FOR YOUR TOMORROW

Volume One

FOR YOUR TOMORROW

A record of New Zealanders who have died while serving with the RNZAF and Allied Air Services since 1915

Volume One: Fates 1915-1942

Errol W Martyn

A Volplane Press Publication

First published in 1998 by
Volplane Press
P O Box 1521
Christchurch, New Zealand
e-mail: errol.martyn@xtra.co.nz

ISBN 0-473-05263-6 (paperback)
ISBN 0-473-05274-1 (hardback)

Typeset by Page Plus Ltd, Christchurch
Printed by Spectrum Print Ltd, Christchurch

"In the springtime of our lives it was an adventure -
not many of us realised what it meant in heartache to those who loved us."

(survivor, Sqn Ldr Bob Spurdle, DFC and bar - *The Blue Arena*)

This work is dedicated to the loved ones of those who did not return from that adventure, and who sacrificed their lives in Freedom's name while serving with the RNZAF and Allied air services, and especially to those left to grieve for those whose final resting place is known only to God.

Contents

COVER DESIGN by Ron Fulstow.

The Maltese Cross with New Zealand fern leaf emblem is widely used on Commonwealth War Graves Commission headstones on the graves of those who died while serving overseas as members of any of the New Zealand armed forces.

Extracts from Carncross letters appearing in Foreword courtesy of the RNZAF Museum.

Foreword

Extracts from a navigator's letters home to his mother

Feltwell, Norfolk

16/7/42

Dear Mum,

It is only a few days since I last wrote but on Tuesday I was thrilled to receive 10 letters from N.Z. so I must write again. There were four letters from yourself (written between 4th April & 5th May), one from Pat R., four from Barbara, and one from Aunt Maud. I also received a letter from Arch, who is on Stirling bombers & a letter from Rob Morton who is on "night" intruders. Rob was out bombing German aerodromes, to distract the fighters, the night we went to Bremen.

I was extremely sorry to hear about Johnny but as he was only "missing" I hope he will have turned up when this arrives.

Don't kick yourself for signing my papers because I've never regretted joining. It has been the greatest experience of my life, not so much on account of the places I have seen but because of the marvellous chaps I have met. They're the tops!

In spite of what critics may say, Bomber Command is the only force that is hitting the German in Germany. In our crew we are determined that we shall do all that is humanly possible to hit the target without taking unnecessary risks. You'll forgive me if I skite and say that we have got team spirit down to a fine art in our crew...

Well Mum I'll sign off here wishing all the family & friends at home all the very best. Best wishes to Bert & yourself.

All my love,

Murray.

P.S. Don't worry, they haven't a hope against a crew like ours,

MEC.

25th July

Dear Mum,

...We have been up since 7A.M. waiting to go on a daylight raid but it has been cancelled. The night before last I went to Duisburg in the Rhur (I hope that's spelt right). I didn't go with my own crew but with a crew who have done fifteen "ops" and whose navigator was sick.

As I think I told you, our crew now has its own plane, "U for Uncle". Well this is the plane we took the other night. Before we started it was a practically new kite but now it is out of commission for at least a week. We got to the target O.K. but then the fun started and for ten minutes solid we got [word censored] pasted out of us by heavy "flak" (excuse my crudeness). I'm afraid I can't describe the barrage with words as it has to be experienced to be believed. We knew the plane had been hit by shrapnel because we could feel the draught blowing through. The captain was grand and every minute when we were in the middle of it he was asking "Everyone O.K? Good show boys."

Well, sufficient to say we got home O.K. Yesterday I went to have a look at the kite and felt quite proud of it. There were holes all over the place and in the nascelle [sic] petrol tank there was a hole big enough to put your fist in. Believe me these "Wimpies" are great planes and will stand up to anything.

The position is now that our plane is unserviceable so our crew hasn't got a plane. When I come home I'll tell you all about Thursday night and it will also make a grand story for the kids...

Well its just about time for lunch so I'll take a breather here.

27/7/42

Well since I left off I have been to Duisburg again & last night we went to the hottest target in Europe - Hamburg. It was by far the most successful show I have been on and the city got a hell of bashing. It was a beautiful clear moonlight and the target was blazing well when we got there. We got a beautiful photo of the city and our plane was the closest to the target in the squadron. The intelligence officer is shouting free beers for our crew because of the success. I did the bombing with a 4,000 lb bomb which wipes out an area of a square mile. They say this is the most successful raid ever made on Germany and there should be quite a few big holes in Hamburg this morning. However my one ambition is still to place a 4,000 lb bomb in the centre of Krupps works. They are going to give each member of our crew a copy of our photo of Hamburg so you'll be able to see it one of these days. There were hundreds of searchlights around the city and we got caught in a cone of twenty and the pilot threw the plane all over the sky for three minutes before we got out of them.

Tonight I am going to a concert in the camp here for a bit of relaxation. We've done three raids in the last four nights so I'll have to catch up on a bit of lost sleep...

I guess I might as well sign off & get this posted or I'll be weeks finishing it.

Best wishes to everyone at home.

All my love,

Murray.

But the weeks had already run their course. There would be no more time for finishing letters, because the next night exuberant nineteen-year old Murray Carncross's youthful adventure came to an abrupt and violent end, when, along with 32 of his countrymen, he died while undertaking another op to 'the hottest target in Europe'.

The 'kids' never would get to hear Murray's grand story 'about Thursday night'.

For him, and thousands like him, there would be no homecoming:

no thankful mother's heartfelt hug;

no grateful lover's impassioned embrace;

no children to raise or grandchildren to cherish.

Victory March cheers would be beyond hearing;

not for them the enjoyment of the fruits of Freedom.

Many today who enjoy those fruits so dearly won would have us 'forgive and forget'. To them, from those who lost so early those they loved so much, the riposte is likely to be: 'Forgive? Perhaps, in time. Forget? Never!'

Preface

THE OBJECT OF THIS THREE VOLUME SERIES is to set down in a comprehensive manner an account of how each of some 4800 New Zealand airmen and women who died while serving with the RNZAF and the Allied Air Services met their end, and to provide brief biographical details on each of them.

It is hoped that such a record will be helpful to those seeking a starting point from which to research or study further any one individual or group of casualties. It is also hoped that it will serve as an enduring memorial to those who sacrificed their lives in times of conflict or while preserving the peace.

Although covering fatal casualties to New Zealanders serving with the military air services since 1915, this work predominantly deals with those who lost their lives during the Second World War. The reason is straightforward - nearly nine out of every ten men and women recorded died during that time, by far the majority being aircrew operating in the European-Mediterranean theatres, a part of the world now receding from the collective memory of a country whose destiny is increasingly linked with the Asia-Pacific region. The Royal Air Force's Bomber Command and its southern sister, 205 Group, accounted for some 60% of these deaths while conducting an arduous and unrelenting bombing campaign from the first day of the war until almost the last - in effect providing the only viable 'Second Front' until the Normandy beaches were stormed in June 1944. Less demanding in terms of manpower, but with losses no less significant, were Coastal and Fighter Commands, the 2nd Tactical Air Force and a myriad of highly mobile commands which operated throughout the Mediterranean.

A modest number died undergoing instruction in the RAF's Training Command (later Flying Training Command). This would have been greater but for the extraordinarily successful Empire Air Training Scheme (EATS), or the British Commonwealth Air Training Plan (BCATP). Agreed to in December 1939, this scheme arranged for the training of RAAF, RAF, RCAF and RNZAF personnel to be undertaken largely in the relatively clear and uncrowded skies of Australia, Canada and New Zealand. Under EATS thousands of New Zealand airmen destined for eventual service with the RAF received training in New Zealand, a large proportion of whom proceeded to Canada to complete their advanced training. Graves and memorials throughout their homeland and Canada bear testimony to the human cost of preparing for battle.

Much more than with land or sea forces death was constantly at the airman's side, whether practising the art of flying in peaceful skies or undertaking an operational flight over enemy territory. Those lost in accidents form a staggering one third of the total who died flying. The air is unforgiving of error, be it through ignorance or an act of bravado. Once in trouble, pulling over to 'sort things out' at leisure is not an option open to those in the air.

Japan's remarkable, although short-lived, success in its sudden thrust into South-East Asia and the Pacific in December 1941 and early 1942 resulted in the RNZAF building up a sizeable operational force in the South and South-West Pacific. Contrary to the common picture postcard portrayal of a tranquil Pacific paradise, tropical weather and geography often posed threats greater than enemy action. RNZAF operations were conducted over a surface area similar in size to Europe, but with little land and vast quantities of water. Over 300 died serving there.

More than 150 New Zealanders died in world-wide service with the Fleet Air Arm and they are accounted for here also, as are those who lost their lives serving as civilians with quasi-military organisations such as the Air Transport Auxiliary.

As WWII demobilisation ended and the RNZAF rapidly shrank in size, the numbers killed while flying dropped accordingly, to the point where from 1958 they were regularly exceeded by deaths from natural or other causes. Aside from the obvious massive decrease in total hours flown, increasingly reliable equipment and improved training have played a significant part in bringing about a steady reduction in fatal flying accidents, to the stage where the measure of accidents per year has been replaced by one of years per accident.

Although few in number, casualties for the First World War and between the wars have proved, in relative terms, the most difficult and time consuming to research. Records are fragmentary and much information is still lacking. Although the coverage is believed to be comprehensive it is likely that a handful of individuals have been overlooked and remain to be identified.

Although the country's WWII air losses were exceeded in number by those incurred by ground forces, the fatal casualty rate, in terms of manpower employed throughout the operational areas, was by far the highest - in the case of the Europe-Mediterranean-Middle East theatres alone, nearly three times so. Small wonder, therefore, that survivors are often perplexed to discover that while our Official Histories are lavish in coverage of the army's endeavours (some 30 dedicated volumes) the air has, by comparison, been treated as a very poor relation (four volumes).

On 28 November 1979 an Air New Zealand DC-10 slammed into the slopes of Mt Erebus, killing 257 in New Zealand's worst ever disaster. The 3.1 million population was shocked and bewildered by the tragedy, the personal losses of which deeply affected a vast number of families and friends throughout the country. Extrapolated into today's terms, the nation's loss of airmen and women during the Second World War is the equivalent of an 'Erebus' every eight and a half weeks. If the land and sea forces losses are added, then the rate rises to one every three weeks. Readers who recall the nation-wide sorrow in November 1979 may care to reflect on the implications for a nation of a mere 1.6 million at war a little over half a century ago. Aside from the enormous scale of personal grief and suffering, society could ill afford this sacrifice of young lives, so many of whom possessed the advanced education and leadership qualities so essential to the growth of a young and developing country.

Much research time has been taken up with resolving the large number of errors, contradictions and omissions uncovered in the records. In some cases their presence is explained by circumstances of the time, e.g. the chaos of retreats during the Battle of France and Japanese invasion of the Far East. Sadly, in others simple neglect or dyslexia often seem to be the cause. For reasons of space, only the more notable contradictions have been commented upon.

Often statements appearing on surviving reports are cryptic or ambiguous and in interpretation I have had to guard against unintentionally adding further confusion or error. Although manuscript drafts have been checked by others there are doubtless occasions where I have still not got it quite right, and for this I take full responsibility.

A work of this nature can never claim to be complete, let alone definitive, and contribution is invited from readers who feel they may have useful material which adds to or corrects the record. Most welcome would be missing take off or crash times, number of operations and hours flown, and especially any information which may lead to the identification of New Zealanders who died other than as members of the RNZAF but who may have not come to my attention (in particular, a number of postwar to present day RAAF and RAF names are thought to be lacking). Assistance is also sought in respect of the location of graves of those whose names are not to be found in the registers of the Commonwealth War Graves Commission (i.e. primarily those who died between 1 September 1921 and 2 September 1939, and after 31 December 1947).

Particularly helpful - in respect of the Fate volumes as well as that to come on Biography - would be the supply of copies or access to any of the following:-

- Letters or documents relating to the circumstances of casualty, searches, and burial.
- Official Record of Service documents.
- Log book pages on which appear last flight entries, postings and Certificates of Qualification.
- Itineraries, dates and names of ships which transported airmen overseas for service with the RAF and RCAF.
- RNZAF Routine Orders for all of 1939.
- RNZAF List of officers, dated June 1940.
- RAF Lists of officers for 1918-1945 (published monthly).

Errol W Martyn, May 1998
P O Box 1521, Christchurch, New Zealand
Fax: #64 3 338 2772 or e-mail: errol.martyn@xtra.co.nz

Acknowledgements

FIRST AND FOREMOST I owe a considerable debt of gratitude to the team of four who graciously agreed to 'volunteer' to review the complete draft of fate narratives for this volume: Arthur Arculus, a long-time respected WWII aircraft and aircrew research specialist; David Duxbury, a virtual encyclopaedia on the subject of the New Zealand and Pacific organisation and operation of the RNZAF in WWII, and pre-war; Colin Hanson, compiler of the forthcoming comprehensive and detailed work, *By Such Deeds - Honours and Awards in the Royal New Zealand Air Force, 1923-1998*; and Trevor Richards, who possesses a wide knowledge of WWI and WWII military history, and a skill at identifying obscure but valuable information often overlooked by others. Trevor has also laboured long hours over the fiche reader checking New Zealand birth registration information for a great many of the casualties. In addition to guiding me away from frequent error these gentlemen have generously contributed additional material from their own researches.

I am especially grateful to David Gunby, author of *Sweeping the Skies*, who, in addition to delving into his files on 40 Squadron and 205 Group, RAF, edited the manuscript and provided invaluable advice and suggestions. I owe thanks also to Richard Robinson of South Croydon for the most professional manner in which he has handled innumerable requests seeking information from London-based offices. My warmest thanks are extended also to Ron Fulstow for the concept and design of the cover, and for his sage advice during a difficult time as this first volume neared completion.

The assistance of the Lottery Grants Board in funding the purchase of a set of RAF Bomber Command Loss Cards on microfilm and of sixteen Commonwealth War Graves Commission registers, and in contributing towards London and Canadian research costs is greatly appreciated. Likewise, the granting of the RNZAF Museum Trust Board's 1995 Study Award, which funded further study into Second World War RAF units and their commands, is gratefully acknowledged.

A work of this kind is dependent in good part on the findings of others who have gone before. In this regard I have gained much from a number of the published works of Air-Britain, the international association of aviation historians. The contribution of authors and editors Jim Halley and Ray Sturtivant, in particular, is gratefully acknowledged. Because of the dearth of records concerning New Zealanders with the FAA, I am much obliged to Ray and another of his Air-Britain colleagues, Ron Pankhurst, for making it possible to present the essential facts of their fates where before there were none. Also due for acknowledgement is the immense amount of research undertaken by Bill Chorley into Bomber Command casualties, as admirably evidenced by his ongoing series, *Royal Air Force Bomber Command Losses of the Second World War*.

My thanks are due also to the following individuals and organisations who assisted with information and in many other ways too numerous to describe in detail:-

Air Historical Branch, London (Sebastian Cox, Air Cdre Henry Probert and Gp Capt Madelin - head and late heads of the Branch respectively; Mike Hatch, Alan Thomas), Archives Office of Tasmania, Ashburton Public Library, Service Historique de Armée de l'Air (Luce Gaume), Australian Archives, Canberra (Margaret Turner), Aviators Gallery (Nigel Wilcox), David J Barnes, Sandra Bird, the late Les Brake, Brentwood Library, Essex (Andrea Cooper), Brian Bridgeman, Canterbury Library, Kent (David S Cousins), Canterbury Public Library, Christchurch, Cheshire Record Office, the late Chris Cole, Clive Collett, Commonwealth War Graves Commission (Tim Brown and Mrs C Connerty-Ketels), Wayne Condon, Coningsby Tattershall Library, Lincolnshire (Dinah Emmitt), Peter Connon, Crawley Reference Library, West Sussex (Charles Kay), Croydon Central Library, Surrey (Oliver Harris), Stan Daniels Photographer, Ted Darcy, Dayton & Montgomery County Public Library, Ohio (Clinton Lowell), Dunedin Pubic Library (Jean Strachan), Edinburgh Library (Andrew Bethune), Mrs A C Elworthy, Owen Gough, Bryant Haigh, Hampshire County Divisional Library (Tina Munro and Philippa Stevens), Harlow Library, Essex (Sheila Sullivan), Father John Harrison, Hatfield Public Library, Hertfordshire (Christine Martindale), The Havering Library Service, London (G H Saddington), Brian Hewson, Indianapolis Municipal Public Library (Dan Fast), Invercargill Pubic Library (Alan Bryce), Inverness Public Library (Norman S Newton), the late Air Cdre P G 'Jamie' Jameson, Phil Jarrett, Cliff Jenks, Kevin Kelly, Carole King, Kingston upon Hull Central Library, Yorkshire (David Smith), Lewes Library, East Sussex (Alan Hibbs), Lincoln Central Library, Lincolnshire (Jane Bailey), the late Ross Macpherson, Marlborough District Library (Walter Elder), Adrienne Martyn, Fred McEvoy, Steve Miezenbeek, Ministère Des Anciens

Combattants et Victimes de Guerre, Paris (Bernard Koelsch), Gerard Morris, National Archives, Washington DC (Mitchell Yockelson), National Archives, Wellington, National Library of New Zealand-Alexander Turnbull Library (Diane Woods), Geoffrey Negus, Nicholas Neill, Wayne Nelson, New Zealand Defence Force (Dr Malcolm McNamara), New Zealand Defence Force Personnel Archives (Walter Guttery and Lynly Yates), New Zealand Fighter Pilots Museum (Ian Brodie and Sir Tim Wallis), Nottinghamshire County Library (Corinne Phillips), Val Nutsford, Ann Orange, Vincent Orange, D Len Peach, David Phillips, Mrs Margot Phillips, Alan Polaschek, Dave Pope, Public Record Office, London, Mrs Veronique Rae, RAF Museum (Andrew Cormack and Peter Elliott), RAAF Historical Section (Mollie Angel and Janet Beck), RAAF Personnel Records (D Pullen), Richmond Library, London (Jane Baxter), RNZAF Wellington (Wt Off Phil Bolton - late of Air Staff), RNZAF Museum (late director Sqn Ldr John Barry, Thérèse Angelo, and Matthew O'Sullivan), RNZAF Public Relations Office (Sqn Ldr Paul Harrison), Gordon Rutherford, Salisbury Library, Wiltshire (Bruce Purvis), Selby Library, Yorkshire (Chris Pearson), Sleaford Library, Lincolnshire (Julia Dabbs), Paul Sortehaug, Southampton City Library, Hampshire (H A Richards), The Southland Times (Clive Lind and Peter Muller - editor and late editor respectively), Stamford Library, Lincolnshire (Veronica Ghosley), Thetford Library, Norfolk (Kathryn Boler), Stuart V Tucker, Alan Tunnicliffe, USAF Historical Research Agency at Maxwell Air Force Base (Marcie Green), USAF Museum at Wright-Patterson Air Force Base (D Menard), Ray Vann, Harold H Vavasour, Wanganui District Library (Lynley Fowler), War Graves Office, Wellington (Jim Dunlop, Daniel Lambert, Margaret Marks, Bernard McGee and Paul Riley), Joe Warne, Larry Webster, Perc Wellington, and Central Library Westgate, Oxfordshire (Mark Lawrence).

Explanatory notes

(On conventions observed and arrangement of data in the body of this work)

Qualification for inclusion

Those who died are recorded here if at time of death they were (a) a member of the New Zealand Forces, (b) New Zealand-born or (c) had received secondary education in or lived in New Zealand for a number of years. In anticipation of any criticism which may be forthcoming in respect of (c), it should be stated that an inclination towards error of commission has been chosen over that of omission. All fate entries falling under this category carry an italicised footnote to identify the nature of the New Zealand link.

Deaths from all causes are included - enemy action, accident, natural causes, suicide, etc - highlighting the fact that death through everyday causes is present in times of war as well as peace.

Some publications refer to a casualty as being a 'native' of New Zealand when this is not necessarily true. The error has been arrived at through an assumption that all next of kin addresses given in the Commonwealth War Graves Commission (CWGC) registers are concurrent with death, the reality being that the information was gathered some years after the war, by which time a family might have migrated from elsewhere.

Coverage includes civilians engaged in military aviation of all kinds and all airmen and women who died while on leave, but not those discharged, held in the Reserve or posted to the Retired List. Territorial personnel appear only if they died whilst engaged on active service following mobilisation or during a refresher or training course run by the regular force.

Height and distance

These are quoted in contemporary terms. Straight line distances are generally given, normally calculated as a measure from a town's central railway station or post office. When, however, these figures have been drawn from official documents it is not always clear if the distance given is taken from the airfield or its associated town.

Place names

Confusion can arise on reading original records. Phonetic spellings, misspellings, or completely wrong locations can be found. Every effort has been made to check these but some have defied resolution - different units arriving at a foreign-named base often had independent interpretations when entering the name in their records. Minor disagreement may also be found in respect of targets throughout Europe. A number of latter day publications opt for present day names (some of which are doubtless at risk of further change - *vide* the Balkans) which can be far removed from the contemporary, and thereby add further confusion. As far as possible this work refers only to names as they would be found in the original records by those undertaking further study of the personnel and fates detailed. In this area the compiler has been largely guided by the definitive works, Jefford's *RAF Squadrons* and Middlebrook and Everitt's *Bomber Command War Diaries*.

The notes below follow the fate description format used throughout. Although concentrating on the Second World War entries, they are generally relevant to the other periods described.

Dates

Fate descriptions appear under the date of the event leading to death; normally both are the same, or within a day. Where a longer interval between the two occurs a cross reference is made under the actual date of death. Occasionally a lack of surviving records has resulted in some doubt as to precisely when an event took place. For example, a death may be known to have occurred on the 18th, but uncertainty could exist as to whether this was in the late evening or early morning hours - at 0100 would indicate the night of 17th/18th, whereas 2300 would fall under the 18th/19th.

Theatre/Command

Grouping by theatre or superior command has been used as a means of gathering together related losses and to help set out fate descriptions in approximate geographical and chronological order within any given date. Starting with NZ and the Pacific, which first see the day, the pattern works westward across the Far and Middle East, to Europe and over the Atlantic.

Given the modest scale of RNZAF operations compared with those of the RAF, there was no requirement for commands in New Zealand, the South Pacific, or South-West Pacific theatres (save in New Zealand, operational command of the RNZAF was in fact vested in the United States Forces).

Within the RAF a Command's title may reflect that of its theatre of responsibility - e.g. the Far East. Occasionally confusion can arise where a unit from one command was detached to operate under another, as when Bomber Command aircraft were detached from time to time to assist Coastal Command.

Duty

The nature of final flights has not always been precisely established. Unit diaries - or Operational Record Books (ORBs), aircraft accident cards, etc - sometimes give only vague or cryptic descriptions, although usually there is sufficient evidence to confirm if the flight was operational or non-operational.

As related in the preface, the greatest demand for aircrew - reflected in the enormous level of casualties- was from Bomber Command. Because it was based in Britain throughout the war, a considerable quantity of detailed records have survived intact. Among some of the most useful publications based on this material are *The Bomber Command War Diaries* and Chorley's *Royal Air Force Bomber Command Losses of the Second World War* volumes. Operations undertaken by the Command gradually developed into large scale undertakings, regularly involving hundreds of aircraft concentrated against a single target. The numbers shown as despatched have been extracted from the *War Diaries*. A shortcoming with this work, however, is that loss figures do not normally include aircraft which were destroyed in Britain or close to its shores, thereby underrating the risk factor in operational flying, particularly in the early years of the war. The numbers shown as lost have therefore been calculated from the more detailed data in *Bomber Command Losses*. Lest the reader be misled by what may seem to be relatively low loss figures, the cumulative effect on aircrew survival rates over a typical first 'tour' of 30 operational flights (or 'ops') needs to be borne in mind (and with a second of 20 ops to follow after a 'rest' period, or, in the case of the Path Finder Force, a continuous one of 45 ops).

Regrettably, it has not been practical to provide this kind of information for the majority of operations undertaken by other organisations. These were of a far more diverse nature and could involve anything from the despatch of a single aircraft to several squadrons operating together as a wing. Here a tour was normally measured in terms of operational hours flown or, in the case of the RNZAF in the Pacific, by the period of time spent in the theatre, rather than in actual number of ops undertaken.

Calculations made by the RAF in late 1942 had established that the chance of surviving a single tour varied from 17.5% for a torpedo bomber crew to 77.5% for one on Catalina flying boats. With heavy and medium bombers crews only 44 out of every 100 were expected to survive.

Unit

The unit shown is that operating the aircraft at time of loss. Usually this coincides with that to which the casualty was posted, a footnote being added when this is not so. Some New Zealand personnel in New Zealand have been difficult to identify in this respect as surviving records simply show the station to which they were posted, when a further 'internal' posting to a specific unit on the base is likely to have taken place.

For the most part, Dominion air forces of the RAAF, RCAF and RNZAF followed RAF procedures and organisation. On the outbreak of war in September 1939 the RAF retitled its Flying Training Schools (FTS) as Service Flying Training Schools (SFTS) and the RAAF and RCAF appear to have followed suit. For some reason the RNZAF did not make the change until several years into the war, at a date as yet undiscovered. As the changeover time is not known, and to avoid confusion with Elementary Flying Training Schools (EFTS), also introduced with the outbreak of war, all FTS losses are shown as SFTS.

The RNZAF was unique amongst the Commonwealth air forces during the Second World War in that it stayed with pre-war RAF practice by subdividing each of its SFTSs into a distinct Intermediate Training Squadron (ITS) and an Advanced Training Squadron (ATS). Thus, after ground training at an Initial Training Wing (ITW), the trainee pilot's flying career began at an EFTS and, all going well, progressed through an SFTS via its ITS and ATS, where he would graduate at the conclusion of his course. Thanks to research undertaken by David Duxbury it is possible to identify the relevant ITS or ATS for almost all SFTS casualties in the RNZAF.

For clarity, role descriptions, such as F (for Fighter) or GR (General Reconnaissance) have been omitted from the titles of numbered squadrons - e.g. 1 GR Squadron appears as 1 Squadron.

Under Article XV of the Empire Air Training Scheme (EATS) it was agreed that each Dominion's contribution should be recognised by squadrons of its own air force in the field. Those formed for New Zealand were 485, 486, 487, 488, 489 and 490 Sqns, RNZAF. As far as possible air and ground crew were drawn from RNZAF sources, with RAF (or other Dominion air force) personnel making up the difference. Although their respective crests include the wording 'Royal New Zealand Air Force' each squadron's administration, equipment issue and operational control were entirely an RAF responsibility. While technically incorrect, they were often informally (and more appropriately) referred to as 485 (NZ) Squadron, RAF, etc, but the only squadron officially entitled to this form of nomenclature is 75 (NZ) Squadron, RAF.

RAF flying units are comprehensively recorded in *RAF Squadrons* and Sturtivant, Hamlin and Halley's *Royal Air Force Flying Training and Support Units*.

Base, superior formations and sub-commands

All bases are identified by country and, in the case of Britain where the largest number are to be found, by pre-1974 county name. Where possible, numbered airfields are provided with their alternative name or a nearby named location, or map reference for the more isolated examples - e.g. LG 221 (Landing Ground 221) is shown as LG 221 [Fayid], Egypt.

It was not uncommon for units to have one or more detachments based at locations other than the airfield on which they were headquartered. These satellite fields or relief landing grounds are mentioned only when it can be established that the aircraft lost left there on its final sortie. Unfortunately, it is often impossible to state with certainty from which airfield departure was made. Short range fighters and aircraft operating in the highly mobile desert and South Pacific campaigns could frequently touch down at an intermediate point for refuelling before entering contested territory. Otherwise, generally speaking and particularly for Bomber Command, it may be said that final take off was most likely to have been from the main base as shown.

For those who wish to carry out further research into a particular loss (or losses), formations or commands which exercised control over the unit are provided in ascending order, i.e. Wing, Group, sub-Command. The Path Finder Force (PFF) was established within Bomber Command on 15 August 1942, to mark and illuminate targets for the Main Force, but did not achieve Group status until 8 January 1943. Until that time the PFF squadrons continued to be administered by their original groups. To identify them as such [PFF] appears immediately before the group number. Readers should also be aware that, in addition to the RCAF's Eastern Air Command, the RAF had two like-named commands - the first from September 1942 to March 1943 for the purposes of *Operation TORCH* in North West Africa and the second from December 1943 in South-East Asia.

The most accurate and comprehensive listing (with location maps) of RAF squadron bases since 1912 is to be found in *RAF Squadrons*. Similar detail (but no maps) on the Fleet Air Arm is given in Sturtivant and Ballance's *Squadrons of the Fleet Air Arm*. Regrettably, no similar work covering the RNZAF has yet been published.

Aircraft

The name, model, individual serial number and, where known, individual code letter are shown. In the interests of space manufacturers' names are not included under entries for WWII casualties - a table of these may be found in the next section, under Abbreviations & Glossary. Some of the American aircraft operated by the Fleet Air Arm were unusual in that they were known within the service by more than one official name, although at different times. British names were used until January 1944 when, for example, the Martlet was renamed Wildcat to conform with US Navy nomenclature.

Where a serial has not thus far been identified this is indicated by '(serial?)' and if there is doubt concerning one quoted it is followed by '(?)'. These problems are usually brought about by lack of information or legibility in the original source, or through a conflict of information when checking one source against another.

The subject of squadron code letters is a complex one and these have not been shown. The reader is referred to Bowyer and Rawlings' *Squadron Codes 1937-56* for information on RAF codes. Individual aircraft code letters are supplied where there is good reason to believe they were being carried on the aircraft at time of loss.

Unless it is obvious or otherwise stated, the aircraft can be taken to have been written off or struck off charge as a result of the event described.

Times

The 24-hour clock system - introduced by the RAF from midnight of 19/20 October 1918 - is used throughout. It has usually been possible to establish a time for take off or when the aircraft crashed, but not often both. Times may be local, Greenwich Mean Time (GMT), British Summer Time (BST), British Double Summer Time (BDST), Central European Time (CET), etc - or even a combination, when both an 'up' and 'down' time is provided. For most events it is now impossible to say which was being referred to as the necessary information rarely accompanies the original entry. It is unwise to draw the conclusion that the flight time of an aircraft taking off from Britain at 1800 and crashing in occupied Belgium at 0100 totalled seven hours. The 1800 may have been in GMT, BST or BDST, and the 0100 CET or GST (German Summer Time), for instance.

Event description

Because of their small number and the limited amount of information generally available about them, as much detail as possible has been given on losses up to August 1939. With rare exceptions, all the crew are named, including non-New Zealanders. Although the fate of the complete crew is still given for Second World War entries, the only non-New Zealand names displayed are those of the pilot (or captain) of crewed machines; a crew identified with its 'Skipper' more than it would with the aircraft. Mention is made of any surviving New Zealander, perhaps taken prisoner, injured, or both, who was directly involved in the event.

Descriptive detail varies from very little in the case of those who simply disappeared without trace or where even the most basic information is lacking, through to the comprehensive where substantially complete records survive. Some of the more lengthy entries serve to portray what may have overtaken others for whom less is known, and also indicate the hazards faced by more fortunate comrades who survived.

One of the greatest difficulties faced by the compiler, aside from deciphering the cryptic and illegible, has been the remarkable number of errors and contradictions which have come to light on trekking through official records. These range from the simple spelling mistake to quoting an incorrect day, month or year, the wrong unit or even country, to something as embarrassing as unknowingly stating as target the very place at which the bomber was based!

Much detective work has been undertaken in trying to establish the facts, not always with complete success, it has to be said. Often too much remains unknown or unresolved to permit the reader to draw many conclusions from the data presented. This should be kept particularly in mind when reviewing flying accidents. Many were clearly caused through ignorance or a moment of overconfidence, but for others a satisfactory explanation was never discovered, given the pressure under which wartime accidents were investigated. The compiler's approach to these casualties has been to set down what was seen or known to have happened to the aeroplane, without venturing into the contentious areas of blame or responsibility.

Where flak or fighter is the known cause for an operational loss this is stated, but aircraft noted as having crashed in enemy territory through a reason not yet established are usually described as having been 'brought down'; mechanical failure and the weather, especially in the early war years, and mid-air collision (more often with friend than foe) also took their toll. With rare exceptions no attempt has been made to identify the particular aircraft, unit or crew that may have been credited with the destruction of those believed to have been shot down. Whilst the lone-wolf hunting tactics of the night-fighter mean their claims can be matched up with a reasonable degree of certainty, where records survive, this is much less so with the day fighters. All sides were guilty of over claiming to varying degrees, and a fighter claim might involve little more than finishing off a flak-crippled victim, while the opportunities for confusion in the fast moving whirl of a dogfight were endless. These aspects are frequently overlooked by many enthusiasts, who sadly seem to limit their perspective of the airwar to adding up who-shot-down-who.

Crash sites, burials and memorials

Note on the Commonwealth War Graves Commission (CWGC): Set up in the United Kingdom by Royal Charter in 1917 as the Imperial War Graves Commission (the name was changed in 1959), it was charged with the task of commemorating by name every serviceman and woman of the Commonwealth who lost his or her life in the Great War of 1914-18. In the course of its work the Commission constructed special war cemeteries into which it concentrated all those graves which were either too widely scattered to be properly maintained, or were not already in an appropriate cemetery. It also organised the erection of special memorials on which were inscribed the names of all those whose graves are unknown. The Commission's role was later extended to include Second World War dead, and it is now responsible for the care of over one and a quarter million war graves throughout the world. A record of the names is available by way of the Commission's register booklets (containing approximately 1000 entries each) which are reprinted or republished in updated editions from time to time. The 'war' periods covered by the Commission's charter are 4 August 1914 to 31 August 1921 and 3 September 1939 to 31 December 1947. These dates extend well beyond the cessation of hostilities and the registers also include the names of some who died after they had ceased to serve. Those New Zealanders who did not die while serving or as the result of war service are not included in the main body of this work. However, their names are noted in appendices.

The random nature of crash sites has led to a considerable effort being put into identifying and relating these to better known nearby towns or cities, in order that the reader may more readily locate the general area through a conventional map or atlas.

When aircraft fell in enemy territory or no-man's land the dead were usually buried close to the crash site, or in the nearest village or town cemetery. After hostilities had ceased most were reinterred in concentrated cemeteries, almost always within the country containing the original grave, as recorded in the CWGC registers. Notable exceptions to this practice, however, occurred throughout the occupied territories in Europe, especially where the liberated people of a town or village elected to maintain existing graves in perpetuity, which they continue to do so with loving care and attention. For some of the reinterred it has been possible to identify and record here their first place of burial. This applies especially to the North African theatre, where strong winds and shifting sands rapidly obliterated all trace of many a temporary wartime desert grave before hostilities ceased or the CWGC could relocate them. While the first place of burial may be given for a New Zealander it cannot automatically be assumed that the rest of his crew was also buried there. They could at times be dispersed, some being killed as the aircraft crashed and others dying later of injuries in a hospital, perhaps many kilometres away.

Overseas personnel who died in Britain or Canada were usually buried near to their base or the crash site. Later in the war several regional cemeteries were established in England for the burial of those who died there. Exceptions to these practices occurred when next of kin or close friends of the deceased living in the same country requested the burial be in their locality. For those who died in New Zealand, burial in their hometown was the norm. The same was true of local personnel who died in Britain and explains why, for instance, a heavy bomber crew killed there in a crash are often divided in death, with Dominion members buried together in one cemetery while the graves of their RAF comrades are perhaps widely dispersed throughout the land.

As full CWGC cemetery, grave and memorial panel details will be contained in the biographical volume, final place of burial information in Volumes I and II is generally limited to identifying a recognisable locality rather than the cemetery or churchyard itself, e.g. those buried in Christchurch's Bromley Cemetery are simply noted as being at Christchurch, which is more readily found on a New Zealand map or in an atlas than is the suburb of Bromley.

About one third of all wartime aircrew casualties were lost without trace and, with a few noted exceptions, they are commemorated on one of the following eight CWGC memorials:-

- Alamein Memorial, Egypt (for those who served in North Africa and the Middle East)
- Arras Memorial, France (for those who served on the Western Front during the First World War)
- Bourail Memorial, New Caledonia (for those who served in the South and South-West Pacific)
- Malta Memorial (for those who served in the Western Mediterranean and Southern Europe)
- Ottawa Memorial, Ontario, Canada (for those who served in North America)
- Runnymede Memorial, Middlesex, England (for those who served in the Azores, Faeroe Islands, Iceland, Northern and Eastern Europe, Northern Ireland, and the United Kingdom)
- Singapore Memorial (for those who served in the Far East)
- The Royal New Zealand Naval Memorial, Devonport, Auckland (includes members of the Royal New Zealand Navy who served world-wide in the Fleet Air Arm)

Provincial memorials are also to be found in Auckland, Wellington, Christchurch and Dunedin for those who served in New Zealand.

In addition to the above there are a number of other official memorials and countless private ones commemorating those with and without graves alike. These range from a roll honouring the collective dead of a school to a large park set aside in memory of a family's lost son. Their very number, widespread location and lack of a collective register of same make it impractical to mention them here. The compiler would, however, be pleased to receive information concerning any of them.

Roles of individual aircrew
The roles shown are those being undertaken at the time of the final flight. When a casualty was travelling as a passenger or acting outside his normal role (a pilot acting as navigator, for instance) brackets are used to draw attention to this - e.g. (Pilot) - with a note of further explanation provided at the foot of the entry. The role terms used here are not necessarily indicative of an airman's 'trade' qualification - he who was acting as mid-upper gunner, for example, may have been a qualified air gunner or wireless operator-air gunner.

The term captain rather than pilot is used for aircraft carrying five or more crew, although there may have only been one pilot on board. On dual training flights pilots often exchanged roles and in these cases it is not usually possible to say who was in control or command at the time of accident. Unless otherwise stated all aircrew listed under a 'School' (e.g. Service Flying Training School or Air Gunnery School) may be taken as being still 'under training'.

In early 1942 Bomber Command, by far the largest customer for aircrew, introduced a dramatic change as to the crew composition of its heavy bombers. The 2nd pilot was dispensed with while other members adopted more specialised roles. Thus the observer became a dedicated navigator (previously he had also been bomb aimer and, if required, an air gunner), while the front gunner, after a short course, took on additional responsibility as bomb aimer (or air bomber as he was officially designated). By May 1942 the standard Wellington bomber crew had been reduced from six to five, although it could increase temporarily to six again when taking along a 'freshman' pilot for operational experience. For some reason Wellington crews operating in the Middle East remained at six until early 1943, and observers were still used as such and in many other types elsewhere. Naturally, there was an introductory phase where observers were still coming through the training system while the first dedicated navigator and air bomber courses got underway. However, as it is not always possible to separate one from the other after introduction of the change, the term navigator replaces observer in entries from May 1942 onward. An exception is the Fleet Air Arm, as that service, with different operating and training procedures to the RAF, continued with the observer through to war's end.

A new specialist member to be taken on board the four engined 'heavies' under the new arrangements was the flight engineer, who took care of fuel and engine management. Since the earliest days of the RAF, qualified technical personnel of airman or NCO rank had been carried as part of the crew of larger flying boats during detachment from their home bases to carry out routine maintenance as well as to operate the wireless equipment, etc. Later they were carried as permanent crew members in aircraft such as the Sunderland and Catalina, although the trade of flight engineer did not exist as such until introduced by Bomber Command.

Attempts to identify the exact role played by those understood to have graduated as air gunners or wireless operator-air gunners (WOpAG) have met with varied success. In four-engined heavy bombers such as the Lancaster a WOpAG could have been acting as rear gunner, mid-upper gunner, bomb aimer (and front gunner) or wireless operator. Where doubt exists these crew are simply shown as 'Air Gunner' or 'WOpAG'.

Service Number
Airmen, or Other Ranks, of the RNZAF were not allotted numbers until September 1938 and direct entry officers until late 1941. Numbers for airmen enlisting after the outbreak of war bore the letter 'A' prefix, while pre-war regulars retained their original non-prefixed number. On or shortly before 21 August 1940 all numbers were prefixed 'NZ' only, although later variations such as NZC, NZT and NZW were introduced. A 'W' prefix was applied to numbers (commencing at 1100) allotted to members of the Women's Auxiliary Air Force (WAAF) when that service was created in early 1941. The first two digits of an airman's number indicated the year in which he enlisted, e.g. NZ421234 attests to enlistment in 1942. Pre-war Territorial Air Force (TAF) airmen provided an exception, their numbers featuring an additional number prefix to indicate membership of the Wellington (4 prefix), Christchurch (5) or Auckland (6) Squadrons - e.g. NZ539084 identified this airman as having enlisted with the Christchurch Territorial Squadron in 1939. When numbers for officers, other than those commissioned from the ranks (who retained their original airman's number), were allotted from September 1941 they started at NZ1001.

This system was rationalised in 1948-49, a new number-only series commencing at 70,000 (130,000 for the TAF) being introduced in January 1949. With a formula-selected single letter prefix (related to computer usage) added in 1971, the series continues today.

Numbers in the Royal Navy's 'FX' series were issued for airmen selected in New Zealand for the FAA, although some of them appear never to have been made aware of this. Officers were not allotted a number.

Space considerations prohibit detailing the systems used by other air services, but briefly it may be said that numbers held by personnel of the RAAF were prefixed 'Aus' and those of the RCAF by letter C, J or R. Numbers for airmen and officers of the RAF (and its antecedents) were not prefixed and where an airman was later commissioned his service number also changed.

Rank

Ranks are often at least one above that which will be found in records created during the time leading up to death. This could come about when confirmation of promotion was 'in the pipeline' but had not been received by the casualty's unit.

Name

Full names are supplied and some of the more significant anomalies noted. In comparing records it is not uncommon to find first names reversed or minor inconsistencies in spellings - Laurence, for instance, is frequently misspelled as Lawrence, or vice versa, and can vary from one record to the next. On checking the index to registration of New Zealand births to verify conflicting age information additional first names have sometimes come to light also, but they are ignored in this work if this is the only place in which they have been found. The degree of these variations is greater than perhaps is generally realised and inspection of several sources, sighting of a birth certificate (not always faultless either), etc, is recommended for those seeking to research the career of any casualty in depth.

Honours and Awards

All known honours and awards, including foreign, are given. For citation and additional information for RNZAF personnel only, Hanson's forthcoming *By Such Deeds - Honours and Awards in the Royal New Zealand Air Force, 1923-1998* is strongly recommended.

Air Arm

This is the air service or arm to which the casualty belonged. It is not necessarily the same as that with which he was serving at time of death - nearly 2900 RNZAF personnel in fact died while on attachment to the RAF.

Prior to the reconstitution of the New Zealand Naval Forces as the Royal New Zealand Navy (RNZN) in October 1941, naval airmen enlisting in New Zealand for service in the FAA nominally joined as members of the New Zealand Division of the Royal Navy, although apparently few of them, if any, knew of this at the time.

Age

Ages have been calculated from birth dates noted in official records. In comparing these with ages recorded in CWGC registers a number of contradictions arose; at times a year over (i.e. presumably referring to age year rather than age at last birthday) and, less logically, a year under. Where the difference is even greater a check (for those born in New Zealand) has been made against the birth registration index, and where confirmed the result noted at the foot of the appropriate entry. Although a few 'fibbed' their way around age restrictions, the majority of these discrepancies appear to be the result of genuine error. It should be noted that where event and death dates are separated by some months or years, the age shown under the former is that at time of death.

Number of hours and operations flown

(For a summary of the operational 'tour' concept see 'Duty' above.)

In late 1960 the British authorities, in what today would be described as an act of mindless vandalism, destroyed tens of thousands of deceased airmen's log books which remained unclaimed after minimal advertising of their impending destruction. New Zealand's good fortune, however, is that RNZAF staff working in London with the Missing Research and Enquiry Service (MRES) during the late 1940s retrieved all log books which could be identified as belonging to New Zealand aircrew who had died while serving in or on attachment to the RAF. While many of these were in due course handed over to next of kin, unclaimed log books held at Wellington are believed to have later suffered a similar fate to those held by the RAF.

About two thirds of the official draft biographies compiled in the late 1940s appear to have been in part based upon information found in these documents. The compilers usually noted (although not always correctly) the total number of hours at the foot of each biography and frequently made reference in the narrative to the number of operational flights carried out. Where hours are not noted, the explanation is usually that the log book was lost - perhaps along with the aircraft in cases of ferry flights - or it had been passed on to next of kin prior to the preparation of the biography.

Flying hour figures for pilots, and sometimes 2nd pilots, often appear on accident cards or files as well, usually broken down into categories of dual, solo or type. Occasionally two sets are provided, in which case one rarely matches with the other.

Operational flight totals quoted have been drawn almost exclusively from the draft biographies. A shortcoming of this source is that it is not always clear if the 'op' on which the crew member was lost is included in the total. An accurate count could also be hindered where the log book had not been updated to show the last flight (or flights). Another area of conflict concerns ferry flights flown through hostile skies and which qualified as ops. Typically such flights would be undertaken from Portreath in Cornwall to the Middle East, via Gibraltar and Malta, but the draft entries are inconsistent, ambiguous or vague in accounting for them. In the Pacific a similar situation applied to RNZAF Hudsons ferrying between New Zealand and the Islands in 1942-43, when all such flights were classified as 'operational' because enemy submarines or shipping could conceivably be intercepted.

Ops, and to a lesser extent flying hours, for those who flew with the RNZAF in the Pacific and New Zealand theatres are generally not noted, presumably because the relevant log book was already with next of kin when compiling the draft.

Flying hours and op figures have been included in the entries because they are available for such a large percentage of the casualties. Given the above comments and earlier remarks about errors, they should be treated with a degree of caution and looked upon only as a guide. The flying hour figures recorded here are usually rounded to the last completed hour and represent the total number flown, unless stated to the contrary. One popular notion which a cursory examination of the ops figures quickly disposes of is that a complete crew usually succeeded in staying together for all of its tour.

Footnotes

Generally speaking, an italicised footnote is used only to clarify a New Zealand connection where the casualty was neither born in New Zealand nor a member of the New Zealand Forces; to name other immediate family members who also lost their lives in military aviation; to highlight the fate of others who may have been involved in one event but died later as the result of another; or to draw attention to the occasional contradiction or error found in related and otherwise dependable works. The names of immediate family who died with the other services have not been included as the resources required to extract this information are presently unavailable to the compiler.

Aircraft manufacturers' names

(for September 1939-December 1942 entries only - pre-war entries show the name in full)

Bell	**Airacobra**	Lockheed	**Hudson**
Fairey	**Albacore**	Hawker	**Hurricane**
Armstrong Whitworth	**Albemarle**	Curtiss	**Kittyhawk**
Avro	**Anson**	Avro	**Lancaster**
Blackburn	**Baffin**	Saunders-Roe	**Lerwick**
Martin	**Baltimore**	Consolidated	**Liberator**
Fairey	**Battle**	Lockheed	**Lodestar**
Bristol	**Beaufighter**	Westland	**Lysander**
Bristol	**Beaufort**	Miles	**Magister**
Bristol	**Bisley**	Avro	**Manchester**
Bristol	**Blenheim**	Martin	**Marauder**
Douglas	**Boston**	Grumman	**Martlet**
Blackburn	**Botha**	Martin	**Maryland**
Brewster	**Buffalo**	Miles	**Master**
Consolidated	**Catalina**	De Havilland	**Mosquito**
Cessna	**Crane**	De Havilland	**Moth**
Boulton Paul	**Defiant**	North American	**Mustang**
De Havilland	**Dominie**	Airspeed	**Oxford**
Short	**Empire Flying Boat**	Blackburn	**Roc**
De Havilland	**Express**	Hawker	**Sea Hurricane**
Boeing	**Fortress**	Supermarine	**Spitfire**
Fairey	**Fulmar**	Short	**Stirling**
Gloster	**Gladiator**	Short	**Sunderland**
Fairey	**Gordon**	Fairey	**Swordfish**
De Havilland	**Gipsy Moth**	De Havilland	**Tiger Moth**
Handley Page	**Halifax**	Curtiss	**Tomahawk**
Handley Page	**Hampden**	Hawker	**Typhoon**
Handley Page	**Harrow**	Lockheed	**Ventura**
Hawker	**Hart**	Vickers	**Vildebeest**
North American	**Harvard**	Vickers	**Vincent**
Douglas	**Havoc**	Supermarine	**Walrus**
Hawker	**Hector**	Vickers	**Wellington**
Hawker	**Henley**	Armstrong Whitworth	**Whitley**
Hawker	**Hind**		

Abbreviations & Glossary

Use of abbreviations throughout this volume has been kept to a minimum in the interests of clarity. Those, however, who wish to carry out further research on their own account may find the following useful in interpreting contemporary documents.

Imperial/metric conversions

1 foot	= 0.305 metres	1 metre	= 3.281 feet
1 gallon	= 4.546 litres	1 metre	= 1.094 yards
1 kilo	= 2.205 pounds	1 mile	= 1.609 kilometres
1 kilometre	= 0.621 miles	1 nautical mile	= 1.853 kilometres
1 kilometre	= 0.540 nautical miles	1 pound	= 0.454 kilograms
1 litre	= 0.220 gallons	1 yard	= 0.914 metres

*	bar
2nd Lt	Second Lieutenant
2nd Off	Second Officer
2NZEF	Second New Zealand Expeditionary Force [WWII]
(A)	Air Branch [Navy]
AA	Anti-aircraft [guns]
AACF/U	Anti-Aircraft Co-operation Flight/Unit
A&AEE	Aeroplane & Armament Experimental Establishment
AAF	Auxiliary Air Force
AAP	Aircraft Acceptance Park
AASF	Advanced Air Striking Force
a/c	aircraft
AC1	Aircraftman 1st Class
AC2	Aircraftman 2nd Class
ACC	Army Co-operation Command
ACH	Aircrafthand
ACM	Air Chief Marshal
ACS	Airfield Construction Squadron
ACS	Aerodrome Construction Squadron [RNZAF]
Act	Acting
ADDL	Aerodrome Dummy Deck Landing
ADF	Automatic Direction Finding
ADF/U	Aircraft Delivery Flight/Unit
ADF/U	Air Defence Flight/Unit
AFC	Air Force Cross
AFC	Australian Flying Corps
AFU	Advanced Flying Unit
AG	Air Gunner
AG u/t	Air Gunner under training
AGS	Air Gunners School
AHB	Air Historical Branch [RAF]
AHQ	Air Headquarters
AI	Airborne Interception [radar]
AID	Aeronautical Inspection Department [later Directorate]
AIF	Australian Imperial Force
Air Cdre	Air Commodore
ALG	Advanced Landing Ground
ALO	Air Liaison Officer
ALT	Attack Light Torpedo
AM	Air Mechanic

AM	Air Marshal
AM	Air Ministry
AMES	Air Ministry Experimental Station [radar station]
AMO	Air Ministry Order
AMP	Air Member for Personnel
ANS	Air Navigation School
AO	Air Observer
AO u/t	Air Observer under training
AOC	Air Officer Commanding
AOC-in-C	Air Officer Commanding-in-Chief
AOS	Air Observers School
AP	Airman Pilot
AP u/t	Airman Pilot under training
AP	Armour Piercing
APO	Acting Pilot Officer
ARC	Aircrew Reception Centre
ARP	Air Raid Precautions
A/S	Anti-shipping
A/S	Anti-submarine
ASD	Administration and Special Duties [Branch]
ASI	Air Speed Indicator
ASR	Air-Sea Rescue
ASV	Air-to-Surface Vessel [radar]
ATA	Air Transport Auxiliary
ATC	Air Training Corps
ATC	Armament Training Camp
ATFERO	Atlantic Ferry Organisation
ATS	Advanced Training Squadron
Att	Attached
AVM	Air Vice Marshal
BAFF	British Air Forces France
BAT Flight	Beam [originally Blind] Approach Training Flight
BC	Bomber Command
BCATP	British Commonwealth Air Training Plan
Bde	Brigade
BDST	British Double Summer Time
BDU	Bomber Development Unit
BEF	British Expeditionary Force
BEM	British Empire Medal
B&GS	Bombing & Gunnery School
Bn	Battalion
BOAC	British Overseas Airways Corporation
BR	Bomber Reconnaissance
BST	British Summer Time
Bullseye	Realistic exercise carried out by bomber OTUs
CAC	Canterbury (NZ) Aviation Company
CACF/U	Coast Artillery Co-operation Flight/Unit
CAM ship	Catapult Armed Merchant ship
Capt	Captain
CAS	Chief of Air Staff
CB	Companion of the Order of the Bath
CBE	Commander of the Order of The British Empire
CC	Coastal Command
Cdt	Cadet

CET	Central European Time
CF	Conversion Flight
CFI	Chief Flying Instructor
CFS	Central Flying School
CGI	Chief Ground Instructor
CGM	Conspicuous Gallantry Medal
CGS	Central Gunnery School
CHANNEL STOP	British code name for air operations against shipping in the Straits of Dover by day
ChB	Bachelor of Surgery
Circus	Large-scale combined fighter and bomber operation intended to bring enemy fighters into action
CNS	Central Navigation School
CO	Commanding Officer
Col	Colonel
COMAIRSOPAC	Commander, Air, South Pacific
COMSOPAC	Commander South Pacific Area
(C)OTU	(Coastal) Operational Training Unit
Cpl	Corporal
CU	Conversion Unit
CUAS	Cambridge University Air Squadron
CWGC	Commonwealth War Graves Commission
DBR	Damaged Beyond Repair
DC	Depth Charge
DCAS	Deputy Chief of Air Staff
DCM	Distinguished Conduct Medal
Det	Detached
D/F	Direction Finding [radio navigation]
DFC	Distinguished Flying Cross
DFM	Distinguished Flying Medal
DH	De Havilland [company]
DI	Daily Inspection
DLCO	Deck Landing Control Officer ['batsman']
DLP	Deck Landing Practice
DLT	Deck Landing Training
DNCO	Duty Not Carried Out
D/R	Dead Reckoning [navigation]
DROs	Daily Routine Orders
DSC	Distinguished Service Cross
DSO	Distinguished Service Order
Dt	Detachment
DWI	Directional Wireless Installation [anti-magnetic mine device]
DYNAMO	British code name for evacuation of Dunkirk
e/a	enemy aircraft
EAC	Eastern Air Command
EATS	Empire Air Training Scheme
E-boat	German motor torpedo boat
EFTS	Elementary Flying Training School
ELG	Emergency Landing Ground
EO	Engineer or Equipment Officer
EPS	Emergency Precautions Scheme
E&RFTS	Elementary & Reserve Flying Training School
ETA	Estimated Time of Arrival
EWS	Electrical and Wireless School

FAA	Fleet Air Arm
FAFAI	Forces Available For Anti-Invasion
F/B	Flying Boat
FC	Ferry or Flying Control
FC	Ferry or Fighter Command
FE	Far East
F/G	Front Gunner
Fg Off	Flying Officer
FIS	Flying Instructors School
F/L	Flight Lieutenant
Flak	Fliegerabwehrkanonen [German anti-aircraft guns]
Flt	Flight
Flt Cdr	Flight Commander
Flt Cdt	Flight Cadet
Flt Lt	Flight Lieutenant
Flt Sgt	Flight Sergeant
Flt Sub-Lt	Flight Sub-Lieutenant
F/O	Flying Officer
FP	Ferry Pool
FPP	Ferry Pilots Pool
Freya	German long-range detection radar
F/S	Flight Sergeant
FSL	Flight Sub-Lieutenant
ft	feet
FTC	Flying Training Command
FTF/U	Ferry Training Flight/Unit
FTR	Failed to return
FTS	Flying Training School
FU	Ferry Unit
FULLER	British code name for counter measures should German warships escape from Brest
Gardening	Sea minelaying by aircraft
GC	George Cross
G/C	Group Captain
GCI	Ground Controlled Interception
GD	General Duties [Branch]
Gee	Radio aid to navigation and target identification
GM	George Medal
GMT	Greenwich Mean Time
GP	General Purpose [bomb or aircraft]
Gp	Group
Gp Capt	Group Captain
GR	General Reconnaissance
GR&ANS	General Reconnaissance & Air Navigation School
GRS	General Reconnaissance School
GRU	General Reconnaissance Unit
GST	German Summer Time
GTS	Glider Training School
GTS	Ground Training School
HARPOON	British code name for covering operations of convoy from UK to Malta, June-July 1942
HC	High Capacity [bomb]
HCU	Heavy Conversion Unit
HE	High Explosive [bomb]

HF	High Frequency [radio]
HMS	His Majesty's Ship
HMNZS	His Majesty's New Zealand Ship
HMT	His Majesty's Transport
hp	horsepower
hrs	hours
HSL	High Speed Launch
Hurribomber	Hurricane aircraft equipped with bombs
IAS	Indicated Air Speed
IF	Instrument Flying
IF	Independent Force
IFF	Identification Friend or Foe
Instr	Instructor
Intruder	Night operations against enemy airfields and aircraft
IO	Intelligence Officer
ITS	Intermediate Training Squadron
ITW	Initial Training Wing
Jimcrow	Shipping reconnaissance for suitable targets for attack by fighters or fighter-escorted bombers
KB	Kite Balloon
KBS	Kite Balloon Section
KIA	Killed In Action
km	kilometre
kt	knot [a nautical mile per hour]
LAC	Leading Aircraftman
LACW	Leading Aircraftwoman
Ldg Airman	Leading Airman
Leigh Light	Searchlight fitted to aircraft for night attack on U-boats
LF	Low Frequency
LG	Landing Ground
Low Ramrod	A Ramrod undertaken at below 5000 feet
Lt	Lieutenant
Lt (A)	Lieutenant (Air Branch)
Lt Cdr (A)	Lieutenant Commander (Air Branch)
Lt Col	Lieutenant Colonel
m	metre
MAC ship	Merchant Aircraft Carrier
Mae West	Lifejacket
Maj	Major
Maj Gen	Major General
MAP	Ministry of Aircraft Production
MB	Bachelor of Medicine
MBE	Member of the Order of the British Empire
M-boat	German minesweeper
MC	Military Cross
MC	Medium Capacity [bomb]
ME	Middle East
MEFC	Middle East Ferry Control
Met	Meteorological
METS	Middle East Training School
MF	Medium Frequency
mg	machine gun
MIA	Missing In Action
mid	mentioned in despatches

MILLENNIUM	British code name for 1000-bomber raid on Cologne 30/31 May 1942
Mk	Mark [number]
MM	Military Medal
MO	Medical Officer
mph	miles per hour
MRAF	Marshal of the Royal Air Force
MRES	Missing Research and Enquiry Service
MSFU	Merchant Ship Fighter Unit
MT	Mechanised Transport
M/U/G	Mid-Upper Gunner
MV	Motor or Merchant Vessel
NA	North Africa
NAS	Naval Air Station
Nav	Navigator
NCO	Non-Commissioned Officer
NEI	Netherlands East Indies
NF	Night Fighter
NFT	Night Flying Test
Nickel	Leaflet dropping raid
Nickels	Leaflets
nm	nautical mile
NWA	North West Africa
NZAF	New Zealand Air Force
NZEF	New Zealand Expeditionary Force [WWI]
NZFS	New Zealand Flying School
NZMC	New Zealand Medical Corps
NZPAF	New Zealand Permanent Air Force
OADU/F	Overseas Aircraft Delivery Unit/Flight
(O)AFU	(Observers) Advanced Flying Unit
OAPU	Overseas Aircraft Preparation Unit
OBE	Officer of the Order of the British Empire
Oboe	Ground controlled blind bombing radar device
Obs	Observer
OC	Officer Commanding
Op/s	Operation/Operations
ORB	Operations Record Book
ORBAT	Order of Battle
OTF	Overseas Training Flight
OTU	Operational Training Unit
OUAS	Oxford University Air Squadron
(P)AFU	(Pilots) Advanced Flying Unit
PEDESTAL	British code name for Malta convoy operations, August 1942
PDRC	Personnel Dispatch and Reception Centre
PDU	Photographic Development Unit
PFF	Path Finder Force [or Pathfinder Force]
Plt Off	Pilot Officer
PMO	Principal Medical Officer
PNB	Pilot, Navigator, Bomb Aimer [aircrew grouping]
P/O	Pilot Officer
POW	Prisoner Of War
PR	Photographic Reconnaissance
PRC	Personnel Reception Centre
PRO	Public Record Office
PRU	Photographic Reconnaissance Unit

psi	per square inch
PTC	Personnel Transit Centre
Pte	Private
RAAF	Royal Australian Air Force
RADAR	Radio Direction and Ranging [American - see RDF]
RAeC	Royal Aero Club
RAF	Royal Air Force
RAF	Royal Aircraft Factory
RAFC	Royal Air Force College [Cranwell]
RAFO	Reserve of Air Force Officers
RAFVR	Royal Air Force Volunteer Reserve
Ramrod	Fighter escort to bombers
RAS	Reserve Aeroplane Squadron
R-boat	Small German minesweeper
RCAF	Royal Canadian Air Force
Recce	Reconnaissance
Retd	Retired
RDF	Radio Direction Finding [original British term for Radar]
RF	Radio Frequency
RFC	Royal Flying Corps
R/G	Rear Gunner
Rhubarb	Small scale fighter or fighter-bomber operations directed against low-flying aircraft or on ground target
RLG	Relief Landing Ground
RN	Royal Navy
RNAS	Royal Naval Air Service
RNVR	Royal Navy Volunteer Reserve
RNZAF	Royal New Zealand Air Force
RNZN	Royal New Zealand Navy
RNZNVR	Royal New Zealand Navy Volunteer Reserve
Roadstead	Fighter escorted bombers or fighter-bombers on operations against ships at sea or in harbour
Rodeo	Fighter-only offensive sweep over enemy territory
ROs	Routine Orders
RS	Radio School
RS	Reserve Squadron
r/t	radio telephony
RTS	Recruit Training School
SAAF	South African Air Force
SAG	School of Aerial Gunnery
SAN	School of Air Navigation
SAP	Semi-Armour Piercing [bomb]
SAP	School of Air Pilotage
SASO	Senior Air Staff Officer
SBA	Standard Beam Approach
Scramble	Emergency take off, usually to intercept enemy raid
SD	Special Duties
SEA LION	German code name for invasion of Britain, 1940
SFF	Service Ferry Flight
SFPP	Service Ferry Pilots Pool
SFTS	Service Flying Training School
SGR	School of General Reconnaissance
SGR&AN	School of General Reconnaissance & Air Navigation
Sgt	Sergeant

S/L	Squadron Leader
SLG	Satellite Landing Ground
SMO	Senior Medical Officer
SOC	Struck Off Charge
SOE	Special Operations Executive
SN&BD	School of Navigation & Bomb Dropping
SSF	School of Special Flying
SPTU	Staff Pilot Training Unit
Sqn	Squadron
Sqn Ldr	Squadron Leader
SR	Special Reserve
SRF	Sea Rescue Flight
SSC	Short Service Commission
Sub-Lt (A)	Sub-Lieutenant (Air Branch)
Sweep	General term for fighters on operation over enemy territory
SWPA	South-West Pacific Area
TAF	Territorial Air Force
TAG	Telegraphist Air Gunner
TAS	True Air Speed
TC	Training Command
TDF/U	Torpedo Development Flight/Unit
TDS	Training Depot Station
TI	Target Indicator
t/o	take off
Torbeau	Beaufighter equipped with torpedo
TORCH	British code name for Anglo-American landings in French North Africa, November 1942
TS	Training Squadron
TSR	Torpedo Spotter Reconnaissance
TT	Target Towing
TTC	Technical Training Centre
TTS	Technical Training School
TTU	Target Towing Unit
TTU	Torpedo Training Unit
Turbinlite	Searchlight fitted to aircraft for night -fighter operations
U-boat	German submarine
u/c	undercarriage
u/s	unserviceable
USAAC	United States Army Air Corps
USAAF	United States Army Air Force
USAAS	United States Army Air Service
USMC	United States Marine Corps
USN	United States Navy
u/t	under training
VC	Victoria Cross
Vegetable	Mine [laid at sea from the air]
VHF	Very High Frequency [radio]
VIGOROUS	British code name for covering operations for convoy from Eastern Mediterranean to Malta, June 1942
VIP	Very Important Person
VLR	Very Long Range [aircraft]
WA	West Africa
WAAF	Women's Auxiliary Air Force
WAG	Wireless Operator-Air Gunner

W/C	Wing Commander
Wg	Wing
Wg Cdr	Wing Commander
WO	War Office
WO	Warrant Officer
WOAG	Wireless Operator-Air Gunner
WOp	Wireless Operator
WOpAG	Wireless Operator-Air Gunner
WS	Wireless School
w/t	wireless telegraphy
Wt Off	Warrant Officer
Würzburg	German ground radar system controlling searchlights, flak and night-fighters

Bibliography

Official Records

Aircraft accident cards, summaries and reports.

Aircraft movement cards for RAF aircraft.

Bomber Command loss cards.

First World War casualty cards for RFC, RNAS and RAF.

First World War casualty files for RFC, RNAS and RAF.

Operations Record Books.

RNZAF [draft] biographies of deceased personnel, 1939-45.

Books

Note: A large number of published school registers and squadron histories have been consulted, although by no means all, but they are not included in this list for reasons of space.

Air Force List, The. London: Air Ministry [various editions from April 1918]

Allison, Les and Hayward, Harry. *They Shall Grow Not Old: A Book of Remembrance.* Brandon, Manitoba: Commonwealth Air Training Plan Museum, 1992.

Barker, Ralph. *The Thousand Plan: The Story of the First Thousand Bomber Raid - on Cologne.* London: Chatto and Windus, 1965.

Bowyer, Michael J F. *2 Group R.A.F: A Complete History, 1936-1945.* London: Faber and Faber, 1974.

Bowyer, Michael J F and Rawlings, John D R. *Squadron Codes, 1937-56.* Cambridge: Patrick Stephens, 1979.

Brew, Alec. *The Defiant File.* Tonbridge: Air-Britain, 1996.

Bruce, J M. *The Aeroplanes of the Royal Flying Corps (Military Wing).* London: Putnam, 1982.

Chappell, F R. *Wellington Wings: An RAF Intelligence Officer in the Western Desert.* London: William Kimber, 1980.

Cheesman, E C. *Brief Glory: The Story of the A.T.A.* Leicester: Harborough, 1946.

Chorley, W R. *Royal Air Force Bomber Command Losses of the Second World War, 1939-1940.* Earl Shilton: Midland Counties Publications, 1992.

————*Royal Air Force Bomber Command Losses of the Second World War, 1941.* Earl Shilton: Midland Counties Publications, 1993.

————*Royal Air Force Bomber Command Losses of the Second World War, 1942.* Earl Shilton: Midland Counties Publications, 1994.

Christie, Carl A. *Ocean Bridge: The History of RAF Ferry Command.* Earl Shilton: Midland Publishing, 1995.

Coulthard-Clark, C D. *The Third Brother: The Royal Australian Air Force, 1921-39.* North Sydney: Allen & Unwin, 1991.

Cutlack, F M. *The Australian Flying Corps in the Western and Eastern Theatres of War, 1914-1918.* Sydney: Angus & Robertson, 1923.

Delve, Ken. *The Source Book of the R.A.F.* Shrewsbury: Airlife, 1994.

Douglas, W A B. *The Creation of a National Air Force: The Official History of the Royal Canadian Air Force, Volume II.* Toronto: University of Toronto Press, 1986.

Duxbury, David; Ewing, Ross and Macpherson, Ross. *Aircraft of the Royal New Zealand Air Force.* Auckland: Heinemann, 1987 (revised ed. - first published as *New Zealand Military Aircraft 1913-1977*)

Eather, Steve. *Flying Squadrons of the Australian Defence Force.* Weston Creek, Australian Capital Territory: Aerospace Publications, 1995.

Franks, Norman. *Valiant Wings.* London: William Kimber, 1988.

Gomersall, Bryce B. *The Stirling File [revised edition].* Tonbridge: Air-Britain, 1987.

Halley, James J. *The K File: The Royal Air Force of the 1930s.* Tonbridge: Air-Britain, 1995.

————*The Lancaster File.* Tonbridge: Air-Britain, 1985.

————*Royal Air Force Aircraft L1000 to N9999.* Tonbridge: Air-Britain, 1995.

————*Royal Air Force Aircraft P1000 to R9999.* Tonbridge: Air-Britain, 1996.

————*Royal Air Force Aircraft T1000 to T9999.* Tonbridge: Air-Britain, 1981.

————*Royal Air Force Aircraft V1000 to V9999, W1000 to W9999.* Tonbridge: Air-Britain, 1983.

————*Royal Air Force Aircraft X1000 to X9999, Z1000 to Z9999.* Tonbridge: Air-Britain, 1984.

————*Royal Air Force Aircraft AA100 to AZ999.* Tonbridge: Air-Britain, 1985.

————*Royal Air Force Aircraft BA100 to BZ999.* Tonbridge: Air-Britain, 1986.

————*Royal Air Force Aircraft DA100 to DZ999.* Tonbridge: Air-Britain, 1987.

————*Royal Air Force Aircraft EA100 to EZ999.* Tonbridge: Air-Britain, 1988.

————*Royal Air Force Aircraft FA100 to FZ999.* Tonbridge: Air-Britain, 1989.

————*Royal Air Force Aircraft HA100 to HZ999.* Tonbridge: Air-Britain, 1989.

————*Royal Air Force Aircraft JA100 to JZ999.* Tonbridge: Air-Britain, 1990.

————*Royal Air Force Aircraft KA100 to KZ999.* Tonbridge: Air-Britain, 1990.

————*Royal Air Force Aircraft LA100 to LZ999.* Tonbridge: Air-Britain, 1991.

————*Royal Air Force Aircraft MA100 to MZ999.* Tonbridge: Air-Britain, 1991.

————*Royal Air Force Aircraft NA100 to NZ999.* Tonbridge: Air-Britain, 1992.

Hamlin, John F. *The Harvard File.* Tonbridge: Air-Britain, 1988.

Hanson, Colin. *By Such Deeds: Honours and Awards in the Royal New Zealand Air Force, 1923-1998.* Publication due 1998.

Hatch, F J. *Aerodrome of Democracy: Canada and the British Commonwealth Air Training Plan 1939-1945.* Ottawa: Directorate of History, Department of National Defence, 1983.

Hayward, Roger. *The Beaufort File.* Tonbridge: Air-Britain, 1990.

Hobson, Chris. *Airmen Died in the Great War, 1914-1918: The Roll of Honour of the British and Commonwealth Air Services of the First World War.* Polstead: J B Hayward, 1995.

Jefford, Wg Cdr C G, MBE RAF. *RAF Squadrons.* Shrewsbury: Airlife, 1987.

Jenks, C F L. *New Zealand Military Aircraft and Serial Numbers.* Wellington: Aviation Historical Society of New Zealand Inc, 1980.

Merrick, K A. *Flights of the Forgotten: Special Duties Operations in World War Two.* London: Arms and Armour, 1989.

Middlebrook, Martin and Everitt, Chris. *The Bomber Command War Diaries: An Operational Reference Book, 1939-1945.* Hamondsworth: Viking, 1985.

Morgan, Eric B and Shacklady, Edward. *Spitfire: The History.* Stamford: Key Publishing, 1987.

Moyle, Harry. *The Hampden File.* Tonbridge: Air-Britain, 1989.

Nesbit, Roy C. *The Armed Rovers: Beauforts & Beaufighters over the Mediterranean.* Shrewsbury: Airlife, 1995.

Prisoners of War, Naval and Air Forces of Great Britain and the Empire, 1939-1945: Alphabetical Nominal Registers...Corrected generally up to 30th March 1945. 1st pub London: HMSO, 1945; facsimile ed. Polstead: J B Hayward & Son, 1990.

Roberts, R N. *The Halifax File.* Tonbridge: Air-Britain & the British Aviation Archaeological Council, 1982.

————*The Whitley File.* Tonbridge: Air-Britain, 1986.

Ross, Sqn Ldr J M S. *Royal New Zealand Air Force [Official History of New Zealand in the Second World War 1939-45].* Wellington: War History Branch, Department of Internal Affairs, 1955.

Shores, Christopher. *Dust Clouds in the Middle East: The Air War for East Africa, Iraq, Syria, Iran and Madagascar, 1940-42.* London: Grub Street, 1996.

Shores, Christopher, *et al. Air War for Yugoslavia, Greece and Crete, 1940-41.* London: Grub Street, 1987.

————*Bloody Shambles: The First Comprehensive Account of Air Operations Over South-East Asia, December 1941-April 1942; Volume One: The Drift to War to the Fall of Singapore.* London: Grub Street, 1992.

————*Bloody Shambles: The First Comprehensive Account of Air Operations Over South-East Asia, December 1941-May 1942; Volume Two: The Defence of Sumatra to the Fall of Burma.* London: Grub Street, 1992.

————*Fledgling Eagles: The Complete Account of Air Operations During the 'Phoney War' and Norwegian Campaign, 1940.* London: Grub Street, 1991.

————*Malta: The Hurricane Years, 1940-41.* London: Grub Street, 1987.

————*Malta: The Spitfire Year, 1942.* London: Grub Street, 1991.

Sturtivant, Ray. *The Anson File.* Tonbridge: Air-Britain, 1988.

Sturtivant, Ray and Ballance, Theo. *The Squadrons of the Fleet Air Arm.* Tonbridge: Air-Britain, 1994.

Sturtivant, Ray and Burrow, Mick. *Fleet Air Arm Aircraft 1939 to 1945.* Tonbridge: Air-Britain, 1995.

Sturtivant, Ray; Hamlin, John and Halley, James J. *Royal Air Force Flying Training and Support Units.* Tonbridge: Air-Britain, 1997.

Sturtivant, Ray & Page, Gordon. *Royal Navy Aircraft Serials and Units, 1911-1919.* Tonbridge: Air-Britain, 1992.

————*The Camel File.* Tonbridge: Air-Britain, 1993.

————*The S.E.5 File.* Tonbridge: Air-Britain, 1996.

Terraine, John. *The Right of the Line: The Royal Air Force in the European War, 1939-1945.* London: Hodder and Stoughton, 1985.

Thetford, Owen. *Aircraft of the Royal Air Force Since 1918.* London: Putnam, 1962 [3rd ed.]

————*British Naval Aircraft Since 1912.* London: Putnam, 1958.

Thomas, C H. *The Typhoon File.* Tonbridge: Air-Britain & the British Aviation Archaeological Council, 1981.

Thompson, Dennis and Sturtivant, Ray. *Royal Air Force Aircraft J1-J9999 and WWI Survivors.* Tonbridge: Air-Britain, 1987.

Thompson, Wg Cdr H L. *New Zealanders with the Royal Air Force: Volume I: European Theatre, September 1939-December 1942 [Official History of New Zealand in the Second World War 1939-45].* Wellington: War History Branch, Department of Internal Affairs, 1953.

————*New Zealanders with the Royal Air Force: Volume III: Mediterranean and Middle East, South East Asia [Official History of New Zealand in the Second World War 1939-45].* Wellington: War History Branch, Department of Internal Affairs, 1959.

War Graves of the British Empire, The: Registers of the names of those who fell in the Great War... London: Imperial War Graves Commission [various dates - see also next entry]

War Dead of the British Commonwealth and Empire, The: Registers of the names of those who fell in the 1939-1945 War... London: Imperial War Graves Commission [various dates - see also next entry]

War Dead of the Commonwealth, The: Registers of the names of those who fell in the 1939-1945 War... Maidenhead: Commonwealth War Graves Commission [various dates - see also above entries]

Wynn, Kenneth G. *A Clasp for 'The Few': New Zealanders with the Battle of Britain Clasp.* Auckland: K G Wynn, 1981.

————*Men of the Battle of Britain: A Who was Who of the Pilots and Aircrew, British, Commonwealth and Allied, who flew with Royal Air Force Fighter Command July 10 to October 31 1940.* Norwich: Gliddon Books, 1989.

Periodicals

Aeromilitaria: The Air-Britain Military Aviation Historical Quarterly. [various issues 1975-1998]

Aeroplane, The. [various issues 1915-1942]

Aeroplane Monthly. [various issues 1973-1998]

Flight. [various issues 1915-1942]

Journal of the Aviation Historical Society of New Zealand. [various issues 1958-1998]

New Zealand Gazette, The. [various issues 1923-1943]

New Zealand Wings. [various issues 1932-1998]

RNZAF Routine Orders. [various issues 1938, 1940-1942]

Southland Times, The. [various issues 1915-1956]

First to Fall - The Great War

(4 August 1914 - 11 November 1918)

MILITARY AVIATION in the First World War developed at a remarkable rate over a brief four years. For the British, a modest 1844 personnel and 272 aircraft on 5 August 1914 had by the Armistice been developed into the world's largest air force, with nearly 300,000 personnel and over 22,000 aircraft on charge. The squadron was the tactical unit, almost 200 of which were engaged on operations by November 1918.

Britain's original pre-war army and navy components, the Royal Flying Corps (RFC) and Royal Naval Air Service (RNAS), were merged into the Royal Air Force (RAF) on 1 April 1918. Australia, the only Dominion to operate its own air arm, possessed the Australian Flying Corps (AFC). Aside from meeting a constant demand to train increasingly greater numbers of men within the air services, the aircraft industry too underwent a major metamorphosis from cottage industry construction to scientific application and mass production.

The immediate effect of aircraft presence over the battlefield was to dramatically elevate and enlarge the view available from the 'high ground'. Thus the earliest activity engaged in was observation or photographic reconnaissance, rapidly followed by 'battery' work - spotting targets and opposing gun batteries for the artillery. Although this was not a decisive factor in the war, no major movement could now be undertaken by armies without their first taking into account aerial observation and attack. Fighters (or 'scouts') and air combat ensued as opposing forces fought to gain control of the skies, to permit their respective 'Art Obs' (Artillery Observation) crews to proceed unhindered. Observation from kite balloon and airship, low and high level bombing, strafing and anti-submarine work were also undertaken.

Adventurous New Zealanders were naturally attracted to aviation from the beginning and many hundreds made their way into the Allied air services. Those already in Britain or Australia, or who had immediately made their own way there, were able to enlist direct. Ongoing expansion resulted in regular calls for volunteers from other units within the Imperial Forces to serve as aircrew, the NZEF also proving a valuable source throughout 1917 and 1918, in Europe and the Middle East.

Two private schools undertook elementary pilot training in New Zealand, the Walsh Brothers' NZ Flying School qualifying pupils on seaplanes at Kohimarama from December 1915 and Henry (later Sir Henry) Wigram's Canterbury (NZ) Aviation Company on landplanes at Sockburn from August 1917. A creditable total of 253 had taken their 'ticket' by the Armistice, the majority with the CAC during 1918. Nearly 90% sailed for Britain to undergo further training with the RFC or RAF, but many reached the RAF late in the war and fewer than 75 had graduated with their respective 'Wings' by its end.

Although a very modest sized contribution (probably some 700 ground and aircrew in all, about half of whom saw operational service), the contribution of New Zealanders was quite disproportionate to the number employed. Casualties were relatively high, however, over 70 losing their lives by the Armistice, 40% of them through flying accidents - a reflection of poor training, general ignorance as to aircraft behaviour and constant pressure to replace operational losses with airmen possessing very limited flying experience.

By a strange twist of fate New Zealand's first military air casualty personified the ANZAC tradition then being forged at such bloody cost on the rugged shores of Gallipoli. Australian born and New Zealand educated, William Burn, as an officer of the regular New Zealand Defence Force, was despatched to Mesopotamia to fly against the same enemy as his comrades - the Turk. His Australian pilot who died with him, George Merz, became the first member of the Australian Flying Corps to lose his life.

Fri 30 Jul 1915

MESOPOTAMIA (IRAQ)

Return to base

Mesopotamia Flight, Indian Flying Corps (Basra, Iraq)

Caudron GIII IFC4 - upon completion of reconnaissance work in connection with the capture of Nasiriyeh on the 25th, took off early morning from Asani piloted by Lt G P Merz, Australian Flying Corps. Another Caudron, flown by the Flight commander, New Zealander Capt **H L Reilly**, followed ten minutes later. Both aircraft, unobserved by one another, suffered engine failure en route and carried out forced landings. Reilly came down near a friendly Arab village, made repairs and flew on to Basra. The less fortunate IFC4 came down some 20 miles away from a filling station at Abu Salibiq and was set upon by a hostile group of nomad Arabs. The Caudron being unarmed, the pilot and observer fought a running battle using their revolvers while retreating towards Abu Salibiq. According to varying reports of two Arab eyewitnesses, the airmen killed one of their attackers and wounded up to five others before themselves being overcome and killed. Reilly located the wrecked Caudron three days later during an aerial search, but no trace of the airmen was found and both are commemorated on the Basra Memorial.

Observer: Lt **William Wallace Allison BURN**, mid(2), New Zealand Staff Corps & RFC - Age 24.

Burn, the first member of the regular New Zealand Defence force to learn to fly, was born at Melbourne, Australia, but emigrated to New Zealand at an early age. His name was originally omitted from the Memorial, the error being discovered and corrected only in recent years.

Wed 1 Sep 1915

WESTERN FRONT

(Duty?)

Escadrille MF2, Aviation Militaire (Clermont-en-Argonne, France)

Maurice Farman MF11(?) (serial?) - believed to have been brought down during an operational sortie and crashed near Parois, 19km west of Verdun. The pilot (and observer?) died and is buried at Vauquois, 11km north of Clermont-en-Argonne.

Pilot: 1370 Sergent Pilot **Claude COUTURIER**, Aviation Militaire - Age 38.

Couturier was born at Le Creusot, France, and came to New Zealand in 1912 or 1913, after qualifying as a pilot at the Wright Brothers school in the United States. He proposed setting up a flying school at Rolleston, near Christchurch, before returning to France upon the outbreak of war. Following his death, a Bleriot machine he had left behind in New Zealand was sold by auction, and later resold to the Canterbury Aviation Company of Sockburn, whose pupils used it as a ground handling trainer.

Sat 29 Apr 1916

MESOPOTAMIA (IRAQ)

Capture upon the fall of Kut el Amara, Iraq

30 Squadron, RFC (Camp Wadi, Iraq)

Kut, under siege since early December 1915, surrendered to Turkish forces. Included among the large number of British and Indian Army personnel taken prisoner during the campaign were some 30 of the Squadron's ground crew. The majority of these airmen did not survive incarceration because of the cruel conditions imposed upon them by their indifferent captors. A New Zealander among them is buried in Baghdad.

Air Mechanic: 44 Air Mechanic **Francis Luke ADAMS**, Australian Flying Corps - Age 23 (if died before 8 October).

The exact death date for Adams is not known. He is reported to have succumbed to malaria sometime during August-November 1916, probably while on a forced march to Turkey.

Fri 2 Jun 1916

ENGLAND

Training?

Unit - possibly 7, 8, 22 or 24 Reserve Squadron, RFC (all Netheravon, Wiltshire)

Morane-Saulnier Type L 5059(?) - started up in one of the hangars, taxied out and proceeded to take off, but on lifting off, banked steeply to the left and side-slipped into the ground. Almost immediately the overturned monoplane burst into flames. The trapped observer, Capt L Prickett, Royal Garrison Artillery, attached RFC, died before he could be extricated from the wreckage. The injured pilot died at the Tidworth Military Hospital on the 7th and is buried at Tidworth. Independent contemporary reports and RFC casualty card confirm the aircraft type, but serial 5059 is that of a Maurice Farman MF11 Shorthorn.

Pilot: Lt **William Archibald BUCHANAN**, 1st Bn Connaught Rangers & RFC - Age 21.

Tue 20 Jun 1916

ENGLAND

Solo training

Most likely 4 or 11 Reserve Squadron, RFC (both Northolt, Middlesex)

Martinsyde S1 696 - while flying from 'an aerodrome in an outlying district of London', fell into a spin at about 3000 feet and dived into the ground. The pilot died later in the day at the RFC Hospital at Bryanston Square, London, and is buried at Ruislip.

Pilot: 2nd Lt **George Edmund Vernon AIMER**, RFC - Age 30.

Wed 21 Jun 1916

WESTERN FRONT

Proceeding to Rouvrel for practice with Le Prieur rockets.

11 Squadron, RFC (Savy, France - 13th Army Wing, 3rd Brigade)

Nieuport 16 A121 - took off at 0900 and entered a right hand bank, which increased to vertical at about 200 feet before the fighter side slipped into the ground. The pilot died of his injuries 15 minutes later without recovering consciousness and is buried at Aubigny, 15km NNW of Arras. In his casualty report, the Squadron's commanding officer was of the opinion that the extra weight and drag of the unfamiliar rocket fittings carried on the wing struts seriously degraded the Nieuport's normal performance. He stressed that very great care should be taken in banking a machine thus fitted. No actual rockets (for attacks against German observation balloons) were being carried at the time.

Pilot: Capt **Herbert Ambrose COOPER**, mid, RFC - Age 32.

The CWGC register incorrectly gives Cooper's age as 26.

Mon 26 Jun 1916

WESTERN FRONT

Bombing raid on Biache railway junction

23 Squadron, RFC (Le Hameau, France - 13th Army Wing, 3rd Brigade)

Royal Aircraft Factory FE2b 6348 - took off at 0850 piloted by Lt **H B Russell**, Royal Field Artillery, attached RFC, and failed to return. This was a replacement aircraft as the engine of the first refused to start. Because of the delay the opportunity of accompanying the rest of the Squadron was lost. After successfully bombing the target independently 6348 was attacked at 5500 feet by three Fokkers and shot down in flames over Biache. The pilot received a wound in his upper right lung and fainted for a time, while his companion was hit three times in the stomach. Despite their injuries, a crash landing was brought off in the German trenches, about a mile short of their own lines. Both crew were taken prisoner, but the observer died of his wounds on 9 August at Ohrdruf, 12km south of Gotha. He was buried in the Ohrdruf Prisoners of War Cemetery, but later reinterred at Niederzwehren, a few kilometres SW of Kassel.

Observer: Lt **James Robert DENNISTOUN**, North Irish Horse & RFC - Age 33. 1st op.

Dennistoun and Russell were cousins. Russell secured a permanent commission in the post-war RAF and attained the rank of Air Vice Marshal during Second World War service.

Sat 15 Jul 1916

WESTERN FRONT

Patrol

22 Squadron, RFC (Bertangles, France - 14th Army Wing, 4th Brigade)

Royal Aircraft Factory FE2b 6980 - took off at 1715 piloted by 2nd Lt R H Sievwright, RFC, and dived to attack a party of enemy infantry in the vicinity of Hébuterne, 20km SSW of Arras. The pilot was unscathed but his observer took a bullet in the thigh from rifle fire. The engine was also hit, necessitating a force landing immediately behind British lines, where 6980 was shelled and completely wrecked by hostile artillery. The observer died the following day and is buried at Couin, 25km SW of Arras.

Observer: 2nd Lt **John Laurie REID**, 18th Bn Northumberland Fusiliers & RFC - Age 22.

Sat 15/Sun 16 Jul 1916

WESTERN FRONT

Bombing raid against Sallaumines

18 Squadron, RFC (Bruay, France - 1st Corps Wing, 1st Brigade)

Royal Aircraft Factory FE2b 5233 - took off at 0030 and brought down an hour later through German infantry machine gun fire damage to the propeller while flying below 1000 feet. 5233 attempted to glide to its own lines at Hill 70 (Loos - map ref Sheet 36c H.25) but struck the parados of the German first line trench some 50 yards short, bounced back into the air and came down in the wire. The observer, Capt J H F McEwen, RFC, was injured on being catapulted out of the bomber and soon captured. His pilot, hit by rifle or machine gun fire which broke out from both sides of the lines as the machine crashed, died a few minutes after the two crew managed to find cover in a nearby trench. According to a message later dropped by the Germans they buried him at Carvin, 26km NE of Arras, on the 17th, but after the war he was reinterred near Souchez, 12km NNW of Arras.

Pilot: Lt **Harold Winstone BUTTERWORTH**, RFC - Age 21.

The CWGC register incorrectly gives the 15th as Butterworth's date of death.

Wed 9 Aug 1916

GERMANY

Lt **J R DENNISTOUN**, a prisoner in Germany, died of wounds received in combat on 26 June.

Sat 26 Aug 1916

WESTERN FRONT

Artillery observation over Contalmaison district

7 Squadron, RFC (Warloy, France - 15th Corps Wing, 5th Brigade[?])

Royal Aircraft Factory BE2d 6743 - brought down by flak at 1100 and crashed on British side of the lines at map reference Sheet 57d X.11.a. The observer, 2nd Lt D Clarke, Honourable Artillery Company, attached RFC, was killed. His pilot, seriously wounded by flak or in the crash, or through a combination of both, died a week later, on 2 September, and is buried at Heilly, near Méricourt-l'Abbé, 40km SSW of Arras.

Pilot: 2nd Lt **Lawrence Dobrée RUSSELL**, RFC - Age 18.

Russell was born in England but emigrated to New Zealand as a child. His brother, Francis Gerald Russell, died on 28 January 1917 while flying with 21 Sqn, RFC.

Fri 15 Sep 1916

ENGLAND

Solo training

7 Reserve Squadron, RFC (Netheravon, Wiltshire - 4th Wing, Training Brigade)

Maurice Farman S11 Shorthorn 546 - appeared to approach the airfield at 0940 for a landing but nosed dived into the ground from about 100 feet. The pilot, on his third solo flight, died and was buried on Monday morning at Upavon, the band from the New Zealand Force at nearby Larkhill being present.

Pilot: 2nd Lt **Victor Stephen Henry ABBOTT**, RFC (Special Reserve) - Age 23.

Thu 19 Oct 1916

ENGLAND

Training

Cranwell, RNAS (Cranwell, Lincolnshire)

Royal Aircraft Factory BE2c 8429 - dived into the ground from 200 feet, severely injuring the pilot, who was admitted to the Royal Naval Lincoln Hospital. He died the following day without regaining consciousness and is buried at Cranwell.

Pilot: Flt Sub-Lt **Arthur Robert GREENWELL**, RNAS - Age 20.

Greenwell lived in New Zealand for at least 12 years prior to travelling to England to enlist. His place of birth is unknown.

Sun 22 Oct 1916

WESTERN FRONT

Offensive patrol east of Gommecourt

11 Squadron, RFC (Izel-le-Hameau, France - 13th Army Wing, 2nd Brigade)

Royal Aircraft Factory FE2b 7684 - took off at 1410 piloted by 2nd Lt A L M Shepherd, 6 Bn King's Royal Rifle Corps, attached RFC, and last seen in combat with enemy aircraft very low over the lines. 7684 was brought down and its two crew captured, the observer dying later in the day from a gunshot wound to the head received in the fight. He was buried in the German Extension of the Gavrelle Communal Cemetery, 9km ENE of Arras. His pilot, also wounded, died 12 days later. After hostilities the observer's grave could not be located and he is now commemorated by a special memorial in the Villers Station Cemetery at Villers-au-Bois, 12km NW of Arras.

Observer: 17094 Air Mechanic 1st Class **Norman Leslie BRAIN**, RFC - Age 24.

Brain's name does not appear in any CWGC register and seems to have been omitted in error.

Thu 26 Oct 1916

WESTERN FRONT

Artillery observation - counter battery work over Mametz Wood-Miraumont with 56th Siege

7 Squadron, RFC (Warloy, France - 15th Corps Wing, 5th Brigade)

Royal Aircraft Factory BE2d 6235 - took off at 1510, but last heard from at 1544 and presumed to have been brought down north of Courcelette, 25km south of Arras. No grave could be located post-war for either the pilot or his observer, 2nd Lt G A Palfreyman, 3 Bn East Kent Rgt, attached RFC, and they are commemorated on the Arras Memorial.

Pilot: 2nd Lt **Forrest Gale PARSONS**, RFC - Age 34.

Parsons had also served in the Boer War.

Thu 7/Fri 8 Dec 1916

ENGLAND

Motor vehicle accident

25 Reserve Squadron, RFC (Thetford, Norfolk - 7th Wing, Training Brigade)

While walking along the Euston Road at about 1730 towards Thetford and pushing a cycle, was struck from behind by a following motor cyclist and seriously injured. Died six hours later at the Cambridge Eastern General Hospital and is buried at Hatfield, Hertfordshire.

(Pilot): 2nd Lt **George Aubyn POWELL**, RFC - Age 27.

Tue 16 Jan 1917

ENGLAND

1 Reserve Squadron, RFC (Fort Grange, Gosport, Hampshire - 17th Wing, Southern Group Command, Training Brigade)

Died shortly after an operation for appendicitis at the RFC Hospital at Bryanston Square, London, and buried at Fulham, London, on the 19th.

(Pilot): Capt **Rudolph Dunstan VAVASOUR**, Royal Field Artillery & RFC - Age 23. Served operationally.

Sun 28 Jan 1917

WESTERN FRONT

Defensive patrol

21 Squadron, RFC (Bertangles, France - 14th Army Wing, 4th Brigade)

Royal Aircraft Factory BE12 6646 - took off at 1310 and seen to be hit by flak, crashing between Guedecourt and le Transloy, about 26km south of Arras. The pilot is buried at Flers, 2km SW of Guedecourt.

Pilot: 2nd Lt **Francis Gerald RUSSELL**, Royal Field Artillery & RFC - Age 28.

Russell, who served with the NZEF prior to transferring to the RFA, had earlier been wounded on 8 March 1916, while flying as an observer with 7 Sqn, RFC. His brother, Lawrence Dobrée Russell, died on 2 September 1916 of wounds received on 26 August while also flying with that squadron.

Tue 3 Apr 1917

WESTERN FRONT

Reconnaissance patrol

11 Squadron, RFC (Izel-le-Hameau, France - 13th Army Wing, 3rd Brigade)

Royal Aircraft Factory FE2b A808 - took off at 1300 piloted by Lt E T C Brandon, RFC, and seen by two other members of the patrol to be hit by flak before falling at map reference Sheet 51B H.27.d. An eyewitness on the ground reported that at about 1530 the aircraft came down into the Scarpe River, near Arras. The two crew are commemorated on the Arras Memorial.

Observer: 2nd Lt **George MASTERS**, mid, NZ Maori (Pioneer) Bn NZEF & RFC - Age 26.

Sat 14 Apr 1917

WESTERN FRONT

Offensive patrol

60 Squadron, RFC (Filescamp Farm, France - 13th Army Wing, 3rd Brigade)

Nieuport 17 B1511 - took off at 0830 with four others and one of four which failed to return. At about 0915 the formation sighted and dived upon two enemy two-seater aircraft east of Douai. It in turn was attacked and badly mauled by enemy fighters, four of the Nieuports being swiftly felled. Three pilots forced landed and were taken prisoner, one of whom died of wounds two days later. The pilot of B1511 died during the combat and is buried at Beaumont, 5km NW of Douai.

Pilot: 2nd Lt **John Herbert COCK**, NZ Rifle Brigade NZEF and RFC - Age 23.

Mon 23 Apr 1917

WESTERN FRONT

Patrol

24 Squadron, RFC (Flez, France - 14th Army Wing, 4th Brigade)

Airco DH2 7909 - took off at 1830 and attacked by enemy aircraft while patrolling over IV Army Front. During the engagement 7909 accidentally collided with a 22 Sqn Royal Aircraft Factory FE2b (6929) and both machines crashed near le Verguier, 12km NW of St Quentin. The two FE2b crew died, as did the pilot of the DH2, who is buried at Jeancourt, just west of le Verguier.

Pilot: 2nd Lt **Melville Arthur WHITE**, Wellington Regiment NZEF & RFC - Age 24.

Tue 24 Apr 1917

ENGLAND

Solo training

13 Reserve Squadron, RFC (Dover, Kent - 6th Wing, Southern Group Command, Training Brigade)

Royal Aircraft Factory RE8 A3187 - stalled and spun into the ground from 500 feet at about 1530, bursting into flames on falling into a lane about a mile from Dover's Swingate Down airfield. The pilot died and was buried at Dover on the 28th.

Pilot: 2nd Lt **David John BARNES**, RFC - Age 22.

Barnes was born in Australia but emigrated to New Zealand about 1904. He is thought to have made his own way to England and had graduated as a service pilot the very day of the accident.

Training

9 Reserve Squadron, RFC (Norwich, Norfolk - 7th Wing, Eastern Group Command, Training Brigade)

Henry Farman F20 A1214 - crashed on failing to flatten out properly during a landing in the evening, the pilot being thrown out and suffering head injuries. He died on the morning of the 27th at the Norfolk War Hospital, Norwich, and was buried at 1500 on the 30th at Ashtead, Surrey.

Pilot: 1313 Cdt **Thomas Meredith ALLEN**, 2nd Infantry Bn AIF & Australian Flying Corps - Age 23. 2hrs 25min solo on Henry Farman.

Allen was born at Apia in Samoa but spent eight or nine years in New Zealand to complete his education and gain early work experience. He returned to the islands a few years prior to the war to work for his New Zealand father's trading company. Allen and one of his five brothers arrived in Sydney on board a company vessel shortly after the outbreak of war, whereupon both enlisted with the Australian Imperial Force. Thomas, who had been wounded serving on Gallipoli, later transferred to the Australian Flying Corps. His tombstone states that he 'died of injuries sustained whilst flying at Norwich, April 26th', while the CWGC register and official records give the 27th as date of death. According to a brother who was at his side when he died, death occurred 52 hours after the accident. It would appear therefore that Allen died very close to midnight 26/27 April.

Sat 19 May 1917

WESTERN FRONT

Balloon observer engaged on ground duty

9 or 32 Kite Balloon Section, RFC (near Franco-Belgian border - 6 Balloon Company, 2nd Balloon Wing, 2nd Brigade)

Attending a late morning briefing at a battery headquarters farm house located between Neuve Eglise and Ploegsteert in Belgium, a few kilometres north of Armentières, France. Shortly after arrival the HQ took a direct hit from a lucky first ranging shot from an enemy battery. The mortally wounded observer died at the scene soon afterwards and is buried at Trois-Arbes, near Steenwerck, 7km west of Armentières. The day's tragedy is graphically described by accompanying fellow observer, Goderick Hodges, in chapter nine of his autobiography, *Memoirs of an Old Balloonatic*. Hodges survived the explosion but lost a leg.

Balloon Observer: 2nd Lt **Carrel Watt FIDLER**, 7th Bn Argyll & Sutherland Highlanders & RFC - Age 30.

Fri 1 Jun 1917

ENGLAND

Dover to Farnborough, Hampshire

13 Training Squadron, RFC (Yatesbury, Wiltshire - 28th Wing, Southern Group Command, Training Brigade)

Royal Aircraft Factory BE2e A1362 - the pilot, accompanied by Air Mechanic 1st Class W Fozard, became lost near Farnborough and made a landing at Wokingham, Berkshire, to enquire as to his whereabouts. He then took off, circled the field once and had half completed a second circuit when A1362 stalled and crashed. The passenger died instantly and the pilot, unconscious, almost immediately afterwards, and is buried at Sutton Coldfield, Warwickshire.

Pilot: 2nd Lt **Nathaniel Fuhrmann CLARKE**, RFC - Age 31.

Wed 6 Jun 1917

WESTERN FRONT

Offensive patrol

6 (Naval) Squadron, RNAS (Flez, France - attached to 14th Army Wing, 4th Brigade, RFC)

Nieuport 17bis N3204 - took off at 1030 and believed to have been shot down by an enemy fighter at approximately 1200-1215. N3204 was last seen diving after the enemy near Bois de Bourlon, but it appeared to be suffering engine trouble as well. The pilot is commemorated on the Arras Memorial.

Pilot: Flt Lt **Fabian Pember REEVES**, RNAS - Age 21. 180+hrs.

Fri 8 Jun 1917

WESTERN FRONT

Special mission

1 (Naval) Squadron, RNAS (Bailleul, France - attached 11th Army Wing, 2 Brigade, RFC)

Sopwith Triplane N5491 - took off at 0940 and believed shot down in combat across the French border NE of Armentières, near Moorslede or Warneton, Belgium. The pilot is commemorated on the Arras Memorial.

Pilot: Flt Lt **Thomas Grey CULLING**, DSC, RNAS - Age 21. 141+hrs. 68+ ops.

Sat 23 Jun 1917

NEW ZEALAND

Workshop accident

New Zealand Flying School (Kohimarama)

Struck by the propeller of a Roberts engine undergoing a bench test. While the engine was running, instructor **Vivian Walsh** noticed a shadow and immediately stopped the engine, but the youth was either too close or the machine had moved forward, since the blades struck him on the right shoulder and head. He died an hour after admission to hospital.

Mechanic: **Charles Percy McGILL**, New Zealand Flying School - Age 18.

Up to the Armistice the NZFS and Canterbury (NZ) Aviation Company between them trained 253 pilots for war service to Royal Aero Club Aviator's Certificate standard. McGill's was the only fatal accident connected with this training. As the employee of a civilian company he does not appear in any CWGC register.

Sat 7 Jul 1917

WESTERN FRONT

Bombing raid on Ramegnies-Chin aerodrome, Belgium

55 Squadron, RFC (Boisdinghem, France - 9th HQ Wing)

Airco DH4 A7493 - took off at 0850 piloted by Lt P W Battersby, West Somerset Yeomanry, attached RFC, and last seen on the return journey going down under control over Lille. The two crew are commemorated on the Arras Memorial.

Observer: Capt **Wyndham Waterhouse FITZHERBERT**, 13th Bn Royal Sussex Regiment & RFC - Age 25.

Escort to photographic reconnaissance

1 (Naval) Squadron, RNAS (Bailleul, France - attached 11th Army Wing, 2 Brigade, RFC)

Sopwith Triplane N6309 - took off on patrol and believed to have been shot down in combat at about 1100 over Dadizeele, Belgium, 25km NE of Armentières, France, to crash near Comines, 12km NE of Armentières. The pilot is buried at Pont-du-Hem, near La Gorgue, 12km WSW of Armentières.

Pilot: Flt Sub-Lt **Kenneth Henry MILLWARD**, RNAS - Age 20. 202+hrs. 76+ ops.

Mon 16 Jul 1917

WESTERN FRONT

Bombing raid on Moorslede, Belgium

27 Squadron, RFC (Clairmarais North, France - 9th HQ Wing)

Martinsyde G100 7499 - successfully returned from the raid but still carrying a single bomb which had failed to drop clear; hanging vane down, it had caught on the landing gear. The pilot chose to ditch in Lake Arques, rather than on the airfield to the north, presumably to lessen the risk of an accidental explosion. Having carried off the ditching, however, he was drowned upon leaving the aircraft and is buried at Longuenesse, to the SW of St Omer.

Pilot: Capt **Melville JOHNSTONE**, RFC - Age 29.

Johnstone's brother, Godfrey Gleeson Johnstone, died on 30 January 1918 while flying with 22 Squadron, RFC.

Sun 22 Jul 1917

MEDITERRANEAN

Practice - first flight on type

F Squadron(?), RNAS (Mudros, Island of Lemnos, Aegean sea - 2 Wing)

Sopwith Camel F1 N6366(?) - spun in immediately after take off, killing the pilot, who is buried at Mudros. The only Camels which appear to match the theatre and date are N6352, N6353, N6354, N6365, N6366 and N6367. Assuming the crash wrecked the machine beyond repair, N6366 becomes the most likely candidate, being surveyed at Mudros in September and struck off as a total loss two weeks later.

Pilot: Flt Sub-Lt **Leslie Henry BRETT**, RNAS - Age 24.

WESTERN FRONT

Offensive patrol

32 Squadron, RFC (Droglandt, France - 22nd Army Wing, 5th Brigade)

Airco DH5 A9412 - shot down from 4000-5000 feet by flak and crashed just behind the enemy's second line east of Hooge, near Ypres, Belgium. The pilot is commemorated on the Arras Memorial.

Pilot: Lt **William Henry COATES**, Auckland Mounted Rifles NZEF & RFC - Age 29.

Mon 30 Jul 1917

WESTERN FRONT

Offensive patrol by four aircraft

32 Squadron, RFC (Droglandt, France - 22nd Army Wing, 5th Brigade)

Airco DH5 B336 - took off at 1750 accompanied by three others, including patrol leader, New Zealander, Capt **A Coningham**. One DH5 returned early but at about 1900 the others engaged five enemy fighters in combat near Langemarck, 7km NNE of Ypres, Belgium. Coningham was wounded but managed to complete the patrol and land at Droglandt. The pilot of B336, however, was more seriously wounded and injured himself further breaking both legs when a forced landing ended in a shell hole north of Vlamertinghe, a few kilometres west of Ypres. He died on 12 August and is buried at Etaples, France.

Pilot: 2nd Lt **Albert William GORDON**, RFC - Age 29. 1st op(?)

Gordon was the first New Zealand Flying School graduate to lose his life on active service.

Sat 11 Aug 1917

ENGLAND

Training

46 Training Squadron, RFC (Catterick, Yorkshire - 19th Wing, Northern Training Brigade, Training Division)

Royal Aircraft Factory BE12 6575 - collided over the airfield at 1045 with a 6 Training Squadron Avro 504A (7967), flown by 2nd Lt O A Browning, RFC. The latter was killed, while the pilot of 6575 died of injuries the following day and is buried at Catterick.

Pilot: Lt **Edward Errington CASTLE**, RFC - Age 33. 19hrs solo.

Sun 12 Aug 1917

WESTERN FRONT

2nd Lt **A W GORDON** died of wounds received in combat on 30 July.

Thu 20 Sep 1917

WESTERN FRONT

Offensive patrol

23 Squadron, RFC (La Lovie, Belgium - 22nd Army Wing, 5th Brigade)

SPAD 7 B3493 - took off at 1245 and later forced to land at map ref Sheet 28 I.18.c, after the pilot had been badly wounded in combat. B3493 came to a halt in a shell hole and was completely wrecked. Although rushed to hospital, the pilot died later in the day and is buried at Ypres.

Pilot: Lt **Frank BULLOCK-WEBSTER**, Canadian Machine Gun Corps & RFC - Age 31.

The CWGC register gives Bullock-Webster's age as 33.

Thu 27 Sep 1917

ENGLAND

Training

198 Depot Squadron, RFC (Rochford, Essex - Southern Home Defence Wing, Home Defence Brigade)

Avro 504J B3111 - took off at 0820 and over the NE corner of the airfield collided with a machine piloted by 2nd Lt J F Dick. Dick and his pupil survived the collision, but B3111's pilot, and his pupil, 2nd Lt G C Malcolm, 3 Bn King's Own Yorkshire Light Infantry, attached RFC, both died and are buried at Rochford.

Pilot Instr: Lt **Arthur Sydney TALBOT**, Australian Imperial Force & RFC - Age 27.

Talbot was born at Sydney, Australia. It is not known when he came to New Zealand, but he lived in Wanganui prior to the war and served with the AIF before transferring to the RFC.

Thu 4 Oct 1917

WESTERN FRONT

Patrol

19 Squadron, RFC (Bailleul, France - 11th Army Wing, 2nd Brigade)

SPAD 7 B1533 - took of at 0815 but crashed immediately afterwards in an adjacent field, killing the pilot, who is buried at Bailleul.

Pilot: 2nd Lt **Harold William DAWSON**, NZ Divisional Signal Company NZEF & RFC - Age 21.

Mon 8 Oct 1917

ENGLAND

Training

40 Training Squadron, RFC (Croydon, Surrey - 6th Wing, Eastern Training Brigade, Training Division)

Avro 504J A9769 - spun into the airfield at 1200 after beginning a left turn at 500 feet, killing the pilot, who is buried at Beddington.

Pilot: 2nd Lt **John Leask FOUBISTER**, RFC - Age 28. 11hrs.

Foubister was a graduate of the New Zealand Flying School.

Fri 19/Sat 20 Oct 1917

ENGLAND

Interception flight during Zeppelin raid against Northern England

33 Squadron, RFC (Gainsborough, Lincolnshire - Northern Home Defence Wing, 6th Brigade)

Royal Aircraft Factory FE2b A5656 - took off at 2000 in slight ground mist, but on reaching 200 feet dived steeply to the right, turned over on hitting the ground, and caught fire. The Observer, 2nd Lt H Preston, was thrown well clear and escaped badly shaken. The pilot died on impact and is buried at Gainsborough. The Squadron's commanding officer described him as a 'very experienced night pilot' and speculated that the accident might have been caused by loss of control through his thigh-length 'fug boots' fouling the controls.

Pilot: 2nd Lt **Hubert Philip SOLOMON**, NZ Cyclist Bn NZEF & RFC - Age 34. 268hrs solo (12 at night).

The 19th was Solomon's 34th birthday.

Wed 24 Oct 1917

WESTERN FRONT

Offensive patrol

23 Squadron, RFC (La Lovie, Belgium - 22nd Army Wing, 5th Brigade)

SPAD 7 B6762 - took off at 0810 and believed brought down out of control near Zonnebeke, 8km east of Ypres. The wrecked B6762 was later found at Korek Farm (map ref Sheet 28, D.9.c.1.7) but had to be abandoned owing to its proximity to the front line. Its pilot had died, but after hostilities his remains could not be found and he is commemorated on the Arras Memorial.

Pilot: 2nd Lt **Frederic James SHARLAND**, RFC - Age 34.

Sharland was a graduate of the New Zealand Flying School.

ENGLAND

Practice shoot

66 Training Squadron, RFC (Yatesbury, Wiltshire - 28th Wing, Southern Training Brigade, Training Division)

Royal Aircraft Factory RE8 A3841 of 'A' Flight - stalled and dived into the ground about half a mile east of the airfield at 1100. The pilot was buried 1330 on the 30th at Brookwood in Woking, Surrey.

Pilot: 2nd Lt **Stanley Howard PILKINGTON**, Australian Flying Corps - Age 28. 39hrs solo (8 on RE8).

Sat 27 Oct 1917

ENGLAND

Training

25 Training Squadron, RFC (Thetford, Norfolk - 26th Wing, Eastern Training Brigade, Training Division)

Maurice Farman S.11 Shorthorn B4759 - dived into the ground at the airfield at 0830, killing both pupil pilot 2nd Lt D McGill (on his first training flight) and his instructor, who is buried at Euston, Suffolk.

Pilot Instr: 2nd Lt **Charles William HOMER**, NZ Medical Corps NZEF & RFC - Age 23.

Homer was born in England. It is not known when he came to New Zealand but he enlisted in Auckland with the NZEF and sailed with the 7th Reinforcements in October 1915.

Tue 30 Oct 1917

MESOPOTAMIA (IRAQ)

Practice?

63 Squadron, RFC (Samarra)

SPAD 7 A8809 - crashed in the morning near Mantares, killing the pilot. Information on the accident is sparse, apart from a Court of Inquiry summary recording that 'the only obvious explanation of the accident seems that [the pilot] inadvertently closed the throttle instead of opening it as the Hispano throttle works in reverse action to most machines.' The Squadron was re-equipping with the Hispano-powered fighter at the time. Samarra is near Baghdad, but detachments of the Squadron were also based elsewhere, although Mantares (location unknown) was not one of them and none lie close to Basra, where Hyslop is buried.

Pilot: 2nd Lt **Ninian Steele HYSLOP**, Wellington Mounted Rifles NZEF & RFC - Age 31.

Tue 6 Nov 1917

WESTERN FRONT

Contact patrol

6 Squadron, RFC (Abeele, Belgium - 2nd Corps Wing, 2nd Brigade)

Royal Aircraft Factory RE8 A3643 - took off at 1455 and assumed to have completed the patrol, as it crashed while firing at a practice target on Lake Dickebusch, 13km east of Abeele. After diving steeply towards the target A3643 pulled out sharply, broke up and dived into shallow water. The pilot and his observer, Lt R H Richardson, 18th Bn London Regiment attached RFC, were both drowned and are buried at Lijssenthoek.

Pilot: 2nd Lt **Geoffrey Walter Gavin Maidens CATO**, RFC - Age 21.

Sun 23 Dec 1917

SCOTLAND

Practice

73 Training Squadron, RFC (Turnhouse, Lothian - 30th Wing, Northern Training Brigade, Training Division)

Albatros DV G56 - struck the water during aerobatics over the Firth of Forth, killing the pilot, who is buried at Edinburgh. A close friend of his was reported as saying that, in the course of completing a loop, the exhaust manifold became detached and struck the pilot on the forehead. Stunned or unconscious, he did not recover in time to regain control. However, a Court of Inquiry summary viewed that the cause was 'not due to any structural breakage,' and was further of the opinion that the pilot may simply have misjudged his height above the water and 'not come out of [a] spin before striking it.' Serial G56 was the RFC identity given to German Albatros DV 1162/17, which had been forced down intact behind Allied lines on 16 July 1917 and later flown to England for testing.

Pilot: 540 Capt **Clive Franklyn COLLETT**, MC*, RFC - Age 31. Over 1179hrs. At least 41 ops.

Coincidentally, Collett died three years to the day from when he had disembarked in England to enlist in the RFC. He made regular use of G56 from mid-October, possibly as personal 'hack', and had made over 30 flights in the machine by end of November. In January 1917, at the experimental station of Orfordness, Suffolk, Collett was the first airman to make a successful parachute jump from an RFC aeroplane.

Mon, Christmas Eve 1917

MIDDLE EAST

Died of exposure

67 (Australian) Squadron, RFC (Julis, Palestine - 40th Army Wing, Palestine Brigade)

Volunteered to escort two visiting officers, who were strangers to the area, part of the way back to their neighbouring camp at 2300. Very heavy rain was falling and shortly afterwards a violent gale blew up, during which 2nd Lt **Matheson** apparently lost his bearings or was cut off by a raging torrent in a small wadi. At 1700 two days later a search party located the body behind a clearing station about 1½ miles from the base. Thought to have died from exposure, he was buried on the 28th near Julis, about 20 miles north of Gaza, but later reinterred at Gaza. Renumbered from 1 Sqn, AFC, to 67 Sqn, RFC, the previous September, the unit regained its original identity in the following February.

Observer: 2nd Lt **Harry MacKay MATHESON**, mid, RFC - Age 30.

Matheson was born in Australia but it is not known when he came to New Zealand. His name appears in the 'Nominal Roll of New Zealanders Serving With the Australian Expeditionary Force' *compiled in 1915. In November 1916 he gained a transfer to the RFC. The CWGC register gives 24/26th as date of death, while Australian files record it as being on the 26th and RFC documents as 26/27th.*

Mon, New Year's Eve 1917

MIDDLE EAST

En route to Middle East

55 Kite Balloon Section, RFC (embarked on *Osmanieh*)

Carrying troops and medical staff to Alexandria, Egypt, the 4041 ton ex-liner *Osmanieh* struck a mine at the harbour entrance and very quickly sank. Of the 199 who died, 48 belonged to the RFC - mostly airmen of 55 and 56 Kite Balloon Sections - including a New Zealander, who is commemorated on the Chatby Memorial at Alexandria.

(?): Air Mechanic 1st Class **Philip Courtenay WAUD**, RFC - Age 18 or 19.

Although he was born in New Zealand it is thought that Waud emigrated (to England?) at an early age.

Tue, New Year's Day 1918

MIDDLE EAST

Training

School of Armament, RFC (Heliopolis, Egypt, Training Group)

Maurice Farman S.11 Shorthorn (serial?) - dived into the ground at Heliopolis after port wings collapsed while passing overhead and to the rear of the Aotea New Zealand Convalescent Hospital. The pilot, Lt **A C Upham**, Duke of Cornwall's Light Infantry attached RFC, was pinned under the wreckage and seriously injured. It took 12 men to lift the wreck clear of him. His passenger was thrown clear but suffered a head injury, from which he died a few minutes afterwards without gaining consciousness. He is buried at Cairo.

(Armament Officer): 2770 Lt **Wesley Neal SPRAGG**, RFC - Age 23. Believed to have served operationally in France.

Upham was born in England and had visited or lived in New Zealand prior to the war. He emigrated to New Zealand in 1920 and served in the RNZAF during the Second World War.

Sat 12/Sun 13 Jan 1918

ENGLAND

Solo night training

33 Squadron, RFC (Gainsborough, Lincolnshire - 46th Northern Home Defence Wing, 6th Brigade)

Royal Aircraft Factory FE2b A5626 - crashed while landing at 2315, killing the pilot, who is buried at Gainsborough.

Pilot: 2nd Lt **Frederick James LIVINGSTONE**, RFC - Age 25.

Tue 22 Jan 1918

WESTERN FRONT

Practice

19 Squadron, RFC (Ste-Marie-Cappel, France - 11th Army Wing, 2nd Brigade)

Sopwith Dolphin C3826 - dived into the ground when wings 'folded up' on completing a loop at 1035 near the Abeele-Watou road by the Belgian border, about 11km NE of the base. The pilot is buried at Bailleul, France. The Dolphin had accumulated a mere 6hrs 15mins of flying time.

Pilot: 2nd Lt **Allan Adolphus VEALE**, Wellington Regiment NZEF & RFC - Age 24.

Veale was born in Australia but it is not known when he came to New Zealand. He had served with the New Zealand Forces in Samoa in 1914, prior to embarking for Egypt with the NZEF in June 1915.

Fri 25/Sat 26 Jan 1918

WESTERN FRONT

Bombing raid on Schelderwindeke, Belgium

101 Squadron, RFC (Clairmarais South, France - 9th HQ Wing)

Royal Aircraft Factory FE2b B480 - took off at 0100, but on returning struck tree tops in thick ground mist and crashed at Le Nieppe, 7km east of the base. The pilot, 2nd Lt R C Lovell, and observer are buried at Hazebrouck, about 9km further east.

Observer: Lt **Seaforth William McKENZIE**, 2nd Bn Wellington Regiment NZEF attached RFC - Age 21.

Seaforth McKenzie and his brother Hector McKenzie briefly owned and flew an American Hamilton biplane at Marton in 1913.

Wed 30 Jan 1918

WESTERN FRONT

Offensive patrol

22 Squadron, RFC (Auchel, France - 10th Army Wing, 1st Brigade)

Bristol Fighter F2B C4832 - took off at 1035 and seen to fall in flames during an aerial engagement in area of map reference Sheet 36 M.6.d. The pilot was killed and his observer, Air Mechanic 3rd Class R A Duff, died of injuries later in the day. Both are buried at Merville, 17km NE of the base.

Pilot: 2nd Lt **Godfrey Gleeson JOHNSTONE**, RFC - Age 22.

Johnstone's brother, Melville Johnstone, died on 16 July 1917 while flying with 27 Sqn.

Sat 23 Mar 1918

ENGLAND

Solo training

School of Instruction, RFC (Hendon, Middlesex)

Caudron 'No.15' - stalled on a climbing turn while operating from Stag Lane relief landing ground at 1815 and dived into the ground from 40 feet. The pilot is buried at Beddington, Surrey. It is likely that the Caudron was not on RFC inventory, but owned and operated by a civilian flying school under contract to the Corps.

Pilot: 23089 2nd Lt **William George MITCHELL**, RFC - Age 34.

Mitchell was a graduate of the New Zealand Flying School.

Thu 4 Apr 1918

SCOTLAND

Training

36 Training Squadron, RAF (Montrose, Angus - 30th Wing, 21 Group, Training Division)

Sopwith Pup B7529 - collided with one of the Squadron's Sopwith Camel F1s (B7338), piloted by Lt A A Payne, RAF. They fell out of control and crashed, killing both pilots, that of B7529 being buried at Montrose.

Pilot: Lt **Ernest Wilfred BURTON**, Australian Flying Corps attached RAF - Age 25.

Burton had served with the Australian Imperial Force prior to transferring to the AFC.

Sun 7 Apr 1918

WESTERN FRONT

Offensive patrol and bombing at low level

4 Squadron, Australian Flying Corps (Bruay, France - 10th Army Wing, 1st Brigade, RFC)

Sopwith Camel F1 B5635 - took off at 1100 and last seen 15 minutes later going down in flames near Illies, 31km north of Arras, after being hit by flak. The pilot died and although his body is understood to have been recovered and buried 3km to the east at Marquillies, it could not be located post war. He is therefore commemorated on both the Arras Memorial and the Australian Villers-Bretonneux Memorial.

Pilot: Lt **John Classon COURTNEY**, Australian Flying Corps - Age 24.

Wed 10 Apr 1918

MIDDLE EAST

Offensive patrol

111 Squadron, RAF (Ramleh, Palestine - 40th Army Wing, Palestine Brigade)

Nieuport 17 B3594 - took off with three others and brought down behind enemy lines near Nablus, about 50km NW of the Dead Sea, following a combat with eight or nine enemy fighters and a two-seater. The pilot is commemorated on the Jerusalem Memorial on Mt Scopus.

Pilot: 2nd Lt **Thomas Lancaster STEELE**, MC, NZ Machine Gun Corps NZEF & RAF - Age 27.

Thu 2 May 1918

WESTERN FRONT

Special mission

56 Squadron, RAF (Valheureux, France - 13th Army Wing, 3rd Brigade)

Royal Aircraft Factory SE5a C1796 - took off at 1040 and last seen at 1130 engaged in combat over Martinpuich, 26km south of Arras. The pilot is buried at Carnoy, 8km south of Martinpuich.

Pilot: Major **Rainsford BALCOMBE-BROWN**, MC mid, Royal Field Artillery & RAF - Age 23. Approx. 880hrs.

Balcombe-Brown was the Squadron's commanding officer and the highest ranking New Zealand airman to lose his life while flying during the war.

Sat 11 May 1918

WESTERN FRONT

Bombing

206 Squadron, RAF (Alquines, France - 11th Army Wing, 2nd Brigade)

Airco DH9 B7587 - took off at 1707 piloted by Lt G A Pitt, RAF, and was attacked by two enemy aircraft over Armentières. Forced to land near Winnezeele, 29km NW of Armentières, B7587's pilot survived uninjured but his observer was killed during the combat and is buried at Longuenesse, on the western outskirts of St Omer.

Observer: 2nd Lt **Charles Edward ANKETELL**, MM, 12 Bn Royal Fusiliers & RAF - Age 20.

Anketell lived in New Zealand prior to the war, but his place of birth and when he immigrated is unknown. Prior to the Royal Fusiliers he was with the Royal Navy Reserve and thus served with all three of the Services - Navy, Army and Air Force. His MM was won during Army service.

Sun 12 May 1918

ENGLAND

Training?

Artillery and Infantry Co-operation School, RAF (Hursley Park, Hampshire - Training Division)

Royal Aircraft Factory RE8 C2473 - crashed and caught fire at about 1500 after stalling 400 yards from the Worthy Down airfield boundary, 3 miles north of Winchester. The pilot was buried at Hursley, 4 miles SW of Winchester, at 1430 on the 15th.

Pilot: Lt **Leslie GEORGE**, Australian Imperial Force & Australian Flying Corps - Age 25.

Airmen Died in the Great War 1914-1918 *indicates that George died in the same crash as 2nd Lt J S W Lord, also of the AFC and serving at the same School. However, court of inquiry evidence states that George took off with ballast in the observer's seat, confirming that he was flying solo. Other records indicate that there were two separate, although similar, fatal accidents that day, both on or near the aerodrome. Lord is also buried at Hursley.*

Mon 20 May 1918

ENGLAND

Practice

2 School of Special Flying, RAF (Redcar, Yorkshire - Training Division)

Sopwith Camel F1 B5693 - entered an intentional spin at 1200 but failed to recover before striking the ground. The pilot is buried at Redcar.

Pilot: Capt **John Alexander McDonald ALLAN**, RAF - Age 23. 120+hrs. At least 15 ops.

Sun 16 Jun 1918

ENGLAND

Training

26 Training Squadron, RAF (Narborough, Norfolk - 7th Wing, 3 Group, Training Division)

Airco DH4 (serial?) - suffered engine failure immediately after take off at 1500 and stalled on turning back towards the airfield, side-slipping into the ground. The pupil pilot, 2nd Lt J S L Oakes, RAF, survived, injured, but the pilot died and is buried at Narborough.

Pilot Instr: 2nd Lt **Trevor Dudley Hall ALDERTON**, RAF - Age 24.

Alderton was a graduate of the New Zealand Flying School.

Tue 18 Jun 1918

WESTERN FRONT

Offensive patrol over Villers-Bretonneux

84 Squadron, RAF (Bertangles, France - 22nd Army Wing, 5th Brigade)

Royal Aircraft Factory SE5a C1923 - took off at 0945 and last seen in combat with two enemy aircraft SE of Abancourt [sic?]. The pilot is buried at Heath Cemetery, Harbonnières, 11km ESE of Villers-Bretonneux. However, he is also commemorated on the Arras Memorial, which suggests that it may have been quite some time after the war before his war grave was finally found. There are two Abancourts in France, both over 30km from Harbonnières; possibly this is a corruption in casualty records of Aubercourt, which is only 9km west of Harbonnières and therefore much more likely to have been the scene of the combat.

Pilot: Lt **Peter NIELSEN**, Wellington Mounted Rifles NZEF & RAF - Age 32.

In the CWGC register entry under Heath Cemetery, Nielsen's name is incorrectly shown as Neilsen, but it appears correctly in that for the Arras Memorial.

Mon 1 Jul 1918

WESTERN FRONT

Artillery observation duty over Coisy

35 Squadron, RAF, (Flesselles, France - 15th Corps Wing, 5th Brigade)

Bristol Fighter F2B C4849 - took off at 1805 and forced landed at Coisy, a few kilometres north of Amiens, after being attacked at 1930 and badly shot about. Pilot Lt A McGregor, RAF, survived wounded but his observer was killed during the fight and is buried at Vignacourt, 15km NW of Amiens.

Observer: Lt **Roy James FITZGERALD**, MC mid, 12th Bn Gloucestershire Regiment & RAF - Age 27.

Prior to attachment to the RAF, Fitzgerald won the MC, was wounded, then captured by the Germans, only to escape five months later. He received the mid for valuable services rendered while a prisoner.

Sat 20 Jul 1918

ENGLAND

Camera gun exercise

Central Flying School, RAF (Upavon, Wiltshire - Training Division)

Avro 504A C696 - dived on to the wrong target while engaged in practice attacks on the firing range at 0930, colliding with a SE5a of the School (B8358) flown by 2nd Lt S J A Dodds, RAF. Both pilots are buried at Upavon.

Pilot: 2nd Lt **William Caithness TALBOT**, 6 Bn Leinster Regiment & RAF - Age 25.

Talbot was born in Scotland. It is not known when he came to New Zealand but he was a bank employee at New Plymouth prior to joining the Wellington Regiment NZEF upon outbreak of war. He transferred to the Leinster Regiment in July 1915.

Mon 22 Jul 1918

MIDDLE EAST

3 Cadet Wing, RAF (Aboukir, Egypt - Training Brigade)

Died of self inflicted wounds and is buried at Alexandria.

Commanding Officer: Lt Colonel **John Henry Morris ARDEN**, DSO, mid(2), Order of Osmanieh 2nd Class, Reserve of Officers, 2nd Bn Worcestershire Regiment & RAF - Age 43.

Arden served in the Anglo-Boer war and in the Sudan before leaving the Army in 1912, and emigrated to New Zealand. He rejoined the British Army upon the outbreak of the war in 1914. It is believed that his awards all predate attachment to the RAF - the DSO being awarded in 1915.

Tue 30 Jul 1918

WESTERN FRONT

Bombing raid on Martinpuich

52 Squadron, RAF (Auxi-le-Château, France - 12th Corps Wing, 3rd Brigade)

Royal Aircraft Factory RE8 E2 - took off at 0925 and brought down east of the lines. The pilot is buried at Ovillers-La Boisselle, about 6km SW of Martinpuich, while his observer, 2nd Lt H G Scott, is commemorated on the Arras Memorial.

Pilot: Capt **Robin Tudor BARLOW**, Canterbury Infantry Bn NZEF & RAF - Age 27.

Barlow was born in India and educated in England. It is not known when he came to New Zealand but he enlisted in the NZEF upon outbreak of war.

Thu 1 Aug 1918

ENGLAND

Training

1 (Observers) School of Aerial Gunnery, RAF (Hythe [Dymchurch], Kent - Training Division)

Bristol Fighter F2B C4798 - collided at 1530 with another of the School's F2Bs (A7262), flown by Sgt F T Hoskins and 2nd Lt J S Morbey, RAF. All four airmen died, including the pilot of C4798 and observer, Flt Cdt G Finney, RAF. The New Zealander is buried at Shorncliffe.

Pilot: 2nd Lt **Henry Stokes RICHARDS**, 12th Bn Essex Regiment & RAF - Age 24.

Tue 3 Sep 1918

ENGLAND

Training

189 (Night) Training Squadron, RAF (Suttons Farm, Essex - 49th Wing, 24 Group, 6th Brigade)

Sopwith Camel F1 D9569 - crashed near the airfield at 1105, killing the pilot, who is buried at Kensington, London.

Pilot: 2nd Lt **Brian Gerald MAHONEY**, Royal Munster Fusiliers & RAF - Age 29.

Mahoney previously served with the Auckland Regiment NZEF before transferring to the Royal Munster Fusiliers.

ATLANTIC

New Zealand trained pupils en route for RAF (embarked on HMT *Tahiti*)

Eight Cadets, six recently graduated as pilots with the Canterbury (NZ) Aviation Company and two with New Zealand Flying School, left New Zealand on 10 July by the *Tahiti* in company with the NZEF 40th Reinforcements. They arrived at Plymouth on 10 September to enter the RAF, but one of their number had died of influenza a week earlier and is thought to have been buried at sea.

(Trainee Pilot): Cdt **Daniel George SPENCE**, CAC (for RAF) - Age 20.

Spence, the first Canterbury (NZ) Aviation Company graduate to die, does not, according to the CWGC, qualify for inclusion in their registers as, not having reached Britain, he had yet to formally enlist in the RAF. However, although records are incomplete, it is believed that all flying school cadets from New Zealand were required to enlist in the NZEF before embarkation for Britain. Failure to continue flying and graduate there did not absolve pupils of their obligation to serve in the military.

Sun 22 Sep 1918

UNITED STATES OF AMERICA

Returning to Indianapolis from Greenfield

British Air Mission, RAF (USA)

Bristol Fighter F2B (serial?) - returning to Indianapolis at about 1730, after giving a flying display at Greenfield in support of the Fourth Liberty Loan War Bond Drive, entered a right hand spin from 600 feet, its left wing striking a tree before crashing in a cornfield of the Marion County Poor Farm near its boundary with the Indianapolis Speedway. One of the passengers, civilian J L Kinder, was killed on impact, while the other, Lt R W Pickett of the US Army Air Service, was seriously injured but recovered. The pilot also died in the crash and was buried at Indianapolis on the 26th.

Pilot: Capt **Joseph Joel HAMMOND**, RAF - Age 32. Approx. 6000hrs(?) Believed to have served operationally in France in 1914.

At the time of his death Hammond was the longest serving New Zealand pilot in the British Air Services. He had been flying regularly since 1910 and much of his career was spent as a test pilot. The 6000 hours is mentioned in an Indianapolis newspaper report of the accident, but no log book or official document has yet been located to corroborate this figure.

Mon 23 Sep 1918

FRANCE

Close offensive patrol

3 Squadron, RAF (Valheureux, France - 13th Army Wing, 3rd Brigade)

Sopwith Camel F1 F6029 - took off at 0635 and last seen at 0800 at 6000 feet before crashing and bursting into flames near Boursies (at map ref Sheet 57C K.2.a), 15km WSW of Cambrai, after combat with enemy aircraft. The pilot is buried at Sains-les-Marquion, 11km west of Cambrai.

Pilot: 2nd Lt **Stuart Herbert RICHARDSON**, RAF - Age 18.

Sun 13 Oct 1918

ENGLAND

Motor vehicle accident

Headquarters 33rd Wing, RAF (Lake Down, Hampshire - 7 [Training] Group)

A car travelling along the London Road towards Salisbury at 1750 skidded and overturned after striking a kerbstone at the end of St Thomas's Bridge. The driver, Acting Sgt A O Weeks of a nearby American Aero Squadron, 2nd Lt C J French, RAF, Lt Francis, and Lt **Nicol** of the Australian Flying Corps were all thrown out. French and Nicol were both injured, the latter being partly trapped under the vehicle and suffering a head wound from which he died at 1730 on the 15th without regaining consciousness. He was buried at Stratford-sub-Castle on the 21st at 1500. Nicol was returning to Lark Hill from Chattis Hill, after an inspection of wireless telephony equipment there, when the accident happened.

Equipment Officer 2nd Class: Lt **Hector NICOL**, Australian Flying Corps - Age 32.

The CWGC register incorrectly gives Nicol's date of death as the 16th, which was also entered in Australian records, but later corrected.

Sat 9 Nov 1918

ENGLAND

Training

51 Training Depot Station, RAF (North Shotwick, Cheshire - 37th Wing, Training Division)

Avro 504K F8719 - crashed while making a left hand turn close to the ground after engine failure. The pilot is buried at Shotwick.

Pilot: 2nd Lt **Francis Athol HINTON**, Otago Infantry Bn NZEF & RAF - Age 29.

WESTERN FRONT

Bombing raid on Denderleeuw, Belgium

108 Squadron, RAF (Bisseghem, Belgium - 80th Army Wing, 10th Brigade)

Airco DH9 E8980 - took off at 1305. The Squadron was attacked by 12 fighters during the raid, E8980 also being hit later by flak. The pilot and his observer, Lt J D Todd, RAF, were both wounded and a forced landing was made in the vicinity of Audenarde, about 16km south of their base. The observer recovered but his comrade died on the 23rd, and is buried at Terlincthun, to the north of Boulogne, France.

Pilot: 2nd Lt **James Duncan SLOSS**, RAF - Age 21.

Sloss was the only Canterbury (NZ) Aviation Company graduate to lose his life as the result of an operational flight.

Mon 11 Nov 1918 (hostilities ceased on the Western Front at 1100)

ENGLAND

Training

1 Training Depot Station, RAF (Stamford, Lincolnshire - 35th Wing, Training Division)

Avro 504K E330 - stalled on a climbing turn while travelling down wind and dived into the ground. Flt Cdt E R Spurrier, RAF, was slightly injured but his pilot died and is buried at Stamford.

Pilot Instr: 2nd Lt **Allan William MACDONALD**, RAF - Age 24.

Macdonald was the first Canterbury (NZ) Aviation Company graduate to lose his life on active service.

First to Fall - supplementary notes

The following entries fall outside the criteria set down for the primary part of this work, but are presented here in chronological order as an addendum.

Fri 17 Mar 1916

ENGLAND

Solo training

Ruffy-Baumann School (Hendon - civil flying school)

50hp Caudron biplane (No.?) - took off at 0750 and climbed steeply to about 250 feet then stalled, side-slipped and dived into the ground. The pilot is buried at Hendon. It is not known if the School numbered its machines.

Pilot: **John Ritchie Caldwell LAIDLAW** - Age 24.

Laidlaw, a civilian, was learning to fly at his own expense with the intention of obtaining a commission in the RNAS. He does not therefore qualify for entry in a CWGC register.

Mon 29 Oct 1917(?)

AT SEA(?)

In transit to New Zealand

RFC (at sea?)

Died en route to New Zealand. No further details known.

Air Mechanic: 664331 [sic] Air Mechanic 3rd Class (Cdt) **Lyonel Clare Fyans NEILL**, RFC - Age 23(?)

Neill enlisted in the RFC on 14 March 1917 but his service details are unknown. His number, as recorded above, contains one digit too many and should fall within the range of 61951-68950. He died on or after 29 October 1917, but no record has been found stating exactly when or where. Curiously, although his name is recorded (as Lyonel Clare Neill) on the roll at the RAF's Central Church of St Clement Danes it is not to be found in any CWGC register.

Wed 2 Jan 1918

ENGLAND

ex-RFC

Discharged earlier for medical reasons, an airman late of the RFC died at 0030 at Walton-on-Thames from tuberculosis and was buried at Brookwood in Woking, Surrey, on the 4th.

(?): 52890 Air Mechanic 2nd Class **Cedric Alfred HUNTER**, RFC - Age 25.

Previously with the NZEF, Hunter's RFC trade is not known but his service number indicates he enlisted as an Aeronautical Inspection Department (AID) examiner. At one time he was reported to have been employed in an 'aeroplane factory' in Birmingham.

Sat 12 Jan 1918

PACIFIC

En route from Britain to New Zealand

ex-RFC (embarked on HMT *Maheno*)

An officer of the RFC who had graduated on 27 November 1916 as a pilot, was obliged to relinquish his commission on account of ill health on 13 October 1917. Permitted to retain honorary rank, he departed Britain on 24 December aboard the *Maheno*, bound for New Zealand, but died en route on 12 January 1918 from tuberculosis. Believed to have been buried at sea in the Pacific, he is commemorated on the Hollybrook Memorial at Southampton, Hampshire, England.

(Pilot): Lt **Herbert Cyrus SIMSON,** late RFC - Age 31(?)

Fri 24 May 1918

ENGLAND

Familiarisation

8 Training Depot Station, RAF (Netheravon, Wiltshire - 4th Wing, 7 Group, Training Division)

Airco DH6 C6518 - piloted by Lt J J Daly and while looping over Salisbury Plain at 1900 suddenly pulled out of a prolonged dive at 1000 feet, whereupon a portion of the aircraft broke away, followed by the collapse of its outer port wings. The pilot and his New Zealand passenger died in the crash, the latter being buried at Codford St Mary. Although visited by accident investigators the same evening, so much of the wreckage had been 'souvenired' before a guard could be mounted that it was impossible to conduct a worthwhile examination of the wreckage.

(Commander): 6/1109 Maj **Benjamin Stevens JORDAN**, Canterbury Regiment NZEF - Age 34.

Jordan had previously served in the Boer War and at the time of his death was second in command of the New Zealand Command Depot at Codford. His flight was apparently a courtesy one and not directly related to his Army duties.

Names of airmen with a New Zealand next of kin connection only

Fates are not given in this work for the following, as they are understood not to have been born or resident in New Zealand. In most of these cases the next of kin lived in New Zealand when the airman died (as dated below), although some may have migrated to the country later, but prior to publication of the relevant CWGC register recording his details.

COOK, Alfred Burton, RFC - 20 Nov 1917
(born ? - father was the 'late A B Cook, auctioneer, Christchurch')

CORRY, Frank Moring, RFC - 12 Dec 1917
(born England or Wales - next of kin in Wellington?)

ELDER, James Jarvis, RAF - 3 Oct 1918
(born Scotland - parents at Waituna, Waimate, Canterbury)

FULLER, Charles Stephen, MC, RFC - 11 Nov 1917
(born Hobart, Tasmania - parent[s] possibly born in New Zealand)

PULLING, Edward Laston, DSO, RNAS - 2 Mar 1917
(born Devonshire, England - mother, remarried, at Raetihi)

RHODES-MOORHOUSE, William Barnard, VC mid, RFC - 27 Apr 1915
(born in England - mother born in New Zealand, father in England)
Rhodes-Moorhouse was the first airman to win the VC.

WHITESIDE, Reginald Cuthbert, RFC - 20 Dec 1916
(born Northern Ireland - parents migrated to Onehunga, Auckland, post-war)

Death Between the Wars

(12 November 1918 - 2 September 1939)

WITH THE ARRIVAL OF PEACE in 1919 (aside from small scale operations continuing in South and North Russia) RAF strength was quickly whittled down to a tenth of its wartime peak, and even less in terms of capability. During the 1920s it had to constantly battle against War Office and Navy intrigue for its very survival as an independent service. One means by which it strengthened its case was through the economical policing of remote and troublesome tribes in the Middle East and North West India; air mobility greatly reduced the need to maintain large and costly army garrisons.

For 15 years manpower levels changed little and the number of squadrons increased but slowly, largely because of the effects of the 'Ten Year Rule' (an assumption there would be no war for at least ten years, leading to minimal defence spending) and the Depression. Haunted by the horrific slaughter of the World War, British protagonists for disarmament held sway until the mid-1930s, when Hitler's rearming of Germany, gathering pace at an alarming rate, was belatedly recognised as a serious threat to national and European security. The Air Estimates for 1936 doubled over those of the previous year, and increased fourfold by 1939. Rebuilding and transforming the RAF from an era of the fabric covered biplane to one of the modern monoplane received highest priority. In the race to catch up before Germany went to war, as now seemed inevitable, it would very nearly be a case of too little too late.

Half a world away the New Zealand Permanent Air Force (NZPAF) was established in June 1923, its tiny staff taking charge of 15 machines of wartime vintage. A territorial component, the New Zealand Air Force (NZAF), with 73 officers, most of whom had seen war service, was created at the same time. Granted a royal title in 1934, the RNZAF became a separate service in 1937, by which time the regular strength totalled almost 200. The recently elected Labour government, eager to avoid repetition of the horrors of trench warfare, enthusiastically promoted air power, perceiving it as a less costly and more effective means of defence. Up to date equipment was still painfully slow to appear, though, and not until late 1938 did modern monoplane trainers, in the form of Airspeed Oxfords, enter service. However, in 1937 a landmark decision was made to order 30 Wellington bombers, at a cost which handsomely exceeded the total of all previous equipment purchases. The first six were scheduled to arrive from Britain in the last months of 1939. Intended as a counter to naval raiding forces, they were expected to range over the South Pacific and, if required, fly to Singapore in support of RAF units there.

The NZAF, without its own aircraft, was effectively little more than a reserve until the reorganisation in 1937, when it was reconstituted as the Territorial Air Force (TAF). By 1939 three TAF squadrons were in operation with a fourth planned, and a Civil Reserve was also created.

The RAF expansion programmes carried with them important implications for New Zealand. A small number of New Zealanders had in earlier years made their way to Britain in the hope of entering the service, but not always with success, as competition was keen and spaces few. To met the increasing demand for pilots, it was agreed that the RNZAF should itself select and examine candidates for four-year short service commissions in the RAF, and ship them to Britain, where they would automatically be accepted for training. A second scheme, introduced in 1937, entailed training officer pilots at Wigram before sending them to the RAF for four years of service, after which it was intended they return and serve a further period on the reserve. Pilots from the two schemes were thus part way through their RAF commitment when war was declared in September 1939.

As war approached, the RNZAF, now over 700 strong (over 1100 including TAF personnel), was thus heavily committed to the training of pilots for the RAF and to building up the infrastructure for impending Wellington operations. A few days prior to declaration of war the Wellingtons and their New Zealand crews, then working up in Britain, were offered to the British Government, an action which would eventually lead to the formation of 75 (NZ) Squadron, RAF. Sadly, although gratefully accepted by the British, the offer was not to be reciprocated with equal alacrity when the Dominion's own time of need arose.

New Zealand deaths for the 21 year interwar period were almost equal in number to those of the previous four years of war. From a dozen or so during the demobilisation period of 1918-1919, fatal casualties fell away to a handful for the 1920s before rising to an average annual rate of a little over three for most of the 1930s. With the expansion programmes well underway in 1938-1939, however, the rate more than tripled. The human cost of learning and practising the art of military flying on an increasingly large scale was beginning to show.

Fri 15 Nov 1918

MIDDLE EAST

Training

18 Training Depot Station, RAF (Moascar, Ismailia, Egypt - 32nd Wing, Training Brigade)

Armstrong Whitworth FK8 D5035 - experienced engine failure shortly after take off and crashed while attempting to return to the airfield. The pilot is buried at Ismailia.

Pilot: Flt Cdt **Oscar Edward OBERHUBER**, Auckland Mounted Rifles NZEF & RAF - Age 27(?)

Sat 23 Nov 1918

FRANCE

2nd Lt **J D SLOSS** died of wounds received in combat on 9 November.

Mon 25 Nov 1918

ENGLAND

Australian Flying Corps (based at?)

Admitted nine days earlier, died at 0515 of influenza-pneumonia in the Military Hospital at Sutton-Veny, Wiltshire, and was buried at Tidworth on 3 December.

(?): 1404 Private **Francis KITTO**, Australian Flying Corps - Age 37.

Kitto was wounded in July 1916 while serving with the AIF in France. He transferred to the AFC in October 1917.

Mon 16 Dec 1918

ENGLAND

Practice(?)

44 Squadron, RAF (Hainault Farm, Essex - 49th Wing, 24 Group, 6th Brigade)

Sopwith Camel F1 6326(sic) - involved in an accident but circumstances unknown. The pilot is buried at Hornchurch. While the aircraft is identified on a casualty record as being 110hp Le Rhone-powered Sopwith Camel 6326, the serial prefix has been omitted from the card. A possible match is F6326 but confirmation is lacking.

Pilot: Lt **Thomas Roland Roth BURNS**, mid, NZ Army Service Corps NZEF & RAF - Age 24.

Sun 22 Dec 1918

FRANCE

Flight from Spa to Paris

57 Squadron, RAF (Spy, Belgium; detachment at Spa - 90th Wing[?], 3rd Brigade[?])

Airco DH4 F2636 - took off at 1205 and crashed after striking a tree or hedge on encountering dense fog on approach to Paris. The passenger, Major General C D Rhodes of the United States Army Air Service, was thrown out and suffered facial injuries, but his pilot died and is buried at Louvres, 24km NE from the centre of Paris.

Pilot: 2nd Lt **Robert Rooke GILPIN**, RAF - Age 25.

Gilpin was born in England but it is not known when he came to New Zealand. Enlisting in the Wellington Mounted Rifles NZEF upon the outbreak of war, he sailed with the Main Body and later served briefly with the RNAS prior to its merger with the RAF.

Wed 22 Jan 1919

AT SEA

En route from Suez, Egypt, to Australia

Australian Flying Corps (embarked on HMT *Wiltshire*)

Fell overboard at 0930 immediately after climbing a railing to retrieve a quoit from behind the starboard central lifeboat and throwing it back to a party playing deck tennis. The ship, which had been rolling in a long swell at the time of the accident, turned about and within 20 minutes a lifeboat was over the spot where he was last seen treading water. His body was not recovered, however, and he is commemorated on the Chatby Memorial at Alexandria, Egypt. Although a good swimmer, on embarkation at Suez on 26 December he was still recovering from severe bouts of malaria, influenza and pleurisy, the effects of which greatly reduced his powers of endurance.

(Pilot): Lt **Carrick Stewart PAUL**, DFC mid, AIF & Australian Flying Corps - Age 25. Served operationally during the war.

Paul's mid was for conspicuous service while serving with the Australian 6th Light Horse Regiment in 1915. He had been twice wounded before attachment to the AFC in August 1917.

Thu 23 Jan 1919

UNITED STATES OF AMERICA

Training

McCook Field, United States Army Air Service (Dayton, Ohio)

Curtiss JN-4 (serial?) - made a right turn on approach for a landing at about 1500 but from 300 feet suddenly dived into the ground. The pupil, Mr Fred Golder, 'flying director at the field', suffered minor injuries and was discharged from hospital later in the day. His pilot, however, was killed instantly and is buried at Hibbing, Minnesota.

Pilot: 1st Lt **Frank BANKS**, USAAS - Age 29.

Fri 31 Jan 1919

ENGLAND

Training(?)

7 Training Depot Station, RAF (Feltwell, Norfolk - 39th Wing, 3 Group, Training Division)

Avro 504 C5281 [sic] - stalled and crashed from 100 feet on turning in to land. The pupil or passenger, 2nd Lt C Hansen, survived, injured. His pilot, more seriously hurt, died the following day, and is buried at Feltwell. The casualty card identifies the aircraft as a 100hp Gnome Monosoupape-powered 'Avro', carrying serial C5281. This identity actually belongs to a Caudron GIII. It seems likely a transposition of figures has occurred and that the Avro, either a J or K model, is one from serial block C5751-6050.

Pilot: 2nd Lt **John Harold CHAMBERS**, NZ Army Postal Service NZEF & RAF - Age 24. 87hrs solo (61 on Avro 504).

Sun 9 Feb 1919

SCOTLAND

Climbing accident

RAF

While on leave, pending embarkation for return to New Zealand, a trainee pilot and a brother officer set out in the afternoon to climb the lower shoulder of Ben Nevis. At a height of about 1000 feet, in attempting to negotiate a snow slide, he lost his footing and fell down the slope. He was dead when his companion reached him and is buried at Kilmallie, 2 miles NW of Fort William, Inverness-shire.

(Trainee Pilot): 179138 2nd Lt **Alexander Campbell CRAIG**, RAF - Age 21.

Craig was a graduate of the Canterbury (NZ) Aviation Company.

Thu 13 Feb 1919

MIDDLE EAST

In transit

RAF

En route to a posting in India died in Port Said, Egypt, after falling dangerously ill on the 8th, and is buried in that city.

Previously Aircraft Park Commander(?): Major **Norman Marshall HALCOMBE**, mid(3), Royal Engineers and RAF - Age 40.

Two of Halcombe's mid were for air work in Palestine and Egypt; the third possibly relates to his service with the Royal Engineers.

Sun 20 Jul 1919

MIDDLE EAST

Swimming accident

111 Squadron, RAF (Ramleh, Palestine - Palestine Brigade)

While bathing, slipped when making a dive and received spinal injuries. Died from same three days later and is buried at Ismailia.

(Pilot): Lt **Richard CROUCHER**, NZ Pack Wireless Troop NZ Engineers NZEF & RAF - Age 25.

Croucher had been 'in and out of hospital' since at least 29 October 1918 but was admitted as 'dangerously ill' on the above date, which is therefore presumed to be that on which he suffered the swimming accident.

Wed 17 Sep 1919

NORTH RUSSIA

Operational

HMS *Vindictive* (cruiser-carrier)

Sopwith 1½ Strutter N9055 or N9056 - both failed to return from flights made this day but it is not known if they were together or engaged on separate sorties. The two pilots lost their lives, although who was flying which machine, and under what circumstances, has not been ascertained. The New Zealander is buried at Koivisto, Finland. As access to the grave became difficult after Koivisto was ceded to Russia in 1940, his name is also commemorated on the Brookwood (Russia) Memorial at Woking, Surrey.

Pilot: Lt **Samuel DAWSON**, DFC, RAF - Age 25. Served operationally during the war.

Mon 14 Jun 1920

MIDDLE EAST

Flight from Ramleh to Ismailia, Egypt

14 Squadron, RAF (Ramleh, Palestine - Palestine Group HQ)

Bristol Fighter F2B E2293 - flew off course and forced landed in the Sinai desert, where the pilot's body was later located 30 miles west of Nekhl, some 70 miles west of Suez. A court of inquiry found he had died of exposure, accepting that this had occurred a week after the landing, on the 21st. The body of his passenger, AC1 P W J Thacker, was found by the Camel Corps on the mountains near Nekhl after the inquiry, his date of death being accepted as that of the day of the flight. Apparently no trace of the aircraft had been found up to the time of

discovery of the two bodies. Both airmen are buried at Suez.
Pilot: Fg Off **Neale FITZGERALD-EAGER**, RAF - Age 32.
Fitzgerald-Eager was born in Australia but emigrated to New Zealand prior to 1912. He enlisted with the NZEF, sailing with the Main Body, before transferring to the RFC and RAF

Sat 18 Sep 1920

MIDDLE EAST

Operational patrol
30 Squadron, RAF (Baghdad West, Iraq; detachment at Kazvin, Persia - Mesopotamian Wing)
Royal Aircraft Factory RE8 D4698 - spun in from 300 feet on approach when returning to Kazvin. The pilot and his observer, LAC L R Dellow, died and were buried at Kazvin, but later reinterred at Tehran. Leo White, in his early New Zealand aviation historical account, *Wingspread*, writes of the pilot being 'shot through the heart by an Arab taking a pot-shot at him while landing in Mesopotamia in 1918 [sic].' Although he is listed as a 'battle casualty', no confirmation of White's version has been located in the records.
Pilot: 2863 Flt Lt **Frank NUTTALL**, MC DFC AFC mid(2), RAF - Age 32. Served operationally during the war.

Mon 16 Jan 1922

ENGLAND

Air Ministry, RAF (London)
Died of double pneumonia after a few days illness at a nursing home in Chester Terrace, London. The funeral took place on the 18th at Brompton Cemetery, Kensington, London.
Secretary to the Secretary of State for Air: Gp Capt **Alan John Lance SCOTT**, CB MC AFC, RAF - Age 38. Served operationally as a pilot during the war.
Scott, although born in New Zealand, is believed to have emigrated to England at a very early age.

Mon 15 Sep 1924

ENGLAND

Practice(?)
17 Squadron, RAF (Hawkinge, Kent - 6 Group, Inland Area)
Sopwith Snipe F2484 - took off from Eastchurch and made a steep climbing turn to the right, then dived into the ground and caught fire, killing the pilot.
Pilot: Plt Off **Arthur Douglas BAILLIE**, RAF - Age 26.
Baillie had served in France with the NZEF during the war. He returned to New Zealand in 1919 but later made his way to Britain to join the RAF.

Mon 1 Jun 1925

INDIA

60 Squadron, RAF (Peshawar, India - 2 Indian Wing)
Died at Nowshera Hospital of enteric fever contracted at Miranshah.
(Pilot): Fg Off **Lawrence William AIKEN**, RAF - Age 27. Served operationally during the war.

Wed 17 Mar 1926

NEW ZEALAND

Type handling demonstration during annual Refresher Course
Wigram, NZ Permanent Air Force (Wigram)
Bristol Fighter F2B H1557 - took off shortly after 1700 and 30 minutes later executed two loops over Papanui from about 2000 feet. Immediately after the second loop H1557 entered a left hand spin which ended in a crash in the Methodist Orphanage's rose garden in Harewood Road, about 12 yards from a room where 60 children were seated at their evening meal. The pilot was killed outright and a passenger died of his injuries en route to the hospital. The officer under instruction, Lt **P A Turner**, NZAF, although seriously injured, later recovered. The pilot was buried at Rangiora on the afternoon of the 19th and the passenger at Bromley the following morning.
Pilot: Capt **Frederick James HORRELL**, mid, NZAF (Territorial) - Age 34. Served operationally during the war.
(Waiter): **Lewis Martin REID -** Age 22.
Horrell served with the NZEF before transferring to the RFC and RAF during the war, where he flew operationally. His was the first military flying death in New Zealand. Ironically, he had just returned from attending his uncle's funeral a few hours prior to the accident.
Reid, a well-known ex-jockey, was temporarily employed as a mess waiter at Wigram in connection with the programme of annual refresher courses.
Turner had served as a driver and observer in the RFC and RAF during the war. This was his first NZAF refresher course.

Mon 9 May 1927

AUSTRALIA

Formation flypast at Canberra, Australian Commonwealth Territory

1 Flying Training School, RAAF (Point Cook, Victoria)

Royal Aircraft Factory SE5a A2-24 - during an afternoon mass flypast at 500 feet, on the occasion of a military review marking the official opening of Parliament House by the Duke of York, inexplicably fell out of formation and dived towards the ground. A2-24 crashed and rolled over on a small rise some 600 yards in front of the House, fatally injuring the pilot, who died in Canberra Hospital a few hours later at 1900. During his funeral service at St John's Church on the 11th, three RAAF aircraft circled the graveside and dropped wreaths in a final tribute.

Pilot: Fg Off **Francis Charles EWEN**, RAAF - Age 27.

Thu 9 Jun 1927

MEDITERRANEAN

Bombing practice

402 Flight, HMS *Eagle* (disembarked at Hal Far, Malta)

Fairey Flycatcher N9887 - dived into the sea and sank while carrying out bombing practice in Marsa Scirocco Bay at 1220. The pilot is buried on Malta.

Pilot: Fg Off (Lt Cdr, RN) **Haydon Marriott Sutherland FORBES**, DSC, RAF - Age 29.

Forbes won his DSC as a Midshipman with the RN during the landings at Gallipoli on 25 April 1915.

Wed 8 Jan 1930

MIDDLE EAST

Training

4 Flying Training School, RAF (Abu Sueir, Egypt)

Avro 504N J8981 - collided at 1500 feet with another of the School's Avro 504Ns (JR8716) piloted by Flt Lt R S Greenslade, RAF. The pilots and their two airmen passengers died.

Pilot: Fg Off **Charles Edison GALPIN**, RAF - Age 28.

Fri 13 Feb 1931

ENGLAND

Training

RAF College (Cranwell, Lincolnshire)

Armstrong Whitworth Atlas (serial?) - piloted by instructor Fg Off A F P Anning, RAF, and dived into the ground during a gliding turn for a practice forced landing. The instructor was seriously injured and died in hospital later in the day, while his pupil died in the crash.

Pupil Pilot: Flt Cdt **Arthur Morland ACTON-ADAMS**, RAF - Age 20.

Fri 30 Oct 1931

ENGLAND

Training

3 Flying Training School, RAF (Grantham, Lincolnshire - 23 Group, Inland Area)

Avro Tutor I K2498 - after completing a loop, fell out of control and crashed two fields away from Hucknall airfield, Nottinghamshire. K2498 fell on to the 14th tee of the Bulwell Hall golf links, missing a golfing party by only 40 yards. Both airmen died.

Pilot: Fg Off **Jack Eugene MARKBY**, RAF - Age 25(?)

Pilot: Fg Off **Louis McHARDY**, RAF - Age 27.

Although killed while flying a 3 FTS aircraft, both New Zealanders were at the time on the strength of 504 Sqn (a Special Reserve unit based at Hucknall). It is not known who was piloting at the time.

Tue 2 Feb 1932

ENGLAND

99 Squadron, RAF (Upper Heyford, Oxfordshire - Wessex Bombing Area, Air Defence of GB)

Taken seriously ill while serving with 502 Squadron at Aldergrove, Northern Ireland, and died at RAF Halton Hospital, Buckinghamshire. Had been flown to Halton by late October 1931 and although posted to 99 Squadron effective 16 November did not in fact get to serve with it.

(Pilot): Fg Off **William Edward GRANT**, RAF - Age 26.

Grant was born in England but emigrated to New Zealand at age eight. He made his way back to England to join the RAF in the late 1920s. Two of his brothers died while flying during the Second World War - Ian Allan Charles Grant on 13 February 1943 with 485 Sqn, RNZAF, and Reginald Joseph Cowan Grant on 28 February 1944 with 65 Sqn, RAF, while commanding 122 Wing.

Wed 27 Jul 1932

ENGLAND

Practice(?)

501 Squadron, RAF (Filton, Gloucestershire - 1 Group, Fighting Area, Air Defence of GB)

Westland Wapiti K1369(?) - took off from Manston, Kent, and shortly after 2000 flew into a tree in front of a house at 'Howletts', north of Bekesbourne. The tail and rudder broke away on impact with the tree and the wrecked Wapiti hit and slid along the ground before bursting into flames. The pilot is buried at Margate. The Squadron was based at Manston for a fortnight's training at summer camp. Wapiti K1369 was struck off Squadron strength on 19 September and is thought likely to have been the aircraft destroyed in this crash.

Pilot: Fg Off **Peter Walter Johnston PHARAZYN**, RAF (Special Reserve) - Age 21.

The wedding of Pharazyn's sister had taken place earlier in the day. His brother, William Forster Pharazyn, died on 29 June 1938 while flying with 72 Sqn, RAF.

Tue 23 Aug 1932

ENGLAND

Motor vehicle accident

RAF Halton HQ (Halton, Buckinghamshire)

Died as the result of an accident when a car driven by a fellow officer collided with a telegraph pole at Twickenham, Middlesex. Appears to have died of injuries some time after the crash, the accident possibly taking place many months earlier.

(Pilot): Plt Off **Henry Edward SLOWEY**, RAF - Age 20.

Wed 12 Oct 1932

ENGLAND

Training

3 Flying Training School, RAF (Grantham, Lincolnshire - 23 Group, Inland Area)

Avro Tutor I K2509 - entered a spin and dived into the ground at Great Ponton, about 3 miles south of Grantham. The instructor was killed in the crash. Although the pupil succeeded in baling out at the last minute, he died also when his parachute failed to fully deploy.

Pilot Instr: Flt Lt **Harold Claude MARETT**, RAF - Age 28(?)
Pupil Pilot: Act Plt Off **Adrian Kinross WHITE**, RAF - Age 19.

Wed 4 Jan 1933

ENGLAND

Test flight - Service trials

111 Squadron, RAF (Hornchurch, Essex - Fighting Area, Air Defence of Great Britain)

Bristol Bulldog IIIA 'R5' - three of the Squadron's pilots arrived at the RAF Aeroplane and Armament Experimental Establishment, Martlesham Heath, Suffolk, to collect prototype Bristol, Gloster, and Vickers fighters for trials at Hornchurch. Flying back to base they dived and pulled up into a Prince of Wales feather manoeuvre, during which the Bulldog disintegrated and crashed at Norton Heath, near Chipping Ongar, killing the pilot. R5 was a manufacturer's designation, no RAF serial having been allocated.

Pilot: Fg Off **John Laurance ARMSTRONG**, RAF - Age 26.

Easter Saturday 15 Apr 1933

MIDDLE EAST

Practice(?)

208 Squadron, RAF (Heliopolis, Egypt)

Armstrong Whitworth Atlas I J9965 - in a flying accident at Ghobit el Bos landing ground, about 25 miles SSW of Suez, the pilot and his passenger, LAC A L Alleway, RAF, were both killed.

Pilot: Plt Off **William Athol Wynne JAMESON**, RAF - Age 20.

Tue 25 Jul 1933

MIDDLE EAST

Bombing practice

8 Squadron, RAF (Khormaksar, Aden)

De Havilland DH60M Moth K1212 - crashed on the Aden bombing range, fatally injuring the pilot, who died later in the day.

Pilot: Fg Off **Eric Arthur COLLYNS**, RAF - Age 21.

Collyns brother, Basil Gordon Collyns, died on 20 August 1944 while flying with 19 Sqn, RAF.

Fri 29 Dec 1933

ENGLAND

Practice(?)

111 Squadron, RAF (Hornchurch, Essex)

Bristol Bulldog IIA K1685(?) - forty minutes after take off crashed into a potato field at Hall Farm, Shenfield, near Brentwood, throwing out the pilot, who was killed. Bulldog K1685 was struck off Squadron strength 12 days later and is thought likely to have been the aircraft involved in this crash.

Pilot: Fg Off **Ronald Rowley MURPHY**, RAF - Age 24(?)

Tue 1 May 1934

ENGLAND

Training

RAF College (Cranwell, Lincolnshire)

Bristol Bulldog Trainer K3928 - took off at 1355 and collided a few minutes later with one of the College's Hawker Harts (K3152) coming from the opposite direction about a mile from Digby village. The wreckage of K3928 fell in Boothby Bottoms field, Lodge Farm, and that of the Hart in Horse Field about 300 yards away. Fg Off J Douthwaite and Flt Cdt J Askell-Rutherford, in the Hart, and Flt Lt J S Tanner and his pupil all died.

Pupil Pilot: Flt Cdt **John Aicken PLUGGE**, RAF - Age 19.

Plugge's brother, Arthur James Philson Plugge, died on 23 Feb 1942 while flying with 832 Sqn, FAA.

Thu 24 May 1934

ENGLAND

Display or practice(?)

Station Flight Andover, RAF (Hampshire - Western Area, Air Defence of Great Britain)

Armstrong Whitworth Atlas (serial?) - entered a spin, crashed and caught fire in a wheat field within sight of visitors to Andover on the occasion of the first annual Empire Air Day. The pilot and his passenger, Cpl R F Brown, RAF, both died. Flt Lt R Hart, in an accompanying machine, tried several times to land and assist but could not safely touch down. A shepherd quickly on the scene tried to extricate the crew but was driven back by the flames. A contemporary New Zealand newspaper report refers to the aircraft as a 'Monxton [sic] machine'; possibly Manston was intended, as Atlases were based there at the time with 2 Sqn, which may have been at Andover for the display.

Pilot: Fg Off **George Elyton Sloan WILLIAMS**, RAF - Age 26.

Mon 22 Oct 1934

MEDITERRANEAN

Melbourne Centenary Air Race (starting from Mildenhall, England)

Fairey Fox I G-ACXX (Air Race No 62) - on the third day of the race took off from Rome at 0355 for Athens, but crashed and burst into flames 100 metres from a provisional landing ground on a small plateau near Palazzo San Gervasio, 70km SE of Foggia. The Fox was believed to be attempting an emergency landing due to engine trouble (which it had also experienced earlier in the race). Rain was falling with a strong NE wind at the time. Both airmen died on impact, their funeral being held at Naples on the morning of the 26th with Italian military honours.

Pilot: Fg Off **Harold Darwin GILMAN**, RAF - Age 28. Approx. 1600hrs.

Pilot: **James Keith Campbell BAINES** - Age 28. Approx. 3900hrs.

Gilman was born in Australia but came to New Zealand prior to 1925. He joined the NZ Staff Corps and was sent to Aldershot in 1928 on attachment to the Suffolk Regiment, prior to transferring to the RAF in 1931. He was granted special leave from 15 Sqn, RAF, to participate in the race.

Baines was born in England and emigrated to New Zealand about 1922. Some reports incorrectly state that he was a member of the New Zealand Air Force.

Baines owned the machine but it is not known who was piloting at the time of the crash.

Wed 21 Nov 1934

ENGLAND

58 Squadron, RAF (Worthy Down, Hampshire - Western Area, Air Defence of Great Britain)

Died at King Edward VII's Hospital for Officers, London, following an operation.

Adjutant: Flt Lt **Charles McMenamen LAING**, MC AFC, RAF - Age 43.

Laing's MC was awarded for service in the British Army, prior to transfer to the RAF, where he qualified as a pilot and served operationally during the war.

Mon 18 Mar 1935

ENGLAND

Aerobatics training

2 Flying Training School, RAF (Digby, Lincolnshire - 23 Group, Inland Area)

Avro Tutor K3318 - took off shortly after 1330 and while executing a loop at 1430 the automatic slat separated

from the top left wing. K3318 began to disintegrate and crashed in Flintham's Field at Martin Moor 150 yards from the main road, a few miles east of Blankney and Scopwick. Both airmen died, the instructor still in his cockpit, while the pupil was found some six yards away with his parachute partially deployed. They were buried at Scopwick.
Pilot Instr: Fg Off **George Edward Bullen NIXON**, RAF - Age 24. 859hrs.
Pupil Pilot: Act Plt Off **Colin KIRKLEY**, RAF - Age 19(?). 32hrs.

Thu 26 Sep 1935

MIDDLE EAST

Practice
4 Flying Training School, RAF (Abu Sueir, Egypt)
Avro Tutor K3314 - stalled at the top of a loop and spun into the ground at Sarafand, Palestine, killing the pilot and his passenger, Fg Off T C Sanders, RAF.
Pilot: Fg Off **William Daniel DENNEHY**, RAF - Age 28. 515hrs.
Dennehy was Adjutant of 2 Armoured Car Company, RAF, based at Ramleh, only a few miles from RAF station Sarafand.

Mon 7 Oct 1935

ENGLAND

Returning to base
32 Squadron, RAF (Biggin Hill, Kent - Fighting Area, Air Defence of Great Britain)
Bristol Bulldog IIA K2171 - left Lee-on-Solent, Hampshire, at about 1030, after an overnight visit for return to Biggin Hill in company with another of the Squadron's Bulldogs. Over the Solent shortly afterwards K2171 dived very steeply out of a loop from about 2000 feet. As it suddenly flattened out a few hundred feet above the water the port wing folded up, causing the Bulldog to crash in the sea 400 yards off Calshot Spit, killing the pilot. High speed RAF launches were on the scene within minutes, the body surfacing as they began searching. Haze at the time made the sea surface almost invisible and may have deceived the pilot as to his height above it.
Pilot: Plt Off **Haldane GEORGESON**, RAF - Age 21. 220hrs.

Thu 14 Nov 1935

NEW ZEALAND

RNZAF Station Wigram
Died of natural causes at Christchurch.
(?): AC1 **Ian Alfred Burtoft MORE**, RNZAF - Age 26.
More was the first regular serving member of the RNZAF to die.

Tue 25 Feb 1936

NEW ZEALAND

Air to ground firing practice during annual Refresher Course
RNZAF Station Wigram
Bristol Fighter F2B 7121 - at about midday, dived from 1000 feet towards a target on the shore of Lake Ellesmere but failed to pull out. On striking the muddy ground, the undercarriage was ripped off and 7121 rolled over several times. The pilot died of his injuries two hours later at the Christchurch Hospital and is buried at Auckland.
Pilot: Fg Off **Graham MacKellar OWEN**, NZAF (Territorial) - Age 23.
Owen, as a member of the Auckland-based territorial No 1 (Army Co-operation) Squadron, NZAF, was at Wigram undergoing a 10 day refresher course.

Sat 31 Jul 1937

NEW ZEALAND

Boating accident
RNZAF Depot Hobsonville
Two off-duty air force photographers, sailing a small Tauranga class yacht between Beach Haven and Hobsonville wharf during the afternoon, were thrown overboard when a gust of wind suddenly overturned their craft 400 yards from the air base. AC2 **H C Hanning** clung to the upturned yacht and was rescued, but his companion, who had struck out for the shore, disappeared and was presumed drowned. A two-day aerial search could not locate his body, however.
Photographer: AC1 **Leonard Orr SMELLIE**, RNZAF - Age 22.
Smellie's was the first death in the RNZAF subsequent to the Service gaining independent status with effect from 1 April 1937.

Sat 20 Nov 1937

CANADA

Training(?)

RCAF Station Trenton (Ontario)

Fleet Fawn II 237 - while landing in front of the seaplane hangar, a wing caught the water causing the Fawn to dive into the bay. The pilot died, his body subsequently being recovered and reportedly transported to England for burial.

Pilot: Flt Lt **Lancelot Francis Jameson TAYLOR**, RAF - Age 28(?)

Taylor had trained as a Pilot with the NZAF at Wigram in 1927-29 before joining the RAF in 1931. He had been on exchange duty with the RCAF since April 1937.

Thu 16/Fri 17 Dec 1937

BOMBER COMMAND

Long distance night bombing exercise; base-Aldergrove-West Freugh-base

102 Squadron, RAF (Honington, Suffolk - 3 Group)

Handley Page Heyford III K5183 - returned on completion of the exercise at 2120, but crashed 100 yards short of the flarepath. The captain lost his life, while Plt Off B G L Betbeder was severely injured and died two days later. A New Zealander, Act Plt Off **L H McFarlane,** and AC1 E Marrows were slightly injured, only AC1 C A F Kettle emerged unscathed.

Captain: 37789 Plt Off **Thomas Arthur BUNTING**, RAF - Age 23. 182hrs.

McFarlane died on 2 July 1940 while flying with 58 Sqn, RAF.

Fri 21 Jan 1938

TRAINING COMMAND

Training

1 Elementary and Reserve Flying Training School, RAF (Hatfield, Hertfordshire - 26 Group)

De Havilland DH82A Tiger Moth G-ADHW - collided with a 66 Squadron Gloster Gauntlet II (K5303) at 800 feet over Smallford, $1^1/_2$ miles SW of Hatfield, and fell into a coal and sand dump at the rear of offices of the Dunbrick Company's brick works, missing a man on the ground by less than 20 yards. Plt Off L R Helps, pilot of the Gauntlet which crashed in a field about 200 yards away from ADHW, was killed, as were the two New Zealanders in the latter. Act Plt Off R F Rimmer, in another Gauntlet accompanying Helps, only just managed to avoid the collision.

Pilot Instr: **Theodore Quentin Paviour SMITH -** Age 30. Approx. 1300hrs.

Pupil Pilot: **Brian William GRIEVES**, RAF - Age 22.

Smith, although a Sgt in the RAF Volunteer Reserve, was employed in a civilian capacity at the time by the de Havilland company, which operated 1E&RFTS under contract to the RAF. His name, sometimes noted as Paviour-Smith, is therefore not recorded on the RAF Memorial Roll at St Clement Danes.

Grieves, selected for a short service commission in the RAF, had only arrived in England on the 17th. He was engaged in the preliminary training required prior to formal acceptance as a commissioned pilot in the service.

Thu 27 Jan 1938

BOMBER COMMAND

Solo single engine flying and landing practice

44 Squadron, RAF (Waddington, Lincolnshire - 5 Group)

Bristol Blenheim I K7078 - took off at 1420 and apparently went out of control at about 1530, when the pilot was seen to abandon it after completing several loops. In baling out of K7078, which crashed near Tattershall Thorpe, a mile or two north of Coningsby, the pilot struck the tail, causing several pieces of it to break away, and fatally injuring himself. He fell 4 miles away to the SE, at Reedham's Lane, New York, but died a few seconds after a nearby witness reached him.

Pilot: 37819 Plt Off **Alan Norman MACFARLANE**, RAF - Age 25. 256hrs.

Tue 22 Feb 1938

FIGHTER COMMAND

Formation flying

1 Squadron, RAF (Tangmere, Sussex - 11 Group)

Hawker Fury I K2061 - while flying at 300-400 feet over the Essex landmark of Black Cap, about 2 miles west of Lewes, unexpectedly broke formation, possibly due to engine trouble, and lost height. Descending through haze at 1445 K2061 struck two high tension cables stretched across a valley 98 feet above the ground, causing the fighter to plunge into a hillside, killing the pilot.

Pilot: 37957 Plt Off **Merton Vivian BAXTER**, RAF - Age 22. 364hrs.

Mon 28 Mar 1938

COASTAL COMMAND

Naval co-operation with the Home Fleet

228 Squadron, RAF (Pembroke Dock, Pembrokeshire - 16 Group)

Supermarine Stranraer I K7288 - took off from Mount Batten, Devonshire, at 0330 in search of a 'convoy' in the Bay of Biscay, but at 0517 crashed into the sea while descending through fog 15 miles SW of Ile d'Ouessant, off Brittany, France. The nearby SS *Consett* immediately steamed to the scene of burning and sinking wreckage, arriving 45 minutes later. None of the five crew could be found, however, only the body of passenger Mr A E Barlow, a journalist for *The Aeroplane*, being recovered. Besides the captain, the others who died were Plt Off J P Owens, Cpl J H Audley, AC1 B D C Lewis and LAC E W Neale.

Captain: Flt Lt **Reginald Henry HOBBS**, RAF - Age 28. 822hrs.

Fri 6/Sat 7 May 1938

TRAINING COMMAND

Solo night training

5 Flying Training School, RAF (Sealand, Flintshire - 23 Group)

Hawker Audax I K5598 - completed a perfect first landing, bounced on the second then stalled and crashed from 500 feet during the overshoot. The pilot died on the 6th but the accident time is not known. It may in fact have occurred on the evening of 5/6 May.

Pilot: Act Plt Off **Errol Desmond GIMSON**, RAF - Age 22. 89hrs.

Sun 29 May 1938

AUSTRALIA

Bombing practice at Greenhills

22 Squadron, RAAF (Richmond, New South Wales)

Avro Anson I A4-27 - after flying over the bombing range ground party to indicate completion of the practice, banked 45 degrees to head for Richmond, but side slipped into the ground. The pilot, Flt Lt M P Allsopp, was killed and Cpl W A Lockwood died of injuries the same day. Three others were also seriously hurt. LAC J A Gordon and AC2 J A Mackey apparently recovered but the New Zealander among them died on the 31st.

Wireless Op(?): AC2 **Raymond Conrad LAWSON**, RAAF - Age 33.

Wed 15 Jun 1938

TRAINING COMMAND

Landing practice

4 Elementary and Reserve Flying Training School, RAF (Brough, Yorkshire - 26 Group)

Miles Magister I L6910 - during the afternoon and about 45 minutes into the flight, stalled off a gliding right turn at 500-600 feet during an approach and spun into the River Humber. The injured and unconscious pilot drowned before a rescue boat could save him.

Pilot: **George Alexander McLeod BAIRD**, RAF - Age 24. 26hrs.

Baird, selected in New Zealand for a short service commission in the RAF, had arrived in England five weeks earlier and was engaged in the preliminary training required prior to formal acceptance as a commissioned pilot in the service. Baird was born in 1913, but it is not known where.

Tue 21 Jun 1938

FAR EAST

Night navigation exercise Seletar-Port Swettenham-Seletar

100 Squadron, RAF (Seletar, Singapore)

Vickers Vildebeest II K2918 or III K6372 - piloted by Fg Off G L A Cooper and Act Flt Lt R C Meares, RAF, respectively. Both machines lost without trace in a storm, their last known position being 10 miles off Cape Rachado, Malaya. A mid-air collision in poor visibility was considered the most likely cause of their demise (squadron practice was to close up on meeting bad weather). Three airmen, LACs W S Lawrence and H Thomas, AC1 G L Walters, and a New Zealand officer also made up the crews, although who was in which aircraft is unknown. All six lost their lives.

(Pilot): Plt Off **Hugh Angus BETHUNE**, RAF - Age 23.

Bethune is likely to have been acting as observer on this flight.

Wed 29 Jun 1938

FIGHTER COMMAND

Fighter attack practice formation

72 Squadron, RAF (Church Fenton, Yorkshire - 12 Group)

Gloster Gladiator I K6139 - collided at 0900 with another of the Squadron's Gladiators (K6138) in rough air conditions over Brayton, near Selby. The latter, flown by Plt Off R J Walker, although damaged, made a successful emergency landing. Rather than bale out, the formation leader evidently tried to regain control of K6139 and make a forced landing, but crashed and was killed.

Pilot: Flt Lt **William Forster PHARAZYN**, RAF - Age 35. 703hrs.
Eight days later Pharazyn's widow, Stella Margaret, gave birth to a daughter. His brother, Peter Walter Johnston Pharazyn, died on 27 July 1932 while flying with 501 Sqn, RAF.

Wed 10 Aug 1938

COASTAL COMMAND

Training - torpedo light attacks
Torpedo Training Unit, RAF (Gosport, Hampshire - 17 Group)
Fairey Swordfish I K8886 - flying in line astern formation with two others shortly after noon when, after torpedo release, it made a steep and sudden turn, apparently to avoid colliding with the leading machine. With inadequate height for recovery, K8886 crashed into the Solent off Stokes Bay Pier, near Gosport, killing the sole occupant, whose body was later recovered.
Pilot: 40262 Act Plt Off **Reginald Alexander ROBERTSON**, RAF - Age 25. 143hrs.

Wed 1 Feb 1939

MIDDLE EAST

Training(?)
33 Squadron, RAF (Ramleh - detachment at Lydda, Palestine)
Gloster Gladiator I K8038(?) - involved in an accident under unknown circumstances near Deir el Ballah, Palestine, about 13km SW of Gaza, resulting in death of the pilot. K8038 was struck off Squadron strength on 1 March and is thought likely to have been the aircraft involved.
Pilot: 40749 Act Plt Off **John Marcus Joseph POYNTON**, RAF - Age 21.

Sat 18 Mar 1939

COASTAL COMMAND

(Duty?)
269 Squadron, RAF (Abbotsinch, Renfrewshire - 18 Group)
Anson I K6321 - dived out of cloud in bad visibility and flew into the ground near Torrance, about 5 miles north of Glasgow. In addition to the New Zealander, those who died were Sgt G C MacDonald, AC1 W G Reid and AC2 D MacDonald.
Pilot(?): Sgt **Hugh Lynn McPHEE**, RAF - Age 25(?)

Wed 5 Apr 1939

BOMBER COMMAND

Low level bombing practice
15 Squadron, RAF (Abingdon, Berkshire - 1 Group)
Fairey Battle I K9359 - took off at 0715 piloted by Fg Off P G Chapman, RAF, and while manœuvring for a fourth bombing run an officer in the rear cockpit, not wearing a parachute or safety harness, was thrown out and fell to his death at Lashford Lane, Wootton, to the north of the airfield.
(Pilot): 40268 Plt Off **Peter SHENNAN**, RAF - Age 23.
Shennan was born in England and came to New Zealand prior to 1926. Selected for a short service commission in the RAF, he embarked for Britain in 1937. His role on the flight is unknown, but he may have been acting as observer.

Thu 11 May 1939

NEW ZEALAND

Night flying exercise - dual instruction
Flying Training School, RNZAF (Wigram)
Vickers Vildebeest III NZ110/B6 - took off at 1925 and on approach at 2015 for a third landing, undershot, struck two trees 150 yards from the southern boundary of the airfield, then dived into the ground 60 yards further on, bursting into flames. The pilot died in the crash, while the instructor succumbed to injuries at 0100 next day. The third occupant, an airman taking the flight for air experience, also died of his injuries on the 12th, at 1810. They were buried at Christchurch, Wanganui and Bluff, respectively.
Pilot Instr: Plt Off **William Frederick DAWSON**, RNZAF - Age 20. 215hrs.
Pilot: Act Plt Off **Reginald James McCRORIE**, RNZAF - Age 22. 104hrs.
(Fitter II): 36137 LAC **George Henry WEST**, RNZAF - Age 25.
According to contemporary newspaper accounts, Southland-born West was the first Maori to be accepted into the RNZAF (on 4 November 1936).

Tue 16 May 1939

NEW ZEALAND

Motor vehicle accident
RNZAF Station Wigram
Two airman clerks from the base were travelling into the city at about 1930 by car when it ran off the road and struck a post in Riccarton, between Konini and Puriri Streets. Cpl **W G C Paterson** recovered from his injuries but his companion died at Christchurch Public Hospital at 2340 that night.
(Clerk): 37104 LAC **Alexander Lachlan CALDER**, RNZAF - Age 25.

Mon 5 Jun 1939

FAR EAST

Camera obscura exercise
230 Squadron, RAF (Seletar, Singapore)
Short Sunderland I L5801 - took off at 0730 captained by Act Flt Lt W W Campbell, RAF, and climbed to 100 feet, but within a few minutes difficulty arose in holding back the control column. The nose dropped, then rose a little but dropped again before L5801 struck the waters of Johore Straits. As the throttles were cut the Sunderland went over onto its back. Four of the crew - the captain, Cpl S H Williams, LAC J J A Smith and AC1 J Colville - survived with varying degrees of injury. The remainder, LAC W B Cook, AC1 G H Leighton, AC1 G W J Mills and the 2nd pilot (also acting as bomb aimer for the exercise) either died in the crash or shortly afterwards from their injuries. The latter's body was recovered from the crash site and buried in Bidadari Cemetery, Singapore, but later reinterred at Kranji.
2nd Pilot: 40352 Plt Off **Allan Henry BARNES**, RAF - Age 24. 190hrs.

Tue 18 Jul 1939

BOMBER COMMAND

Formation flight from Waddington to Tangmere, Sussex
44 Squadron, RAF (Waddington, Lincolnshire - 5 Group)
Handley Page Hampden I L4046 - turned away in cloud from accompanying aircraft and not seen again by them. In low cloud and rain L4046 dived into the ground and exploded at Corby, about 7 miles SE of Grantham, killing the pilot and his three crew, Sgts J A Hawes and E W Jones, and AC2 R J Andrews.
Pilot: 37814 Fg Off **David Ivan JOBSON**, RAF - Age 25. 513hrs.

Tue 8 Aug 1939

MIDDLE EAST

Flight to ?
203 Squadron, RAF (Basra, Iraq)
Short Singapore III K4584 - hit sea wall on take off and caught fire at Aboukir, Egypt. AC1 W H Scott and a New Zealander were killed, while AC1 A A Shaw died of injuries two days later at the Ras el Tin Military Hospital, Alexandria. The other occupants were Sqn Ldr J R Scarlett-Streatfield, Plt Off P R M Williams, Cpls P Ingrey and J R Smith, and AC1s W N Arrowsmith and W H Ard, RAF, all of whom were slightly injured.
2nd Pilot(?): Sgt **Leonard Clunie STEED**, RAF - Age 26(?)
Steed had just been accepted for transfer to the RNZAF in connection with the proposed delivery of Wellington bombers to New Zealand.

Wed 30 Aug 1939

TRAINING COMMAND

Dual instruction - RAF examining officer's five hour test.
19 Elementary and Reserve Flying Training School, RAF (Gatwick, Surrey - 26 Group)
Miles Magister I L8214 - entered a left hand flat spin from over 2000 feet while carrying out a spinning test and failed to recover before crashing near Horley, killing both airmen. The testing officer, Flt Sgt T R W Owens, had over 2000 flying hours to his credit.
Pupil Pilot: **Roger John GILLINGHAM**, RAF - Age 21.
Gillingham, a New Zealander selected for a short service commission in the RAF, had arrived in England three weeks earlier and was engaged in the preliminary training required prior to formal acceptance as a commissioned pilot in the service.

Death Between the Wars - supplementary notes

The following entries fall outside the criteria set down for the primary part of this work, but are presented here in chronological order as an addendum.

Tue 5 Nov 1918

ENGLAND

Training

10 Training Depot Station, RAF (Harling Road, Norfolk - 39th Wing, Training Division)

Avro 504K D7127 - a casualty record for this accident states that the Avro 'spun into the ground'. (The reverse of the record card notes a Court of Inquiry finding that 'while taxiing at a moderate speed, [struck] an unseen object in the water,' but this would appear to relate to another accident altogether.) Serious injury was caused to Flt Cdt C I Casserly and his pilot instructor, Casserly dying two days later. Over eight months later, on 20 July 1919, the pilot was obliged to relinquish his RAF commission 'on account of ill health', and died shortly afterwards.

Pilot Instr: Lt **John Edward HUNT**, RAF - Age 29.

Although Hunt's exact date of death is unknown, it must have taken place between 20 July 1919 and 10 February 1920, and was most likely related to this accident. He is understood to have embarked on the SS Ruapehu *for repatriation to New Zealand on 7 June 1919, but it is not known if he reached his homeland. RAF records stating that Hunt relinquished his commission on 20 July 1919 make no reference as to his fate. On 10 February 1920 his name was entered into the* New Zealand War Deaths Register *but, unusually, no entry appears under the column for place and date of death. His name is not to be found in any CWGC register nor on the roll at the RAF's Central Church of St Clement Danes.*

Thu 13 Feb 1919

ENGLAND

Low flying over camp area

(unit?), RAF (base?)

Type and serial unknown - piloted by 2nd Lt W S Featherstonhaugh, RAF. While flying at 100 feet over Army Headquarters No 5 Camp at Larkhill, Wiltshire, a mechanical failure occurred necessitating a forced landing. As the aircraft came down a wing struck a New Zealand soldier on the head. He died instantly and was buried at Tidworth on the 17th.

(Soldier): 16/113 Private **Joseph TAIRUA**, NZ Maori (Pioneer) Bn, NZEF - Age 27.

Tairua had served on Gallipoli and in France during the war.

Mon, Anzac Day 1938

ENGLAND

Army co-operation searchlight exercise

Wrightways Ltd (Croydon, Surrey)

De Havilland DH84 Dragon I G-ACHX - took off at about 1900 but on returning to Croydon $2^1/_2$ hours later the starboard engine failed during approach from the southwest. ACHX crashed into a row of trees near the Rose Walk and Promenade de Verdun intersection, Purley, a mile to the SW. The New Zealand wireless operator, **Ernest Ratcliffe**, received a broken right thigh, but his fellow countryman was more severely injured. It took a dozen firemen to lift the wreckage and finally extricate them, but the pilot died at Purley and District War Memorial Hospital three days later and was buried south of Croydon, at Wallington on 2 May.

Pilot: **Colin Calder BYAR** - Age 25.

Byar served with the RAF 1933-37 as a short service commission officer and was on the Reserve at the time of the accident. Army co-operation searchlight flights by civil aircraft were a regular event in 1938 as war appeared increasingly inevitable.

A doubtful identified casualty

On page 494 of his 1928 work, *Some Records of Officers of the New Zealand Expeditionary Force (unofficial but compiled from official records)*, Lt Col Studholme lists 2nd Lt **L[aurence] R[ichard] HENRYS** as 'killed while flying' but provides no date of death. Shown as a graduate of the New Zealand Flying School, he was actually trained by Canterbury (NZ) Aviation Company. Henrys returned safely to New Zealand in May 1919 and no evidence has been found to suggest he could have died while flying between then and 1928.

Names of airmen with a New Zealand next of kin connection only

nil

List of inter-war Territorial and Reserve Air Force pilots killed in civil aircraft accidents

The following personnel also fall outside the criteria set down for the primary part of this work. They are listed here, however, to highlight the high casualty rate, given the modest amount of flying undertaken during the period, amongst airmen who died in civil aircraft accidents while members of territorial or reserve air forces. The number of territorial officers in New Zealand between 14 June 1923 and 3 September 1939 amounted to a little over 250, fourteen of whom appear below (honours and awards are omitted).

ARMSTRONG, Hamish Francis (NZAF) 21.7.35
BINLEY, Alan Keith Hilton (RAFO) 18.11.34
CAMERON, Maurice John (NZAF) 7.4.30
DUSTIN, David Louis (RAFO) 25.6.38
DUTHIE, Cyril Mercer (NZAF/TAF) 10.5.38
GREEN, Stanley William George (NZAF) 13.9.34
HARKNESS, Donald Ernest (NZAF) 12.12.29
HOOD, George (NZAF) 10.1.28 #
KIGHT, Ivan Louis (NZAF) 8.2.31
KINGSFORD SMITH, Charles Edward (NZAF/Reserve) 8.11.35 *
McGREGOR, Malcolm Charles (NZAF/Reserve) 19.2.36
MONCRIEFF, John Robert (NZAF) 10.1.28 #
RENTON, Jack Bertram (NZAF) 20.11.33
ROAKE, Richard Noel (NZAF) 19.1.35
STEWART, Eric George Duncan (RAFO) 21.1.37
ULM, Charles Thomas Phillipe (NZAF/Reserve) 4.12.34 *
WAITE, Clarence Clifford (NZAF) 28.11.34

Hood and Moncrieff died during the first attempt to cross the Tasman Sea from Australia.
* Kingsford Smith and Ulm were commissioned into the NZAF and placed on the Reserve during their visit to New Zealand in 1928 after carrying out the first successful trans-Tasman flight.

The following three ex-1914-1918 airmen were likewise killed prior to formation of the NZAF:-
HILL, Cecil McKenzie (Canterbury [NZ] Aviation Company Flying Instructor) 1.2.19
GROUT, Harry Cranston (ex-RAF) 30.12.21
RUSSELL, Richard (ex-RAF) 11.11.20

Sacrifice of the Citizen Air Force
The Second World War

(3 September 1939 - 2 September 1945)

Out of every ten New Zealand airmen and women who died during the Second World War at least nine were civilians who had volunteered for 'Duration of War' service only, with the balance largely made up of short service commission officers appointed to the RAF during 1937-1940.

Part I: September 1939 to December 1942

1939-1940

ON 1 SEPTEMBER 1939 Germany invaded Poland. Aided by Russia from the 17th, the Wehrmacht's *Blitzkrieg* ('Lightning War') campaign quickly defeated Poland in little over a month. Three days after the attack was launched, with no response from Germany to an ultimatum, Britain declared war against Hitler at 1100 (BST). The Dominion of New Zealand, at one with the 'Home' country in this moment of need, simultaneously pronounced itself also to be at war (at 2130 local time) with Germany.

Already mobilising, the country now called for volunteers to enlist for war service. Brought up under the shadow of the Depression, when car ownership and overseas travel (then by sea) were largely an indulgence of the wealthy, and when flying, although enjoying a popular profile through highly publicised record breaking and pioneering flights, was a luxury beyond reach of the average person, volunteers were not found wanting for the air. This was decades before the television age in New Zealand, with radio only just beginning to appear in a large number of homes and the telephone even less common. For those in their late teenage or early adult years the war opened up irresistible once-in-a-lifetime opportunities; the excitement and adventure of flying, accompanied by the thrill and freedom of travel, were suddenly all on offer. Even better, one would be paid for the privilege, and the flying would not be limited to the low powered aero club machine of little over 100hp, since already in RAF service were high performance Hurricanes and Spitfires with 1000hp plus under their cowlings. These attractions notwithstanding, it had been tacitly accepted well before September 1939 that the nation's youth would inevitably be required to fight and deter Hitler. True, flying, even in peaceful skies, was not an occupation without risk, as media reports on almost every crash, from fatal to the slightest brush with death, stressed so clearly. But then death, in the minds of confident young men, was something that overtook others - not them.

Britain now braced itself against anticipated air raids and despatched land and air forces to France. Although the Luftwaffe frequently carried out minor attacks on coastal targets over the next seven months, no major attack eventuated. While many came to look on this as the 'Phoney War', those at the 'sharp end' at sea and in the air were understandably of a different view. Bomber Command in particular paid a heavy price for persisting for several months, against the evidence, with daylight raids on well defended targets in the belief that its bombers could 'see off' the fighters. When the reality was finally acknowledged night raids had to be resorted to, with all the attendant difficulties of navigating to the target and back, not to mention pin-pointing it in the dark. For the Command it was the beginning of a hard and costly road to perfect a weapon which could deliver heavy and accurate blows against the enemy.

Meanwhile RNZAF recruits (and particularly aircrew) were building up in such numbers that a bottleneck was soon reached and the training system unable to cope with them - at least not all at once - although the need was great. Foreseeing this, on 17 December the New Zealand Government, together with those of Australia, Canada and Britain, signed an agreement at Ottawa which brought into being the Empire Air Training Scheme. Under the Scheme, in their spacious and unthreatened skies, Britain's three partners would furnish the bulk of the RAF's training requirements. From October 1940 large numbers of aircrew, after undergoing preliminary training in New Zealand, would proceed to Canada to gain more advanced skills, after which they would be posted to Britain, or to where ever else the RAF might have need of them.

For each New Zealand casualty in operational flying between September 1939 and early April 1940, two others occurred during training and preparation for war service. This was not to be for long, however, since the 'Phoney' phase abruptly ended on 9 April with Germany's invasion of Denmark and Norway, followed a month later, on 10 May, by a stunning *Blitzkrieg* assault on Belgium, Luxembourg, France and the Netherlands. The Allies could produce no effective counter and by the end of June all four had surrendered. The British forces only just succeeded in evacuating France through Dunkirk four weeks earlier, but left with little but their lives. The casualty equation now swung to four operational deaths for each by accident. By April's end total losses had risen past 50. Within two months it had doubled and by the year's end would double yet again.

During June 1940 Italy also entered the war, an event which brought conflict into the Mediterranean and across North Africa. This would lead to the 2nd New Zealand Expeditionary Force (2NZEF) fighting the whole of its northern hemisphere war in those theatres, beginning with disastrous and costly battles in Greece and on Crete in early 1941. A further threat also revealed itself closer to home when the

German raider *Orion*, in laying mines near Auckland on the night of 13 June, sank the SS *Niagara* on the 19th. Two days later the first fatality on an operational flight from New Zealand took place at Taieri, near Dunedin. Ironically, given that New Zealanders were dying on active service with the RAF in the northern hemisphere, the airman killed was an English RAF officer who happened to be on loan to the RNZAF. A dozen more would lose their lives in the course of patrolling New Zealand waters.

In July Hitler launched the aerial phase of *Operation SEALION*, the planned invasion of Britain, which would be vigorously contested over the next four months. The initial campaign by the Luftwaffe against Channel convoys was followed by others against airfields and London, the last subsequently being switched from daylight to night as the RAF, although still recovering from its heavy losses in France, exacted an increasing toll. Sustaining losses in the air of almost two to one against, Hitler in October was finally compelled to postpone the invasion. Although over 20 New Zealanders died serving in Fighter Command during the Battle more than twice that number fell while engaged on Bomber and Coastal Command operations, many while directly or indirectly countering the invasion threat. Regrettably, several otherwise worthy works on the Battle, while providing comprehensive lists of Fighter Command aircrew and Luftwaffe aircrew of *all* types, demean the contribution of the other RAF Commands by failing to treat their casualties with similar respect. Critical though the Battle was, its cost was relatively small compared with later losses. It must be remembered, however, that size of the RAF was also then modest and thinly spread.

Like Bomber Command the Luftwaffe discovered that their bombing raids had to be undertaken at night in order to avoid unacceptable casualties. The *Blitz*, as it was called, would cost over 43,000 British lives by the time it ended with a big raid on London in mid-May 1941, as Germany began withdrawing its forces eastwards in preparation for *Operation BARBAROSSA* - the invasion of Russia. The difficulties of locating and engaging enemy aircraft at night were immense, compared to daylight operations, and the development of the equipment and tactics to combat them took time to bring to fruition.

With the immediate threat of invasion removed and winter's approach, the intensity of combat flying lessened and casualties fell accordingly, the ratio of non-operational losses rising once again. During the year over 600 RNZAF air and ground personnel had embarked for service in Britain and the Middle East, but they arrived gradually and 80% of the New Zealand casualties to date in those theatres were borne by the short service commission officers, over 150 of whom paid the supreme price for being first in the line. It was a qualitative loss which could not quickly be replaced.

On 9 July, 20 volunteers from the list of candidates impatiently awaiting call up into the RNZAF, taking advantage of an opportunity to transfer to the Fleet Air Arm, sailed direct to England and immediately began training at HMS *St Vincent* (Gosport, Hampshire). By the time Scheme 'F', as it was called, ended in June 1945 over 1000 others had similarly been drafted. Meanwhile, 70 RNZAF trainee air gunners had arrived in Canada at the end of October, the forerunner of over 7000 aircrew who would train there. By the year's end nearly 60% of the RNZAF fleet consisted of Tiger Moths and Oxfords, a clear indication that the country was earnestly setting about fulfilling its EATS obligations.

Thu 7 Sep 1939

COASTAL COMMAND

Night patrol - North Sea

224 Squadron, RAF (Leuchars, Fifeshire - 18 Group)

Hudson I N7247 - took off at 1710 piloted by Fg Off H D Green, RAF, and later seen off the coast NE of Fife Ness to dive into the sea with a loud explosion at 'position 315', 3 miles NE of *North Carr Light Vessel 3*. All four crew are commemorated on the Runnymede Memorial.

2nd Pilot: 36184 Plt Off **Cedric Nicholas WHITTINGTON**, RAF - Age 21. 205hrs. 4th op.

Whittington is believed to have been the first New Zealander of any of the armed services to lose his life on operations during the Second World War.

Fri 8 Sep 1939

BOMBER COMMAND

Air firing exercise

9 Squadron, RAF (Honington, Suffolk - 3 Group)

Wellington I L4320 - captained by Plt Off Rosofsky, RAF. After completing the exercise at Berners Heath, flew low over the range and collided with trees, crashing and catching fire near Elveden, 4 miles SW of Thetford, Norfolk. All five crew died, the 2nd pilot being buried at Honington.

2nd Pilot: 36170 Plt Off **Bruce Innes CLIFFORD-JONES**, RAF - Age 22.

Sat 9 Sep/Sun 10 Sep 1939

COASTAL COMMAND

Anti-submarine patrol - southern North Sea

48 Squadron, RAF (Thorney Island, Sussex - 16 Group)

Anson I L7058 of 'B' Flight - took off at about 1800, but became lost on return and ran short of fuel. At 2230 Sgt G N Wilkinson, RAF, took control for a forced landing in a field on the summit of Cap Gris Nez, France, but the starboard engine cut through lack of petrol during a right hand turn on approach. Control was lost and L7058 crashed. Two of the crew survived, injured, but the third died and is buried at Terlincthun, near Wimille, 5km north of Boulogne.

Pilot: 41267 Plt Off **John James DRUMMOND**, RAF - Age 25. 226hrs. 4th op.

Tue 12 Sep 1939

TRAINING COMMAND

Instruction on quarter attacks

10 Service Flying Training School, RAF (Tern Hill, Shropshire; detachment at Penrhos, Caernarvonshire - 23 Group)

Harvard I N7105 - piloted by instructor Flt Lt J P Marriott, RAF. In carrying out an attack at 1420 on a towed target at 1200-1300 feet N7105 turned to starboard then tightly to port. In executing the latter manœuvre the Harvard side slipped, rolled over and spun into the sea off Pwllheli, Caernarvonshire. Both pilots are buried at Llanengan, Caernarvonshire.

Pilot: 41764 Plt Off **Alan George WHITEHEAD**, RAF - Age 22. 147hrs.

Sun 17/Mon 18 Sep 1939

COASTAL COMMAND

Anti-submarine patrol

210 Squadron, RAF (Pembroke Dock, Pembrokeshire - 15 Group)

Sunderland I L2165 - took off at 1130 captained by Flt Lt T I Davies, RAF, but on return was unable to locate the base because of poor visibility and the blackout. Two hours later, at 0030, after obtaining a bearing, L2165 exhausted its fuel on approach to the base and crashed, breaking up and sinking in the sea off Milford Haven. All nine crew died, the 2nd pilot's body being recovered from the wreckage and buried at Pembroke Dock. It was thought the Sunderland probably stalled while trying to prolong its glide after the engines cut.

2nd Pilot: 40647 Plt Off **Gordon William SUTTON**, RAF - Age 26. 327hrs.

Tue 19 Sep 1939

BOMBER COMMAND

Direction finding bearings & homing practice

207 Squadron, RAF (Cranfield, Bedfordshire - 6 Group)

Battle I K9448 - in poor visibility crashed into a hillside near Winchcombe, 6 miles NE of Cheltenham, Gloucestershire. Both crew died, the pilot being buried at Little Rissington, Gloucestershire.

Pilot: 36206 Plt Off **James Edward HULL**, RAF - Age 24 . 135hrs.

Sat 23 Sep 1939

TRAINING COMMAND

Formation practice

1 Service Flying Training School, RAF (Netheravon, Wiltshire - 23 Group)

Harvard I P5791 - collided with another of the School's Harvard Is (P5792), both of which caught fire and crashed at Collingbourne Kingston, 5 miles NE of the airfield. At least three airmen died, the New Zealander and one other being buried at Netheravon.

Pilot: Midshipman (A) **John Clutha CASEY**, RN, of HMS *Pembroke* (FAA) - Age 21.

Casey was the first New Zealand member of the Fleet Air Arm to lose his life in WWII, albeit in an RAF aircraft.

Tue 3 Oct 1939

NEW ZEALAND

RNZAF Station Ohakea
Died of natural causes while on parade at Ohakea and is buried at Gisborne.
(Aircrafthand): A39587 AC1 **Charles Haig TAYLOR**, RNZAF - Age 21.
Taylor, the first RNZAF airman of WWII to die, had been in the RNZAF but 14 days, having enlisted on 20 September.

Wed 4 Oct 1939

BOMBER COMMAND

Wireless training
49 Squadron, RAF (Scampton, Lincolnshire - 5 Group)
Anson I N5096 - spun in off a steep turn at 1115 while circling Newton aerodrome, Nottinghamshire, and crashed near Radcliffe-on-Trent, 2 miles SW of the airfield. Both crew are buried at Lincoln.
Pilot: 40620 Plt Off **Brian Leslie Gordon HARKER**, RAF - Age 21.

Tue 10 Oct 1939

COASTAL COMMAND

Patrol
233 Squadron, RAF (Leuchars, Fifeshire - 18 Group)
Hudson I N7227/Y - returning after nightfall in poor visibility and low cloud struck high ground and caught fire $1^1/_2$ miles south of Freuchie, 15 miles SW of Leuchars. All four crew died, the New Zealand pilot being buried at Leuchars.
Pilot: 39272 Fg Off **Charles Hampton CLARK**, RAF - Age 27.

Thu 19/Fri 20 Oct 1939

TRAINING COMMAND

Night navigation training
School of Air Navigation, RAF (St Athan, Glamorgan - 23 Group)
Anson I N5055 - starboard engine failed at 5000 feet when 9 miles west of Trevose Head. A return course was set for base but 20 minutes later, unable to maintain height, a forced landing was made in darkness off the beach at Bude, Cornwall. All four crew successfully took to a dinghy but found nothing available by way of a paddle. The pilot therefore made his way back into the Anson to seek something suitable. At that moment a large wave swept over N5055 and carried the dinghy and its three injured occupants towards the shore. Nothing further was seen of the pilot and he is commemorated on the Runnymede Memorial.
Pilot: 39650 Plt Off **Joseph Wendell BOREHAM**, RAF - Age 24 . 698hrs.

Mon 30 Oct 1939

BOMBER COMMAND

Practice formation flying
9 Squadron, RAF (Honington, Suffolk - 3 Group)
Wellington I L4288/A - one of three in the formation and at 1015 collided with another of the Squadron's Wellington Is (L4363) while running in towards Honington at 800 feet under low cloud. This occurred during a crossover evasive manœuvre, when the propellers of L4288 sheered off the tail of L4363 positioned below it. The tail-less aircraft then reared up and struck L4288. Locked together, they broke up as they fell and crashed in marshy ground at the Water Mill, Sapiston, less than a mile from the runway. The five crew of L4288 and four of L4363 all died, the New Zealander in each being buried at Honington.
Captain: 29171 Sqn Ldr **Lennox Stanley LAMB**, RAF - Age 29. 1506hrs. At least 1 op.

9 Squadron, RAF (Honington, Suffolk - 3 Group)
Wellington I L4363 - captained by Plt Off J F Chandler, RAF, and collided with L4288 as described above.
2nd Pilot: 40987 Plt Off **Colin Charles CAMERON**, RAF - Age 22. 287hrs.

Mon 6 Nov 1939

NEW ZEALAND

Drowning accident
RNZAF Station Hobsonville
The body of a drowned airman of the Station was discovered in the upper Waitemata by a public works ganger crossing the harbour in a dinghy and is buried at Auckland.
(Aircrafthand): A39658 AC1 **Eric Joseph CLINCE**, RNZAF - Age 21.

Wed 8 Nov 1939

COASTAL COMMAND

Experience in use of oxygen

220 Squadron, RAF (Thornaby, Yorkshire - 18 Group)

Hudson I N7290 - after descending from a climb to 20,000 feet, stalled at 600 feet on approach to Thornaby and spun into a house in Cambridge Road, Middlesbrough, catching fire. All four crew died, the pilot being buried at Stoke-on-Trent and his fellow New Zealander at Thornaby-on-Tees.

Pilot: 41324 Plt Off **Augustus Harold Jervis RYAN**, RAF - Age 21.

(Pilot): 41469 Plt Off **Douglas Haig ROBERTSON**, RAF - Age 21.

Robertson was probably acting as observer on the flight.

Ryan's brother, Leonard Terence Jervis Ryan, died on 26 November 1940 while flying with 105 Sqn. They are buried in the same cemetery.

BOMBER COMMAND

Night flying practice - circuits and bumps

97 Squadron, RAF (Abingdon, Berkshire - 6 Group)

Whitley II K7225 - captained by Sgt A V Strong, RAF, and on approach at 2140 undershot, striking a tree 400 yards out from the airfield boundary, followed by a house roof 100 yards further on, before crashing at St Helen Without. All three crew died, the 2nd pilot being buried at Drayton.

2nd Pilot: 40334 Plt Off **Ian St John BEERE**, RAF - Age 25.

Wed 8/Thu 9 Nov 1939

FIGHTER COMMAND

Night formation flying practice

64 Squadron, RAF (Church Fenton, Yorkshire - 13 Group)

Blenheim IF L8731 - struck trees 700 yards from the airfield and crashed after flaps were prematurely raised. The pilot and another were killed, while the third crew member, seriously injured, died on the 12th. The pilot is buried at Kirkby Wharfe, 2 miles north of the base. An observation made during the subsequent investigation noted concern at the general level of inaccuracy of Blenheim altimeters and condensation found on L8731's windscreen.

Pilot: 40604 Plt Off **Desmond Charles Reid CARTER**, RAF - Age 26. 447hrs.

Wed 15 Nov 1939

FIGHTER COMMAND

Local flying - demonstration of take off and landing procedure of type

92 Squadron, RAF (Tangmere, Sussex - 11 Group)

Blenheim IF L1127 - piloted by Plt Off Hinds, RAF, and seen at about 1700 on approach at 300 feet with port engine throttled back, or failing, the starboard engine being under full power. On turning slowly away to the left, losing height as the undercarriage began to go up, L1127 stalled and crashed. The pilot and the two officers on board as 'passengers' to observe Blenheim handling technique died, the New Zealander among them being buried at Tangmere. It was thought that the pilot, wearing an Irvin flying suit with very thick cuffs, may have inadvertently pushed forward the port mixture control with the throttle and thereby choked the engine.

(Pilot): 41834 Plt Off **Hugh Kendrick Arthur DRUMMOND**, RAF - Age 24. 150hrs.

Although shown as 92 Sqn in the CWGC register, official records give Drummond's last postings as 11 Service Flying Training School to 222 Sqn (in the process of reforming with the Blenheim) and, on the same date, 11 Group Pool for temporary duty. His presence at 92 Sqn appears to have been purely in the nature of a familiarisation visit.

Sun 19 Nov 1939

COASTAL COMMAND

Air firing practice at sea markers

220 Squadron, RAF (Thornaby, Yorkshire - 18 Group)

Hudson I N7284 - completed the practice 10 miles off shore from Thornaby but during return to base crashed in the sea $4^1/_2$ miles east of Seaham Harbour, County Durham. N7284 was seen to sink by the mine sweeper *Dalmatian*, but no bodies could be found and the two crew are commemorated on the Runnymede Memorial.

Pilot: 36173 Plt Off **Hubert John KELLER**, RAF - Age 20.

Keller's brother, Ronald Frederick Keller, died on 11 April 1943 while flying with 232 Sqn, RAF.

Fri 24 Nov 1939

TRAINING COMMAND

Test flight

9 Air Observers School, RAF (Penrhos, Caernarvonshire - 25 Group)

Battle I L5255 - collided soon after take off with another of the School's Battle Is (L5256), which was engaged in a left hand circuit preparatory to landing. Both crashed 4 miles NNW of Penrhos and caught fire. All three crew of L5255 died, the pilot being buried at Pwllheli.

Pilot: 41287 Plt Off **Keith Percival HAMILTON**, RAF - Age 20. 228hrs.

Sat 25 Nov 1939

FIGHTER COMMAND

Cloud reconnaissance

Station Flight Northolt, RAF (Northolt, Middlesex - 11 Group)

Gladiator I K8012 - wings broke off after entering a spin over Hadley Common, causing it to crash near Barnet. The pilot is buried at Northwood, 5 miles NW of Northolt. Ice was considered to have been a factor leading to loss of control and the accident.

Pilot: 40341 Plt Off **Bryan GRAHAM**, RAF - Age 26. 343hrs.

Graham was serving with 65 Sqn, also at Northolt, at the time. He may have been rostered for the weather test flight in the Station Flight aircraft as an additional duty.

Thu 14 Dec 1939

BOMBER COMMAND

Search for German naval vessels in Heligoland Bight area, Germany (by 12 aircraft - 6 lost)

99 Squadron, RAF (Newmarket, Suffolk - 3 Group)

Wellington IA N2986 - took off at 1143 captained by Sgt R H J Brace, RAF. On attacking the light cruiser *Nurnberg* and its escort in the Schillig Roads, north of Wilhelmshaven, came under fire from flak and was shot down into the sea by fighters a few minutes later, at 1435. All six crew are commemorated on the Runnymede Memorial.

2nd Pilot: 564505 Sgt **Charles Denis CALDWELL**, RAF - Age 26.

99 Squadron, RAF (Newmarket, Suffolk - 3 Group)

Wellington IA N2957 - took off at 1143 on same raid as the above. Badly damaged by fighters during the attack, N2957 limped back to Newmarket, jettisoning bombs over the sea before crossing the coast. However, on approach to base at 1805 it entered a steep spiral and crashed into a field near the racecourse. Three of the six crew died, including the captain, who is buried at West Row, 2 miles NW of Mildenhall.

Captain: 39026 Flt Lt **Eugene John HETHERINGTON**, RAF - Age 25. 699hrs. 2nd op.

Wed 27 Dec 1939

COASTAL COMMAND

Routine patrol south of Orkneys

269 Squadron, RAF (Wick, Caithness-shire - 18 Group)

Anson I K6286/E - took off at 1026 and during the patrol twice signalled that it had suffered a starboard engine failure, after which nothing further was heard. Believed to have ditched east of the Orkneys, K6286's four crew are commemorated on the Runnymede Memorial.

Pilot: 36176 Plt Off **Angus Joseph MACDONALD**, RAF - Age 25. 371hrs. 41st op.

Sat 6 Jan 1940

NEW ZEALAND

Drowning accident

4 Technical Training Centre, RNZAF (Hillside, Dunedin)

During the afternoon got into difficulties in the surf at Smaill's beach, Dunedin, and drowned. His body was not recovered and he is commemorated on the Provincial Memorial at Dunedin.

(Flight Mechanic - Instructor): A39030 Sgt **Arthur WILLS**, RNZAF - Age 41.

Wills had served four years with the RFC and RAF during the First World War.

Tue 9 Jan 1940

FAR EAST

Routine patrol off Colombo to identify shipping, etc

273 Squadron, RAF (China Bay/Trincomalee, Ceylon)

Vildebeest III K4156 - took off at 0555 piloted by Sgt D B Ross, RAF, and crashed 18 miles off Colombo at 0620 while circling an unidentified ship at 50 to 100 feet. K4156's starboard wing apparently stalled, causing the Vildebeest to spin into the sea and sink. One of the crew survived, injured, but the other two, including a New Zealander, are commemorated on the Singapore Memorial.

(Pilot): 40583 Plt Off **John Denys WILLIAMS**, RAF - Age 20. 341hrs. 17th op.

Although a pilot, Williams was acting as observer on this flight.

Fri 26 Jan 1940

BOMBER COMMAND

102 Squadron, RAF (Driffield, Yorkshire - 4 Group)

Admitted to York Hospital two days earlier suffering from suspected cerebro-spinal meningitis, died and is buried at Driffield.

(Pilot): 40241 Plt Off **William Ralph McCOMB**, RAF - Age 27. 522hrs. 3 ops.

Sun 4 Feb 1940

MIDDLE EAST

Motor vehicle accident

70 Squadron, RAF (Helwan, Egypt)

Returning to Helwan from Heliopolis at 2330, was killed when his motor car collided with a lamp post and a telegraph pole near the Egyptian Army HQ at Abbassia, near Cairo. The car was completely wrecked but his passenger, fellow Squadron member Plt Off **B S M Jones**, RAF, escaped with minor injury and shock. Buried at Cairo the following afternoon.

(Pilot): 39852 Fg Off **Brian Mortimer BROWNING**, RAF - Age 23. 624hrs.

Jones died on 29 April 1941 while flying with the Squadron.

Mon 5 Feb 1940

NEW ZEALAND

Message dropping practice

Christchurch General Reconnaissance Squadron, RNZAF (Wigram)

Baffin NZ152 - at 1210 climbed to 700 feet, but stalled coming out of a turn and spun into the surf 500 yards from the shore a mile south of the Waimakariri River mouth, killing the three crew. The bodies of pilot and wireless operator were recovered at the scene, and the third two days later on the beach at South Brighton, 7 miles to the south. They are buried at East Eyreton, Christchurch, and Roxburgh, respectively. The wrecked NZ152 was hauled ashore on the evening of the crash by nearly 200 men pulling on ropes.

Pilot: (NZ2523) Flt Lt **Arthur Darvall POULTON**, RNZAF - Age 41. 261hrs solo.

(Wireless Electrical Mechanic): A539003 AC1 **Douglas Haig BIRBECK**, RNZAF - Age 23.

Wireless Op: A539071 AC1 **William James Reynold STRACHAN**, RNZAF - Age 34.

At the time of Poulton's death service numbers had not yet been assigned to officers of the RNZAF. NZ2523 was a posthumous allocation made at war's end.

Tue 6 Feb 1940

COASTAL COMMAND

Convoy patrol - OBM85G

217 Squadron, RAF (St Eval, Cornwall - 15 Group)

Anson I K8746/O - took off at 0740 captained by Fg Off J A C White, RAF, and at about 1120 lost control in a turn at 350 feet, diving into the sea off Bishop's Rock. Three of the crew were killed, the sole survivor being picked up uninjured by HMS *Vimy*. The New Zealander is commemorated on the Runnymede Memorial.

(Pilot): 41643 Plt Off **Arthur Hugh Moncrief WRIGHT**, RAF - Age 23.

Wright was probably acting as observer on this flight.

Mon 12 Feb 1940

NEW ZEALAND

1 Electrical & Wireless School, RNZAF (Wigram)

After a short illness, died at Christchurch Public Hospital and is buried at New Plymouth.

(WOp): A39881 AC1 **Alfred Henry MOORES**, RNZAF - Age 21.

Sat 17 Feb 1940

FIGHTER COMMAND

Map reading and forced landing practice

222 Squadron, RAF (Duxford, Cambridgeshire - 12 Group)

Magister I R1827 - collided with a tree at 1630 while low flying and crashed near Ashwell, Berkshire. One of the two crew was killed, while the New Zealander received injuries from which he died next day, and is buried at Whittlesford.

Pilot: 33500 Plt Off **Arthur Frank DELAMORE**, RAF - Age 19. 244hrs solo.

Delamore was born in Kuala Lumpur, Malaya. When he came to New Zealand is unknown, but he was educated at Marton and Wanganui before making his way to England and entering Cranwell as a cadet in 1938.

Thu 29 Feb 1940

FIGHTER COMMAND

Night circuits & landings

19 Squadron, RAF (Duxford, Cambridgeshire - 12 Group)

Spitfire I K9809 - took off at about 2025 and shortly afterwards made a sudden turn to port at 1200 feet, lost altitude, then dived into the ground 1$^1/_2$ miles NE of Duxford. The pilot is buried at Whittlesford.

Pilot: 40271 Plt Off **Horace Arthur TRENCHARD**, RAF - Age 28. 776hrs.

Sat 2 Mar 1940

BRITISH AIR FORCES FRANCE

Patrol of right front

1 Squadron, RAF (Vassincourt, France - 67 Wing, 14 Group, Advanced Air Striking Force)

Hurricane I L1971 - took off at 1045 with two others as 'Red 3' but during the patrol one was obliged to force land after its propeller disintegrated. The others continued and found an enemy bomber which they attacked. On the third pass, L1971 was hit in the engine by return fire and forced to descend, the pilot's vision being partially obscured by smoke. Sgt F J Soper, RAF, in the other Hurricane, abandoned the fight and went down to pick a field out for an emergency landing by his companion, but lost sight of him in the final stages. Conflicting eye witness reports on the ground later indicated that the Hurricane, which crashed at noon, either spun in at the edge of the field or hit a tree while on final approach a mile south of Sarre-Union. The pilot is buried at Metz.

Pilot: 41607 Plt Off **John Stewart MITCHELL**, RAF - Age 21. 11th op.

Mitchell's brother, Stuart Stewart Mitchell, died on 31 May 1942 while flying with 112 Sqn, RAF.

Thu 7/Fri 8 Mar 1940

BOMBER COMMAND

Security patrol Sylt-Borkum-Norderney, East and North Frisian islands

61 Squadron, RAF (Hemswell, Lincolnshire - 5 Group)

Hampden I L4111 - took off at 2018, but on returning from the patrol found England covered in mist. L4111 flew for three hours over Yorkshire and Lincolnshire seeking a landing place but difficulty was experienced by base staff in trying to make wireless contact. The Hampden eventually arrived over Digby airfield after a flight of 9hrs 15mins but appeared incapable of turning to the left, limiting itself to right handed circuits only. It finally ran out of fuel on approach to land shortly after 0530 and crashed. The four crew died, the pilot being buried at Scotter. The wireless operator had suffered severe bullet wounds, indicating an engagement with the enemy during the patrol.

Pilot: 39855 Fg Off **Derek Charles Gray CLINKARD**, RAF - Age 24. 475hrs. 3rd op.

Tue 12 Mar 1940

NEW ZEALAND

Dual instruction

Flying Instructors School, RNZAF (Mangere)

Tiger Moth II NZ724 - took off at 0915 with Act Plt Off **P G H Newton** in the front cockpit. During slow rolls at 0930 over Black Ridge at 2800 feet, 2 miles from the airfield, the instructor's Sutton harness fastener accidentally disengaged, causing him to fall to his death into a paddock about 100 yards on the eastern side of the main highway. It was some moments before Newton realised his instructor was no longer present, after which he promptly returned to Mangere to report. Flt Lt **R A Kirkup** of the School then began an aerial search, quickly spotting the missing airman's body on the ground at 1020. The body was later cremated at Auckland. Steps were taken to modify the flange on the control stick of Tiger Moth aircraft to avoid a reoccurrence - the safety cord for the fastener pin having apparently been fouled by the stick when the instructor 'checked forward.'

Pilot Instr: (NZ2524) Sqn Ldr **David Malyon ALLAN**, RNZAF - Age 43. Approx. 6000hrs.

Allan, who served with the RNAS and RAF in the First World War, was a much revered figure in New Zealand aviation throughout the 1930s, as a flying instructor and display pilot. Attendance at his funeral on the 14th included some 400 members of the RNZAF. At the time of his death service numbers had not yet been assigned to officers of the RNZAF. NZ2524 was a posthumous allocation made at war's end.

Wed 13 Mar 1940

FIGHTER COMMAND

Formation practice

219 Squadron, RAF (Catterick, Yorkshire - 13 Group)

Magister I P2375 - spun off a stall turn or loop at 1510 while breaking formation and crashed to the east of Catterick at Bolton-on-Swale, killing the two pilots, the New Zealander being buried at Catterick.

Pilot: 40949 Plt Off **Arthur James SKIDMORE**, RAF - Age 25. 461hrs.

It is not known which pilot was in command at the time.

Wed 27 Mar 1940

COASTAL COMMAND

Navigation reconnaissance exercise, interception of launch

School of General Reconnaissance, RAF (Thorney Island, Sussex - 17 Group)

Anson I K6249 - entered a spin off a stall turn at 700 feet while circling the launch at 1550, crashed into the sea and sank. Two other aircraft circled the area for some time but no trace of the four crew was found, including the pilot who is commemorated on the Runnymede Memorial.

Pilot: 37885 Fg Off **Ewen Campbell LILBURN**, RAF - Age 29. 1310hrs.

Mon 1 Apr 1940

FIGHTER COMMAND

Target towing practice

1 Anti-Aircraft Co-operation Unit, RAF (Cleave, Cornwall - 22 Group)

Henley I L3245 of 'D' Flight - experienced engine failure a mile north of Bude and ditched 50 yards off shore in heavy seas at 1515. Both crew members got clear of the sinking machine but were drowned in the surf, the pilot being buried at Morwenstow, 6 miles north of Bude.

Pilot: 40933 Plt Off **Roderick Campbell MATHIESON**, RAF - Age 22. 331hrs.

Sat 6/Sun 7 Apr 1940

BOMBER COMMAND

Security patrol of German seaplane bases

83 Squadron, RAF (Scampton, Lincolnshire - 5 Group)

Hampden I L4054 - took off at 1915 and ran out of fuel returning after nine hours in the air. Abandoned at 0415 over Whitley Bay, Northumberland, L4054 crashed into the sea near St Mary's lighthouse. The bodies of three crew members were later recovered from the sea, while the wireless operator is commemorated on the Runnymede Memorial. Both pilots are buried at East Chevington, Northumberland.

Pilot: 40319 Plt Off **Wilfred ROBERTS**, RAF - Age 25. 357hrs.

(Pilot): 41251 Plt Off **Keith BROOKE-TAYLOR**, RAF - Age 21. 266hrs. 5th op.

Roberts was born at Brisbane, Australia, but educated in New Zealand. Selected for a short service commission in the RAF, he embarked for England in 1937. His number of operational flights is unknown but the first was undertaken on the first day of the war.

Brooke-Taylor was probably acting as observer on this flight. His brother, Rex Brooke-Taylor, died on 30 December 1940 while flying with 3 Service Flying Training School, RNZAF.

Mon 8/Tue 9 Apr 1940

BOMBER COMMAND

Circuits and landings

12 Operational Training Unit, RAF (Benson, Oxfordshire - 6 Group)

Battle I P2273 - took off at 2310, climbed to 500-600 feet, but swung away to the right before losing height to crash and catch fire at Ewelme, a mile east of Benson. The pilot is buried at Benson.

Pilot: 36235 Plt Off **Michael Edward Fulton BARNETT**, RAF - Age 18. 176hrs (29 on Battle).

Barnett was born in London, England, but educated in New Zealand. Selected for a short service commission in the RAF, he embarked for England in 1939 after graduating as a pilot with the RNZAF.

Thu 11/Fri 12 Apr 1940

BOMBER COMMAND

Raid on Stavanger/Sola, Norway (by 6 aircraft - 1 lost)

115 Squadron, RAF (Marham, Norfolk - 3 Group)

Wellington IC P9284/J - took off at 1800 captained by Plt Off F E Barber, RAF, and shot down over the target area by flak or fighters while bombing at low level. All six crew are buried at Stavanger.

2nd Pilot: 36180 Plt Off **David Alexander RANKIN**, RAF - Age 25.

Fri 12 Apr 1940

COASTAL COMMAND

Special operation - to intercept naval vessels off Norwegian coast

233 Squadron, RAF (Leuchars, Fifeshire - 18 Group)

Hudson I N7258 - took off at 0728 to search for enemy warships off coast of Norway but shot down into the sea by fighters SW of Kristiansand while shadowing the battlecruiser *Scharnhorst*. All four crew died, the pilot being commemorated on the Runnymede Memorial.

Pilot: 36185 Plt Off **George James Denzil YORKE**, RAF - Age 25. 414hrs. 23rd op.

BOMBER COMMAND

Attack on warships off Norwegian coast (by 12 aircraft - 7 lost)

44 Squadron, RAF (Waddington, Lincolnshire - 5 Group)

Hampden I P1173 - took off at 0820 captained by Fg Off H Robson, RAF, but unable to locate any of the warships at sea in the poor weather, with others attacked instead two ships in Kristiansand Harbour shortly after noon. On heading out to sea in clearing weather P1173 came under attack by fighters, being seen to catch fire and dive into the sea about 100 miles SW of Kristiansand. All four crew died, the captain's body later being recovered on or near the coast of Denmark, where he is buried. His three comrades are commemorated on the Runnymede Memorial.

(Pilot): 40229 Fg Off **Kenneth John Alan JOHNSTONE**, RAF - Age 24.

Johnstone was born at Suva, Fiji, but educated in New Zealand. Selected for a short service commission in the RAF he embarked for England in 1937. He was probably acting as observer on this flight.

Attack on warships off Norwegian coast (by 12 aircraft - 3 lost)
149 Squadron, RAF (Mildenhall, Suffolk - 3 Group)
Wellington IC P9266 - took off at 1153 captained by Sgt G E Goad, RAF, but upon turning south with the formation on reaching the Norwegian coast at about 1600 was attacked by fighters. P9266 was last seen some ten minutes later at position 5845N:0530E, still being pursued while flying east at a height of 50 feet. The captain's body was recovered on or near the coast of Norway, where he is buried, while his five comrades are commemorated on the Runnymede Memorial.
WOpAG: A37156 Cpl **Jack Harry LANGRIDGE**, RNZAF - Age 21. 262hrs. 10th op.
Langridge was the first member of the RNZAF to lose his life on an operational flight and on overseas service. He was one of the first six wireless operators sent to Britain in January 1939 in connection with the proposed delivery of Wellington bombers to New Zealand.

Sat 13 Apr 1940

COASTAL COMMAND

Raid on Stavanger, Norway
233 Squadron, RAF (Leuchars, Fifeshire - 18 Group)
Hudson I N7323/Y - took off at 1342 and intercepted by a fighter at 1725, being shot down to crash near Stavanger. All four crew are buried at Sola, 10km SW of Stavanger.
Pilot: 580171 Sgt **Reginald Stephen CALLINAN**, RAF - Age 31. 1007hrs. 51st op.

Sat 13/Sun 14 Apr 1940

BOMBER COMMAND

Night reconnaissance of Vaernes (Trondheim) aerodrome, Norway
38 Squadron, RAF (Marham, Norfolk - 3 Group)
Wellington I L4339 - took off from Wick, Caithness-shire, captained by Plt Off G L Crosby, RAF, and believed to have been damaged in a fighter attack during the reconnaissance. L4339 crashed into the sea shortly after sending an emergency message at 0004 when about 22 miles NE of Whitby, Yorkshire. All six crew died, the body of only one being recovered. The others, including the 2nd pilot, are commemorated on the Runnymede Memorial, with the exception of a naval officer who is commemorated on the Fleet Air Arm Memorial at Lee-on-the-Solent.
2nd Pilot: 41860 Plt Off **Arthur Guy Wyndham MEREWETHER**, RAF - Age 23.

Sun 21/Mon 22 Apr 1940

BOMBER COMMAND

'Gardening' - minelaying in the Baltic sound
44 Squadron, RAF (Waddington, Lincolnshire - 5 Group)
Hampden I L4088 - captained by Flt Lt F G Dutton, RAF, successfully released its mine in the Kattegat before attacking a fleet of 'fishing boats', which turned out to be flak ships. With an engine put out of action, a forced landing was made on the coast at Heiligendamm, 15km WNW of Rostock, Germany. All four crew, suffering only minor injuries, were captured. Tragically, the New Zealander among them was to lose his life on 19 April 1945, less than three weeks prior to VE Day, while on a forced march with some 500 Allied prisoners from Stalag 357 at Fallingbostel to Lübeck. He was one of 30 who died at Gresse, 14km NE of Lauenburg, when RAF fighter-bombers unwittingly attacked the column, mistaking them for enemy troops. The 30 bodies were buried in the parish churchyard at Gresse, but in July 1947 all, except that of an American, were reinterred at Berlin.
WOp: 623703 Wt Off **William Philip Jeffery WATSON**, RAF - Age 30.
Watson was born in England but emigrated to New Zealand in 1926. After completing his education he saw service with the Royal New Zealand Artillery before returning to England in 1938, where he joined the RAF.

Tue 23/Wed 24 Apr 1940

BOMBER COMMAND

Raid on Stavanger airfield, Norway (by 6 aircraft - 1 lost)
107 Squadron, RAF (Wattisham, Suffolk; detachment at Lossiemouth, Morayshire - 2 Group)
Blenheim IV L8750 - took off at 0120 from Lossiemouth, Morayshire, and believed to have been shot down by a night fighter while attacking at low level, crashing into the sea west of Stavanger. All three crew are commemorated on the Runnymede Memorial.
Pilot: 36177 Plt Off **John Dawson MURPHY**, RAF - Age 21. 287hrs. 9th op.

Wed 24 Apr 1940

COASTAL COMMAND

Escorting French destroyers to Skagerrak, Norway
220 Squadron, RAF (Thornaby, Yorkshire - 18 Group)
Hudson I N7286/Z - after a diversionary landing away from base the previous evening, took off at 0432 from Leuchars, Fifeshire, with two others. The formation was intercepted by two enemy fighters shortly after 0746, with N7286 last seen at position 'MBJV 3812' being pursued over the sea and trailing black smoke from one

engine. The Hudson was reported to have crashed near three French destroyers, but the four crew were lost without trace and are commemorated on the Runnymede Memorial.
Pilot: 40425 Plt Off **Murray Charles PETRIE**, RAF - Age 21. 621hrs. 38th op.
(Pilot): 41301 Plt Off **Ronald Owen LAWRY**, RAF - Age 20. 392hrs. 29th op.
Lawry was probably acting as observer on this flight.

Tue 30 Apr/Wed 1 May 1940

BOMBER COMMAND

Raids on airfields in Denmark and Norway (by 50 aircraft - 7 lost)
51 Squadron, RAF (Dishforth, Yorkshire - 4 Group)
Whitley IV K9039 - took off at 1900 captained by Plt Off E Cotton, RAF, to attack Fornebu airfield, Oslo, but became lost in poor weather on return to England and ran out of fuel. While carrying out a forced landing K9039 struck a coll between Burnside Fell and Dunsop Fell, 3 miles WNW of Slaidburn on the Yorkshire moors. An RAF Sgt in the crew was killed and the other four members injured. The 2nd pilot, however, later died on 2 May at the Queen Victoria Hospital, Lancashire, and is buried at Dishforth.
2nd Pilot: 41690 Plt Off **John Martin GILMER**, RAF - Age 23. 204hrs.

Wed 1 May 1940

BOMBER COMMAND

Ferry flight Kinloss-Leuchars-Driffield
102 Squadron, RAF (Driffield, Yorkshire; detachment at Kinloss, Morayshire - 4 Group)
Whitley V N1500 - strayed off course in bad weather and at 1059 struck the northern slopes of the Hill of Foudland, 6 miles SE of Huntly, Aberdeenshire, and caught fire. Of the crew and passengers on board, five died, including the captain, who is buried at Dyce, while three others were injured, two of whom died later in the month. The crash site is sometimes quoted as Bains Hole on the north side of Glens of Foudland.
Captain: 40342 Fg Off **Kenneth Neil GRAY,** DFC, Czech War Cross, RAF - Age 25. 712hrs. 7 ops.
Gray was the twin brother of Colin Falkland Gray, who became the top scoring New Zealand fighter ace of the war.

Fri 10/Sat 11 May 1940

BOMBER COMMAND

Night circuit and landing practice (first night solo on type)
17 Operational Training Unit, RAF (Upwood, Huntingdonshire - 6 Group)
Blenheim IV L4889 - during an approach at 2240, after some three hours of practice, suddenly dived from 1000 feet into the ground at Green Dyke, near the airfield. The pilot is buried at Bury, 2 miles NE of Upwood.
Pilot: 42413 Plt Off **Grenville James McFARLAND**, RAF - Age 20. 127hrs solo.

Sat 11 May 1940

FIGHTER COMMAND

Investigating X-Raid
54 Squadron, RAF (Hornchurch, Essex - 11 Group)
Spitfire I N3187 - took off at 1825 with one other on a patrol over the English Channel. N3187's last position report indicated it was 5 miles north of Margate, but shortly afterwards the section leader realised his 'No 2' was no longer present. A search was instigated but no trace could be found and the pilot is commemorated on the Runnymede Memorial.
Pilot: 40726 Plt Off **John Douglas Bruorton McKENZIE**, RAF - Age 25. 393hrs.

Sun 12 May 1940

COASTAL COMMAND

Solo reconnaissance over German coast
206 Squadron, RAF (Bircham Newton, Norfolk - 16 Group)
Hudson I N7353 - took off at 0445 and shot down by fighters into the sea north of the German island of Juist. All four crew were lost, including the pilot, who is commemorated on the Runnymede Memorial.
Pilot: 36188 Plt Off **Ian Lowrie GRAY**, RAF - Age 21. 400hrs. 13th op.

BOMBER COMMAND

Attack on bridges at Maastricht, Netherlands (by 12 Blenheims - 6 lost)
15 Squadron, RAF (Alconbury, Huntingdonshire - 2 Group)
Blenheim IV P6912 - took off at 0750 and shot down during the attack by flak at about 0915, crashing in flames in the Kattevennen, near Genk, Belgium, 14km NW of the target. All three crew were buried near Diepenbeek, 8km SW of Genk, but later reinterred at Leopoldsburg.
Pilot: 40616 Plt Off **Claude Randolph FRANKISH**, RAF - Age 25.

15 Squadron, RAF (Alconbury, Huntingdonshire - 2 Group)
Blenheim IV L8847 - took off at 0750 on same raid as the above and damaged by flak over the target at about 0916, crashing in flames in the Gemeente of Borgharen, adjacent to Maastricht. All three crew are buried at Maastricht.
Pilot: 40781 Fg Off **Thomas George BASSETT**, RAF - Age 22. 475hrs. At least 6 ops.

Further attack on bridges at Maastricht, Netherlands (by 12 Blenheims - 4 lost)
107 Squadron, RAF (Wattisham, Suffolk - 2 Group)
Blenheim IV P4914 - took off at 0810 and shot down by fighters, crashing near Bierset, Belgium, 8km west of Liege. All three crew are buried a kilometre further west at Voroux-Goreux.
Pilot: 41427 Plt Off **Osborne Harvey KEEDWELL**, RAF - Age 26. 265hrs. At least 8th op.

Attack on road in Tongres, Belgium (by 9 Blenheims - 1 lost)
21 Squadron, RAF (Watton, Norfolk - 2 Group)
Blenheim IV L8739 - took off at 1900 and while 7000 feet over the target at about 2040 was hit by flak or fighter, the rear fuselage breaking away. All three crew are commemorated on the Runnymede Memorial.
Pilot: 39249 Flt Lt **Arthur David WATSON**, RAF - Age 25.
It is not known where Watson was born but he came to New Zealand in 1933 and left several years later for Britain, where he joined the RAF in November 1936.

BRITISH AIR FORCES FRANCE

Attack on bridge near Bouillon, Belgium
103 Squadron, RAF (Betheniville, France - 75 Wing, Advanced Air Striking Force)
Battle I P2193 - took off at 1630 with two others and one of two which failed to return. Shot down and crashed in the vicinity of Noirefontaine and Sensenruth, Luxembourg, 3km north of Bouillon, P2193's two crew are buried at Sensenruth.
Pilot: 40635 Plt Off **Edgar Elliot MORTON**, RAF - Age 25.

Tue 14 May 1940

BRITISH AIR FORCES FRANCE

Reconnaissance over the Sedan area, France
59 Squadron, RAF (Poix, France - 52 Wing, British Expeditionary Force Air Component)
Blenheim IV N6173 - lost without trace, all three crew being commemorated on the Runnymede Memorial. While some records indicate that the Blenheim took off and was lost on the 13th others, including the CWGC register, confirm that the crew died on the 14th. Actual take off time is unknown but, being a reconnaissance operation, the sortie would almost certainly have been undertaken in daylight, hence it is unlikely to have taken off late on the 13th and been brought down in the early hours of the 14th.
Pilot: 40508 Plt Off **Croydon Jelfs Edinborough CHAMBERLAIN**, RAF - Age 20.
Valiant Wings *suggests that Chamberlain, of the 'RNZAF'[sic], 'later turned up in Luxembourg, although he had been wounded,' but this account cannot be corroborated and may have been based on hearsay evidence.*

Patrol of Louvain - Gembloux area
87 Squadron, RAF (Seclin, France - 60 Wing, 14 Group, British Expeditionary Force Air Component)
Hurricane I L1612 - took off on a morning patrol with others and one of two which failed to return. L1612's pilot was seriously wounded when shot down while intercepting an incoming bombing raid near Mainvault, 4km WNW of Ath. He is reported to have died of his injuries in the Lille Hospital and is buried at Fenain, 34km SSE of Lille.
Pilot: 41620 Plt Off **Gordon Cranleigh SAUNDERS**, RAF - Age 23.
The CWGC register and Air Ministry records give Saunders date of death as the 14th. Several unsubstantiated accounts have stated that he died on another date, perhaps as late as the 19th.

Attack on a pontoon bridge near Mouzon on the Meuse, SE of Sedan, France
103 Squadron, RAF (Betheniville, France - 75 Wing, Advanced Air Striking Force)
Battle I L5516 - took a direct flak hit during the attack and exploded in mid-air over the west bank of the Meuse in the vicinity of Mouzon and Remilly-et-Aillicourt, SE of Sedan. Both crew members are buried at Remilly-et-Aillicourt.
Pilot: 36212 Plt Off **Vernon Allan CUNNINGHAM**, RAF - Age 24.

Wed 15 May 1940

BRITISH AIR FORCES FRANCE

Operational
13 Squadron, RAF (Douai, France - 50 Wing, British Expeditionary Force Air Component)
Lysander II L6885 - took off about noon and brought down over Belgium, crashing in the Forest of Meerdaal, 7km south of Louvain. The two crew are buried at Louvain.
Pilot: 41729 Plt Off **Alan Christie OLLERENSHAW**, RAF - Age 19.

BOMBER COMMAND

Attack against troops on the Dinant-Celles road (by 12 aircraft - 3 lost)

40 Squadron, RAF (Wyton, Huntingdonshire - 2 Group)

Blenheim IV P4913 - took off at 0930 and shot down by fighters shortly after bombing the target near Ecaussinnes d'Enghien, 18km NE of Mons, Belgium. All three crew are buried at Ecaussinnes d'Enghien.

Pilot: 36151 Fg Off **John Ernest EDWARDS**, RAF - Age 26.

Sun 19 May 1940

BRITISH AIR FORCES FRANCE

Ferry flight from Bertangles to Abbeville

16 Squadron, RAF (moving from Bertangles, France to Lympne, Kent this day - 50 Wing, British Expeditionary Force Air Component)

Lysander I or II (serial?) - piloted by Plt Off A P Reed, RAF, took off with seven others as enemy ground forces were surrounding the airfield. Intercepted by enemy fighters at 1500 feet over Glisy, to the west of Amiens, the Lysander was shot down, killing its two occupants. The wreckage fell immediately south of Glisy but no remains or grave could be located post-war and both airmen are commemorated on the Runnymede Memorial.

(Pilot): 42461 Plt Off **Ian DROMGOOLE**, RAF - Age 24. 200hrs.

Dromgoole's brother, Sydney Houston Dromgoole, died on 22 April 1942 while flying with 75 (NZ) Sqn, RAF.

FIGHTER COMMAND

Photographic reconnaissance of Hamburg and Bremen, Germany

Photographic Development Unit, RAF (Heston, Middlesex)

Spitfire PRIII P9308 - took off at 1120 and on the return journey was intercepted, being shot down near Templeuve, Belgium, 15km east of Lille, France. The pilot is buried at Templeuve. The stripped, unarmed and specially modified P9308 included fuel tanks which extended its range to 2000 miles, about four times that of a normal Spitfire.

Pilot: 43692 Plt Off **John Herbert COLEMAN**, RAF - Age 27.

Mon 20 May 1940

COASTAL COMMAND

Practice fighter attack

248 Squadron, RAF (Gosport, Hampshire - 17 Group)

Blenheim IVF P4837 - one of three 248 Squadron Blenheims carrying out practice attacks on a 42 Sqn Beaufort I (L4501), piloted by New Zealander Plt Off **K W Trigance**, RAF. During the exercise P4837 collided with Blenheim L9455, badly damaging the port aileron and mainplane of the latter, but continued the attack. It then appeared to get caught in the slipstream of the target Beaufort at about 1000-1500 feet and dived into the sea between Thorney creek and Selsey Bill. Both crew members are commemorated on the Runnymede Memorial. Blenheim L9455 was also written off but it is not clear if it ditched into the sea or succeeded in landing back at Gosport; both of its crew survived uninjured.

Pilot: 42100 Plt Off **Arthur Edmond Quentin BOURGEOIS**, RAF - Age 21. 199hrs.

Tue 21/Wed 22 May 1940

BOMBER COMMAND

Attack on bridge at Dinant, Belgium

75 (NZ) Squadron, RAF (Feltwell, Norfolk - 3 Group)

Wellington IC R3157/H - took off at 2120 and hit by flak at 2000-3000 feet during run in to the target, crashing near Kain, 4km NNW of Tournai, Belgium. Both pilots died but the other three crew survived and were captured. The 2nd pilot is buried at Kain, but the body of the captain was not recovered and he is commemorated on the Runnymede Memorial.

Captain: (NZ2513) Flt Lt **John Noel COLLINS**, RNZAF - Age 23. 850hrs. 11th op (?)

This was the 'New Zealand' Squadron's first loss. Collins graduated as a pilot while engaged on a short service commission with the RAF. He transferred to the RNZAF in connection with the proposed delivery of Wellington bombers to New Zealand, and became the first RNZAF pilot to lose his life while serving overseas. At the time of Collin's death service numbers had not yet been assigned to officers of the RNZAF. NZ2513 was a posthumous allocation made at war's end.

Thu 23/Fri 24 May 1940

BRITISH AIR FORCES FRANCE

Raid on Bingen-on-Rhine, Germany

88 Squadron, RAF (Les Grandes Chappelles, France - 75 Wing, Advanced Air Striking Force)

Battle I P2356 - brought down at the village of Springen, 21km NW of Wiesbaden. All three crew were buried at Springen, but later reinterred at Durnbach, 16km east of Bad Tölz.

Pilot: 36200 Plt Off **Albert Edgar WICKHAM**, RAF - Age 22. 243hrs.

Fri 24 May 1940

BOMBER COMMAND

Cross country - Frome-Tiverton-Holsworthy-Exeter-Lyme Regis-Frome-Benson

12 Operational Training Unit, RAF (Benson, Oxfordshire - 6 Group)

Battle I L4977 - in low cloud and poor visibility at 1120, flew into a hill at Bowd Croff on the Exeter-Sidmouth road, Devonshire. All three crew died, the pilot being buried at Plymouth.

Pilot: 36276 Plt Off **Peter Knox SIGLEY**, RAF - Age 24. 216hrs.

12 Operational Training Unit, RAF (Benson, Oxfordshire - 6 Group)

Battle I K9481 - took off on same cross country flight as the above and at 1140, in similar conditions, flew into a hill near Sidmouth, Devonshire. All three crew died, the pilot being buried at Plymouth.

Pilot: 36268 Plt Off **William David FINLAYSON**, RAF - Age 25. 200hrs (103 solo -12 on Battle).

Sigley and Finlayson had graduated together as pilots on the same RNZAF course in New Zealand before transferring to the RAF in Britain.

Sun 26 May 1940

FIGHTER COMMAND

Reconnaissance over northern France

53 Squadron, RAF (Andover, Hampshire - 22 Group)

Blenheim IV L8863 - brought down near Bayenghem-lès-Seninghem ,14km SW of St Omer, France. All three crew are buried at Bayenghem-lès-Seninghem. Twenty-two Group was an Army Co-operation formation but came under Fighter Command until 24 June, when it became an independent group with Command status.

Pilot: 42330 Plt Off **Gerald Montague BAILEY**, RAF - Age 24.

Mon 27 May 1940

FIGHTER COMMAND

Supply drop to Calais, France

26 Squadron, RAF (Lympne, Kent - 22 Group)

Lysander II L6863 - took off at 0935 from Hawkinge and was one of two of the Squadron's Lysanders seen to crash in flames near the target area. Both crew members died and are buried at Calais. Twenty-two Group was an Army Co-operation formation but came under Fighter Command until 24 June, when it became an independent group with Command status.

Pilot: 42129 Plt Off **Ernest Elliot HOWARTH**, RAF - Age 20. 207hrs.

Wed 29 May 1940

FIGHTER COMMAND

Patrol over Dunkirk, France

229 Squadron, RAF (Digby, Lincolnshire - 12 Group)

Hurricane I P3489 - took off as leader of a section. The patrol encountered enemy fighters and bombers over the beach and P3489 was last seen joining an attack on two of the enemy which were dive bombing shipping. The pilot's body later washed ashore on the German island of Borkum, near the Dutch border, and was buried there, but later reinterred at Sage, 24km south of Oldenburg.

Pilot: 40990 Flt Lt **Falcon Nelson CLOUSTON**, RAF - Age 27. 488hrs. 48+ ops.

Fri 31 May 1940

FIGHTER COMMAND

(duty?)

605 Squadron, RAF (Drem, East Lothian - 13 Group)

Hurricane I L2117 - took off from Hawkinge, Kent, but experienced engine failure shortly after take off. In attempting to force land, L2117 rolled over at low altitude and dived into a road 6 miles north of Folkestone. The pilot is buried at Hawkinge, near Folkestone. It was thought that he may have been overcome by glycol fumes. The Squadron had left Hawkinge and moved north to Scotland three days earlier but it is not known why this aircraft was still at the former base.

Pilot: 40609 Plt Off **Antonio Simmons DINI**, RAF - Age 22.

Sun 2 Jun 1940

FIGHTER COMMAND

Reconnaissance patrol over Narvik area, Norway

263 Squadron, RAF (Bardufoss with detachment at Bodo, Norway - Fighter Command)

Gladiator II N5893 - last seen at 1430 after an engagement with enemy aircraft near Narvik while in company with another of the Squadron's Gladiators, flown by fellow New Zealander, Plt Off **L R Jacobsen**. N5893 had been brought down east of Rombaken Fjord but, as the crash site was not discovered for some years after the war, the pilot's name appeared on the Runnymede Memorial before his body was recovered and buried at Narvik.

Pilot: 42170 Plt Off **James Leon WILKIE**, RAF - Age 20. 230hrs. 5th+ op.
Wilkie was born at Perth, Australia, but came to New Zealand at an early age. Selected for a short service commission in the RAF, he embarked for England in 1939. Jacobsen died with the week, when HMS Glorious *was sunk.*

Mon 3/Tue 4 Jun 1940

BOMBER COMMAND

Raid on Emmerich, Germany
44 Squadron, RAF (Waddington, Lincolnshire - 5 Group)
Hampden I P1340 - captained by Sgt E J Spencer, RAF, and at 3800 feet in bad visibility on return, struck a barrage balloon cable, diving steeply out of control to crash onto the mud flats of the Orwell River near Harwich, Essex. The pilot was rescued after parachuting into the river. His three comrades, however, were never found and they are commemorated on the Runnymede Memorial.
(Pilot): 36202 Plt Off **Reginald William ROOTS**, RAF - Age 20. 316hrs. 7th op.
Roots was probably acting as observer on this flight. He was born in England and came to New Zealand at an early age. Selected for a short service commission in the RAF, he graduated as a pilot with the RNZAF prior to embarking for England in 1939.

Thu 6 Jun 1940

FIGHTER COMMAND

Transit from West Hartlepool, County Durham, to Catterick, Yorkshire
152 Squadron, RAF (Acklington, Northumberland; detachments at Leconfield, Yorkshire and Sumburgh, Shetlands - 13 Group)
Spitfire I K9896 - thought to have been returning from an operational flight - crashed while landing at dusk at Catterick when its undercarriage struck a dispersal pen. The pilot died from severe spinal injuries in Catterick Hospital next day and is buried at Catterick.
Pilot: 41653 Plt Off **Bewsher Eric BELL**, RAF - Age 20. 281hrs. 8 ops.

Fri 7 Jun 1940

BRITISH AIR FORCES FRANCE

Attack on enemy troop columns in Abbeville area
150 Squadron, RAF (Houssay, France - 71 Wing, Advanced Air Striking Force)
Battle I L5288 - took off from Échemines and shot down while attacking enemy motor transport near the village of Vergies, 18km south of Abbeville. All three crew are buried at Vergies.
Pilot: 36145 Fg Off **James Edward VERNON**, DFC, RAF - Age 24.

Flight to Le Mans
73 Squadron, RAF (Échemines, France - 67 Wing, Advanced Air Striking Force)
Hurricane I L1826 - took off at about 1115 but, in concluding a departure 'beat up' of the airfield, stalled and spun into the ground off a third left-hand flick roll. The pilot died instantly on being thrown clear as the fighter broke up and burst into flames. At noon on the 8th he was buried at Troyes, 20km ESE of Échemines, but later reinterred at Choloy.
Pilot: 39534 Fg Off **Edgar James KAIN**, DFC mid, RAF - Age 21. 636hrs. 80+ ops.
Kain was the first Allied air ace of the war. He was due to fly to England on leave later in the day for his wedding there the following week.

Sat 8 Jun 1940

FIGHTER COMMAND

Offensive patrol over northern France
32 Squadron, RAF (Biggin Hill, Kent - 11 Group)
Hurricane I N2406 - took off at 0820 with others and one of three which failed to return. At approximately 0915-0930 some 28 enemy bombers and fighters were engaged by the patrol near Le Treport, N2406 being shot down to crash at La Ganerre, near Houville-en-Vexin. The pilot is buried at the latter location.
Pilot: 72526 Plt Off **Kenneth KIRKCALDIE**, RAFVR - Age 28. 240hrs.
Kirkcaldie's brother, John Allan Kirkcaldie, died on 2 December 1944 while serving in Wellington, New Zealand, with the RNZAF.

FLEET AIR ARM AND FIGHTER COMMAND

In transit from Norway to Britain
HMS *Glorious*
The 22,500-ton aircraft carrier *Glorious*, returning with RAF aircraft and personnel of 46 and 263 Squadrons, escorted by the destroyers *Acasta* and *Ardent*, was surprised at 1545 by the German battlecruisers *Gneisenau* and *Scharnhorst* in company with several other warships about 300 miles west of Narvik at position 6845N:0430E. The battlecruisers opened fire at 1630 and within 90 minutes all three British ships had been destroyed. Of 1559 men in total only 40 survived, including New Zealander Flt Lt **P G Jameson** of 46 Sqn. Three other New Zealand

airmen were lost without trace, the FAA pilot being commemorated on the Fleet Air Arm Memorial at Lee-on-Solent and his two RAF comrades on the Runnymede Memorial.
Of **802 Squadron, FAA (a Sea Gladiator unit embarked on HMS *Glorious*)**
(Pilot): Lt (A) **Reginald Joseph Braithwaite MILES**, RN - Age 24.
Of **263 Squadron, RAF (a Gladiator unit in transit on board HMS *Glorious*)**
(Pilot): 41423 Plt Off **Louis Reginald JACOBSEN**, DFC, RAF - Age 25.
(Pilot) 39143 Fg Off **Harold Edward VICKERY**, RAF - Age 26.
Although the Glorious *was sunk on the 8th, men were known to still be alive in the water on the following day and the 9th is therefore is accepted as the official date of death for many or all of those lost.*
Miles had served as an officer in the RAF prior to transferring to the FAA in October 1938.

Mon 10/Tue 11 Jun 1940

BOMBER COMMAND

Short cross-country
12 Operational Training Unit, RAF (Benson, Oxfordshire - 6 Group)
Battle I K9463 - took off at 0205 but suffered engine failure immediately afterwards. K9463 tried to return to the airfield but entered a steep gliding high speed turn, spun, recovered, then dived into the ground and caught fire a quarter mile SW of Benson. All three crew died, the pilot being buried at Benson.
Pilot: 36264 Plt Off **Basil Percy THOMSON**, RAF - Age 20. 188hrs (43 on Battle).

Tue 11/Wed 12 Jun 1940

BOMBER COMMAND

Attack on target north of Ameins, France
10 Squadron, RAF (Dishforth, Yorkshire - 4 Group)
Whitley V P4954/T - took off at 2047 captained by Sgt L A Keast, RAF, and shot down 6km north of Abbeville. The five crew are buried at Abbeville.
WOp: 624643 LAC **Raymond Robin Henry NUTTALL**, RAF - Age 28. 94hrs.

Wed 12 Jun 1940

ENGLAND

Test flight from Filton to Farnborough, Hampshire(?)
Bristol Aeroplane Company (Filton, Gloucestershire)
Beaufort L4443 (3rd prototype Beaufort) - after an engine caught fire, the pilot, having instructed his two 'passengers' to bale out, headed away from a residential area towards open country. He died when L4443 dived into the ground at 1556 near South Hinksey, Oxfordshire, and is buried at Berrow, Somerset.
Pilot: **Charles Alan WASHER**, Bristol Aeroplane Co. - Age 29.
During 1929-1934 Washer served in the RAF as a short service commission officer. Afterwards he joined Bristol, serving first as an instructor and later as a test pilot. His civilian status means that his name is not recorded in a CWGC register.

Thu 13 Jun 1940

BOMBER COMMAND

Local instrument flying practice
16 Operational Training Unit, RAF (Upper Heyford, Oxfordshire - 6 Group)
Hampden I P4297 - captained by Fg Off J E S MacAlister, RAF, stalled on a turn at 1000 while circling round a Boys' School at Iwerne Minster, near Blandford, Dorset, and crashed. All three crew died, the 2nd pilot being buried at Middleton Stoney, Oxfordshire. It was thought that the crew may have lost their bearings and been searching for a landmark.
(Pilot): 42443 Plt Off **Edgar Albert Ernest SEDGLEY**, RAF - Age 22. 70hrs solo (on Hampden?)
Sedgley may have been piloting at the time as part of the practice was for the pilots to change seats, an awkward and extremely hazardous manoeuvre which was later banned.

Fri 14 Jun 1940

BRITISH AIR FORCES FRANCE

Attack on a target near Dreux
226 Squadron, RAF (Faux-Villecerf, France - 76 Wing, Advanced Air Striking Force)
Battle I (P2335)/N - took off from Sougé and brought down at 1830 near Breux-sur-Avre, 15km west of Dreux. One crew member survived to be captured but died of his wounds less than two months later. His two comrades two died on the day of the crash, the pilot being buried at Dreux, while the other is commemorated on the Runnymede Memorial.
Pilot: 40260 Fg Off **Kenneth Noel REA**, RAF - Age 27.

Sat 15 Jun 1940

NEW ZEALAND

Air to ground firing exercise

1 Service Flying Training School (ATS), RNZAF (Wigram)

Oxford I NZ279 - stalled during a steep left hand turn while low flying over Akaroa township at 1130. NZ279 spun into buildings in the main street and burst in flames, causing extensive property damage, in addition to injuring two civilians. The two crew members are buried at Nelson and Greymouth, respectively. The injured were **Valmai Dann**, age 12, and **Nathalie Ramsay**, 13.

Pilot: A391840 LAC **Francis Maurice McFARLANE**, RNZAF - Age 24. 182hrs (115 solo - 38 on Oxford)

2nd Pilot: A391839 LAC **John Lindsay McFADYEN**, RNZAF - Age 22.

Two of McFarlane's brothers, Laurence Herbert McFarlane and John Osbourne McFarlane, also died while flying; Laurence only a few weeks later with 58 Sqn, RAF, on 2 July, and John with 16 Sqn, RNZAF, on 17 December 1943.

COASTAL COMMAND

Attack on ammunition dump at Stavanger, Norway

224 Squadron, RAF (Leuchars, Fifeshire - 18 Group)

Hudson I N7217/W - took off with two others, the entire formation being shot down by flak and fighters near the target area. All four crew of N7217 are commemorated on the Runnymede Memorial.

Captain: 41838 Plt Off **Niel EWART**, RAF - Age 23. 296hrs. 16th op.

(Pilot): 41558 Plt Off **Francis Sirius CRAWFORD**, RAF - Age 20. 393hrs. 36th op.

Crawford was probably acting as observer on this flight.

Sun 16 Jun 1940

FIGHTER COMMAND

Reconnaissance of SW of Seine River, France

59 Squadron, RAF (Odiham, Hampshire - 22 Group)

Blenheim IV R3817 - took off from Le Mans, France, at 0945 with the intention of landing at Rennes airfield upon completion of the sortie, but brought down near Nonancourt. The three crew were buried in the La Madeleine-de-Nonancourt Cemetery, but later reinterred at Dreux. Twenty-two Group was an Army Co-operation formation but came under Fighter Command until 24 June, when it became an independent group with Command status.

Pilot: 41769 Plt Off **Clive Robinson WYLIE**, RAF - Age 26. 220hrs.

Tue 18 Jun 1940

NEW ZEALAND

Accident(?)

1 Service Flying Training School, RNZAF (Wigram)

During the morning was found dead at the foot of Scarborough cliff, Sumner, and later buried at Invercargill. He had earlier injured his wrist while working at Wigram and was on a month's leave at the time.

(Flight Rigger): A39838 AC1 **Eric Anthony BULMAN**, RNZAF - Age 30.

MAINTENANCE COMMAND

Ferry flight to Malta, via France and Tunisia (by 12 Blenheims and 12 Hurricanes)

4 (Continental) Ferry Pilots Pool, RAF (Kemble, Gloucestershire - 41 Group)

Blenheim IV L9314 - took off from Tangmere between 0630 and 0645. The formation became split up over France in thick cloud 110km south of the Loire River. Five of the Blenheims crashed, killing all fifteen crew members, L9314 coming down at Prunières, 37km NNW of Mende. The three crew were buried at the Prunières parish, but later reinterred at Marseilles.

Pilot: 41426 Plt Off **Douglas Stewart JOHNSTON**, RAF - Age 24.

Tue 18/Wed 19 Jun 1940

BOMBER COMMAND

Raid on Castrop-Rauxel, Germany

58 Squadron, RAF (Linton-on-Ouse, Yorkshire - 4 Group)

Whitley V N1460/R - took off at 2115 captained by Fg Off J T MacInnes, RAF, and last heard from at 2335, before crashing at 0200 at Leveringhausen, about 7km NE of the target area. All five crew were buried a few kilometres north at Waltrop, but later reinterred at Reichswald Forest, 5km SW of Kleve.

2nd Pilot: 42690 Plt Off **Clement Neville BUIST**, RAF - Age 20. 258hrs.

Buist was born at Sydney, Australia, but educated in New Zealand. Selected for a short service commission in the RAF, he embarked for England in 1939.

Fri 21 Jun 1940

NEW ZEALAND

Reconnaissance patrol

1 Service Flying Training School, RNZAF (Wigram)

Oxford I NZ275 - took off from Taieri after two others on the same patrol at 0725, each machine carrying a 250lb bomb, but was unable to gain height. After clearing the airfield boundary NZ275 entered a slow right hand turn in order to clear trees ahead, but stalled and hit the ground while still turning, and caught fire. Plt Off **P R McNab**, the observer, and AC1 **R E Kilpatrick**, wireless operator, were injured but the pilot died within an hour or two of being extricated from the wreckage. He is buried at Christchurch. A layer of frost on the Oxford's wings had significantly degraded its take off performance. Investigation also revealed that the throttle control lock wire had not been broken, indicating that the pilot had failed to push the throttles through the gate when the emergency arose. For their courageous efforts in rescuing the three crew LAC **D T Herrick** and Cpl **E A Stratton** were each awarded the George Medal, and AC2 **D M Nichol** presented with an Air Board letter of commendation

Pilot: 34160 Sqn Ldr **John Hugh KITSON**, RAF - Age 28. 1373+hrs (412+ on Oxford)

Kitson, a flying instructor with the ITS at Wigram, was an English RAF officer on loan to the RNZAF. He was the first fatal casualty on an operational flight undertaken from a New Zealand base (a German raider was active in New Zealand waters at the time - the SS Niagara *striking a mine and sinking only two days earlier).*

Herrick died of wounds on 30 June 1941 after being shot down over France while flying with 53 Sqn, RAF.

Wed 26/Thu 27 Jun 1940

BOMBER COMMAND

Raid on Langenhagen airfield near Hannover, Germany

50 Squadron, RAF (Waddington, Lincolnshire - 5 Group)

Hampden I L4078 - took off at 2217 and brought down by flak, crashing in the early morning near Rethen, about 10km SSE of Hannover. All four crew were buried in the Hannover-Limmer civil cemetery, but later reinterred in the Hannover War Cemetery, also in the suburb of Limmer.

Captain: 40628 Plt Off **Anthony Seton LUXMOORE**, RAF - Age 26. 520hrs. 15th op.

Thu 27 Jun 1940

COASTAL COMMAND

Special reconnaissance of Norwegian coast

269 Squadron, RAF (Wick, Caithness-shire - 18 Group)

Hudson I N7330/C - took off at 0600 with one other and shot down by fighters off the Norwegian coast, killing the four crew. No trace of the captain was found and he is commemorated on the Runnymede Memorial.

Captain: 36183 Plt Off **Peter Noel TROLOVE**, RAF - Age 25. 658hrs. 31st op.

Two of Trolove's brothers, Derek Roy Trolove and William Alan Trolove, died while serving with the RNZAF; Derek while flying in New Zealand with 2 OTU on 11 August 1944, and William while flying in the South Pacific with 31 Sqn on 31 May 1944.

Reconnaissance of Zuider Zee, Netherlands

235 Squadron, RAF (Bircham Newton, Norfolk - 16 Group)

Blenheim IVF L9447/Y - took off at 1255 with five others and one of two which failed to return. The patrol was heavily engaged by flak over the target and later by fighters, L9447 being brought down at 1500 by the latter at Waverveen, 2km west of Vinkeveen. Two of the crew died but the RAF observer survived to be taken prisoner. The pilot was buried in the Vinkeveen Cemetery, 17km south of Amsterdam, but later reinterred at Bergen-op-Zoom.

Pilot: 42109 Plt Off **John Richard CRONAN**, RAF - Age 21.

Cronan was born at Simla, India, but came to New Zealand at an early age. Selected for a short service commission in the RAF, he embarked for England in 1939.

Fri 28 Jun 1940

FIGHTER COMMAND

Escort to six Blenheims on photographic reconnaissance Calais-Boulogne, France

151 Squadron, RAF (North Weald, Essex; detachment at Manston, Kent - 11 Group)

Hurricane I P3322 - after refuelling at Manston, took off at 1730 in company with two others to provide escort in company with Hurricanes of 56 Sqn. P3322 was shot down by enemy fighters near Boulogne, its pilot baling out into the sea. A destroyer was despatched to make a rescue, but he was not seen again and is commemorated on the Runnymede Memorial.

Pilot: 40785 Fg Off **Kenneth Edward NEWTON**, RAF - Age 24. 539hrs.

Newton was born in London, England, and came to New Zealand at an early age. Selected for a short service commission in the RAF, he graduated as a pilot with the RNZAF prior to embarking for England in 1938.

Mon 1/Tue 2 Jul 1940

BOMBER COMMAND

Attack on *Scharnhorst* at Kiel

58 Squadron, RAF (Linton-on-Ouse, Yorkshire - 4 Group)

Whitley V N1461 - took off at 2100 captained by Plt Off C J T Jones, RAF, and brought down near the target, killing four of the crew. A fifth member, an RAF Sgt, survived to become a prisoner of war. The dead were all buried in the Garrison Cemetery at Kiel, but later reinterred in the Kiel War Cemetery.

2nd Pilot: 39545 Fg Off **Laurence Herbert McFARLANE**, RAF - Age 25. 581hrs.

WOp: 542430 Sgt **Henry Eric Archibald CRAVEN**, RAF - Age 24.

McFarlane had earlier survived the crash which killed Plt Off T A Bunting on 16/17 December 1937. Two of his brothers, Francis Maurice McFarlane and John Osbourne McFarlane, died while flying also; Francis only a few weeks earlier, with 1 Service Flying Training School, RNZAF, on 15 June, and John with 16 Sqn, RNZAF, on 17 December 1943.

Craven was born at Burntisland, Scotland, but came to New Zealand at an early age. He returned to Britain in 1937 and enlisted in the RAF in July.

Wed 3 Jul 1940

NEW ZEALAND

Practice bombing

1 Air Observers School, RNZAF (Ohakea)

Vincent NZ305 - took off shortly after 0940, piloted by Flt Lt **L A Robertson**, for practice over the Raumai bombing range. Four bombing runs were made at 3000 feet but the release gear failed to operate on each. One of the trainee observers was called upon to check the faulty mechanism from the prone bombing position after the first run, and he was not seen again after completing this action. The Vincent returned to the airfield with the gear still not in working order and while taxiing across the airfield it was noticed that the observer was missing. His body was later found near the range and it was presumed he had fallen to his death at about 1000 during the practice while his safety cable (or 'monkey strap') was unsecured (the strap would have been unhooked while examining the faulty bomb release equipment). He is buried at Onerahi. The other two crew members, including trainee observer LAC **O W Knight**, returned safely with the aircraft.

Observer: A40739 LAC **David Handforth HAWKES**, RNZAF - Age 24.

Wed 3/Thu 4 Jul 1940

BOMBER COMMAND

'Gardening' - minelaying to south of Langeland island, Denmark

44 Squadron, RAF (Waddington, Lincolnshire - 5 Group)

Hampden I P4352/J - took off at 2122 and last made contact at 0130 to advise that an engine had failed, its position being estimated as 45 miles NW of the Dutch island of Terschelling. Two 206 Sqn Hudsons and their crews were lost searching for the aircraft and its crew early in the morning (see below). The body of the Hampden's air gunner washed up three weeks later on the west coast of Denmark, at Strandlehen. The other three crew members are commemorated on the Runnymede Memorial.

Captain: 42162 Plt Off **Desmond Arthur TODD**, RAF - Age 21. 319hrs. 16th op.

Thu 4 Jul 1940

COASTAL COMMAND

Search for missing 44 Sqn aircraft

206 Squadron, RAF (Bircham Newton, Norfolk - 16 Group)

Hudson I P5162 - took off at 0550 captained by Plt Off J E McKinnon, RAF, in company with another Hudson to search for missing Hampden P4352 (see above). Both failed to return, their last report indicating no sign of the missing bomber had been found and that they were heading north of the Dutch island of Texel to search further. Both four-man crews are commemorated on the Runnymede Memorial.

(Pilot): 41434 Plt Off **Stanley John LESTER**, RAF - Age 26.

Lester was a pilot but acted as navigator on this flight.

Tue 9 Jul 1940

BOMBER COMMAND

Raid on Sola airfield, Stavanger, Norway (by 12 aircraft - 7 lost)

21 Squadron, RAF (Lossiemouth, Morayshire - 2 Group)

Blenheim IV L8872/M - took off at 0800 and crashed into the sea when attacked by fighters after bombing the airfield. All three crew are commemorated on the Runnymede Memorial.

Pilot: 39559 Flt Lt **John David William MURRAY**, RAF - Age 27. 701hrs.

21 Squadron, RAF (Lossiemouth, Morayshire - 2 Group)
Blenheim IV R3732/L - took off at 0800 on same raid as above, and escaped into cloud after the attack, but a message sent at 1145 advised that because of damage inflicted by fighters it might have to ditch. Contact was maintained until 1231, after which nothing further was heard. Two of the crew are commemorated on the Runnymede Memorial, but the pilot's body was later recovered and is buried at Lonstrup, just south of Ribe, Denmark.
Pilot: 16216 Wg Cdr **Leslie Clive BENNETT**, RAF - Age 34. 2453hrs.

COASTAL COMMAND

'SA2' - anti-shipping strike off coast of southern Norway
233 Squadron, RAF (Leuchars, Fifeshire - 18 Group)
Hudson I N7377/J - took off at 1534 and shot down by fighters over the North Sea at 1720, all four crew being commemorated on the Runnymede Memorial.
Captain: 41274 Plt Off **Loris Jardine Ernest EWING**, mid, RAF - Age 22. 578hrs. 58th op.

Fri 12 Jul 1940

FIGHTER COMMAND

Escort for Convoy *Booty* off Orfordness, Suffolk
151 Squadron, RAF (North Weald, Essex - 11 Group)
Hurricane I P3275 - took off at 0835. Three enemy bombers were attacked by the Squadron, following which P3275 was seen to glide down with a dead engine into the sea about 20 miles east of Orfordness. Another pilot circled the area for five minutes but no trace of the Hurricane or its pilot could be seen. He is commemorated on the Runnymede Memorial.
Pilot: 39957 Fg Off **James Henry Leslie ALLEN**, RAF - Age 25.

Sat 13/Sun 14 Jul 1940

COASTAL COMMAND

Anti-submarine patrol
220 Squadron, RAF (Thornaby, Yorkshire - 18 Group)
Hudson I N7231 - took off at 1845, but during the patrol was advised of deteriorating weather and instructed to land at Dishforth, Yorkshire, upon return. On making landfall N7231 appeared uncertain of its position and circled for 30 minutes, by which time Dishforth had closed. After receiving further instructions to proceed to Leuchars, Fifeshire, the Hudson came in over the coast at about 3000 feet, but struck a balloon barrage cable at Boldon Colliery, County Durham, just south of South Shields. All four crew died, the two pilots being buried at Hylton, near Sunderland.
Captain: 36169 Plt Off **Garth Wells Fuller CAREY**, DFC, RAF - Age 24. 632hrs. 71st op.
(Pilot): 41632 Plt Off **Phillip Knox VARTAN**, RAF - Age 22. 462hrs. 47th op.
Vartan was probably acting as observer on this flight.

Fri 19 Jul 1940

FIGHTER COMMAND

Patrol 20 miles south of Folkestone
141 Squadron, RAF (West Malling, Kent - 11 Group)
Defiant I L6974 - took off from Hawkinge at 1233 with eight others and one of six lost. The formation was bounced at about 1245 by enemy fighters, L6974 being one of four Defiants shot down within a minute into the sea, three in flames. The two crew are commemorated on the Runnymede Memorial.
Pilot: 41850 Plt Off **John Richard KEMP**, RAF - Age 25. 303hrs.

141 Squadron, RAF (West Malling, Kent - 11 Group)
Defiant I L7015 - took off from Hawkinge at 1233 on same patrol as the above and lost under the same circumstances. The two crew are commemorated on the Runnymede Memorial.
Pilot: 41297 Plt Off **Rudal KIDSON**, RAF - Age 26. 303hrs.

Sat 20/Sun 21 Jul 1940

BOMBER COMMAND

Raid on Horst, near Gelsenkirchen, Germany
75 (NZ) Squadron, RAF (Feltwell, Norfolk - 3 Group)
Wellington IC R3165 - brought down at Weseke, about 40km NNW of the target. All five crew were buried in a communal grave nearby, but later reinterred at Reichswald Forest, 5km SW of Kleve.
Captain: 77026 Fg Off **Samuel Miles Mackenzie WATSON**, mid, RAFVR - Age 27.
2nd Pilot: 36237 Plt Off **Edward Colin Joseph CAMERON**, RAF - Age 19.
Front Gunner: A36139 Flt Sgt **Ronald Alexander John ANDERSON**, RNZAF - Age 26.
Rear Gunner: A391332 Sgt **John Lewis OWEN**, RNZAF - Age 24. 46hrs. 7th op.
This was the first occasion on which a combination of RNZAF airmen and New Zealander members of the RAF had lost their lives on the same flight.
Owen was the first New Zealand-trained air gunner to lose his life overseas.

Mon 22 Jul 1940

NEW ZEALAND

Ferry flight from Harewood to Birdlings Flat for air firing practice

1 Service Flying Training School (ATS), RNZAF (Wigram; temporarily based at Harewood)

Oxford I NZ262 - took off at about 0900 and 15 minutes later, in thick fog, hit the side of a 300 foot hill near Kaituna, Ataahua, opposite Lake Ellesmere, on the property of Mr I C Gray. All four occupants died and are buried at Havelock North, Christchurch, Invercargill and Christchurch, respectively. Three of five other Oxfords making the flight were safely recalled to Harewood as the fog bank rolled in, while the others managed to land at Wigram and at the Flat.

Pilot: (NZ2521) Plt Off **William Oscar George KROGH**, RNZAF - Age 28. 243hrs solo (41 on Oxford)
(Pilot): A40232 LAC **Archibald John HULL**, RNZAF - Age 24.
(Pilot): A40247 LAC **Herbert John von TUNZELMAN**, RNZAF - Age 22.
(Flt Mech): A39849 AC1 **Albert Bartholomew DUFFELL**, RNZAF - Age 25.

At the time of Krogh's death service numbers had not yet been assigned to officers of the RNZAF. NZ2521 was a posthumous allocation made at war's end.
Hull and von Tunzelman were due to graduate at week's end.

FIGHTER COMMAND

Returning to base airfield after operational flying at Martlesham Heath

85 Squadron, RAF (Debden, Essex; detachment at Martlesham Heath, Suffolk - 11 Group)

Hurricane I P3895 - spun off a half roll at 2000 feet while approaching Debden at 2105 and crashed a quarter mile SW of the airfield. The pilot is buried a few miles to the north at Wimbish.

Pilot: 36266 Plt Off **John Laurance BICKERDIKE**, RAF - Age 21. 166hrs solo. 32 ops.

Tue 23 Jul 1940

COASTAL COMMAND

Operational

269 Squadron, RAF (Wick, Caithness-shire - 18 Group)

Hudson I P5152/(P or J) - took off at 0415 captained by Flt Lt Price, RAF, and immediately collided with a 3 Sqn Hurricane I (P2862), the bomb load exploding after it crashed and caught fire. All four crew died, the 2nd pilot being buried at Wick. The Hurricane pilot survived but was seriously injured.

2nd Pilot: 41277 Plt Off **James Tayne Taine FLEMING**, RAF - Age 20. 737hrs. 96th op.

Thu 25 Jul 1940

COASTAL COMMAND

'S.A.9' - anti-shipping patrol off the Dutch coast

53 Squadron, RAF (Detling, Kent - 16 Group)

Blenheim IV R3836/X - took off at 0602 and crashed on the Koffieboon Plaat off Ameland, one of the West Frisian islands. The RAF observer's body later washed ashore at Friesland, while his two comrades are commemorated on the Runnymede Memorial.

Pilot: 41752 Plt Off **David Bayntun STARKY**, RAF - Age 20. 304hrs. 10th op.

BOMBER COMMAND

Raid on oil storage depots at Cherbourg, France

101 Squadron, RAF (West Raynham, Norfolk - 2 Group)

Blenheim IV N6174 - lost without trace, its three crew being commemorated on the Runnymede Memorial.

Pilot: 42770 Plt Off **Rupert Edward SHORT**, RAF - Age 23. 228hrs. 2nd op.

Thu 25/Fri 26 Jul 1940

BOMBER COMMAND

Raid on Kassel, Germany

75 (NZ) Squadron, RAF (Feltwell, Norfolk - 3 Group)

Wellington IC R3235 - believed to have been brought down into the Ijsselmeer Sea, Netherlands, where the bodies of all six crew later washed ashore, that of the captain on 3 August. All are buried at Amsterdam.

Captain: (NZ2526) Fg Off **William Harcourt COLEMAN**, DFC, RNZAF - Age 23. 1063hrs. 27th op.
2nd Pilot: 42722 Plt Off **Frank Twain POOLE**, RAF - Age 24. 291hrs. 9th op.
Rear Gunner: A391377 Sgt **William Donald Francis ANNAN**, RNZAF - Age 20.

At the time of Coleman's death service numbers had not yet been assigned to officers of the RNZAF. NZ2526 was a posthumous allocation made at war's end.

Sun 28/Mon 29 Jul 1940

BOMBER COMMAND

Raid on Brussels airfield, Belgium

142 Squadron, RAF (Binbrook, Lincolnshire - 1 Group)

Battle I L5584 - took off at 0005 and lost without trace, its three crew being commemorated on the Runnymede Memorial.

Pilot: 36254 Plt Off **Matthew John Alexander KIRDY**, RAF - Age 19.

Sat 3/Sun 4 Aug 1940

BOMBER COMMAND

Raid on 'A71' (Horst), Germany

75 (NZ) Squadron, RAF (Feltwell, Norfolk - 3 Group)

Wellington IC R3176 - took off at 2115 but encountered bad weather on returning to England. Diverted to Mildenhall, Suffolk, R3176 suffered an engine failure and executed a wheels-up forced landing at 0420 at Barton Mills, a few miles to the south. The Wellington slid into a ridge and caught fire, injuring four of the six crew, including the captain who died later in the day. He is buried at Feltwell.

Captain: 34232 Sqn Ldr **Wilfred Ira COLLETT**, RAF - Age 28.

Mon 5 Aug 1940

COASTAL COMMAND

Air-sea-rescue search for lost dinghy

206 Squadron, RAF (Bircham Newton, Norfolk - 16 Group)

Hudson I P5133 - took off at 1505 but on returning to base at 1800 stalled off a vertical turn, flicked on to its back, and crashed at Ringer's Farm, Syderstone. The wreckage caught fire, exploding the bomb load and killing the four crew, the captain of which is buried at Hunstanton. It would appear that the search formed part of an anti-submarine patrol.

Captain: 41295 Plt Off **Raymond Thomas KEAN**, DFC, RAF - Age 22. 606hrs solo. 83rd op.

Tue 6 Aug 1940

BOMBER COMMAND

Solo night flying training

12 Operational Training Unit, RAF (Benson, Oxfordshire - 6 Group)

Battle I L4943 - took off shortly before 2315 and on climbing to 800-1000 feet swung to starboard then port, before flying into the ground to catch fire at Watlington, 5 miles ENE of Benson. The pilot is buried at Benson.

Pilot: 43702 Plt Off **Arthur Vivian FISHER**, RAF - Age 23. 160hrs.

Wed 7 Aug 1940

NEW ZEALAND

Solo night flying training

2 Service Flying Training School (ATS), RNZAF (Woodbourne)

Oxford I NZ271 - took off at 2245 and flew level at 200 feet before stalling off a steep turn to crash near the base. The pilot is buried at Masterton.

Pilot: A40644 LAC **Thomas Edwin DENBEE**, RNZAF - Age 22. 147hrs (28 solo on Oxford - 45min at night)

Sat 10 Aug 1940

FIGHTER COMMAND

Training

5 Operational Training Unit, RAF (Aston Down, Gloucestershire - 10 Group)

Blenheim IF L6799 - struck while on the ground by one of the OTU's Spitfire Is (L1063) which was in the process of taking off. Both occupants of L6799 died in the collision, the instructor being buried at Brimscombe, near Minchinhampton. The Spitfire pilot was injured and later died also.

Pilot Instr: 41673 Plt Off **George Gordon CRAWFORD**, RAF - Age 23. 369hrs. 7 ops.

Sun 11 Aug 1940

FIGHTER COMMAND

Interception of enemy aircraft attacking convoy *Booty*

74 Squadron, RAF (Hornchurch, Essex - 11 Group)

Spitfire I R6757 - took off at 1145 with 10 others from Manston, Kent, being one of two which failed to return. R6757 was shot down in combat over the convoy off the Harwich-Clacton coast at 1204. The pilot's body was later recovered and is buried at Ostende, Belgium.

Pilot: 41552 Plt Off **Donald Gordon COBDEN**, RAF - Age 26. 439hrs. 84th op.

11 August was Cobden's 26th birthday.

Tue 13 Aug 1940

BOMBER COMMAND

Training - circuits and landings

13 Operational Training Unit, RAF (Bicester, Oxfordshire - 7 Group)

Blenheim IV L9038 - took off at 1430, but immediately experienced an engine failure and dived into the ground after a stall turn as it attempted to clear nearby trees. Only one of the crew was killed in the crash, but the pilot died of injuries a few hours later and is buried at Caversfield, just north of Bicester.

Pilot: A39929 Sgt **Henry Algernon Gascoyne MILLER**, RNZAF - Age 25. 256hrs.

Training

11 Operational Training Unit, RAF (Bassingbourn, Cambridgeshire - 6 Group)

Wellington I L4387 - while taking off at 2145 captained by Fg Off F A H Lambart, RAF, struck a dispersed aircraft and burst into flames. All six crew died, the observer being buried in the Bassingbourn cum Kneesworth Cemetery.

Observer: A391869 Sgt **Douglas Oswald REWA**, RNZAF - Age 19. 99hrs.

Rewa was the first New Zealand-trained observer to lose his life overseas.

Tue 13/Wed 14 Aug 1940

BOMBER COMMAND

Raid on Fiat aero engine works at Turin, Italy

10 Squadron, RAF (Leeming, Yorkshire - 4 Group)

Whitley V P4965/H - took off from Abingdon, Berkshire, at 1327 and in a fighter attack over the target received damage to its starboard aileron, and had an engine put out of action. On struggling back to the English coast in the early morning an emergency landing was attempted on the beach near Dymchurch Redoubt, Kent. However, the weakened aileron broke off during the approach and the Whitley fell into the sea a mile off shore. Three of the crew survived but both pilots died, their bodies later washing up on the French coast to be buried at Boulogne.

Captain: 43370 Plt Off **Ernest Ian PARSONS**, DFC, RAF - Age 27.

Parsons was the first New Zealander to lose his life while flying on operations against Italy.

Wed 14 Aug 1940

COASTAL COMMAND

Night training(?)

206 Squadron, RAF (Bircham Newton, Norfolk; satellite airfield at Docking - 16 Group)

Hudson I N7401 - took off at 0315, but control lost on a turn five minutes later while in the circuit at Docking, 2 miles north of Bircham Newton, and crashed at Sunderland Farm. Three crew members died, while the fourth was injured, both pilots being buried at Great Bircham. It was thought that the cockpit lights were on too brightly, disguising the fact that the Hudson was over banking.

Captain: 36265 Plt Off **Herbert Gregory BALLANTYNE**, RAF - Age 22. 317hrs. 17 ops.

Pilot: 41623 Plt Off **John Oscar Lloyd STEPHENSON**, RAF - Age 25. 208hrs.

BOMBER COMMAND

Cross country training flight, base-Salisbury-Barnstaple-Lundy island and return

15 Operational Training Unit, RAF (Harwell, Berkshire - 6 Group)

Wellington I L4308 - captained by Sgt R L Peacock, RAF. Seen at 1125 by the crew of another Harwell aircraft to lose height and fall towards the sea with its port engine on fire, crash and sink about 5 miles east of Lundy Island, in the mouth of the Bristol Channel. An Admiralty yacht picked up the body of one crew member but no trace could be found of the other five, who, including a New Zealand air gunner, are commemorated on the Runnymede Memorial.

Air Gunner: A39913 Sgt **John Edward FULLER**, RNZAF - Age 24. 41hrs.

Thu 15 Aug 1940

FIGHTER COMMAND

Interception of incoming raid on Portland, Dorset

87 Squadron, RAF (Exeter, Devonshire - 10 Group)

Hurricane I P3215 - took off at 1700, leading the Squadron into combat with an enemy raiding force of 120 aircraft. P3215 was hit during the engagement and caught fire, coming down from 15,000 feet towards Warmwell. It headed for and circled the Abbotsbury area before crash landing there at about 1800 in a copse and striking an oak tree. The pilot was thrown clear but was already dead when help arrived. He is buried at Warmwell.

Pilot: 29224 Sqn Ldr **Terence Gunion LOVELL-GREGG**, RAF - Age 27. 2178hrs. 9th op.

234 Squadron, RAF (Middle Wallop, Hampshire - 10 Group)

Spitfire I R6988 - took off at 1705 with 11 others to intercept the same raid as above. R6988 was damaged and the pilot wounded during the ensuing combat. He baled out over Bournemouth but his parachute failed to open. His body fell into the garden of 'Hambledon', a house in Leven Avenue, while the Spitfire crashed further along the same avenue at the corner of Walsford Road. He is buried at Boscombe, near Bournemouth.

Pilot: 41924 Plt Off **Cecil Henry HIGHT**, RAF - Age 22. 68th op(?)

Sat 17 Aug 1940

TRAINING COMMAND

Instrument flying practice

7 Service Flying Training School, RAF (Peterborough, Northamptonshire - 23 Group)

Hart Trainer K5799 - spun in at 1100 from a low height and crashed at Deeping St Nicholas, near Spalding, Lincolnshire. Both pilots died, the New Zealander being buried at Peterborough.

Pilot: FX78421 Leading Airman **Edward George CULLEY**, RN, of HMS *Daedalus* (FAA) - Age 18.

Wed 21 Aug 1940

NEW ZEALAND

Solo flying practice

2 Service Flying Training School (ITS), RNZAF (Woodbourne)

Vincent NZ325 - at 1220, stalled on steep turn at low altitude and dived into a farm shed, and caught fire near Dillons Point Road, near Blenheim. The pilot is buried at Feilding.

Pilot: NZ40655 LAC **Raymond Edward George MASON**, RNZAF - Age 22. 71hrs solo (19 on Vincent)

Mason's death was the first of an RNZAF airman to be recorded in contemporary documents featuring an 'NZ' service number prefix. Up to this time numbers for pre-war regular airmen normally carried no prefix, while those of airmen enlisted for 'Duration of War', including Mason, had been prefixed 'A'.

Wed 21/Thu 22 Aug 1940

FIGHTER COMMAND

Flarepath duty

23 Squadron, RAF (Wittering, Northamptonshire - 12 Group)

On late evening flarepath duty at the station, died when struck by a landing Blenheim and is buried at Wittering.

(Air Gunner): NZ391875 Sgt **Jack Stephen BRENNAN**, RNZAF - Age 22. 36hrs.

Thu 22/Fri 23Aug 1940

BOMBER COMMAND

Motor vehicle accident

15 Operational Training Unit, RAF (Harwell, Berkshire - 6 Group)

Accidentally knocked down by a truck on the main Oxford-Newbury road at 2215. Admitted to the Radcliffe Infirmary, he died there the next day and is buried at Harwell.

(Pilot): NZ391341 Sgt **Robert Henry CLIFFORD**, RNZAF - Age 24.

Sun 25/Mon 26 Aug 1940

BOMBER COMMAND

Raid on Berlin, Germany

49 Squadron, RAF (Scampton, Lincolnshire - 5 Group)

Hampden I P4416/L - believed to have ditched on return journey when strong headwinds were encountered, its four crew being commemorated on the Runnymede Memorial. When flying to Berlin a Hampden was operating at the limit of its range.

Captain: 41005 Plt Off **Nicoll Brian FAWCETT**, RAF - Age 20. 485hrs. 22nd op.

Mon 26 Aug 1940

FIGHTER COMMAND

Air fighting practice

5 Operational Training Unit, RAF (Aston Down, Gloucestershire - 10 Group)

Hurricane I N2606 - seen at 0705 to dive very low before pulling up in a violent steep climbing turn, then stalled and spun into the ground, and caught fire at South Cerney. The pilot is buried at Woodchester, 2 miles SW of Stroud.

Pilot: 36256 Plt Off **Ronald Stanley MAGEE**, RAF - Age 25. 9hrs on Hurricane.

MAINTENANCE COMMAND

Ferry flight from Debden, Essex, to Kemble

4 Ferry Pilots Pool, RAF (Kemble, Gloucestershire - 41 Group)

Hurricane I V6563 - approaching to land at 1500, began a tight turn across wind at low altitude, stalled and struck the ground in a left hand spin, and caught fire. The pilot is buried at Kemble.

Pilot: 41820 Plt Off **Leonard William BENNETT**, RAF - Age 24. 235hrs solo.

Wed 28 Aug 1940

FIGHTER COMMAND

Training
7 Operational Training Unit, RAF (Hawarden, Flintshire - 10 Group)
Spitfire I L1060 - stalled and crashed into the water or onto the beach at Hoylake, 3 miles west of Birkenhead, Cheshire. The pilot is buried at Hawarden.
Pilot: 42136 Plt Off **Michael Ernest Brian MACASSEY**, RAF - Age 23. 371hrs. 2 ops.

BOMBER COMMAND

Training - solo night landings
17 Operational Training Unit, RAF (Upwood, Huntingdonshire - 7 Group)
Blenheim IV L8843 - appeared to lose control while turning immediately after take off at 2145 in very dark night conditions. Lost altitude and crashed, catching fire and killing the pilot, who is buried at Bury, 2 miles NE of Upwood.
Pilot: NZ39919 Plt Off **Henry Michael HOADLEY**, RNZAF - Age 20. 214hrs.

Fri 30 Aug 1940

COASTAL COMMAND

Practice
235 Squadron, RAF (Bircham Newton, Norfolk - 16 Group)
Blenheim IVF L9262 - spun while practising circuits at 1115 and crashed at Barwick Farm, near the airfield. Both crew members died, the pilot being buried at Great Bircham.
Pilot: NZ39934 Plt Off **John Sinclair PRIESTLEY**, RNZAF - Age 27. 156hrs. 2 ops.

Sat 31 Aug 1940

NEW ZEALAND

RNZAF Station Harewood
Admitted to the Christchurch Public Hospital four days earlier, died of natural causes and is buried in that city.
(Clerk): NZ402796 AC1 **Kenneth William SANDO**, RNZAF - Age 27.

Flying practice
Flying Instructors School, RNZAF (Hobsonville)
Tiger Moth II NZ722 - took off at 0915 and practised aerobatics over the Rangitoto channel, but at 1000 spun at low altitude and crashed into the sea about 600 yards off Narrow Neck. Plt Off **B V le Pine** was quickly rescued from the rapidly sinking machine by a nearby boat and rushed to hospital injured. He recovered but his companion died in the crash. The body was recovered during salvage of the wreckage three days later and is buried at Wanganui.
Pilot: NZ39914 Plt Off **Robert Michael GOLDSTONE**, RNZAF - Age 21. 179hrs solo (24 on Tiger Moth)

Sun 1/Mon 2 Sep 1940

BOMBER COMMAND

Raid on Hannover, Germany
37 Squadron, RAF (Feltwell, Norfolk - 3 Group)
Wellington IA N2992 - took off at 2050 captained by Plt Off M S Burberry, RAF, and signalled a position message at 0250, placing it some 50 miles out from the English coast. Shortly afterwards N2992 apparently flew into the sea. Burberry, the sole survivor, was picked up by a warship after being afloat in a rubber dinghy for two days. The 2nd pilot's body later washed up on the Dutch coast on the 28th, but no trace was found of the other four crew and they are commemorated on the Runnymede Memorial.
Air Gunner: NZ40201 Sgt **Thomas Chamberlain Molineux BROWNE**, RNZAF - Age 19. 105hrs. 5th op.

Wed 4 Sep 1940

FIGHTER COMMAND

Night interception patrol
264 Squadron, RAF (Kirton-in-Lindsey, Lincolnshire - 12 Group)
Defiant I N1628 - took off at 2205 piloted by Fg Off D K C O'Malley, RAF, and headed away at low altitude but struck the ground near Northorpe, 3 miles west of Kirton-in-Lindsey. Both crew members are buried at Kirton-in-Lindsey.
Air Gunner: NZ391868 Sgt **Laurie Andrew Woodney RASMUSSEN**, RNZAF - Age 18. 19hrs.

Fri 6 Sep 1940

NEW ZEALAND

Dual instruction
2 Elementary Flying Training School, RNZAF (Bell Block)
Tiger Moth II NZ745 - while on the ground at 0755, with engine running, was struck by one of the School's Moth Majors (NZ521) coming into land, flown by trainee pilot LAC **J S Vibert**. Seeing the impending danger at the last minute, Fg Off **F J McDonald**, the instructor in NZ745, endeavoured to taxi out of the way but was unable to

avoid the collision. Although he was not himself hurt, his pupil was seriously injured and died in the New Plymouth Public Hospital later in the day, at 1845, and is buried at Devonport. Vibert, flying solo, who had not sighted the other aircraft in his landing path as it was in his blind spot, was slightly injured.
Pupil Pilot: NZ402171 LAC **George William de VERE**, RNZAF - Age 23.

Fri 6/Sat 7 Sep 1940

BOMBER COMMAND

Raid on Krefeld, Germany
44 Squadron, RAF (Waddington, Lincolnshire - 5 Group)
Hampden I P2087/M - piloted by Plt Off S R Taunton, RAF. Severely damaged by heavy flak near Munster on the return, P2087 fell with starboard engine and wing afire and crashed, killing two of the crew who are now buried at Reichswald Forest, 5km SW of Kleve. The pilot and fourth crew member successfully parachuted to safety and were captured.
Air Gunner: NZ40199 Sgt **James BRACEGIRDLE**, RNZAF - Age 21. 81hrs. 5th op(?)

Sat 7 Sep 1940

FIGHTER COMMAND

Interception patrol
504 Squadron, RAF (Hendon, Middlesex - 11 Group)
Hurricane I L1615 - took off at 1620 on a patrol over the Thames estuary and towards Margate. A few minutes after 1700 the patrol's rear section was bounced by five or six enemy fighters which sent down its leader's Hurricane out of control to crash at Sandbanks Farm, Graveney, near Faversham, Kent. It is not known if the pilot had abandoned L1615 or was still in the cockpit when it crashed. He was badly burned, however, and died of injuries later in the day, being buried at Faversham.
Pilot: 40651 Fg Off **Kenneth Victor WENDEL**, RAF - Age 24.

Sun 8 Sep 1940

COASTAL COMMAND

'CAL 2' - attack on shipping convoy outside Calais Harbour, France
53 Squadron, RAF (Detling, Kent - 16 Group)
Blenheim IV T2042/H - took off at 1149 and lost without trace, its three crew being commemorated on the Runnymede Memorial.
Pilot: 42749 Plt Off **Robert Gladstone HALL**, RAF - Age 19. 330hrs. 7th op(?)

BOMBER COMMAND

Photographic reconnaissance of Belgian, Dutch and French ports
82 Squadron, RAF (Watton, Norfolk - 2 Group)
Blenheim IV R3730/U - piloted by Plt Off J M McCausland, RAF, being one of three Blenheims which departed independently on the reconnaissance between 1200 and 1340. Their sorties were ordered to be completed with or without the benefit of cloud cover. Two, including R3730, were lost without trace, their three-man crews being commemorated on the Runnymede Memorial.
Observer: NZ39971 Sgt **John Bernard PHILPOTT**, RNZAF - Age 30.

Mon 9 Sep 1940

BOMBER COMMAND

Raid on Ostend, Belgium(?)
107 Squadron, RAF (Wattisham, Suffolk - 2 Group)
Blenheim IV T1831 - took off at 0325 and believed brought down after bombing the target, crashing near Wambercourt, 6km north of Hesdin, France. All three crew are buried at Wambercourt.
Pilot: 43704 Plt Off **Charles de Vic HALKETT**, RAF - Age 24. 188hrs.

Wed 11 Sep 1940

NEW ZEALAND

Engine starting accident
1 Air Observers School, RNZAF, (Ohakea)
DH86 Express NZ553 - while starting the port outer engine a few minutes after 1800, a flight mechanic was killed instantly when he walked into the already spinning propeller of the port inner engine. He is buried at Rongotea.
(Flight Mech): NZ391106 AC1 **Arthur Laurence JESSOP**, RNZAF - Age 22.

Fri 13/Sat 14 Sep 1940

NEW ZEALAND

Motor vehicle accident

RNZAF Station Wigram

Returning from Christchurch City shortly after midnight when his motor cycle collided head on with a car at the intersection of the Main South Road and Watts Road in Upper Riccarton. The car driver was only slightly injured but the airman died instantly and is buried at Christchurch.

(Carpenter): NZ40270 AC1 **Maurice Ingram LOW**, RNZAF - Age 34.

The CWGC register gives the 13th as Low's date of death.

Wed 18 Sep 1940

MIDDLE EAST

HQ Aden, RAF

Taken ill earlier with peritonitis while on sick leave at Mukeiras Hill station, 100 miles from Aden, died this day and was buried at Mukeiras, but later reinterred at Aden.

(Pilot): 27061 Wg Cdr **Alan Cunningham MITCHELL**, RAF - Age 35. 1997hrs.

Thu 19 Sep 1940

COASTAL COMMAND

'Dundee 2' - reconnaissance to Berck-sur-Mer, France

53 Squadron, RAF (Detling, Kent - 16 Group)

Blenheim IV T2045/F - took off at 1455 and lost without trace, its three crew being commemorated on the Runnymede Memorial.

Pilot: 43050 Plt Off **Clarence Francis TIBBITTS**, RAF - Age 23. 262hrs. 1st op.

Fri 20 Sep 1940

FIGHTER COMMAND

Interception patrol - Gravesend

92 Squadron, RAF (Biggin Hill, Kent - 11 Group)

Spitfire I X4417 - took off with nine others at 1115 to patrol over Gravesend. The Squadron first flew SW to Dungeness but north of there was attacked from above and behind by enemy fighters at 1155. Two Spitfires were shot down, including X4417 which, operating at the rear of the formation as part of 'Blue' Section, crashed into high tree tops at West Hougham. The wrecked fighter and its dead pilot were not discovered until a month later. He is buried at Hawkinge, near Folkestone.

Pilot: 41847 Plt Off **Howard Perry HILL**, RAF - Age 20. 299hrs. 45th op.

Hill's brother, Peter John Hill, died on 12 January 1945 while flying with 5 Lancaster Finishing School, RAF.

Fri 20/Sat 21 Sep 1940

BOMBER COMMAND

Raid on Ostend, Belgium

75 (NZ) Squadron, RAF (Feltwell, Norfolk - 3 Group)

Wellington IC T2463 - took off at 2115 captained by Plt Off M R Braun, RAF, and shot down by coastal flak batteries, crashing near Leffinge, Belgium, 8km SSW of Ostend. The captain is buried at Ostend but the other five crew members were buried at Leffinge. Sometime between 1957 and 1991 the graves of the latter were concentrated into the Adegem Canadian War Cemetery, 19km east of Bruges.

Rear Gunner: NZ391321 Sgt **Lindsay Douglas ANDERSON**, RNZAF - Age 20. 119hrs. 15th op.

Mon 23/Tue 24 Sep 1940

BOMBER COMMAND

Raid on Berlin, Germany (by 129 aircraft - 3 lost)

83 Squadron, RAF (Scampton, Lincolnshire - 5 Group)

Hampden I L4049/A - captained by Sqn Ldr A O Bridgman, RAF, and hit by flak, crashing near Bethen (Behteln?). The pilot survived to be taken prisoner but the other three crew died and are buried at Becklingen, 17km south of Soltau.

Air Gunner: NZ40185 Sgt **Alfred Frederick BLATCH**, RNZAF - Age 26. 94hrs. 10th op.

The CWGC register gives Blatch's age as 34. His last posting is recorded as 106 Sqn, but he may have been transferred or detached to 83 Sqn shortly before the raid.

Tue 24/Wed 25 Sep 1940

BOMBER COMMAND

Raid on Calais, France

107 Squadron, RAF (Wattisham, Suffolk - 2 Group)

Blenheim IV T1883 - brought down at Guemps, 12km to the SE of the target, after attacking the docks. The three crew are buried at Guemps.

Pilot: 43711 Plt Off **Winton Selwood SHANN**, RAF - Age 23. 240hrs. 6th op.
The CWGC register entry incorrectly shows Shann as RAFVR.

Wed 25 Sep 1940

FIGHTER COMMAND

Night interception patrol
23 Squadron, RAF (Ford, Sussex; detachment at Middle Wallop, Hampshire - 12 Group)
Blenheim IF L8369/B - took off from Middle Wallop at 2045, but a short time afterwards the port engine failed and at 2110 the fighter crashed at Broughton, 3 miles south of the airfield. All three crew died, the pilot and air gunner being buried at Over Wallop. It was believed that L8369 stalled during a turn as it broke away to make a wide approach into wind.
Pilot: 36272 Plt Off **Eric ORGIAS**, RAF - Age 25. 262hrs. 6th op.

Fri 27 Sep 1940

FIGHTER COMMAND

Interception patrol over Maidstone
92 Squadron, RAF (Biggin Hill, Kent - 11 Group)
Spitfire I X4422 - took off at 0845 with rest of the Squadron to intercept an incoming raid. As the bombers were engaged over Sevenoaks at about 0930 enemy fighters came to their aid, downing two of the Squadron, including X4422 which crashed and burned out at Sparepenny Lane, Farningham. The pilot is buried at Orpington.
Pilot: 36193 Flt Lt **James Alfred PATERSON**, MBE, RAF - Age 20. 410hrs. 149th op.

Mon 30 Sep 1940

NEW ZEALAND

Delivery of new aircraft from Hobsonville to Ohakea
RNZAF Station Ohakea
Oxford I NZ1251 - took off at 1145 and went missing in bad weather. An intensive air search was undertaken and a reward of £50 offered by Air Department for information which would lead to discovery of its fate. It was not until five months later, however, that NZ1251 was found when two pig hunters reported on 2 March 1941 that they had stumbled upon the site in dense bush on the seaward side of the main road between Tongapurutu and the Mohakatino River (some sources, though, state the site as being near Pukearuhe, to the north). The pilot and his two passengers had died instantly, the dense bush into which the Oxford crashed closing in over the wreckage to obscure it from the air. After retrieval the three bodies were conveyed to Auckland, two being buried there and that of the pilot cremated.
Pilot: (NZ2522) Fg Off **Harry Nalder MAJOR**, RNZAF - Age 29.
(Fitter AE) NZ34167 Flt Sgt **Sheldon Benton SMITH**, RNZAF - Age 25.
(Equip Assist): NZ38217 Cpl **Raymond Bernard SMITH**, RNZAF - Age 22.
Raymond Smith's brother, Harry Francis Burton Smith, died two years later, on 24 September 1942 while flying with 9 Sqn, RAF.
At the time of Major's death service numbers had not yet been assigned to officers of the RNZAF. NZ2522 was a posthumous allocation made at war's end.
The CWGC register incorrectly gives 3 October as date of death for Major and Raymond Smith.

Tue 1/Wed 2 Oct 1940

BOMBER COMMAND

Raid on Berlin, Germany
9 Squadron, RAF (Honington, Suffolk - 3 Group)
Wellington IC R3282/G - took off at 1820 captained by Flt Lt C D Fox, RAF, and on the return journey ran out of fuel, ditching in the sea off Lowestoft at 0246. All six crew are commemorated on the Runnymede Memorial.
Air Gunner: NZ40726 Sgt **Charles Hugh Le Blanc NEWBERRY**, RNZAF - Age 22.

Raid on synthetic oil installations in Holten district, Sterkrade, Germany
78 Squadron, RAF (Dishforth, Yorkshire - 4 Group)
Whitley V P4964 - took off at 1906 and believed to have been shot down over the Netherlands by a night fighter, crashing at 0130 on the property of Mr Demmink of Hessenweg, Hummelo, 21km east of Arnhem. All five crew are buried at Hummelo.
Captain: 36252 Plt Off **Neville Halsey ANDREW**, RAF - Age 21. 464hrs. 18th op.
2nd Pilot: 43708 Plt Off **Harry Wright MORGAN**, RAF - Age 27. 366hrs. 9th op.

Thu 3 Oct 1940

FIGHTER COMMAND

Operational patrol
600 Squadron, RAF (Redhill, Surrey - 11 Group)
Blenheim IVF L4905/M - took off at 0325 piloted by Plt Off C A Hobson, RAF, and signalled that an engine was not running well soon after reaching the patrol line but continued patrolling. At about 0345 L4905 further reported

that the engine had failed and a return to base was being made. Ten minutes later, in heavy rain, the Blenheim ploughed through tree tops on high ground to crash at Broadstone Warren, Forest Row, Sussex, 13 miles SE of Redhill. All three crew died, the air gunner being buried at Whyteleafe, near Warlingham, Surrey.
Air Gunner: NZ40920 Sgt **David Ernest HUGHES**, RNZAF - Age 27. 29hrs. 3rd op.

RHODESIAN AIR TRAINING GROUP

Dual instruction
20 Service Flying Training School, RAF (Cranborne, Southern Rhodesia)
Harvard I P5963 - lost control in cloud and only partially recovered before striking the ground 10 miles east of Marandallas Road, near Salisbury. Both pilots are buried at Salisbury.
Pilot Instr: 40505 Fg Off **Willis Stewart BROWN**, RAF - Age 25. 855+hrs.
Brown was the first New Zealand airman to lose his life while flying in Africa. His brother, Ivor Morrison Brown, died on 20 September 1941 while flying with 602 Sqn, RAF.

Fri 4 Oct 1940

FLYING TRAINING COMMAND

Training
2 Flying Instructors School, RAF (Cranwell, Lincolnshire - 21 Group)
Oxford I R5950 - took off from Cranwell and while flying in a southwesterly direction at 500 feet collided with a 1 Signals School Wallace II (K6030). R5950's starboard wing struck the port wing of the Wallace, causing both aircraft to go out of control and crash near Sleaford, 6 miles to the SE of Cranwell. The two airmen in each aircraft died, the pilot instructor of R5950 being buried at Cranwell.
Pilot Instr: 40877 Fg Off **Michael Joseph BERRY**, RAF - Age 23. 873hrs.

Tue 8 Oct 1940

COASTAL COMMAND

Attack on motor transport and shipping concentrations - Gravelines, France
53 Squadron, RAF (Detling, Kent - 16 Group)
Blenheim IV T2036/K - took off at 1845 and brought down by flak, crashing at about 2245 near Fréthun, 5km SW of Calais. Two of the crew died on the day of the crash, while the pilot was seriously injured. Captured and admitted to hospital at Guînes, 6km SE of Fréthun, he died there three days later. All three airmen are buried at Guînes.
Pilot: 42719 Plt Off **John Charles MALLON**, RAF - Age 24. 382hrs. 43rd op.
The CWGC register incorrectly gives date of death for Mallon as the 8th. His brother, Thomas Alexander Mallon, died on 12 March 1945 while flying with 488 Sqn, RNZAF.

FIGHTER COMMAND

Night patrol
264 Squadron, RAF (Kirton-in-Lindsey, Lincolnshire; detachment at Northolt, Middlesex - 12 Group)
Defiant I N1627 - took off at 2150, probably from Northolt, piloted by Plt Off H I Goodall, RAF, and while patrolling at 10,000 feet from Malton to Maidenhead reported that a suspected enemy aircraft was being investigated. No further reports were received, but at 2230 the Defiant crashed at Marlow, Berkshire, 4 miles north of Maidenhead. The two crew died, the air gunner being buried at Northwood, 5 miles NW of Northolt. Suspected bullet holes were found on the wreckage and the identifying colour cartridge of the hour had been fired, suggesting that the enemy may have been engaged.
Air Gunner: NZ40197 Sgt **Robert Bett Mirk YOUNG**, RNZAF - Age 22. 76hrs. 20th op.

Tue 8/Wed 9 Oct 1940

BOMBER COMMAND

Raid on Gelsenkirchen, Germany
58 Squadron, RAF (Linton-on-Ouse, Yorkshire - 4 Group)
Whitley V T4137/K - took off at 1825 and completed a successful attack, but at 0023 on the return journey the port engine caught fire. Although the fire was extinguished, while on approach to Bircham Newton, Norfolk, for an emergency landing at 0115 T4137 crashed and caught fire at Docking, just to the north of the airfield. All five crew died, the captain being buried at Great Bircham. *Bomber Command Losses* lists this event under 7/8th but is in error.
Captain: 79511 Plt Off **Ronald Arnold HADLEY**, RAFVR - Age 20. 375hrs. 15th op.
Hadley's brother, Trevor Harold Hadley, also a Whitley pilot with 58 Squadron, had been brought down and captured with the rest of his crew four weeks earlier during a raid on Bremen.

Fri 11 Oct 1940

COASTAL COMMAND

Plt Off **J C MALLON** died of injuries received when his aircraft was shot down on the 8th.

BOMBER COMMAND

Training

Reserve Flight of 214 Squadron, RAF (Stradishall, Suffolk - 3 Group)

Wellington IC L7840 - took off at 1445 captained by Plt Off C W A H Rafter, RAF, but swung, struck a hangar, crashed, and caught fire. Two crew survived, injured, but the other four died, including the 2nd pilot, who is buried at Stradishall.

2nd Pilot: NZ39922 Sgt **George William HUNT**, RNZAF - Age 22. 208hrs. At least 1 op.

Sun 13/Mon 14 Oct 1940

BOMBER COMMAND

Raid on Kiel, Germany

99 Squadron, RAF (Newmarket, Suffolk - 3 Group)

Wellington IC P9243 - took off at 1900 captained by Fg Off W E N Keller, RAF, and lost without trace, its six crew being commemorated on the Runnymede Memorial.

2nd Pilot: NZ391346 Sgt **Charles Harold DOUGLAS**, RNZAF - Age 20. 244hrs. 2nd op.

Sun 20 Oct 1940

MIDDLE EAST

Kamaran Island to Aden

203 Squadron, RAF (Khormaksar, Aden)

Blenheim IVF T2112 - had just concluded a long patrol, which included an attack on an enemy bomber, and landed at Kamaran Island to refuel. Took off again at 1500 but shortly afterwards the port engine failed. When a landing from a left hand circuit was attempted 30 minutes later the port wing struck the corner of a 30 foot-high mosque. One crew member, although injured, survived, but the pilot either died in the crash or of injuries later that day. The third member died the following day. The pilot is buried on Kamaran but, because of difficulty in maintaining the remote site, is also commemorated on Maala Memorial No 2 at Aden.

Pilot: 40981 Fg Off **Heslop Miles Frederick BARNITT**, DFC, RAF - Age 22. 533hrs.

Barnitt's number of ops is not known but he had flown over 100hrs of fighter patrols and completed at least two reconnaissance flights.

Mon 21 Oct 1940

FIGHTER COMMAND

Transit Stradishall, Suffolk, to Wittering

266 Squadron, RAF (Wittering, Northamptonshire - 12 Group)

Spitfire I X4265 - ran short of fuel after an unsuccessful pursuit out to the east coast of enemy aircraft which had been attacking Cambridge, and at 1250 landed at Stradishall. After refuelling X4265 took off to return to base but immediately stalled and crashed, killing the pilot, who is buried at Stradishall.

Pilot: 42173 Plt Off **Wycliff Stuart WILLIAMS**, mid, RAF - Age 20. 409hrs.

Wed 23/Thu 24 Oct 1940

BOMBER COMMAND

Raid on Potsdam, Germany

75 (NZ) Squadron, RAF (Feltwell, Norfolk - 3 Group)

Wellington IC P9292 - took off at approximately 2230-2300 and brought down in the North Sea, killing the six crew. An RAF air gunner's body later washed up on the island of Fanø off the SW coast of Denmark. That of the 2nd pilot was recovered on the 31st by a Royal Navy trawler and buried at sea near the *Shipwash Light Vessel*, 17 miles off the Essex coast. He is also commemorated on the Runnymede Memorial, along with his four comrades who were lost without trace.

Captain: 42727 Plt Off **Rex Martyn SANDERSON**, RAF - Age 24. 373hrs. 12th op.
2nd Pilot: NZ39911 Plt Off **William John FINLAYSON**, RNZAF - Age 23. 270hrs.

Thu 24/Fri 25 Oct 1940

BOMBER COMMAND

Raid on Air Ministry buildings in Berlin, Germany

102 Squadron, RAF (Linton-on-Ouse, Yorkshire - 4 Group)

Whitley V P5073/D - took off at 2202 captained by Plt Off A G Davies, RAF, and shot down in flames by an enemy night-fighter less than ten minutes later. P5073 fell near Tolthorpe, 4 miles NNW of the base, killing the 2nd pilot and observer, who are both buried at Newton-upon-Ouse. The other three crew were all wounded, two dying within the next nine days.

2nd Pilot: 42759 Plt Off **Thomas Russell MURFITT**, RAF - Age 24. 360hrs. 15th op.
Observer: NZ391870 Sgt **Ian Cowie SCOULAR**, RNZAF - Age 25. 133hrs.

Sat 26 Oct 1940

FIGHTER COMMAND

Patrol

229 Squadron, RAF (Northolt, Middlesex - 11 Group)

Hurricane I W6669 - took off at 1050 as leader of 'Blue' section. The Squadron sighted enemy fighters and pursued them towards the French coast, but lost contact. As they were about to return to England, a low flying He59 float plane was sighted near Boulogne, attacked by Blue section and forced to alight on the water. The three Hurricanes then came under attack by enemy fighters and fire from shore based flak. Only one Hurricane survived to return to base, the others being shot down. The pilot of one was taken prisoner but the section leader was lost without trace and is commemorated on the Runnymede Memorial.

Pilot: 41481 Fg Off **Geoffrey Mervyn SIMPSON**, RAF - Age 21. 492hrs. 99th op.

Night training

151 Squadron, RAF (Digby, Lincolnshire - 12 Group)

Hurricane I V7434 - took off at 2130 from satellite airfield 'L2' (Coleby Grange, 7 miles south of Lincoln) and, on climbing to 500 feet, began a gentle left hand turn then flew into the ground and caught fire. The pilot was extricated from the wreckage but died the following day at the Lincoln County Hospital, and is buried at Scopwick.

Pilot: NZ39944 Sgt **Douglas Owen STANLEY**, RNZAF - Age 24. 215hrs.

Night training

151 Squadron, RAF (Digby, Lincolnshire - 12 Group)

Hurricane I R4184 - took off from 'L2' a few minutes after V7434 above and appeared to lose control before diving into the ground from 200 feet, and catching fire. The pilot is buried at Bidford-on-Avon, Warwickshire.

Pilot: NZ391356 Sgt **Robert HOLDER**, RNZAF - Age 23.

Mon 28 Oct 1940

COASTAL COMMAND

Night patrol over convoy in the Atlantic

204 Squadron, RAF (Sullom Voe, Shetland Isles - 18 Group)

Sunderland I P9620 - took off at 1640 captained by Flt Lt S R Gibbs, RAF. On returning to base over 13 hours later bad weather was encountered and at 0615 an emergency landing had to be made on the open sea off the west coast of Scotland as fuel began to run low. P9620 was wrecked after landing in the rough sea, and most of its 13 crew were thrown into the water. Nine were rescued by the cruiser HMAS *Australia* but four others died, including the observer, who is commemorated on the Runnymede Memorial.

Observer: NZ39866 Sgt **Stuart Hayden MACDONALD**, RNZAF - Age 21.

Mon 28/Tue 29 Oct 1940

BOMBER COMMAND

Raid on Mannheim, Germany

105 Squadron, RAF (Watton, Norfolk - 2 Group)

Blenheim IV T2229 - took off at 1830 piloted by Plt Off I Prosser, RAF, and for reasons unknown, although weather conditions were poor, had to jettison its bomb load near Trier, Germany. T2229 later crashed at Cranmer Hall near Sculthorpe, 2 miles WNW of Fakenham, during the final stages of its return journey. The pilot and observer died and are buried at Great Bircham. The air gunner was injured and later died also.

Observer: NZ39964 Sgt **Arthur Fraser DALLAS**, RNZAF - Age 22. 192hrs. 15th op.

Wed 30 Oct 1940

BOMBER COMMAND

Operational patrol 'M61' (Antwerp, Belgium or Flushing, Netherlands?)

101 Squadron, RAF (West Raynham, Norfolk - 2 Group)

Blenheim IV T2246 - took off at 1714 and upon returning at 2300 crashed in bad visibility into Dunster Wood, near Coleby, 5 miles north of Scunthorpe, Lincolnshire. All three crew died, the pilot being buried at Waddington.

Pilot: 43700 Plt Off **Jack Milton CAVE**, RAF - Age 21. 273hrs. 17th op.

Cave's brother, Verner Grenville Cave, died on 31 March 1945 while flying with 37 Sqn, RAF.

Wed 6/Thu 7 Nov 1940

BOMBER COMMAND

Raid on Hamburg, Germany

58 Squadron, RAF (Linton-on-Ouse, Yorkshire - 4 Group)

Whitley V P4943/P - took off at 1710 captained by Sgt F R Elton, RAF, and signalled an SOS on the return journey at 0018 but nothing further was heard. P4943 is believed to have come down in the sea some ten miles off Lowestoft, Suffolk, its five crew being commemorated on the Runnymede Memorial.

2nd Pilot: NZ39936 Sgt **Reginald Wilfred PROUT**, RNZAF - Age 21. 253hrs.

Wed 13 Nov 1940

NEW ZEALAND

New Zealand General Reconnaissance Squadron, RNZAF (Whenuapai)
Admitted to the Auckland Public Hospital on 7 October, died of natural causes and is buried at Wellington.
(Aircrafthand): NZ639160 AC1 **Kent JOHNSTON**, RNZAF - Age 21.

Thu 14 Nov 1940

NEW ZEALAND

Solo night cross country, Ohakea-Levin-Palmerston North-Ohakea
3 Service Flying Training School (ATS?), RNZAF (Ohakea)
Oxford I NZ1277 - failed to return by 2130, the estimated time of arrival. NZ1277 became lost in low cloud between Levin and Palmerston North, striking trees and crashing into dense bush while flying up a blind gully in the hills behind Tokomaru at 2330. The pilot is buried at Auckland.
Pilot: NZ40649 Plt Off **Thomas Everton KIRK**, RNZAF - Age 26. 179hrs solo (31 on Oxford)

Sat 16 Nov 1940

MIDDLE EAST

Operation WHITE - reinforcement ferry flight from aircraft carrier HMS *Argus* to Malta
261 Squadron, RAF (Luqa, Malta - Mediterranean)
Hurricane I (serial?) - took off at about 0715 as part of the second of two reinforcement flights, each of six Hurricanes escorted by a Fleet Air Arm Skua. Two of the first Flight's fighters ran out of fuel and crashed into the sea before reaching Malta, one of the two pilots being saved. The second flight, however, developed into a complete disaster as a series of events, including a strengthening headwind, saw the six Hurricanes exhaust their fuel and fall into the sea, one by one. All six pilots were lost and they are commemorated on the Runnymede Memorial. The Flight's Skua succeeded in crash landing on Sicily after its fuel also ran out. The *Argus* had sailed from Gibraltar but could not itself take the risk of delivering the aircraft direct to Malta. The New Zealander's Hurricane serial is unidentified but is believed to be one of the following - V7373, V7475, V7487, V7488, V7494 or V7549.
Pilot: 42128 Fg Off **Patrick Wilmot HORTON**, mid, RAF - Age 20.
Horton had been posted in Britain to 261 Sqn on Malta and was making his way there by means of the ferry flight. Several post-war publications state that the Hurricanes flew off Argus *on the 17th, but the 16th is recorded on official wartime documents and in the CWGC register.*

Sun 24 Nov 1940

COASTAL COMMAND

Convoy escort to 'SC11'
272 Squadron, RAF (Aldergrove, County Antrim - 15 Group)
Blenheim IVF Z5734/D - took off at 1620 and while on patrol engaged in a practice dogfight with a Fulmar near the destroyer *Vesper*. While turning at a low altitude Z5734 stalled and crashed into the sea, sinking immediately. Rescue craft were on the scene within three minutes but only a rubber dinghy and a petrol tank could be found. The three crew are commemorated on the Runnymede Memorial.
Pilot: 42003 Fg Off **Brian Henry HERRICK**, RAF - Age 25. 498hrs.
Two of Herrick's brothers, Dennis Trevelyan Herrick and Michael James Herrick, lost their lives while serving as members of the RNZAF and RAF respectively. Dennis was brought down while flying with 53 Sqn, RAF, on 26 June 1941 and died of his wounds four days later as a prisoner. Michael was killed on 16 June 1944 while flying with 305 Sqn, RAF.

Tue 26 Nov 1940

COASTAL COMMAND

Raid on docks at Lorient, France
59 Squadron, RAF (Thorney Island, Sussex - 16 Group)
Blenheim IV R3889/G - took off at 1550 piloted by Plt Off K W Jerwood, RAF, and lost without trace, its three crew being commemorated on the Runnymede Memorial.
Observer: NZ40626 Sgt **Frederic Jack NEWMAN**, RNZAF - Age 26. 140hrs. 9th op.

Tue 26/Wed 27 Nov 1940

BOMBER COMMAND

Raid on port of Cologne, Germany
105 Squadron, RAF (Swanton Morley, Norfolk - 2 Group)
Blenheim IV T1886 - took off at 1850 but on return to base discovered weather had deteriorated, with low cloud and poor visibility. While circling at 2323, preparatory to landing, T1886 flew into the ground at high speed in a shallow dive at Foxley Wood, 4 miles NE of Swanton Morley. All three crew died, the pilot being buried at Stoke-on-Trent, Staffordshire.
Pilot: 41744 Plt Off **Leonard Terence Jervis RYAN**, RAF - Age 21. 289hrs solo.
Ryan's brother, Augustus Harold Jervis Ryan, died on 8 November 1939 while flying with 220 Sqn, RAF. They are buried in the same cemetery.

Wed 27/Thu 28 Nov 1940

BOMBER COMMAND

'Gardening' - minelaying off Le Havre, France
102 Squadron, RAF (Topcliffe, Yorkshire - 4 Group)
Whitley V P5077/B - took off at 1800 captained by Sgt R Collier, RAF, and lost without trace, its five crew being commemorated on the Runnymede Memorial.
2nd Pilot: 43644 Plt Off **Edward Read OSBORN**, RAF - Age 21. 300hrs. 8th op.

Thu 28 Nov 1940

NEW ZEALAND

RNZAF Depot Hobsonville
Died as a result of illness at the Auckland Public Hospital and is buried in that city.
Medical Off: Capt **Robert James MAUNSELL**, MB ChB, NZ Medical Corps att RNZAF - Age 45.
Maunsell, on attachment to the RNZAF for the preceding six months, had served at Gallipoli during the First World War.

MIDDLE EAST

Ferry flight from Stradishall, Suffolk, to Egypt via Malta
In transit, RAF
Wellington IC T2894 - from Malta, having arrived on the 25th, took off at 0901 for Egypt, captained by Wg Cdr C G Hogler, RAF. Immediately afterwards an air raid warning was sounded, preventing accompanying aircraft from departing. T2894 was last seen setting course towards a convoy approaching Malta, which was then being attacked by Italian aircraft. Extensive searches revealed nothing as to the fate of the bomber and its six crew are commemorated on the Runnymede Memorial.
Air Gunner: NZ401223 Sgt **John Charles CORNISH**, RNZAF - Age 19.
Cornish, previously with 214 Sqn, RAF, had been posted from Bomber Command to the Middle East and was making his way there by means of the delivery flight.

Fri 29/Sat 30 Nov 1940

BOMBER COMMAND

Special flight to Malta
Reserve Flight, 214 Squadron, RAF (Stradishall, Suffolk - 3 Group)
Wellington IC T2893 - took off captained by Sqn Ldr B J McGinn, RAF, but later experienced an engine failure and could not maintain height. At 0100 T2893 struck trees, crashed and caught fire at Elmdon, Essex. The pilot and an air gunner both died, the latter being buried at Haverhill, Suffolk. Three others of the crew were hurt and it is thought that one of these died of his injuries the following April. McGinn was a highly experienced pilot with more than 4200 hours to his credit.
Air Gunner: NZ401220 Sgt **Claude Horace BAIN**, RNZAF - Age 23.

Thu 5 Dec 1940

FIGHTER COMMAND

Ferry flight from Drem, Lothian, to Montrose
111 Squadron, RAF (Dyce, Aberdeenshire; detachment at Montrose, Angus - 14 Group)
Hurricane I P3470 - took off at 1527 and later crashed in bad weather into a hill 10 miles NW of Edzell, Angus. The pilot is buried at Montrose.
Pilot: NZ391343 Sgt **Jack Burall COURTIS**, RNZAF - Age 26. 224hrs.

Fri 6 Dec 1940

NEW ZEALAND

Pedestrian accident on airfield
2 Elementary Flying Training School, RNZAF (Bell Block)
Tiger Moth NZ660 - on coming in to land at 1115, one of the wheels struck an airman as he was walking with LAC **R S Bruce** along the eastern boundary of the airfield. The trainee pilots were returning from the rifle range at the time. On realising the danger as the aircraft approached, Bruce shouted a warning and dropped to the ground. However, his companion was struck on the head and killed outright. He is buried at Bombay, Auckland. The slightly damaged Tiger Moth landed safely with both of its occupants unharmed.
(Pupil Pilot): NZ403542 LAC **Maurice Sydney BEST**, RNZAF - Age 27.

Sat 7/Sun 8 Dec 1940

BOMBER COMMAND

Raid on Dusseldorf, Germany
50 Squadron, RAF (Lindholme, Yorkshire - 5 Group)
Hampden I X3004 - took off at 1810 and lost without trace, its four crew being commemorated on the Runnymede Memorial.
Pilot: 41864 Fg Off **Robert Trevor MULLIGAN**, RAF - Age 21. 460hrs. 28th op.

Sun 8/Mon 9 Dec 1940

BOMBER COMMAND

Raid on Bordeaux, France

115 Squadron, RAF (Marham, Norfolk - 3 Group)

Wellington IC T2520/A - took off at 1715 captained by Plt Off A Tindall, RAF, and became lost on the return journey in low cloud, rain and strong winds. T2520 unknowingly entered mountainous country where it crashed and caught fire at Blaen Dyffryn quarries near Tredegar, Monmouthshire, 20 miles north of Cardiff, Glamorgan. The wreckage remained undiscovered for two days. All six crew died, three of whom, including the front gunner, are buried at Llantwit Major, Glamorgan.

Front Gunner: NZ40913 Sgt **Reginald BROWN**, RNZAF - Age 22. 102hrs. 7th op.

Mon 9 Dec 1940

BOMBER COMMAND

Night training

114 Squadron, RAF (Oulton, Norfolk - 2 Group)

Blenheim IV N3617 - crashed at 2130 at Bridge Farm, Hemsby, 4 miles NNW of Caister-on-Sea, reportedly after being hit by anti-aircraft fire from an Allied convoy. N3617 struck telegraph wires before hitting the ground and disintegrated on colliding with a hedge in the centre of a field, then caught fire. All three crew died, the pilot being buried at Norwich.

Pilot: 41503 Fg Off **William Ross WATSON**, RAF - Age 23.

Wed 11 Dec 1940

NEW ZEALAND

Cross country instrument training

1 Service Flying Training School (ATS), RNZAF (Wigram)

Gordon NZ632 - flew low over some tree tops and banked steeply, then dived into the ground at 1000 in a small paddock on the property of Mr A McLean, located on the northern side of Leithfield Beach Road near the main north road junction, catching fire and careering on into the next paddock. The two pilots died and are buried at Gisborne and Hamilton, respectively.

Pilot: NZ401785 LAC **Basil Rawdon Hastings SHARP**, RNZAF - Age 22. 121hrs (35 solo on Gordon)

Safety Pilot: NZ401799 LAC **Leslie WRIGHT**, RNZAF - Age 22.

Fri 13/Sat 14 Dec 1940

BOMBER COMMAND

Raid on Kiel, Germany

78 Squadron, RAF (Dishforth, Yorkshire - 4 Group)

Whitley V N1485 - took off at 1636 but had to be abandoned over The Wash on return because of engine failure. Three of the five crew died. Two bodies were later recovered but not that of the captain, who is commemorated on the Runnymede Memorial.

Captain: 36263 Plt Off **Michael Lomax STEDMAN**, RAF - Age 20. 545hrs. 28th op.

Stedman's father, Gerald Lomax Stedman, was a pilot in the First World War, a prominent aviator in the interwar years, and a serving officer in the RNZAF at the time of his son's death.

Sat 14 Dec 1940

BOMBER COMMAND

Daylight cross country training

17 Operational Training Unit, RAF (Upwood, Huntingdonshire - 7 Group)

Blenheim IV L4835 - piloted by Sgt I M Curror, RAF, and while flying at 1200 feet in bad visibility at 1245, collided with 1750 foot-high Clee Hill, 6 miles east of Ludlow, Shropshire, and caught fire. All three crew died, the observer being buried at Bury, 2 miles NE of Upwood.

Observer: NZ40743 Sgt **Alfred James Victor SECKER**, RNZAF - Age 27.

Sun 15/Mon 16 Dec 1940

BOMBER COMMAND

Raid on Berlin, Germany

77 Squadron, RAF (Topcliffe, Yorkshire - 4 Group)

Whitley V T4292/P - took off at 1708 but reported a severe engine problem at 1933 and turned back. At about 2200 T4292 ditched in the North Sea 30 miles east of Great Yarmouth, Norfolk. A search failed to find any trace of the five crew, all of whom are commemorated on the Runnymede Memorial.

Captain: 43049 Plt Off **Neil Anzac SVENSON**, RAF - Age 24. 356hrs. 14th op.

Mon 16 Dec 1940

COASTAL COMMAND

Low level reconnaissance of Brest, France
1 Photographic Reconnaissance Unit, RAF (St Eval, Cornwall - 15 Group)
Spitfire I X4350 - took off at 1415 and shot down into the sea by flak at Lannilis, 20km north of Brest. The pilot's body later washed ashore near Tréflez, 30km NE of Brest, and is buried there.
Pilot: 41047 Fg Off **Samuel James MILLEN**, DFC, RAF - Age 26. 723hrs. 50th op.

Wed 18 Dec 1940

COASTAL COMMAND

Raid on Wilhelmshaven, Germany
22 Squadron, RAF (North Coates, Lincolnshire - 16 Group)
Beaufort I L4516/W - took off at 2010 captained by Sgt D G How, RAF, but stalled after completing a left hand turn, struck the ground and exploded half a mile from Marsh Chapel, just SE of the base. All four crew are buried at North Coates.
Observer: 580978 Sgt **Paul Victor RENAI**, RAF - Age 22.

BOMBER COMMAND

Raid on 'B.55' (Ludwigshafen, Germany)
99 Squadron, RAF (Newmarket, Suffolk - 3 Group)
Wellington IC R1333 (a presentation machine named *The Broughton Wellington*) - took off at 2128 captained by Flt Lt G S Ogilvie, RAF, lifted clear of the runway, but failed to climb and struck the 20 foot-high Devil's Dyke on the airfield perimeter, and caught fire. Two of the crew were injured but the other four died, including the 2nd pilot and observer, who are buried at Beck Row, 2 miles NW of Mildenhall.
2nd Pilot: NZ391365 Plt Off **Arthur Patnell PRITCHARD**, RNZAF - Age 20. 213hrs.

Sun 22/Mon 23 Dec 1940

BOMBER COMMAND

Raid on Mannheim, Germany
75 (NZ) Squadron, RAF (Feltwell, Norfolk - 3 Group)
Wellington IC T2474 - took off at approximately 1645-1720 captained by Sgt E R Chuter, RAF, and brought down over France near Thérouldeville, 42km NE of Le Havre. The rear gunner is buried at Thérouldeville. The other five crew members survived to be taken prisoner, including Sgt **C F Scott**, RNZAF, who successfully escaped from a French hospital in early 1941, and Sgt **H M McG English**, RNZAF, who was repatriated in July 1943.
Rear Gunner: NZ40207 Sgt **Alfred Henry RITCHIE**, RNZAF - Age 22. 55hrs. 7th op.

Mon 23 Dec 1940

BOMBER COMMAND

Cross country
11 Operational Training Unit, RAF (Bassingbourn, Cambridgeshire - 6 Group)
Wellington IA N3014 - captained by Plt Off D R F Haviland, RAF. During the flight the weather deteriorated considerably, causing the windscreen to ice up. An air raid warning also came into force, with landing lights being dimmed at the base. N3014 made an approach to land shortly before 2020 but then appeared to go round again. It stalled after climbing to 500 feet and crashed near Bassingbourn village, where it caught fire. All six crew died, the captain and New Zealand air gunner being buried in the Bassingbourn cum Kneesworth Cemetery.
Air Gunner: NZ401519 Sgt **Errol James PETERS**, RNZAF - Age 20.

Mon 23 Dec/Christmas Eve 1940

BOMBER COMMAND

Raid on Mannheim, Germany
58 Squadron, RAF (Linton-on-Ouse, Yorkshire - 4 Group)
Whitley V T4159/J - took off at 1710 and believed to have crashed into the sea off Torbay, Devonshire, its five crew being commemorated on the Runnymede Memorial.
Captain: NZ39942 Sgt **Stronach Vivian SMITH**, RNZAF - Age 24.

Tue, Christmas Eve 1940

MAINTENANCE COMMAND

Ferrying from 20 Maintenance Unit (at Aston Down) to 247 Sqn (Roborough, near Plymouth, Devonshire)
HQ Service Ferry Pools , RAF (Kemble, Gloucestershire - 41 Group)
Hurricane I P3083 - late in the morning, on its delivery flight, was heard circling in cloud and rain over the Mendip hills. The engine throttled back and opened up from time to time, the pilot apparently gliding down and climbing again while trying to make a landfall. It was thought that he saw rising ground immediately ahead, pulled up steeply, stalled and entered an incipient spin before crashing near Rodney Stoke or Draycott, about 3 miles from the Cheddar Gorge, Somerset. He is buried at Wells.
Pilot: 40909 Fg Off **John Warren HAMILL**, RAF - Age 24. 400+hrs.

Fri 27 Dec 1940

COASTAL COMMAND

Night convoy escort

210 Squadron, RAF (Oban, Argyllshire - 15 Group)

Sunderland I N9022 - took off at 1154 captained by Act Flt Lt I H Meggitt, RAF, and on return at 2108 failed to flatten out sufficiently, breaking up on striking the water in the Firth of Lorne, a mile off Oban Harbour. One crew member was injured but his ten comrades died. The New Zealand air gunner is commemorated on the Runnymede Memorial.

Air Gunner: NZ40926 Sgt **Noel Edgar THOMSON**, RNZAF - Age 24.

Sat 28 Dec 1940

NEW ZEALAND

Drowning accident

3 Service Flying Training School (ITS), RNZAF (Ohakea)

Caught in the undertow while swimming at Tangimoana during the afternoon. Two fellow officers tried in vain to effect a rescue but he was swept away and drowned. His body was recovered a few days later and is buried at Eltham.

(Pilot Instr): 40573 Fg Off **John Rowley Miller STEWART**, RAF - Age 26.

Stewart had joined the RAF via the New Zealand Locally Selected Candidate Scheme in 1937. He had been loaned to the RNZAF as an instructor in early 1940.

Sun 29 Dec 1940

BOMBER COMMAND

Night training

17 Operational Training Unit, RAF (Upwood, Huntingdonshire - 7 Group)

Blenheim IV L4868 - shortly after take off the observer drew the pilot's attention to what he believed was the rear light of an aircraft ahead. The pilot's attention was thus momentarily diverted from his instruments while engaged in an emergency turn at about 1000 feet. Control was lost and L4868 stalled, crashed and caught fire near the base at 1935. Only the air gunner survived, seriously injured. His two comrades are buried at Birmingham. The 'light' seen by the observer was most likely that of a star.

Pilot: NZ40742 Plt Off **Vernon Sinclair REYNOLDS**, RNZAF - Age 25. 241hrs.

Mon 30 Dec 1940

NEW ZEALAND

Cross country on instruments

3 Service Flying Training School (ATS), RNZAF (Ohakea)

Oxford I R6226 - took off on a triangular cross country flight with pupil pilot under the hood. At 0745 R6226 was seen at 100 feet and to lose height, and carry out a steep left hand turn, still at high speed. It then carried away power lines and scythed through the middle of a two-storey wooden house on Mr C M Kebbell's property near Ashhurst, at the entrance to the Manawatu Gorge. The disintegrated Oxford came to rest 100-150 yards further on from the dwelling. Both pilots died and are buried at Auckland and Wellington, respectively. Miraculously, none of the building's occupants was injured, a nurse and the four Kebbell children having vacated the nursery some ten minutes before the machine struck and demolished it.

Pilot: NZ402154 LAC **Thomas Arthur Hughes ALEXANDER**, RNZAF - Age 26.

Safety Pilot: NZ439002 Sgt **Rex BROOKE-TAYLOR**, RNZAF - Age 25. 174hrs solo (112 solo – 71 on Oxford)

Brooke-Taylor's brother, Keith Brooke-Taylor, died on 7 April 1940 while flying with 83 Sqn, RAF.

1941

THE SECOND FULL YEAR OF WAR saw fighting intensify and spread. By its end, with the entry of Japan, the conflict was truly world-wide. For the Allies there would be little return for the casualties they were to endure during a grim year.

With Hitler forced to postpone his planned invasion of Britain, RAF's Fighter Command began a new offensive in January operating large formations of its short-range fighters on 'sweeps' over Belgium and Northern France. Light bombers were frequently used as bait to tempt Luftwaffe fighters into battle, but despite highly optimistic claims for aircraft destroyed it was the RAF which suffered most heavily. It was akin to fighting the Battle of Britain in reverse, with most advantages held by the enemy, even more so when later in the year the new and startlingly superior Focke-Wulf 190 appeared. The antidote, in the shape of the improved Spitfire Mk IX, would seem a long time coming.

Formed on 1 March, No 485 Sqn, RNZAF, had also begun by June to operate over enemy occupied Europe, participating in the 'Circus', 'Ramrod', 'Rhubarb', and 'Roadstead' variations of this 'leaning forward into France' policy. After 22 June, when Germany invaded Russia, these operations were intensified in an attempt to prevent the former transferring further forces to the east. Yet increasing combat risks notwithstanding, more than half the 67 New Zealanders who died with the Command during the year did so while engaged on non-operational flights.

Critically dependent upon supplies from abroad, and with nearly 900 ships sunk since the fall of France, Britain in early 1941 directed much of its bombing campaign against the causes - long range aircraft bases, shipbuilding yards and ports hosting the U-boats and warships. The French port of Brest in particular was singled out for frequent attack, the need to damage the capital ships there being so great that even risky daylight raids were occasionally resorted to. During a July raid on Munster Sgt J A Ward won New Zealand's first VC of the air. It was presented to him in August, but he died on another raid in September.

Coastal Command sorties varied from seeking out the U-boat during long and watchful convoy patrols to carrying out highly dangerous low-level anti-shipping strikes along the enemy coast, where the chance of survival if brought down was virtually nil. More than 30 New Zealanders lost their lives during the year with the Command.

Bomber Command frequently despatched over 100 aircraft in a night but only rarely exceeded 200. Timing was not yet tightly co-ordinated nor bombing concentrated, aircraft often arriving over the target individually, if they had the good fortune to first find and identify it. The new four-engined Stirling and Halifax bombers began operations in February and March, respectively, but teething troubles and slow production meant that their contribution was minimal, numbers despatched rarely exceeding double figures on any raid throughout the year. Aside from a considerable number of Blenheims on daylight bombing with 2 Group, the Command's mainstay lay in the twin-engined Wellington. Still present in numbers, but increasingly obsolete, were the Hampden and Whitley - simply completing a round trip in one of these, particularly in poor weather to more distant targets, could often be considered an achievement in itself.

The continual improvement of enemy defences, including a substantial night-fighter force now based in the Netherlands, caused casualties to rise to alarming levels, July-October losses alone equalling the front line strength of the Command. Some 200 New Zealanders died serving in Bomber Command during 1941, a heavy price repaid only in small measure in terms of results. An independent investigation into the bombing of June and July depressingly revealed that only one in four crews claiming to have found the target actually dropped their bombs within 5 miles of it, and that a third of the force could not locate the target at all. When the largest force yet despatched (392 aircraft) suffered a disastrous 9% loss on a three-target raid in poor weather on 7/8 November, the bombing offensive was virtually halted for the winter months while the Command's future was reconsidered. Fighter Command too was badly bruised by heavy losses on the following day and for a time was obliged to let up the pressure and conserve its strength.

Early British success against the Italians in North Africa led to support arriving from the latter's German ally early in the year. With protagonists now more evenly balanced the theatre witnessed a series of see-saw battles over the next two years, Libyan territory changing hands five times before the Axis powers were finally driven into Tunisia and defeated in May 1943. An Italian debacle in Greece led to equally disastrous battles for the Allied forces there in April, and on Crete in May, after the Germans once more were obliged to come to their ally's assistance, conquering Yugoslavia en route. The island fortress of Malta had been raided by Italian aircraft as early as June 1940, but with the arrival of strong Luftwaffe contingents in Italy and Sicily the following January it came under increasing attack. Malta was a key Allied base from which to carry out an offensive against enemy supply ships sailing south across the Mediterranean, for logistics were the critical factor in sustaining North African campaigns. But Malta too depended on supply by sea and the survival of its convoys would also be bitterly contested. Some 60 New Zealand airmen died while fighting over North Africa's desert or the 3500km stretch of Mediterranean waters.

A similar number died serving in New Zealand, the majority during the latter half of 1941, 23 of them in Oxford and Hudson crashes during the last 11 weeks. The first Hudsons (for patrol work) and Harvards (for advanced training) arrived by sea in April. Diverted from British orders, they were the first all-metal and American aircraft to be acquired by the RNZAF. The Hudson proved particularly valuable in that it provided the Air Force with efficient operational equipment for the first time in the war.

On 8 December (NZ and Far East time) the Japanese shocked the United States into the war with a surprise Sunday morning assault on Pearl Harbour, while shortly before, 11,000km to the west, their troops stormed ashore on the northeast coast of Malaya. Within hours bombers of the Japanese Navy Air Force had also carried out their first raid on Singapore and two days later its torpedo bombers rapidly sent the British capital ships *Prince of Wales* and *Repulse* to the bottom of the South China Sea. Over the following fortnight Borneo, Hong Kong and the Philippines were also invaded. How far south lay the enemy's final destination could only be guessed at, but Australia, New Zealand and the South Pacific looked far from secure.

To meet this threat the RNZAF had but five operational squadrons, two of them in Fiji. Not one of its nearly 650 aircraft was a fighter and only the 30 or so Hudsons could be considered capable of offering much in the way of resistance. Of its 16,000 personnel, over 4700 were serving with the RAF or undergoing training in Canada, while an aerodrome construction squadron and over 150 freshmen pilots were already caught up in the Far East fighting. Most of the pilots manning 67 Sqn and 243 Sqn, RAF, were New Zealanders, as were all the air and ground crew of 488 Sqn, RNZAF, which had formed in Wellington as recently as September.

As the year ended there seemed to be little going right for the Allies, with the likelihood of more disasters on the horizon. Yet, two significant events did offer hope in the medium and longer term: Germany's ambitious campaign in eastern Europe had failed in its primary objective, the capturing of Moscow, and the German forces now found themselves woefully unprepared for operating through a Russian winter, while the full weight of America's manpower and industrial might was now committed to the defeat of the Axis powers.

Mon 6 Jan 1941

NEW ZEALAND

Flying practice

Flying Instructors School, RNZAF (Hobsonville)

Tiger Moth II NZ723 - on overshooting at 1525 to make another circuit, spun off a steep downwind turn, overcorrected, and failed to recover from a second spin. NZ723 crashed into the Waitamata Harbour a mile off Whau Creek. Although both pilots were seriously injured, Plt Off **P S Worsp** was able to rescue his injured companion, who was trapped in the wreckage. After three months in the Auckland Public Hospital Plt Off **Long** was transferred by air to Wellington, but died there on 27 September 1942 and is buried in that city.

Pilot: NZ401496 Plt Off **Arthur Clendon LONG**, RNZAF - Age 29.

For his unselfish action in rescuing Long, Worsp received a letter of commendation from the Minister of Defence. Sadly, he died on 24 December 1943 while flying with 17 Sqn, RNZAF.

Thu 9 Jan 1941

NEW ZEALAND

Cross country to Ohakea and return

2 Service Flying Training School (ATS), RNZAF (Woodbourne)

Vincent NZ326 - after refuelling at Ohakea left during the afternoon to return to Woodbourne. While flying low over the Manawatu River NZ326 struck power lines strung across the water at Whirokino. The left wing disintegrated, causing the Vincent to flick over and dive into the river. An onlooker, **Maurice A Parsons** of

Palmerston North, dived into the river and as he reached the wreckage the injured pilot, LAC **C W Holmes**, floated free of the semi-submerged machine. After helping Holmes ashore, Parsons swam out again but could not locate the other pilot, whose body was later recovered from the wreckage by an RNZAF salvage party and is buried at Nelson. Parsons was subsequently awarded the Royal Humane Society of New Zealand's Bronze Medal in recognition of his rescue work 'under extremely dangerous conditions'.
Pilot: NZ402181 LAC **Kennard James HAASE**, RNZAF - Age 22.

Fri 10 Jan 1941

BOMBER COMMAND

Night Flying Test, then to Bassingbourn, Cambridgeshire, to collect a new pilot
75 (NZ) Squadron, RAF (Feltwell, Norfolk - 3 Group)
Wellington IC T2550/L - took off at about 1130 and headed for Bassingbourn on completion of the test. While descending to pinpoint its position in poor weather, T2550 flew into trees on a hill top at Heath Farm near Stapleford, 4 miles south of Cambridge. Of the six crew, the only survivor was rear air gunner Sgt **R E Ashby-Peckham**, RNZAF, who was admitted to Addenbrooke Hospital seriously injured. The two pilots are buried at Feltwell.
Captain: 42856 Plt Off **Brian Patrick McNAMARA**, RAF - Age 26. 408hrs. 24+ ops.
2nd Pilot: NZ391367 Plt Off **Alexander James RYAN**, RNZAF - Age 25. 279hrs. 10 ops.

Mon 13 Jan 1941

BOMBER COMMAND

Measured climb and fuel consumption test at 12,000 feet
35 Squadron, RAF (Linton-on-Ouse, Yorkshire - 4 Group)
Halifax I L9487 - took off at 1120 captained by Fg Off M T G Henry, DFC, RAF, and seen at 8000 feet with port undercarriage down, then on fire on descending through 2500 feet. By 1153 flames had consumed the tail control surfaces, causing L9487 to suddenly cartwheel and dive into the ground at Howefield House, just SW of Baldersby St James, 4 miles north of Dishforth. All six crew died, the 2nd pilot being buried at Dishforth. Cause of the fire was put down to a cap of one of the port fuel tanks being unsecured after refuelling.
2nd Pilot: 79513 Plt Off **Leslie Joseph McDONALD**, RAFVR - Age 23. 531hrs. 28 ops.

Tue 21 Jan 1941

NEW ZEALAND

Reconnaissance report writing exercise
2 Service Flying Training School (ATS), RNZAF (Woodbourne)
Vincent NZ354 - took off at 0830, but encountered low cloud and fog and crashed at about 0945 some 100 feet from the top of a 1000 foot hill, in the vicinity of Redwood Pass Road about 8 miles SE of Blenheim. The two airmen are buried at Wellington and Auckland, respectively.
Pilot: NZ403123 LAC **Guy Wilfrid SUISTED**, RNZAF - Age 22. 102hrs (27 solo on Vincent)
(Pilot): NZ402457 LAC **Angus Victor Macleod GILMOUR**, RNZAF - Age 21.
Gilmour was acting as navigator for the exercise.

2 Service Flying Training School (ATS), RNZAF (Woodbourne)
Vincent NZ340 - took off at 1000 on the same exercise as the above and encountered similar weather. NZ340 crashed and caught fire at about 1030 some 25 feet from the top of a 2000 foot hill in the vicinity of Weld Pass, about 7 miles SW of Blenheim. The two airmen are buried at Wellington and Dargaville respectively.
Pilot: NZ402898 LAC **William Frederick READER**, RNZAF - Age 21. 115hrs (24 solo on Vincent)
(Pilot): NZ402878 LAC **Rodney Albert LLEWELLYN**, RNZAF - Age 20.
Llewellyn was acting as navigator for the exercise.
All four of the above pilots had been awarded their Flying Badges with effect from 18 January.

MIDDLE EAST

Raid on Tobruk, Libya
55 Squadron, RAF (Amseat, Libya - 202 Group)
Blenheim IV T1879 - proceeded to take off at 0755 but when almost airborne four bombs fell away. Two exploded, causing the port engine to cut as T1879 climbed to 100 feet. It then turned to port and dived into the ground, and caught fire. All three crew were buried at Bardia, later being reinterred at Halfaya Sollum.
Pilot: 40986 Fg Off **Frank Reidhaar BULLOT**, RAF - Age 24.

Wed 22 Jan 1941

BOMBER COMMAND

Raid on Antwerp, Belgium
139 Squadron, RAF (Horsham St Faith, Norfolk - 2 Group)
Blenheim IV T2435 - took off at 1020 but on approaching the Suffolk coast on return was hit by 'friendly' naval coastal defence anti-aircraft fire. Shortly afterwards, at 1229, the tail broke away and T2435 dived into a field 250 yards from the police house at Oulton, 2 miles NW of Lowestoft. All three crew died, the pilot being buried at

Catton. The Blenheim had failed to make recognition signals, but it was thought doubtful if the pilot had sighted the coast at the time because of poor weather conditions and was therefore unaware of the need to do so.
Pilot: 41604 Flt Lt **Guy Jerrold MENZIES**, DFC, RAF - Age 20. 369hrs solo.
Guy Jerrold Menzies should not be confused with the Australian, Guy Lambton Menzies, who gained fame in January 1931 by completing the first solo flight across the Tasman Sea from Australia to New Zealand. The latter also died while flying with the RAF, during an air-sea-rescue flight over the Mediterranean on 1 November 1940.

Thu 30 Jan 1941

NEW ZEALAND

RNZAF Station Whenuapai
Died of natural causes at the Auckland Public Hospital, after being admitted there on 10 December 1940, and is buried at Papakura.
(Patrolman): NZ402532 LAC **James Martin Stanton SCHWARTFEGER**, RNZAF - Age 48.

Fri 31 Jan 1941

NEW ZEALAND

First solo night flight
1 Service Flying Training School (ITS), RNZAF (Wigram)
Oxford I NZ256 - overshot on first approach and went round again. On second attempt at 0310 approached too high, began to climb away but lost height and dived into the ground at a shallow angle and high speed 2 miles from the airfield. The pilot is buried at Wellington.
Pilot: NZ403466 LAC **Allan Henry McIVER**, RNZAF - Age 26. 111hrs (20 solo on Oxford - 3.45 dual at night)

Tue 4 Feb 1941

BOMBER COMMAND

Cross country to Penrhos, Caernarvonshire
18 Operational Training Unit, RAF (Bramcote, Warwickshire - 6 Group)
Wellington IC R1298 - captained by Fg Off D Warburton, RAF, and drifted 12 miles off course en route. On failing to clear a balloon barrage in poor visibility, R1298 struck a cable at 1250 and crashed near Middlewich, Cheshire, and caught fire. All ten occupants died, seven of whom, including a New Zealand air gunner, are buried at Nuneaton, Warwickshire.
Air Gunner: NZ39859 Sgt **Walter Lionel GOSDEN**, RNZAF - Age 24.

Cross country and bombing training
20 Operational Training Unit, RAF (Lossiemouth, Morayshire - 6 Group)
Wellington IC R1367 - stalled in an over banked turn at low altitude while on a circuit of the airfield at 1040, crashed and caught fire half a mile SE of Lossiemouth. Two of the crew survived, injured, but the other four died, including the pilot, who succumbed to his injuries later in the day and is buried at Lossiemouth. It was thought that R1367 was attempting an emergency landing, but whether it was returning from the exercise or setting out is unknown.
Pilot: NZ40668 Sgt **Daniel Munro WALKER**, RNZAF - Age 21. 195hrs solo.

Cross country and bombing training
20 Operational Training Unit, RAF (Lossiemouth, Morayshire - 6 Group)
Wellington IC R1005 - took off at about 1830. Given a 'green' to land on returning from the exercise at 2205, but then a 'red' when flying the cross wind leg, owing to another aircraft obstructing the flightpath. R1005 turned to port, lost height rapidly and crashed into the sea near Lossiemouth. Two of the crew were injured but the other four died, including the pilot, whose body was later recovered and is buried at Lossiemouth.
Captain: NZ40754 Plt Off **Basil Ivan CLAUSEN**, RNZAF - Age 24. 220hrs.

Fri 7 Feb 1941

NEW ZEALAND

Solo training
1 Elementary Flying Training School, RNZAF (Taieri)
Tiger Moth II NZ717 - had just completed a right hand turn at 0730 when it collided at 300 feet with another of the School's Tiger Moths (NZ873), which had also just executed a right hand turn preparatory to landing. NZ717 dived straight into the ground killing the pilot, who is buried at Ashburton. The other aircraft began a spin from which it recovered and forced landed in a hedge, injuring flying instructor Fg Off **L F Poore** and his pupil LAC **P W D Stewart**. At the time of the collision the sun was low on the horizon and directly in the eyes of the pilots of each aircraft.
Pilot: NZ404869 LAC **Desmond Thomas Francis HANRAHAN**, RNZAF - Age 23. 14hrs (4 solo)

Sat 8 Feb 1941

BOMBER COMMAND

Air to ground firing practice

12 Squadron, RAF (Binbrook, Lincolnshire - 1 Group)

Wellington II W5365 - captained by Act Sqn Ldr P C Lawrence, RAF, and stalled at 500 feet while turning into wind at 1220 on approach to Tollerton, Nottinghamshire, diving into the ground at Cotgrave, 2 miles east of the airfield. Two of the nine occupants were injured but their comrades died, including the New Zealand air gunner who succumbed to his injuries later in the day and is buried at Cotgrave. When it stalled, W5365 was flying over a small 150 foot-high hill fringed with a row of trees. It was considered that this may have created disturbed air conditions which contributed to the accident.

Air Gunner: 903054 Sgt **Neil Colin BOYLE**, RAFVR - Age 24.

Mon 10 Feb 1941

ARMY CO-OPERATION COMMAND

Tactical reconnaissance training

1 School of Army Co-operation, RAF (Old Sarum, Wiltshire - 70 Group)

Lysander II L4778 - piloted by Plt Off E Lunde, RAF, flew low along a railway line and hit telegraph wires before diving towards a train. A wing tip struck the tender before L4778 crashed some 400 yards further on, near Tisbury, 12 miles west of Salisbury. The pilot was killed and the air gunner died of injuries later in the day at the Tower House Emergency Hospital at Salisbury. He is buried in that city.

Air Gunner: NZ401719 Sgt **Trevor Norman BASSETT**, RNZAF - Age 21. 23hrs.

Tue 11 Feb 1941

COASTAL COMMAND

'Nomad' Patrol

206 Squadron, RAF (Bircham Newton, Norfolk - 16 Group)

Hudson I T9350 - took off at 2100 as one of three engaged on the patrol, all of which were abandoned to crash upon returning late at night, because of a combination of widespread fog and fuel exhaustion. T9350 made several unsuccessful attempts to land at Leuchars, Fifeshire, before instructions were given to abandon it. It was thought that the crew were under the impression of being over land, but the aircraft was lost in the sea without trace. All four crew died, the captain being commemorated on the Runnymede Memorial.

Captain: 42417 Plt Off **Miles Arthur George MASON**, RAF - Age 24.

Tue 11/Wed 12 Feb 1941

BOMBER COMMAND

Raid on aircraft factory at Bremen, Germany

83 Squadron, RAF (Scampton, Lincolnshire - 5 Group)

Hampden I AD722/X - took off at 2010 and because of thickening fog over base on returning, diverted towards Lindholme, Yorkshire. However, on sighting en route a flare path at Finningley, Nottinghamshire, a decision was made to land there instead. While circling the airfield at 0122 at low altitude, because of a low cloud base and poor visibility, AD722 crashed into a house and caught fire. An RAF Sgt crew member survived, injured, but the other three died, including the pilot, who is buried at Finningley.

Pilot: 70043 Flt Lt **Desmond Wallace Ferguson BARKER**, DFC, RAF - Age 27. 522hrs. 25th op.

Barker's brother, Richard Percival William Barker, died on 22 December 1941 while flying with 28 Conversion Flight, RAF.

Fri 14 Feb 1941

COASTAL COMMAND

Cross country training at night

235 Squadron, RAF (Bircham Newton, Norfolk - 16 Group)

Blenheim IVF Z5970 - piloted by Sgt W Mason, RAF, and seen to execute a gliding turn at 2130 after engines began misfiring, then cut out completely. Z5970 straightened out but continued to lose height before striking the ground a mile north of Great Croxton. The time of the crash was consistent with exhaustion of fuel in the outer tanks. All three crew died, the observer succumbing to injuries the next day and being buried at Rugby, Warwickshire.

Observer: NZ39943 Sgt **Noel Mizpah STANGER**, RNZAF - Age 24. 196hrs.

Sat 15 Feb 1941

COASTAL COMMAND

Bombing raid on the heavy cruiser *Admiral Hipper* at Brest, France.

217 Squadron, RAF (St Eval, Cornwall - 19 Group)

Beaufort I W6493 - took off at 1640 on the raid with two others, all three being shot down when intercepted by fighters about 20km north of the target. W6493 crashed at Plouguin, 14km SW of Lannilis, the four crew being buried at Lannilis, 20km north of Brest.

Pilot: 36171 Fg Off **Ronald William GAIR**, RAF - Age 21. 778hrs. 131st op.

Thu 20 Feb 1941

NEW ZEALAND

Solo training

3 Elementary Flying Training School, RNZAF (Harewood)

Tiger Moth II NZ886 - while in the circuit at 800 feet, converged and collided at 0930 with another of the School's Tiger Moths (NZ883), entering a spiral dive from which it failed to recover. NZ886 crashed onto Burnside Road, a mile east of Harewood, and caught fire. The pilot is buried at Christchurch. LAC **W J Willis**, piloting the other trainer, retained sufficient control to effect a forced landing. Striking a ditch and tearing through a gorse fence, NZ883 overturned as it hit rough ground, Willis suffering only slight injuries.

Pilot: NZ404422 LAC **William Gordon STEPHENS**, RNZAF - Age 26. 18hrs (2 solo)

Fri 21 Feb 1941

NEW ZEALAND

RNZAF Station Wigram

Died of natural causes and is buried at Christchurch.

(Pilot): NZ40954 Plt Off **William Keith ANDERSON**, RNZAF - Age 26.

Fri 21/Sat 22 Feb 1941

BOMBER COMMAND

Raid on Wilhelmshaven, Germany

75 (NZ) Squadron, RAF (Feltwell, Norfolk - 3 Group)

Wellington IC T2503 - took off at 1830 and believed to have been brought down into the sea. All six crew died, the bodies of two later washing ashore, including at Heligoland that of the rear gunner, who was buried at Wilhelmshaven, but later reinterred at Sage, 24km south of Oldenburg. The others, including captain and observer, were lost without trace and are commemorated on the Runnymede Memorial.

Captain: NZ39910 Plt Off **Arthur James FALCONER**, RNZAF - Age 23. 384hrs. 23rd op.
Observer: NZ40195 Plt Off **Anthony Vincent MUIR**, RNZAF - Age 29. 205hrs. At least 9 ops.
Rear Gunner: NZ391378 Sgt **Andrew Moore BRODIE**, RNZAF - Age 25. 173hrs. 23rd op.

Sat 22 Feb 1941

COASTAL COMMAND

Sweep over Denmark

235 Squadron, RAF (Bircham Newton, Norfolk - 16 Group)

Blenheim IVF T1803 - took off at 1120, piloted by Sgt D S Wallis, RAF, and one of three which intercepted an enemy bomber 30km west of Borkum at about 1500. Their attacks proved unsuccessful but return fire shot T1803 down into the sea. All three crew are commemorated on the Runnymede Memorial.

Observer: NZ40186 Sgt **Richard Waller BROOKMAN**, RNZAF - Age 28. 223hrs. 5th op.

Anti-submarine convoy patrol at approximate position 5500N:1600W

209 Squadron, RAF (Stranraer, Wigtownshire - 15 Group)

Lerwick I L7263/L - took off during daylight captained by the Squadron's commanding officer, Wg Cdr J E M Bainbridge, RAF, and failed to return. The following day a naval tug belatedly signalled at 0940 that it had seen a flare at 5340N:1240W at 2000 on the 22nd. This may have related to the Lerwick's fate, no trace of which was ever found. All 12 crew died, all or most of whom are commemorated on the Runnymede Memorial.

Air Gunner(?): 911664 LAC **Albert Edward McQUITTY**, RAFVR - Age 24. 201hrs. 20th op.

LAC McQuitty enlisted in the RAF as a Flight Rigger. Records state that he flew the above hours as an air gunner, which role normally meant rank of Sergeant or above, although he does not seem to have remustered as such.

Sun 23 Feb 1941

FIGHTER COMMAND

Scramble patrol

151 Squadron, RAF (Wittering, Northamptonshire - 12 Group)

Defiant I N3388 - took off at 1835, piloted by Sgt J Hopewell, RAF, but ran low on fuel during return. The flare path at Feltwell, Norfolk, was lit for an emergency landing, but by the time its lights were functioning N3388 was too far away to see them. At 2050 the Defiant was abandoned as the fuel tanks ran dry, the pilot descending uninjured. The air gunner's parachute, however, caught on the tail and he was killed when the aircraft crashed near Watton, and is buried at Honington, Suffolk.

Air Gunner: NZ401735 Sgt **Jack Fisher WALLACE**, RNZAF - Age 20. 29hrs.

Sun 23/Mon 24 Feb 1941

BOMBER COMMAND

Raid on Boulogne, France

115 Squadron, RAF (Marham, Norfolk - 3 Group)

Wellington IC L7810/R - took off at 1943 captained by Sgt M F Lloyd, RAF, and believed to have been brought down by flak, crashing at 0013 onto the beach 1km east of the Calais lighthouse, near the village of Les Hemmes de Marck. All six crew were buried in dunes by the sea, but later reinterred at Pihen-lés-Guînes, 10km SW of Calais.

Observer: NZ40915 Sgt **Cyril Edgar DONEHUE**, RNZAF - Age 27. 142hrs. 1st op.

Tue 25 Feb 1941

BOMBER COMMAND

Practising stalls with and without flaps and with undercarriage down

11 Operational Training Unit, RAF (Bassingbourn, Cambridgeshire - 6 Group)

Wellington I L4276 - captained by Act Flt Lt A F Riddlesworth, RAF, and stalled after recovery from an involuntary spin at 1440, crashed, and caught fire near Newnham (at position 5201N:0011W), Northamptonshire, 7 miles SW of Bassingbourn. All five crew died, the 2nd pilot being buried in the Bassingbourn cum Kneesworth Cemetery.

2nd Pilot: Aus400032 Plt Off **William John SHIRTCLIFFE**, RAAF - Age 25. 90hrs.

Wed 26 Feb 1941

MIDDLE EAST

Interception patrol

261 Squadron, RAF (Ta Kali, Malta - Mediterranean)

Hurricane I V7474 - took off at about 1300 with seven others to intercept an incoming raid of 90-100 enemy aircraft and one of three which failed to return. The Hurricanes climbed to 28,000 feet before diving on bombers below at 10,000 feet, but were followed down in turn by enemy fighters. Three pilots of the Squadron were shot down during the engagement, all being commemorated on the Malta Memorial.

Pilot: 43705 Plt Off **Charles Edward LANGDON**, RAF - Age 22. 353hrs. 32nd op.

Tue 4 Mar 1941

NEW ZEALAND

Formation practice

3 Service Flying Training School (ATS), RNZAF (Ohakea)

Hind Trainer NZ1507 - while flying between Shannon and Levin in the morning, led another of the School's Hinds (NZ1526) through isolated cloud at 1800 feet. On breaking cloud it was apparently not realised that NZ1526 was now slightly ahead and above. They collided and went down out of control, the pilot of NZ1526, LAC **G G Thomas**, carrying out a forced landing without injury to himself. NZ1507's pilot, however, crashed to his death and is buried at Palmerston North.

Pilot: NZ403114 LAC **Evan Roy ARCHER**, RNZAF - Age 23. 105hrs (32 solo on Hind)

Thomas died on 9 April 1943 while flying with the Central Gunnery School, RAF.

Wed 5 Mar 1941

FIGHTER COMMAND

Flight from Wittering to Digby, Lincolnshire

25 Squadron, RAF (Wittering, Northamptonshire - 12 Group)

Blenheim IF L6602 - collided with a tree during a low level daylight flight in intermittent fog conditions and crashed into the ground near Cottesmore, Rutland. Both occupants died, the passenger being buried at Stamford, Lincolnshire.

(Signals Officer): 84162 Fg Off **John Austin STRONG**, RAFVR - Age 31.

Thu 6 Mar 1941

NEW ZEALAND

Solo training

4 Elementary Flying Training School, RNZAF (Whenuapai)

Tiger Moth II NZ680 - while carrying out aerobatics at 3000 feet a loop was completed at 1610. After levelling out, part of the wing was seen to disintegrate and fold back, another portion separated and landed some 2 miles away. NZ680 then spun into the ground and burst into flames in an open space between two houses in Kaka Street near the railway workshops at Otahuhu, killing the pilot, who is buried at Papatoetoe. The cause of the collapse was traced to failure of a front flying wire through faulty manufacture in Britain.

Pilot: NZ402529 LAC **Frank Robert TRAYNOR**, RNZAF - Age 19. 52hrs.

Fri 7 Mar 1941

MIDDLE EAST

Non-operational flight from Egypt to Athens, Greece

37 Squadron, RAF (Shallufa, Egypt - 257 Wing, 202 Group)

Wellington IC R3239 - captained by Plt Off A de L Thomas, RAF, and on approach to Menidi airfield at Athens was obliged to make another circuit due to Greek Air Force aircraft taking off. On the second approach at 1530 a propeller flew off, causing R3239 to lose height and crash into the edge of a wood at Kyphyssia, in the NE outskirts of Athens. One occupant survived, injured, but six others died, including a New Zealand air gunner, who is buried at Athens.

Air Gunner: NZ40912 Sgt **James William BOLTON**, RNZAF - Age 24.

Five of those killed are identified as 37 Sqn aircrew but the other appears to have been a non-Air Force passenger.

Sun 9 Mar 1941

NEW ZEALAND

Motor vehicle accident

3 Elementary Flying Training School, RNZAF (Harewood)

As driver of a Ford V8 car, was involved in an accident shortly after 0200 while returning to Harewood from Christchurch City. Instead of turning into Fendalton Road from Deans Avenue the vehicle continued straight ahead, crossed a footpath, ran down a bank into the Avon River and turned upside down. The three occupants in the front seats were drowned but four others in the back owed their lives to a small reservoir of air being trapped beneath the floor, and prompt rescue efforts by Constable **G Hart** and Flt Lt **V G H Gee** (of RNZAF Station Wigram) who were in the vicinity at the time. In addition to the driver, who is buried at Christchurch, the deceased included Mrs **S S Magee** and Miss **I Martin**. Rescued uninjured were Mrs **B Pearson** and three airmen - LACs **N G Packard**, **N R Meyer** and **R F Stone**, who were undergoing pilot training at the EFTS.

(Pilot): NZ405270 LAC **Gerald Livingstone HEWITT**, RNZAF - Age 24.

All four airmen had enlisted over the weekend of 21-22 December 1940. Only Packard survived the war - Meyer died on 25 August 1942 while flying with 61 Sqn, RAF, and Stone on 30 July 1942 while flying with 102 Sqn, RAF.

Wed 12/Thu 13 Mar 1941

BOMBER COMMAND

Raid on Berlin, Germany (by 72 aircraft - 4 lost)

102 Squadron, RAF (Topcliffe - 4 Group)

Whitley V T4326/K - took off at 1940 and brought down over the Netherlands by Flak at 0005, crashing at Noord-Deurningen, 3km north of Denekamp. Four of the crew are buried at Denekamp. The fifth member, an RAF Sgt, survived but died later while a prisoner of war.

Captain: 36190 Flt Lt **Frank Hugh LONG**, DFC, RAF - Age 24.

Raid on Boulogne and Calais, France (by 8 aircraft - 1 lost)

40 Squadron, RAF (Alconbury, Huntingdonshire - 3 Group)

Wellington IC T2515/U - took off at 1950 to raid Boulogne and brought down in flames near Wimille, 5km north of Boulogne. All six crew are buried at Wimille.

Captain: NZ40764 Sgt **Douglas William GOUGH**, RNZAF - Age 19. 277hrs. 1st op.

2nd Pilot: NZ40670 Sgt **Thomas Geoffrey WEBB**, RNZAF - Age 19. 245hrs. 1st op.

Gough and Webb had both graduated on No 17 course at Wigram.

Thu 13 Mar 1941

FIGHTER COMMAND

Ferrying pilot to Ford, Sussex, for collection of an aircraft

85 Squadron, RAF (Debden, Essex - 11 Group)

Havoc I BJ500 - took off at about 1620, piloted by Act Flt Lt G Allard, RAF. Almost immediately afterwards the top gun panel detached and lodged against the base of the tail fin, forcing the aircraft to pitch upwards. BJ500 climbed, stalled, spun and crashed, catching fire on impact at Wimbish, 2 miles NE of the base. All three on board were killed, the New Zealander being buried at Saffron Walden, Essex.

(Pilot): 36269 Plt Off **William Henry HODGSON**, DFC, RAF - Age 20. 503hrs. 146 ops.

Hodgson, a veteran of the Battle of Britain, had been classified as medically unfit for flying and went on the flight for air experience.

Thu 13/Fri 14 Mar 1941

BOMBER COMMAND

Raid on Hamburg, Germany (by 139 aircraft - 7 lost)

214 Squadron, RAF (Stradishall, Suffolk - 3 Group)

Wellington IC N2746/M - took off at 2000 captained by Sgt A G Elder, RAF, and shot down over the Netherlands by a night fighter shortly after 2300, crashing near Groningen. An RAF Sgt survived to be taken prisoner but his five comrades died and are buried at Groningen.

Observer: NZ40621 Sgt **Sol GLAZER**, RNZAF - Age 29. 219hrs. 5th op.

Fri 14 Mar 1941

MIDDLE EAST

Cross country training
70 Operational Training Unit, RAF (Ismailia, Egypt - 250 Wing)
Blenheim I L6664 - piloted by Sgt R Ibbotson, RAF, and on returning after the exercise found Ismailia to be closed by a sandstorm. At 1230 L6664 attempted a landing at an alternative airfield at Port Said, but ran out of fuel during an overshoot and crashed near the prison. All three crew are buried at Ismailia.
Observer: NZ401726 Sgt **John Elliot HASLOP**, RNZAF - Age 22. 68hrs.

BOMBER COMMAND

Air firing at sea, bombing practice and cross country
10 Operational Training Unit, RAF (Abingdon, Berkshire - 6 Group)
Whitley V P5087 - intended route was from base to Worcester-Strumble Head-Bardsey Island-Jurby-Great Ormes Head-Ludlow and back to base, but nothing heard after take off. P5087 is believed to have crashed into the Irish Sea sometime between 1045 and 1930. All six crew died, including the captain, who is commemorated on the Runnymede Memorial.
Captain: 36267 Plt Off **Allan BRIDSON**, DFC, RAF - Age 22. 475hrs solo. 22 ops.

Fri 14/Sat 15 Mar 1941

MIDDLE EAST

Ferry flight from Stradishall, Suffolk, to Egypt via Gibraltar and Malta
In transit to RAF Middle East (Egypt)
Wellington IC W5644 - set out in the evening on the Gibraltar-Malta leg with three other Wellingtons and an escorting Maryland. Fifteen minutes prior to estimated arrival time of 0800, a message was received stating that W5644 was under fighter attack some 70 miles out from Malta. Nothing further was heard, and all seven occupants, including an RAF Gp Capt travelling as a passenger, are commemorated on the Runnymede Memorial.
Captain: NZ391335 Sgt **Richard Hugh ALINGTON**, RNZAF - Age 26. 29 ops.
Alington, previously with 99 Sqn, RAF, had been posted from Bomber Command to the Middle East and was making his way there by means of the delivery flight.

Tue 18 Mar 1941

MIDDLE EAST

Reconnaissance of enemy airfields at Sirte and Mechina, Libya
55 Squadron, RAF (Maraua, Libya - 202 Group)
Blenheim IV T1995 - took off at about 0730 and last seen by a Hurricane pilot at 0800 in the vicinity of Benina. All three crew are commemorated on the Alamein Memorial.
Pilot: 40769 Fg Off **Thornton Osborne WALKER**, mid, RAF - Age 25. 653hrs. At least 35 ops.

Thu 20 Mar 1941

BOMBER COMMAND

Attack on shipping off the Dutch coast (by 6 aircraft - 1 lost)
82 Squadron, RAF (Watton, Norfolk - 2 Group)
Blenheim IV R3604 - took off at 1109 piloted by Sgt J H Kelly, RAF, and shot down into the sea off the Dutch coast by flak from a German minesweeper operating off Den Helder. The bodies of two crew members were recovered but not that of the air gunner, who is commemorated on the Runnymede Memorial. The observer was first buried in the Huisduinen Cemetery at Den Helder, but later reinterred at Bergen-op-Zoom.
Observer: NZ40953 Sgt **Graham Robert WILSON**, RNZAF - Age 20. 121hrs. 1st op(?)

Fri 21/Sat 22 Mar 1941

BOMBER COMMAND

Raid on Lorient (by 66 aircraft - 4 lost)
18 Squadron, RAF (Great Massingham, Norfolk - 2 Group)
Blenheim IV T2038/M - took off at 2339 from Boscombe Down, Wiltshire, piloted by Plt Off G H Cowings, RAF. Squadron aircraft were unable to locate the target due to poor weather and landed at various bases on return to England. Blenheim T2038 became lost and shortly after 0500 flew into the ground at high speed near Barkston, 4 miles NNW of Grantham, Lincolnshire. All three crew died, the observer being buried at Newcastle-upon-Tyne, Northumberland. Of the four bombers destroyed, none were lost until the return journey - only one out of the 16 crew members surviving.
Observer: NZ40747 Plt Off **Robert DANIEL**, RNZAF - Age 24. 139hrs. 5th op.

Sat 22 Mar 1941

NEW ZEALAND

Electrocution accident

RNZAF Station Ohakea

Accidentally electrocuted at about 1710 while tuning the transmitting set in the control room of the base's transmitting station, and is buried at Oamaru.

(Wireless Electrical Mechanic): NZ391148 LAC **Alan Harris HALL**, RNZAF - Age 23.

Wed 26 Mar 1941

MIDDLE EAST

Aerodrome patrol

112 Squadron, RAF (Yanina, Greece; detachment at Paramythia - 'W' Wing, 201 Group, Greece)

Gladiator II N5910 - took off from Paramythia at 1720 to relieve another pilot on defensive patrol. The patrol was instructed to land 30 minutes later, at which point N5910 broke off independently at 20,000 feet and dived to 12,000 feet. It then entered another dive and made an aileron turn before coming straight down to crash near the airfield perimeter. The pilot, who it was thought had blacked out during the dive, is buried at Athens.

(Pilot): 34169 Sqn Ldr **Richard John Courtney NEDWILL**, AFC, RAF - Age 27. 2003hrs.

Nedwill had only three or four days beforehand been appointed commanding officer of the Blenheim-equipped 211 Sqn, RAF, which was also operating from Paramythia.

Fri 28 Mar 1941

NEW ZEALAND

Solo training

4 Elementary Flying Training School, RNZAF (Whenuapai)

Tiger Moth II NZ790 - at 1120, stalled off a steep turn at 150 feet, crashed and burst into flames in a paddock of Dr Aitkenhead's farm at Reweti, ten miles west of Whenuapai. The pilot is buried at Auckland.

Pilot: NZ41349 LAC **Edwin Chaytor MOORE**, RNZAF - Age 23. 18hrs (3 solo)

Sat 29 Mar 1941

MIDDLE EAST

Reconnaissance - searching for remnants of Italian fleet after Battle of Cape Matapan, Greece

69 Squadron, RAF (Luqa, Malta - Mediterranean)

Maryland I AR727 - took off at 1100 to reconnoitre results of the previous night's naval battle. Shortly after passing some unidentified naval vessels an engine failed and a forced landing had to be made close to a beach on the Ionian island of Zante. As AR727 touched down the Maryland flipped over on to its back in shallow water. The navigator and air gunner were both injured, and found that the pilot had drowned by the time they were able to extricate him from the wreckage. He is buried at Athens.

Captain: 40873 Fg Off **Frederick Redfearn AINLEY**, RAF - Age 26

Ainley was born at Preston, England but came to New Zealand in 1934. Selected for a short service commission in the RAF, he embarked for England in 1938.

Tue 1 Apr 1941

MIDDLE EAST

Reconnaissance to Tripoli, Libya

55 Squadron, RAF (Maraua, Libya - 204 Group)

Blenheim IV V5423 - took off at 1525 and was forced to ditch in the sea SW of Benghazi. The three crew, which included a Fleet Air Arm officer on attachment, were at one stage sighted in a rubber dinghy. However, all three were lost without trace, the two Air Force airmen being commemorated on the Alamein Memorial and their Navy comrade on the Fleet Air Arm Memorial at Lee-on-Solent.

Pilot: 40523 Fg Off **Maxwell Stewart FERGUSON**, RAF - Age 25. 539hrs. 39th op.

Wed 2 Apr 1941

NEW ZEALAND

Formation practice

2 Squadron, RNZAF (Nelson)

Vincent NZ357 - as the formation returned from Omaka at 1500 experienced engine failure and made a forced landing in bush at the head of the Pelorus River within a few miles of a sawmill. Another machine later dropped a message at Mr G Acton's residence, located about a mile from the crash site, appealing for a local rescue party to be organised. Met part way in by the least injured of NZ357's passengers, the party was taken to the scene and found the pilot and one passenger dead. Both are buried at Nelson. The other passengers, Cpls **E Whitely** and **J G Watt**, were injured, the former with a broken leg. They were brought out over a bridle track to the Acton residence, thence to Nelson Public Hospital. Subsequent investigation revealed that a blockage in the feed pipe from the port fuel tank had caused the engine to fail.

Pilot: NZ402213 Plt Off **Vernon Wellington RAYMOND**, RNZAF - Age 25. 150hrs solo (49 on Vincent)

(Wireless Electrical Mechanic): NZ401325 AC1 **James Edward MURPHY**, RNZAF - Age 31.

Sun 6/Mon 7 Apr 1941

MIDDLE EAST

Ferry flight from Stradishall to Malta

3 Group Training Flight, RAF (Stradishall, Suffolk - 3 Group)

Wellington IC N2818 - deviated from track in bad weather and at 0300 crashed 5km east of Ras el Akba or 4km south of Ain Amara, near Guelma, Algeria. N2818 had been seen descending through cloud with landing lights on and engines throttled back. As the ground came into view the engines were opened up, but the bomber crashed into a knoll situated between two large hills and exploded. It is not known if N2818 was seeking to make an emergency landing or simply trying to establish its position. All six crew were buried at Guelma, but later reinterred at Bone.

Captain: NZ391359 Plt Off **Gilbert Theodore KIMBERLEY**, RNZAF - Age 24.

Air Gunner: NZ401449 Sgt **Walter Alexander WATTS**, RNZAF - Age 27.

Kimberley, previously with 75 (NZ) Sqn, RAF, and with the Group Training Flight since early March, had just been posted to the Middle East. Watts had also been posted to the Middle East, directly from 75 (NZ) Sqn. Both were therefore leaving Bomber Command and making their way to Middle East Command by way of the flight.

Mon 7 Apr 1941

COASTAL COMMAND

Convoy escort

1 Coastal Artillery Co-operation Flight, RAF (Detling, Kent - 17 Group)

Anson I N5203 - took off at 1310 piloted by Sgt C H Blake, RAF, but unsteady, yawing with port engine coughing and smoking. N5203 began a left hand turn when airborne but crashed on the downwind leg into rising ground 3 miles from Detling. One crew member survived, injured, but his three comrades died, the air gunner being buried at Maidstone. The accident report, attributing the accident to mishandling, noted that the pilot had flown a total of only 190 minutes since May 1940.

Air Gunner: NZ402138 Sgt **Harold Lagor BEACH**, RNZAF - Age 23. 22hrs.

Mon 7/Tue 8 Apr 1941

BOMBER COMMAND

Raid on Kiel, Germany (by 229 aircraft - 5 lost)

214 Squadron, RAF (Stradishall, Suffolk - 3 Group)

Wellington IC R1380 - captained by Sgt R A Williams, RAF, and believed to have been shot down by a night fighter, its six crew being commemorated on the Runnymede Memorial.

2nd Pilot: NZ40654 Sgt **Kenneth MANSON**, RNZAF - Age 20. 233hrs. 5th op.

Tue 8 Apr 1941

FIGHTER COMMAND

Convoy patrol

232 Squadron, RAF (Elgin, Morayshire - 14 Group)

Hurricane I V7159 - returned at 0340 but was obliged to circle for 15mins while a duty pilot and electrician were woken to light the flarepath. On landing, overshot slightly and bounced heavily, and thought to have then burst a tyre while braking at high speed. The pilot died when V7159 overturned and is buried at Montrose. He had asked to be placed on 15 minutes standby prior to the fatal flight, notwithstanding that he had already completed four daylight and two night sorties over the preceding 20 hours. V7159 was later salvaged and continued in service until June 1944.

Pilot: NZ40217 Sgt **John Herbert ALLEN**, RNZAF - Age 19. 308hrs.

Allen's brother, Alistair William Allen, died on 15 September 1942 while flying with 4 Glider Training School, RAF.

Training

256 Squadron, RAF (Squires Gate, Lancashire - 9 Group)

Defiant I N3424 - took off about 0450, piloted by Sgt J D H Cunningham, RAF, but lost height and crashed near the airfield at Manton, Blackpool. Both crew members died, the air gunner being buried at Lytham St Anne's. It was considered that the engine may have cut because take off was made without first running up the engine to clear any accumulation of fuel, after idling for some time.

Air Gunner: NZ401819 Sgt **Albert Douglas WOOD**, RNZAF - Age 24. 72hrs. 3 ops.

Tue 8/Wed 9 Apr 1941

BOMBER COMMAND

Raid on Kiel, Germany (by 160 aircraft - 4 lost)

149 Squadron, RAF (Mildenhall, Suffolk - 3 Group)

Wellington IC X3167/H - captained by Sgt J B C Jago, RAF, and brought down near Holtenau, on the northern outskirts of Kiel. Two of the crew were lost without trace and are commemorated on the Runnymede Memorial. The bodies of their four comrades were recovered, including that of the observer, who was buried in the Garrison Cemetery at Kiel, but later reinterred in the Kiel War Cemetery.

Observer: NZ401207 Plt Off **John Alastair GRAHAM**, RNZAF - Age 24. 180hrs. 3rd op.

214 Squadron, RAF (Stradishall, Suffolk - 3 Group)
Wellington IC T2542 - took off at 2025 captained by Sgt J P Cusworth, RAF, on same raid as the above and brought down at Melsdorf, in the SW outskirts of Kiel. All six crew are buried at Kiel.
Observer: NZ40932 Sgt **James Clark CATTELL**, RNZAF - Age 21. 168hrs. 3rd op.

Wed 9 Apr 1941

NEW ZEALAND

Ferry flight from Nelson to Blenheim
2 Squadron, RNZAF (Nelson)
Vildebeest III NZ117 - took off at about 1045, but after circuiting returned to the airfield with black smoke issuing from the exhaust. Over the airfield the engine was opened up but NZ117 was not turned into wind. It then executed a steep right hand turn at 50-100 feet, stalled and dived onto the golf course adjacent to the airfield, and burst into flames. Two of the occupants were killed and are buried at Hamilton and Marton, respectively. A corporal passenger died of injuries in the Nelson Public Hospital the following morning and is buried at Invercargill.
Pilot: (NZ2525) Flt Lt **Lionel Mountjoy SQUIRE**, RNZAF - Age 33. 90hrs solo on Vildebeest.
(Clerk): NZ39662 Cpl **Ernest Ivan JOHNSON**, RNZAF - Age 24.
(WOp): NZ391981 AC1 **Lindo James THOMSON**, RNZAF - Age 20.
At the time of Squire's death service numbers had not yet been assigned to officers of the RNZAF. NZ2525 was a posthumous allocation made at war's end.
Johnson belonged to 2 Service Flying Training School at Blenheim.

MIDDLE EAST

Ferry flight from Egypt to Malta
148 Squadron, RAF (LG 213 [Kabrit], Egypt; detachment at Luqa, Malta - 257 Wing, 202 Group)
Wellington IC T2952/H - captained by Sgt H P Adams, RAF, and one of three Wellingtons making the flight - all of which are believed to have refuelled en route at LG29 (Amiriya), 16 miles SSW of Alexandria. Nearing Malta in the mid-afternoon T2952 exhausted its fuel and ditched in the sea. All six crew were lost without trace, four of whom, including a New Zealand air gunner, are commemorated on the Alamein Memorial. Unusually, it would appear from the CWGC registers that the names of the other two are recorded on the Malta Memorial.
Air Gunner: NZ40725 Sgt **Joseph Lewis McNAMARA**, RNZAF - Age 22.

Thu 10 Apr 1941

COASTAL COMMAND

Photographic reconnaissance of warships at Brest, France
1 Photographic Reconnaissance Unit, RAF (Benson, Oxfordshire)
Spitfire I P9307 - one of two Spitfires which took off independently during the day from St Eval, Cornwall, to reconnoitre Brest. Each was intercepted and shot down into the sea, both pilots being commemorated on the Runnymede Memorial.
Pilot: 36242 Fg Off **Laurence David LOASBY**, RAF - Age 23. 651hrs. 31st op.

Thu 10/Good Friday 11 Apr 1941

BOMBER COMMAND

Raid on Brest, France (by 53 aircraft - 1 lost)
218 Squadron, RAF (Marham, Norfolk - 3 Group)
Wellington IC R1442/D - took off at 1930 captained by Sgt A G Plumb, RAF, and believed to have been hit by flak, crashing into the sea off Brest after sending a message requesting assistance. All six crew died, the body of an RCAF Sgt later being recovered near Brest, but the other five were lost without trace and are commemorated on the Runnymede Memorial.
Air Gunner: NZ402106 Sgt **Tahu William DABINETTE**, RNZAF - Age 21.

Easter Saturday 12 Apr 1941

MIDDLE EAST

German assault on Tobruk, Libya
202 Group HQ, RAF (Heliopolis, Egypt)
Killed by light enemy shell fire outside the perimeter of Tobruk on the Tobruk-El Adem road . Body recovered by an Australian patrol during the night of the 18th/19th and is buried at Tobruk.
(Armament Officer): 29199 Sqn Ldr **Ronald James BENNETT**, RAF - Age 32.
Bennett entered the RAF in 1930 and served as a pilot prior to qualifying as an armament specialist. A footnote entry on page 18, Volume III, of the Official History, New Zealanders in the Royal Air Force, *is incorrect in stating that he was 'killed on air operations.'*

Easter Sunday 13 Apr 1941

MIDDLE EAST

Attack on enemy columns on the Bitolj-Prilep road, Serbia

211 Squadron, RAF (Paramythia, Greece - 'W' Wing, 204 Group, Greece)

Blenheim I L1434 - took off at about 1500 with five others, all of which failed to return. No Allied fighter escort was available and the formation was intercepted on its way in to the target and shot down by enemy fighters. Battle-damaged, L1434 ditched or crashed into shallow water in the SE corner of Lake Prespa, where Albania shares a border with Greece. All three crew died and are commemorated on the Alamein Memorial. In 1993 a Greek Air Force recovery team was reported to have salvaged most of the remains of the Blenheim from the lake.

Pilot: 40885 Flt Lt **Lindsay Basil BUCHANAN**, DFC, RAF - Age 23. 44+ops.

Air War for Yugoslavia, Greece and Crete *refers to an unconfirmed report that Buchanan and another of his crew later died of injuries in an Albanian hospital. It also states that only two airmen survived from a total of 18 aboard the six Blenheims - yet only 10 names are recorded in CWGC registers, suggesting perhaps that the 'hospital' report may relate in some way to others of those who survived.*

Easter Monday 14 Apr 1941

MIDDLE EAST

Interception of incoming raid on Tobruk, Libya

73 Squadron, RAF (LG 145 or LG 146 [both El Gubbi/Tobruk], Libya - 258 Wing, 204 Group)

Hurricane I V7766 - took off in the late afternoon to intercept a raid by 30 enemy aircraft but was shot down during the engagement. The pilot was buried in an isolated grave near the Tobruk-Bardia road, but later reinterred in Knightsbridge War Cemetery, a few miles from Acroma.

Pilot: 42411 Plt Off **Owen Edward LAMB**, RAF - Age 23. 304hrs. 36th op.

Tue 15 Apr 1941

MIDDLE EAST

Reconnaissance of Halfaya-Sollum area, Egypt

14 Squadron, RAF (Heliopolis, Egypt - 204 Group)

Blenheim IV Z5863 - piloted by Plt Off I Ormiston, RAF, and failed to return. The RAF wireless operator-air gunner survived to be taken prisoner, but the pilot and observer died and are buried in a joint grave at Halfaya Sollum.

Observer: NZ402242 Sgt **Allan Hugh Lovatt FRASER**, RNZAF - Age 30. 84hrs. 7th op.

Wed 16/Thu 17 Apr 1941

ENGLAND

Enemy bombing raids on London

Between 2050 and 0518 London experienced its heaviest raids to date, when 759 sorties were launched against the city to claim 1180 lives. Four New Zealand airmen were amongst those killed, the three RNZAF members being buried at Brookwood in Woking, Surrey, and the naval airman at Gillingham, Kent. Two members of the Fleet Air Arm - Leading Airman **V H Hatchard**, RN, and Naval Airman 2nd Class **J A Pickard**, RN - were slightly injured.

Of **Bomber Command - 15 Operational Training Unit, RAF (Harwell, Berkshire - 6 Group)**

(WOp): NZ437007 LAC **Robert Lance CONGREVE**, RNZAF - Age 25. 170hrs.

Of **Bomber Command - 75 (NZ) Squadron, RAF (Feltwell, Norfolk - 3 Group)**

(WOp): NZ438024 LAC **Horace Dean SIMONSEN**, RNZAF - Age 31. 40hrs.

Of **Fighter Command - 256 Squadron, RAF (Squires Gate, Lancashire - 9 Group)**

(Air Gunner): NZ402116 Sgt **Reginald Lee ROBINSON**, RNZAF - Age 24.

Of the **Fleet Air Arm - HMS *Daedalus***

(Pilot): FX84614 Leading Airman **Kenneth Neil CAMPBELL**, RN - Age 20.

According to their CWGC register entries, Congreve and Robinson died on the evening of the 16th, and Campbell and Simonsen the following morning. Robinson and Simonsen were both staying at the Overseas Forces Victoria League Club in Malet Street which took two directs hits during the raid.

Pickard died on 26 June 1942 while flying with the FAA.

Hatchard, who two months after the raid transferred to the RAF, died on 14 May 1943 while flying with 102 Sqn, RAF.

Thu 17/Fri 18 Apr 1941

BOMBER COMMAND

Night training

20 Operational Training Unit, RAF (Lossiemouth, Morayshire - 6 Group)

Wellington IC N2820 - a sudden deterioration in weather caused the pilot to make a very low circuit of the airfield. While doing so N2820 struck an Aircraft Repair Section shed, crashed and caught fire. The two crew died, the captain being buried at Maresfield, Sussex. The weather change was unexpected as it was sudden, occurring within five minutes. Two other aircraft operating in the area at the time were considered to have narrowly escaped disaster.

Captain: NZ401274 Sgt **Frederick Simpson HOBDEN**, RNZAF - Age 28. 154hrs.

Fri 18 Apr 1941

NEW ZEALAND

Training

Flying Instructors School, RNZAF (Hobsonville)

Tiger Moth II NZ724 - at 1420 began a climbing turn at 1000 feet, but fell into a spin from which there was insufficient height to recover. NZ724 crashed in to the sea 400 yards off shore near Castor Bay, at the northern end of the Rangitoto channel. One pilot was killed while the other, Plt Off **M H Hoy**, slightly injured, was rescued with the assistance of three gunners of the Royal New Zealand Artillery and a civilian who witnessed the crash and swam out to him. The pilot's body was recovered from the wreckage later in the day and is buried at Invercargill. Two of the gunners, **James O'Meara** and **John B Hipkins**, were awarded Certificates of Merit by the Royal Humane Society of New Zealand in recognition of their actions.

Pilot: NZ403421 Plt Off **Bertrand William BUTLER**, RNZAF - Age 25. 96hrs solo (32 on Tiger Moth)

Sat 19 Apr 1941

FIGHTER COMMAND

Patrol

74 Squadron, RAF (Manston, Kent - 11 Group)

Spitfire IIA P7381 - took off at 1710, and was intercepted and shot down over the English Channel, an accompanying Spitfire surviving the encounter to make a safe return. No trace of the downed pilot was found and he is commemorated on the Runnymede Memorial.

Pilot: NZ39900 Plt Off **Edward Walter Gillies CHURCHES**, RNZAF - Age 19. 416hrs.

Sun 20 Apr 1941

COASTAL COMMAND

Navigation exercise

2 (Coastal) Operational Training Unit, RAF (Catfoss, Yorkshire - 17 Group)

Blenheim IV V5796 - crashed into the sea 50 miles east of Flamborough Head at 1135 in poor weather - other aircraft in the area reporting clouds down to sea level. The three crew died. After an extensive search the pilot's body was recovered and buried at Great Bircham, Norfolk.

Pilot: NZ40646 Sgt **James Hanley HAMILL**, RNZAF - Age 24. 152hrs solo.

Sun 20/Mon 21 Apr 1941

BOMBER COMMAND

Raid on Cologne, Germany (by 61 aircraft - 4 lost)

99 Squadron, RAF (Waterbeach, Cambridgeshire - 3 Group)

Wellington IC T2997 - took off at 2047 captained by Plt Off F H Cook, RAF. On returning from the raid at 0120, collided with a barrage balloon over Harwich Harbour, caught fire and crashed into the sea 3000 yards to the SE of Landguard, Suffolk. All six crew died, including two New Zealanders, whose bodies were later recovered and buried at Felixstowe.

Observer: NZ401212 Plt Off **Rex Mulligan MORROW**, RNZAF - Age 30. 183hrs. 5th op.

Air Gunner: NZ40284 Sgt **Frederick George HINDRUP**, RNZAF - Age 23. 110hrs. 11th op.

The CWGC register records date of death for Morrow as the 21st but incorrectly gives the 20th for Hindrup.

Tue 22/Wed 23 Apr 1941

BOMBER COMMAND

Raid on Brest, France (by 26 aircraft - 2 lost)

218 Squadron, RAF (Marham, Norfolk - 3 Group)

Wellington IC L7798/S - brought down near Milizac, 11km NW of Brest, where the five crew who died are buried. The sole survivor was air gunner Sgt **J Clark**, RNZAF, who was taken prisoner.

Captain: NZ40665 Sgt **William Henry SWAIN**, RNZAF - Age 23. 305hrs. 5th op.

Observer: NZ40755 Sgt **Malcolm Bruce CROOKS**, RNZAF - Age 25. 224hrs. 5th op.

Wed 23 Apr 1941

MIDDLE EAST

Bombing raid on enemy aircraft at Gazala Landing Ground North, Libya

55 Squadron, RAF (LG 15 [Maaten Bagush satellite], Egypt - 204 Group)

Blenheim IV T1873 - took off at 1114 piloted by Sgt T Fullarton, RAF, with five others. After bombing at 1330 and proceeding out to sea, the formation was intercepted 24 miles west of Tobruk by two fighters. Blenheim T1873 fell behind the rest of the flight (some its bomb load had failed to release) and in an engagement of some 15 minutes was shot down in flames into the sea. All three crew are commemorated on the Alamein Memorial.

Observer: NZ402133 Sgt **Henry Speery LATTA**, RNZAF - Age 32. 97hrs. 8th op.

FIGHTER COMMAND

Patrol over shipping off Falmouth, Cornwall

247 Squadron, RAF (Roborough, Devon; detachment at St Eval, Cornwall - 10 Group)

Hurricane I V7016 - one of two Hurricanes returning from patrol at 0620 in poor weather. They descended together in a shallow dive through a gap in the cloud near Portreath, but on levelling off at 400 feet, about 500 yards off the coast, the leader discovered his No 2 was missing. V7016 had in fact ditched a few miles further north off St Agnes Head. At 1100 a 152 Sqn Spitfire pilot saw the Hurricane pilot still swimming. Sadly, rescue efforts were delayed through a series of errors and he was dead of exposure when picked up after midday. He is buried at Illogan.

Pilot: NZ40771 Sgt **Alan Cameron KELLY**, RNZAF - Age 23. 203hrs. 10th op.

Thu 24 Apr 1941

COASTAL COMMAND

Reconnaissance of Norwegian coast

107 Squadron, RAF (Leuchars, Fifeshire - 18 Group)

Blenheim IV Z5795 - took off at 1800 piloted by Plt Off C P H Maclaren, RAF, and shot down by enemy fighters into the sea. The pilot's body later washed ashore near Kristiansand, but no trace of his two comrades could be found and they are commemorated on the Runnymede Memorial. 107 Sqn, normally under Bomber Command, was on detachment to Coastal Command from 3 March to 10 May 1941.

Observer: NZ40738 Sgt **Anthony James HANNAH**, RNZAF - Age 20. 164hrs. 12th op.

Fri, Anzac Day/Sat 26 Apr 1941

MIDDLE EAST

Communication flight from Suda Bay, Crete, to Kalamata, Greece

228 Squadron, RAF (Alexandria, Egypt - 201 Group)

Sunderland I T9048/N 2130 - took off at 2130 captained by Flt Lt H W Lamond, RAF, to deliver a message from Air Vice Marshal J H D'Albiac, Air Officer Commanding Greece, to Gp Capt A G Lee at Kalamata, and to evacuate 50 RAF personnel to Suda Bay. While approaching to land at 2315 on a moonless night, the height above a perfectly calm sea was misjudged and the aircraft flew into the water. Six of the ten crew died, including the 2nd pilot, but their bodies were not recovered and they are commemorated on the Alamein Memorial. The survivors drifted on an upturned wing until picked up by fishermen a few hours later.

2nd Pilot: 33460 Fg Off **John Clement Julius LYLIAN**, RAF - Age 21. 816hrs.

BOMBER COMMAND

Raid on Kiel, Germany (by 69 aircraft - 1 lost)

218 Squadron, RAF (Marham, Norfolk - 3 Group)

Wellington IC R1507/V - took off at 2054 captained by Fg Off G B S Agar, RAF. On the return flight an SOS was sent at 2356, at which time R1507 was about 80 miles off the enemy coast, but nothing further was heard. Extensive searches found no trace of the six crew and all are commemorated on the Runnymede Memorial.

2nd Pilot: NZ40242 Plt Off **Gilbert Peter Lewis REDSTONE**, RNZAF - Age 24. 300hrs. 9th op.

Sat 26/Sun 27 Apr 1941

BOMBER COMMAND

Raid on Hamburg, Germany (by 50 aircraft - 2 lost)

83 Squadron, RAF (Scampton, Lincolnshire - 5 Group)

Hampden I AD796/D - took off on the raid at 2110 and sent three signals, the last being an SOS when near the mouth of the River Elbe. AD796 was brought down at Eddelak, 6km NNW of Brunsbüttel. The four crew were buried at Brunsbüttel, but later reinterred at Kiel.

Pilot: 43027 Plt Off **Richard Hope CRUSH**, RAF - Age 26. 364hrs. 4th op.

Mon 28/Tue 29 Apr 1941

MIDDLE EAST

Raid on Benghazi Harbour, Libya

70 Squadron, RAF (LG 213 [Kabrit], Egypt - 257 Wing, 202 Group)

Wellington IC T2727 - took off at 1945 from Fuka, bombed shortly after 2300 and on return journey at 0118 signalled to seek a bearing. At about 0200 T2727 descended through low cloud to establish its position, but crashed into an escarpment at Sidi Omar, south of Sollum. Both pilots were killed, while the other four crew, together with a Sqn Ldr from Middle East HQ also on board, were injured. The rear gunner walked for $1^1/_2$ days in search of assistance. The captain's body was either not recovered or the grave later lost, as he is commemorated on the Alamein Memorial.

Captain: 40919 Fg Off **Bernard Stanley Maurice JONES**, RAF - Age 23. 898hrs.

At 23 years of age, Jones was the oldest aircrew member of the Squadron at the time. It will be recalled that he was with fellow Squadron member, Plt Off B M Browning, when the latter was killed in a motor car accident at Abbassia on 4 February 1940.

Tue 29/Wed 30 Apr 1941

BOMBER COMMAND

Raid on Mannheim, Germany (by 71 aircraft - 1 lost)

99 Squadron, RAF (Waterbeach, Cambridgeshire - 3 Group)

Wellington IC T2721 - took off at 2111 captained by Sgt F Hewitson, RAF. An SOS message was sent at 0314 and it is believed that T2721 was shot down into the sea off the English coast by an enemy night-fighter. An extensive search found no trace of the aircraft or its six crew, who are all commemorated on the Runnymede Memorial.

Rear Gunner: NZ401447 Sgt **Richard Raymond THOMAS**, RNZAF - Age 28. 115hrs. 10th op.

Wed 30 Apr 1941

MIDDLE EAST

Ground strafing mechanised transport and dumps east of Gazala, Libya, to Sollum, Egypt

274 Squadron, RAF (LG 10 [Gerawala], Egypt - 258 Wing, 204 Group)

Hurricane I V7555 - from 0600 the Squadron sent off a Hurricane every 15 minutes on strafing strikes, two of the aircraft failing to return. Contrary to some previously published accounts, V7555 was accidentally shot down by Allied anti-aircraft fire, not by enemy fighters. The pilot is buried at Tobruk.

Pilot: 42445 Fg Off **Douglas James SPENCE**, RAF - Age 20.

Wed 30 Apr/Thu 1 May 1941

COASTAL COMMAND

Anti-submarine patrol

612 Squadron, RAF (Wick, Caithness-shire - 18 Group)

Whitley V P5070 - took off for the patrol captained by Plt Off F L Arney, RAF, but recalled early due to a very sudden deterioration in weather, with sea fog and low cloud rapidly forming. There was some doubt as to whether P5070's wireless was working properly and at 0030 it flew into the side of a mountain at 2000 feet on Scaraben, 6 miles WSW of Dunbeath. All six crew died, the observer being buried at Wick.

Observer: NZ401213 Sgt **Douglas John McPHERSON**, RNZAF - Age 23. 151hrs. 10th op.

Sun 4 May 1941

COASTAL COMMAND

Free firing practice at splash targets

1 (Coastal) Operational Training Unit, RAF (Silloth, Cumberland - 17 Group)

Hudson I N7304 - captained by Plt Off M S Smith, RAF. At 1630, on recovering from a shallow dive, climbed very steeply, stalled and spun into the Solway Firth off Silloth, and caught fire. All six crew died, including the 2nd pilot, whose body was not recovered and thus is commemorated on the Runnymede Memorial. The accident report observed that there was a tendency for the bank on a Hudson to increase rapidly when executing tight turns, and in this case the pilot may have been caught unawares in a moment of overconfidence.

2nd Pilot: NZ40781 Sgt **Roderick Barker TAYLOR**, RNZAF - Age 22. 199hrs.

Taylor's brother, John Barker Taylor, died on 28 May 1943 while flying with 45 Group, RAF.

Sun 4/Mon 5 May 1941

MIDDLE EAST

Raid on Benina, Libya

38 Squadron, RAF (Shallufa, Egypt; detachment at Fuka - 257 Wing, 202 Group)

Wellington IC W5681 - took off from Fuka and at 0500, on the return flight, crashed at El Qasaba, 35 miles west of Fuka. The six crew are buried at El Alamein.

Captain: NZ391345 Sgt **Norman Franklin DIXON**, RNZAF - Age 25. 425hrs. 23rd op.
2nd Pilot: NZ40782 Sgt **Basil Alexander WALKER**, RNZAF - Age 21. 286hrs. 6th op.
Rear Gunner: NZ401467 Sgt **Herbert Owen TURNER**, RNZAF - Age 19. 121hrs. 9th op.

Tue 6 May 1941

BOMBER COMMAND

'Beat 7' - anti-shipping strike off the northern Dutch coast

110 Squadron, RAF (Wattisham, Suffolk - 2 Group)

Blenheim IV R3600 - took off with others at 1115 and upon locating a convoy, which included three destroyers, the Blenheims ran into heavy defensive fire from 3 miles out. During the unsuccessful attack R3600 was shot down into the sea. The air gunner's body later washed ashore onto the German coast, but no trace of the other two crew was found and they are commemorated on the Runnymede Memorial.

Pilot: 40955 Flt Lt **Edward Nation STEEL**, RAF - Age 24. 608hr. 14th op.

Tue 6/Wed 7 May 1941

BOMBER COMMAND

Raid on Hamburg, Germany (by 115 aircraft - 1 lost)

75 (NZ) Squadron, RAF (Feltwell, Norfolk - 3 Group)

Wellington IC R3169 - took off at approximately 2215-2235. On returning to England in 10/10ths cloud at 1000 feet, R3169's position was pinpointed as being at the mouth of the Humber River, after a variation in forecast winds had driven it off course. Turning south at 0445, it immediately struck one or two barrage balloon cables. The wrecked Wellington fell into the Humber near Trinity Sands, 6 miles ENE of Grimsby, Lincolnshire. An injured RAF Sgt rear gunner was the sole survivor of the six crew, the two pilots of which are buried at Grimsby.

Captain: NZ39930 Sgt **David Leo NOLA**, RNZAF - Age 26. 382hrs (185 solo on Wellington). At least 16 ops.

2nd Pilot: NZ40656 Sgt **Alexander Coutts MEE**, RNZAF - Age 23. 323hrs. 8th op.

Thu 8 May 1941

FIGHTER COMMAND

Training(?)

111 Squadron, RAF (Dyce, Aberdeenshire - 14 Group)

Spitfire I R6619 - took off at 1015, but at 1105 collided with another of the Squadron's Spitfires (X4931) and crashed SW of Aberdeen, killing the pilot, who is buried at Dyce. The pilot of X4931 was injured but landed safely.

Pilot: Aus402000 Sgt **Frederick Agnew Vance Mansfield DRUMMOND**, RAAF - Age 18.

Drummond's brother, Vance Drummond, died on 17 May 1967 while flying with 3 Sqn, RAAF.

Thu 8/Fri 9 May 1941

BOMBER COMMAND

Raid on Hamburg, Germany (by 188 aircraft - 5 lost)

214 Squadron, RAF (Stradishall, Suffolk - 3 Group)

Wellington IC R1226/L - took off at 2210 captained by Sqn Ldr F L H Eddison, DFC, RAF, and shot down over the Netherlands by a night-fighter at 0248, crashing at Anna Paulowna, 12km SE of Den Helder. The six crew were buried in a communal grave in the Huisduinen Cemetery, just to the west of Den Helder, but later reinterred at Bergen-op-Zoom.

2nd Pilot: NZ401315 Plt Off **Norman Mervyn Kebbell KIRKCALDIE**, RNZAF - Age 23. 235hrs. 3rd op.

83 Squadron, RAF (Scampton, Lincolnshire - 5 Group)

Hampden I X3062/Z - took off at 2225 on same raid as above, and shot down into the sea off Mablethorpe by a night-fighter at 2330 while on the outbound journey. All four crew are commemorated on the Runnymede Memorial.

Captain: 43034 Plt Off **Frederick Kennett GILL**, RAF - Age 26. 402hrs. 11th op.

Sat 10 May 1941

MIDDLE EAST

Returning with aircraft from Maintenance Flight to operational landing ground

55 Squadron, RAF (LG 95 [Ikingi], Egypt; detachment at LG 15 [Maaten Bagush satellite] - 202 Group)

Blenheim IV V5514 - one of two Squadron Blenheims, carrying two pilots and six passengers between them, which touched down at an intermediate landing ground because of engine trouble in one machine. The personnel from the unserviceable machine transferred to V5514, which then left for the intended destination. Unable to land there because of a dust storm, they diverted to Burg el Arab (LG 28) where dust and approaching darkness greatly reduced visibility. Plt Off J L van Breda, RAF, made an unsuccessful approach to land, whereupon the other pilot took over for another attempt. As the Blenheim finally touched down at 2020, it collided with a parked camouflaged Magister (R1974) of the Western Desert Communications Flight and burst into flames. Only one of the passengers survived. The seven who died are buried in a collective grave at Alexandria.

Pilot: 40938 Fg Off **Ralph Holden NICOLSON**, mid, RAF - Age 23. 606hrs solo.

Sat 10/Sun 11 May 1941

BOMBER COMMAND

Raid on Berlin, Germany (by 23 aircraft - 3 lost)

7 Squadron, RAF (Oakington, Cambridgeshire - 3 Group)

Stirling I N6010 - took off at 2211 and collided with a night-fighter over Germany, crashing near Cloppenburg, 50km SW of Bremen. The seven crew were buried at Cloppenburg, but later reinterred at Becklingen, 17km south of Soltau. The air gunner in the night-fighter also died, although his pilot survived.

Captain: NZ1068 Flt Lt **Neville WILLIAMS**, DFC mid, RNZAF - Age 25. 1179hrs. 37th op.

15 Squadron, RAF (Wyton, Huntingdonshire - 3 Group)
Stirling I N3654/B - took off captained by Wg Cdr H R Dale, RAF, on same raid as above, and shot down over the Netherlands by a night-fighter, crashing at 0030 onto a farm 500 metres west of Opmeer, 16km NE of Alkmaar. All seven crew died, the captain's body being recovered and buried at Bergen. The wrecked Stirling, however, buried itself deep in the soft soil, making recovery and identification of the other bodies impossible. They are therefore commemorated on the Runnymede Memorial. After the war villagers covered the crash site with a mound of earth, chained it off and marked the 'grave' with a cross bearing the names of all the crew.
Observer: NZ40189 Plt Off **Daniel McLean CAMPBELL**, RNZAF - Age 25. 303hrs. 16th op.
Air Gunner: NZ391860 Sgt **Eric Russell LUCAS**, RNZAF - Age 26. 178hrs. 29th op(?)

Sun 11/Mon 12 May 1941

BOMBER COMMAND

Raid on Hamburg, Germany (by 92 aircraft - 3 lost)
40 Squadron, RAF (Alconbury, Huntingdonshire - 3 Group)
Wellington IC R1330/H - took off at 2225 and shot down over Germany by a night-fighter, crashing at 0140 near Süderstapel, 19km NNW of Heide. An injured RAF Flt Sgt survived to be taken prisoner, but his five comrades died and are buried at Kiel.
Captain: NZ39923 Sgt **Roderick William FINLAYSON**, RNZAF - Age 24. 432hrs. 27th op.

40 Squadron, RAF (Alconbury, Huntingdonshire - 3 Group)
Wellington IC R1461/Z - took off at 2230 captained by Sgt F T Luscombe, RAF, on same raid as the above. Shot down into the North Sea at 0245, the bodies of three crew members later washed ashore onto the German coast. The other three, including the rear gunner, were lost without trace and are commemorated on the Runnymede Memorial.
Rear Gunner: NZ40623 Sgt **John David Campbell LONG**, RNZAF - Age 21. 70hrs.

75 (NZ) Squadron, RAF (Feltwell, Norfolk - 3 Group)
Wellington IC R1589 - took off at approximately 2320-0045 captained by Sqn Ldr R P Widdowson, RCAF, on same raid as above, and attacked from astern by a night-fighter on the way to the target. The rear gunner being mortally wounded, the Wellington abandoned the operation and returned to Feltwell, where he died shortly afterwards and was later buried. The other five crew were unhurt, including navigator Sgt **L A Lawton**, RNZAF, the only other New Zealander on board.
Rear Gunner: NZ402110 Sgt **Eric Francis GANNAWAY**, RNZAF - Age 21. 111hrs. 10th op.

Wed 14/Thu 15 May 1941

FIGHTER COMMAND

Practice night reconnaissance
256 Squadron, RAF (Squires Gate, Lancashire - 9 Group)
Defiant I T3955/R - piloted by Sgt P J Taylor, RAF, control apparently lost in deteriorating weather while flying in 9/10ths cloud at 6000 feet. The two crew died when at 0145 they abandoned the falling Defiant too low. Both are buried at Mansfield, Nottinghamshire. The aircraft crashed and caught fire at Nether Kellet, Bolton-le-Sands, 2$^1/_2$ miles SSW of Carnforth, Lancashire.
Air Gunner: NZ402108 Sgt **Erle Rutherford FREMLIN**, RNZAF - Age 19.

Thu 15 May 1941

NEW ZEALAND

Formation exercise
2 Service Flying Training School (ITS), RNZAF (Woodbourne)
Oxford II NZ1276 - collided with another of the School's Oxfords (NZ1246) while changing leader position at 1455, and dived steeply into the sea between White Bluffs and the Awatere River mouth, about 11 miles ESE of Blenheim. Floating wreckage was located shortly afterwards by an RNZAF launch which chanced to be proceeding through Cook strait at the time. However, the body of the pilot, the sole occupant, was not recovered and he is commemorated on the Provincial Memorial at Wellington. Oxford NZ1246 forced landed on a beach, its pilot, LAC **J J Fox** being injured.
Pilot: NZ41326 LAC **Alec HAIGH**, RNZAF - Age 25. 141hrs (23 solo on Oxford)
Fox died on 1 June 1942 while flying with 114 Sqn, RAF.

Thu 15/Fri 16 May 1941

BOMBER COMMAND

Raid on Hannover, Germany (by 100 aircraft - 3 lost)
103 Squadron, RAF (Newton, Nottinghamshire - 1 Group)
Wellington IC R1494 - took off at 2245 and port engine failed after being hit by flak over Emden. R1494 later crashed at Oude Pekala, 9km east of Veendam, Netherlands. Four of the crew were injured and taken prisoner, but the other two died and are buried in a joint grave at Groningen.
Captain: NZ40759 Plt Off **Raymond Gordon ECCLES**, RNZAF - Age 26. 382hrs. 11th op.

Sat 17 May 1941

FIGHTER COMMAND

Patrol

54 Squadron, RAF (Southend, Essex - 11 Group)

Spitfire IIA P7833(?) - took off at 1725 as one of a flight to patrol over Maidstone, Kent at 20,000 feet. Shortly afterwards the flight was ordered south to intercept enemy aircraft. As two pairs of the latter came into view below, the Spitfire leader ordered the formation to break and dive upon them. In the ensuing engagement P7833 was shot down, crashing at Biddenden, 11 miles SE of Maidstone. The pilot is buried at Hornchurch, Essex.

Pilot: NZ401390 Sgt **Earl Joseph KEAN**, RNZAF - Age 28. 24th op.

Sun 18 May 1941

MIDDLE EAST

Bombing raid on Derna airfield, Libya

45 Squadron, RAF (Fuka, Egypt - 204 Group)

Blenheim IV T2056(?) - took off at 0400 piloted by Fg Off J Beveridge, RAF, as one of two on the raid but became separated and failed to return. The three crew are commemorated on the Alamein Memorial.

Observer: NZ401737 Plt Off **Allan Herbert WISE**, RNZAF - Age 25. 85hrs. 2nd op(?)

Wed 21 May 1941

BOMBER COMMAND

Anti-shipping strike off Heligoland, Germany

107 Squadron, RAF (Great Massingham, Norfolk - 2 Group)

Blenheim IV (serial?) - took off at 1400 piloted by K Wolstenholme, RAF, with eight others. During the strike the observer was killed by flak. The Blenheim returned safely to base and the dead crew member was later buried nearby at Little Massingham.

Observer: NZ40746 Sgt **John Colin WILSON**, RNZAF - Age 27. 177hrs. 13th op.

Wilson's brother, Arthur Roy Wilson, died on 17 January 1942 while flying with 12 Operational Training Unit, RAF.

Thu 22 May 1941

FAR EAST

Oblique photographic exercise

100 Squadron, RAF (Seletar, Singapore)

Vildebeest III K6369 - took off at about 0830, but stalled off a turn during the exercise and crashed into the sea off the island of Palau Bukom, just to the SW of Singapore. All three crew are commemorated on the Singapore Memorial.

Pilot: NZ401782 Sgt **Cecil James McLaughlin SALT**, RNZAF - Age 28.

Fri 23 May 1941

MIDDLE EAST

Raid on Máleme airfield, Crete

45 Squadron, RAF (Fuka, Egypt - 204 Group)

Blenheim IV V5624 - took off at 1330 piloted by Plt Off P J Vincent, RAF, leading three others. The formation bombed the target but V5642 was last seen to dive and turn towards the mountains, as though taking evasive action. Its three crew are commemorated on the Alamein Memorial

Observer: NZ39970 Plt Off **Stuart Charles NIVEN**, RNZAF - Age 25. 351hrs. At least 21 ops.

Mon 26 May 1941

COASTAL COMMAND

Transit from Melgerdi to Kaldadarnes

98 Squadron, RAF (Kaldadarnes, Iceland - 18 Group)

Battle I P2330 - crashed into a cloud covered mountain at Bakkasel en route to base, killing the two crew and two passengers on board. All four were buried at the site of the crash, the pilot also being commemorated on the Runnymede Memorial, presumably because of the inaccessible nature of the grave site.

Pilot: 36201 Fg Off **Arthur Kavhan ROUND**, RAF - Age 26. 563hrs.

Round's brother, Heathcote George Round, died on 27 August 1944 while flying with 1661 Heavy Conversion Unit, RAF.

Tue 27 May 1941

MIDDLE EAST

Raid on Máleme airfield, Crete

45 Squadron, RAF (Fuka, Egypt - 204 Group)

Blenheim IV Z5896 - took off from Fuka North at 0300 piloted by Fg Off N W Pinnington, RAF, leading two others. Z5896 climbed away steeply but stalled, crashed and caught fire. The pilot and observer died and are buried at El Alamein. Although rescued from the wreckage, the air gunner was badly burned and died of injuries a fortnight later. The raid was cancelled as a result of the crash.

Observer: NZ402132 Plt Off **Howard Fleming IRVING**, RNZAF - Age 25. 117hrs. 9th op.

Bomb troops between Suda Bay and Máleme, Crete
14 Squadron, RAF (LG 21 [Qotafiya III], Egypt - 204 Group)
Blenheim IV T2338 - took off at 1430 with two others. One returned early with engine trouble, while the others bombed the airfield at Máleme after failing to locate the original target. They became separated and lost on the return journey and crashed in the Egyptian desert, two of the crew of T2338 baling out beforehand over the Qattara Depression, some 70 miles south of El Daba. Searchers found the crashed Blenheim three days later, but it was the morning of the sixth day before the observer, Sgt **M B Fearn**, RNZAF, was rescued. Later that afternoon the air gunner, Sgt McConnell, was also picked up, suffering badly from exposure. However, a further three days searching by aircraft and motorcycle revealed no trace of the pilot, although landing strips laid out near the wrecked Blenheim clearly indicated he survived the crash, and he is commemorated on the Alamein Memorial.
Pilot: 40931 Fg Off **Murray MacKENZIE**, RAF - Age 27. 717hrs. 50th op(?)
The Sgt McConnell above is incorrectly identified on page 36 of Volume Three of New Zealanders with the Royal Air Force, *as being Sgt J N McConnell, RNZAF, but he was actually in Canada at the time, and died on 15 June 1942 while flying with 601 Sqn, RAF.*

Raid on Máleme airfield, Crete
55 Squadron, RAF (LG 95 [Ikingi], Egypt; detachment at LG 15 [Maaten Bagush Satellite] - 202 Group)
Blenheim IV T2175 - piloted by Sgt J H Chesman, RAF, took off from Maaten Bagush shortly after 1430 with five others. Two of the bombers collided shortly after take off, leaving four to complete the raid. Two of these returned safely to base while the others became separated and lost over the desert. The crew of one baled out and safely reached their base two days later. The three of T2175 were less fortunate, however, their wrecked aircraft and bodies later being discovered by a searching 6 Sqn Lysander. They are now buried at El Alamein.
Observer: NZ401721 Sgt **Douglas Gordon CALLENDER**, RNZAF - Age 24.

Wed 28 May 1941

FIGHTER COMMAND

Ford to Manston, Kent
23 Squadron, RAF (Ford, Sussex - 11 Group)
Havoc I BB893 - piloted by Fg Off D A Willans, DFC, RAF, and entered an unintentional spin while carrying out a mock dogfight at low altitude at 1929. BB893 crashed and caught fire 400 yards SSW of Manston, killing the three crew, the pilot and air gunner of which are buried at Clymping, Sussex.
Air Gunner: NZ391859 Sgt **Gerald Bruce JOHNSON**, RNZAF - Age 21. 135hrs. 20 ops.

Thu 29 May 1941

NEW ZEALAND

Solo cross country from base to Stephens Island, Otaki and return
2 Service Flying Training School (ITS), RNZAF (Woodbourne)
Oxford II NZ1302 - took off at 0907 in good weather and was due back at 1115, but failed to return. Four others on a similar exercise all returned safely. Six aircraft searched for almost a week but found no trace of NZ1302 or its pilot, who was later commemorated on the Auckland Provincial Memorial. The trainer's fate was finally discovered over an Easter weekend almost 39 years later by Mr **Andy Cox** while deer hunting with three friends near Otaki Forks in the Tararua Ranges on 5 April 1980. It appeared that a wing had struck the side of Oriwa ridge near the summit, causing the Oxford to cartwheel and disintegrate in dense bush. The pilot's remains were buried at the site on the 15th in a brief service carried out by the RNZAF's principal chaplain, the Rev **Theo Carpenter**.
Pilot: NZ41945 LAC **Selwyn Allan SINCLAIR**, RNZAF - Age 20. 80hrs (34 solo - 14 on Oxford)

Thu 29/Fri 30 May 1941

MIDDLE EAST

Bomb raid on either Máleme, Crete, or Scarpanto on the Dodecanese island of Karpathos
37 Squadron, RAF (Shallufa, Egypt - 257 Wing, 202 Group)
Wellington IC W5622 - on beginning its take off run another of the Squadron's Wellington's (W5685) crossed its flightpath. They collided and immediately caught fire, their bomb loads starting to explode shortly afterwards. Of the two crews, four airmen died the night of the crash and another of his injuries the following day. The other seven were all injured to varying degrees, including the captain of W5622, who died on 5 June and is buried at Ismailia with the others.
Captain: NZ391350 Sgt **Henry Noel GOODALL**, RNZAF - Age 21. 396hrs. 25th op(?)

Sat 31 May 1941

FIGHTER COMMAND

Non-operational
23 Squadron, RAF (Ford, Sussex - 11 Group)
Havoc I BJ485 - piloted by Plt Off G H Robinson, RAF. When 3 miles off Rye the port wing and nose dropped violently on turning in cloud. During these manœuvres the rear gunner, neither secured to the aircraft nor wearing a parachute, was thrown out and killed. His body was later recovered and buried at Hawkinge, near Folkestone, Kent. The Havoc and its remaining crew returned to base and landed safely.
Air Gunner: NZ401460 Sgt **Anzac Piripi Tamihana LEWIS**, RNZAF - Age 23. 96hrs. 6 ops.

Sun 1 Jun 1941

FIGHTER COMMAND

Landing practice

41 Squadron, RAF (Catterick, Yorkshire - 13 Group)

Spitfire IIA P8163 - took off at 2040 and in poor visibility 40y minutes later crashed on the side of a hill 5 miles east of Northallerton. The pilot was taken to Catterick Military Hospital with severe head injuries, but died the following morning at 0600 and is buried at Catterick.

Pilot: NZ391826 Sgt **Eric Eugene CROKER**, RNZAF - Age 24. 325hrs. 41 ops.

Wed 4 Jun 1941

BOMBER COMMAND

Raid on De Kooy airfield, Netherlands

139 Squadron, RAF (Horsham St Faith, Norfolk - 2 Group)

Blenheim IV R3903/R - piloted by Plt Off W Baser, RAF, took off at 1900 with five others and one of two shot down into the sea by enemy fighters at 2008 west of Bergen aan Zee. All three crew are commemorated on the Runnymede Memorial.

Observer: NZ40949 Sgt **Arthur William SIMPSON**, RNZAF - Age 32. 201hrs. 19th op(?)

Thu 5 Jun 1941

MIDDLE EAST

Sgt **H N GOODALL** of 37 Sqn, RAF, died of injuries received in a take off crash on 29 May.

Sat 7 Jun 1941

FIGHTER COMMAND

Flight from Heston to Ford, Sussex, for an air-to-air firing practice

53 Operational Training Unit, RAF (Heston, Middlesex - 81 Group)

Spitfire I K9852 - took off as 'No 2' in the second of two sections of three. On descending towards Ford the airfield was found to be obscured by low cloud. Climbing away through cloud, the two sections became separated and the pilot of K9852 was last seen by his leader at that time. The missing fighter was later found to have crashed 3 miles SE of Tangmere, about 4 miles west of Ford. The pilot is buried at Tangmere.

Pilot: NZ402157 Sgt **Clifford Barnaby BARTON**, RNZAF - Age 24. 229hrs.

Sun 8 Jun 1941

COASTAL COMMAND

Circuits and landings at Langham

500 Squadron, RAF (Bircham Newton, Norfolk - 16 Group)

Blenheim IV V5689/F - in misty weather at 1635 flew into the ground at Holme-next-the-sea, 7 miles NW of Bircham Newton, and caught fire. All three on board died, the pilot being buried at Great Bircham.

Pilot: NZ2515 Fg Off **Frederick William HALL-JONES**, RNZAF - Age 21. 340hrs. 7 ops.

BOMBER COMMAND

Night cross country training exercise

11 Operational Training Unit, RAF (Bassingbourn, Cambridgeshire - 6 Group)

Wellington IC R1728 - took off at 1510, most probably from the satellite airfield at Steeple Morden, Hertfordshire, and failed to return. The route to be followed was from base to Helmsley, Yorkshire-5415N:0145E-5320N:0200E-Manby, Lincolnshire, with return to base by 2010. On 21 June the bodies of two of the six crew were found in a dinghy off the east coast of England but no trace was found of the others, including the captain, who is commemorated on the Runnymede Memorial.

Captain: 36210 Fg Off **William Alan FOSTER**, RAF - Age 21. 570hrs (518 solo - 41 on Wellington)

Lost on the same day, Hall-Jones and Foster had graduated together on No 6 course at Wigram.

Wed 11 Jun 1941

FIGHTER COMMAND

'Rhubarb' - low level patrol over France

258 Squadron, RAF (Redhill, Surrey - 11 Group)

Hurricane IIB Z3269/E (a presentation machine named *Conqueror*) - took off at 1250 with five others, operating in three sections of two, and believed shot down into the sea during an engagement with enemy fighters over the northern coast of Pas de Calais. The pilot is commemorated on the Runnymede Memorial.

Pilot: NZ39909 Fg Off **Wilfred Thomas EVERIST**, RNZAF - Age 23. 333hrs. 49th op.

Thu 12/Fri 13 Jun 1941

BOMBER COMMAND

Raid on Osnabrück, Germany (by 61 aircraft - 2 lost)

103 Squadron, RAF (Newton, Nottinghamshire - 1 Group)

Wellington IC T2996/C - took off at 2300 and shot down over the Netherlands by a night-fighter, crashing at 0037 onto the property of Mr Langendijk at Wervershoof, 10km west of Enkhuizen. All six crew died, but as the wreckage had buried itself deep in the ground only the body of the rear gunner was able to be recovered. The captain and the others are therefore commemorated on the Runnymede Memorial. Dutch municipal authorities established a common grave on the site, where a memorial was erected and, after the war, ceremoniously inaugurated.

Captain: 79509 Plt Off **Robert Stanley CHISHOLM**, RAFVR - Age 26. 489hrs. 19th op.

Sun 15 Jun 1941

MIDDLE EAST

Reconnaissance over Quatana-Kissqué-Damascus, Syria

208 Squadron, RAF (Gaza, Palestine - Palestine & Transjordan Command)

Hurricane I N2626 of 'A' Flight - took off at 1545 to repeat a reconnaissance undertaken earlier in the day, but from which that aircraft had failed to return. Continuing the flight at 1650, after refuelling at Rosh Pinna, N2626 also failed to return. While attempting to drop a message the Hurricane was brought down by Allied anti-aircraft fire (rather extraordinarily, a repeat of the fate which had befallen the earlier aircraft). According to *Dust Clouds in the Middle East*, as the pilot left the Hurricane after force-landing near the southern Lebanon town of Merajayoun, the day's tragedies were further compounded by his being shot dead by Free French colonial troops. It is not known where he was at first buried, on the following day, but his body was reinterred at Damascus during the first week of September.

Pilot: 40913 Fg Off **Cedric West HOLDSWORTH**, RAF - Age 25. 516hrs. 25th op.

Sun 15/Mon 16 Jun 1941

BOMBER COMMAND

Raid on Cologne, Germany (by 91 aircraft - 1 lost)

106 Squadron, RAF (Coningsby, Lincolnshire - 5 Group)

Hampden I AD863 - took off at 2300 and lost without trace, all four crew being commemorated on the Runnymede Memorial.

Captain: 43029 Plt Off **David Gray DICKIE**, RAF - Age 21. 353hrs. 6th op.

Ferry flight from Harwell to Malta via Gibraltar

15 Operational Training Unit, RAF (Harwell, Berkshire - 6 Group)

Wellington IC R3293 - took off from Gibraltar, but crashed into the sea near Kalafrana at 0215 as it was about to land on Malta. All six crew died, the captain's body and four others being recovered and buried on Malta. However, that of the New Zealand air gunner could not be found and he is commemorated on the Runnymede Memorial.

Captain: 964587 Sgt **Evan Bruce BEATTIE**, RAFVR - Age 21. 2nd op.

Air Gunner: 1261133 Sgt **Francis Ernest DRAKE**, RAF - Age 20. 2nd op.

Beattie was born in Singapore but came to New Zealand at an early age. He travelled to Britain in 1938 and enlisted in the RAF in 1940.

The other pilot on the flight was Sgt J R Bolton, RAF, and it is possible that he might have been in command, rather than Beattie. The crew may have been posted to the Middle East and making their way there by means of the delivery flight, although surviving records show both New Zealanders and their aircraft as being on the strength of 15 OTU at the time.

Mon 16 Jun 1941

BOMBER COMMAND

'Beat 7' - anti-shipping strike off East Frisian islands, Germany

21 Squadron, RAF (Watton, Norfolk - 2 Group)

Blenheim IV V6034/D - took off at 1319 piloted by Sgt E A R Leavers, DFM, RAF, and during the attack off Borkum with two others collided with the mast of a 'squealer' (radio-equipped early warning trawler), shearing off the outer starboard wing. V6034 cartwheeled into the sea, killing the three crew. Two bodies later washed ashore, but that of the observer was never found and he is commemorated on the Runnymede Memorial.

Observer: NZ40727 Sgt **Ian OVERHEU**, DFM, RNZAF - Age 23. 290hrs. 28th op.

Wed 18 Jun 1941

FIGHTER COMMAND

Aerobatic display

485 Squadron, RNZAF (Leconfield, Yorkshire - 12 Group)

Spitfire IIA P7758 - during an aerobatic display at 1600 for the benefit of the pilot's visiting brother, dived into the ground from 800 feet while demonstrating a slow roll. The pilot is buried at Leconfield.

Pilot: NZ401257 Sgt **Kevin Desmond COX**, RNZAF - Age 20.

Fri 20/Sat 21 Jun 1941

MIDDLE EAST

Raid on Benghazi Harbour, Libya

38 Squadron, RAF (Shallufa, Egypt - 257 Wing, 204 Group)

Wellington IC T2748/K - took off at 2000 captained by Sgt C G E Newton, RAF. Aircraft 'Q' picked up a series of messages from the Wellington between 0245 and 0305, indicating a starboard engine failure. Nothing further was heard and a Maryland on reconnaissance could later find no trace of the aircraft. Subsequently it was discovered that T2748 had crashed about 100 miles (kilometres?) east of the target. Two crew members succeeded in baling out and were captured, but their comrades perished. After hostilities no sign of the wreckage could be found, nor any trace of the four who died, including the observer, and they are commemorated on the Alamein Memorial. Surviving records show the Wellington's serial as C2478, but this belongs to an RE8 of the First World War and is believed to be an error for T2748.

Observer: NZ401457 Sgt **David Aubrey CONNOLLY**, RNZAF - Age 23. 160hrs. 6th op.

BOMBER COMMAND

Raid on Kiel, Germany (by 115 aircraft - 2 lost)

218 Squadron, RAF (Marham, Norfolk - 3 Group)

Wellington IC R1339/J - took off at 2220, the last contact being a message sent at 0335 calling for help when 60 miles away from the enemy coast. Assumed to have gone down in the North Sea, all six crew are commemorated on the Runnymede Memorial.

Captain: NZ40233 Sgt **Gordon Grant JILLETT**, RNZAF - Age 23. 345hrs. 19th op.

218 Squadron, RAF (Marham, Norfolk - 3 Group)

Wellington IC R1713/V - took off at about 2220 on same raid as the above. Lost without trace after sending a message asking for help, its six crew are commemorated on the Runnymede Memorial.

Captain: NZ40224 Sgt **Mason John FRASER**, RNZAF - Age 23. 427hrs. 22nd op.
Observer: NZ391327 Sgt **Desmond Aubrey DACRE**, RNZAF - Age 26. 398hrs. 22nd op.

Sat 21 Jun 1941

NEW ZEALAND

Solo cross country

1 Service Flying Training School (ITS), RNZAF (Wigram)

Oxford II NZ1296 - took off at 1010, but the port engine failed at 150 feet and the port wing dropped as it dived into the ground near Halswell. The pilot is buried at Dunedin. Subsequent investigation revealed that the fuel cock to the port tank had been left in the off position.

Pilot: NZ405849 LAC **Albert MAHER**, RNZAF - Age 22. 74hrs (31 solo - 11 on Oxford).

Mon 23 Jun 1941

FIGHTER COMMAND

Patrol of Hardelot-Le Touquet area, France, to cover withdrawal of a bomber force

485 Squadron, RNZAF (Leconfield, Yorkshire - 12 Group)

Spitfire IIA P7975 - took off at 1300 from West Malling, Kent, with four others and failed to return. Seen to emit smoke after being attacked, roll over on to its back and spin into the sea 10 miles SW of Boulogne. A patrol over the crash site some ten minutes later failed to find any sign of P7975 or its pilot, although a greenish oil was discernible on the sea. The pilot is commemorated on the Runnymede Memorial.

Pilot: NZ40753 Sgt **Richard John BULLEN**, RNZAF - Age 29. 277hrs.

Wed 25 Jun 1941

FIGHTER COMMAND

'Circus' - escort to Blenheim bombers on raid to St Omer, France

616 Squadron, RAF (Westhampnett, Sussex - 11 Group)

Spitfire IIA P7327 - took off at 1549 with 10 others and, as 'Red 4', failed to return. Believed to have been shot down by fighters on becoming detached from the formation after an engagement with the enemy inland from Le Touquet. P7327 was lost without trace, the pilot being commemorated on the Runnymede Memorial.

Pilot: NZ401249 Sgt **Ronald Leslie BREWER**, RNZAF - Age 29.

Thu 26 Jun 1941

COASTAL COMMAND

'C' Patrol - anti-shipping sweep off Brest, France

53 Squadron, RAF (St Eval, Cornwall - 19 Group)

Blenheim IV V6087/O - took off at 1110 with one other, both of which failed to return. Brought down into the sea, the observer and air gunner being lost without trace and commemorated on the Runnymede Memorial. The pilot was seriously wounded and taken prisoner. He died four days later in the Marine Hospital at Lorient and is buried in that city.

Pilot: NZ40974 Plt Off **Dennis Trevelyan HERRICK**, GM, RNZAF - Age 28. 320hrs. 13th op.
It will be recalled that Herrick was awarded the George Medal for his attempts to rescue the crew of a crashed Oxford at Taieri on 21 June 1940. Two of his brothers, Brian Henry Herrick and Michael James Herrick, were killed while serving as members of the RAF. Brian died while flying with 272 Sqn, RAF, on 24 November 1940 and Michael while flying with 305 Sqn, RAF, on 16 June 1944.

Thu 26/Fri 27 Jun 1941

BOMBER COMMAND

Raid on Kiel, Germany (by 41 aircraft - 2 lost)
61 Squadron, RAF (Hemswell, Lincolnshire - 5 Group)
Manchester I L7304 - took off at 2301 captained by Fg Off K G Webb, RAF, and brought down near Brunsbüttel. The six crew died, the air gunner, and possibly some of the others, being buried there, but all were later reinterred at Kiel.
Air Gunner: NZ401807 Sgt **Francis Sydney HASLEMORE**, RNZAF - Age 21. 81hrs. 4th op.

97 Squadron, RAF (Coningsby, Lincolnshire - 5 Group)
Manchester I L7374 - took off at 2336 on same raid as the above, a message later advising that an RAF Sgt in the crew had been killed during a night-fighter attack. L7374 later crashed into the sea off Westerhever, Germany, killing the remaining six crew. The bodies of three were later recovered, including that of Sgt **C K McKenzie** which washed ashore near Flensburg. He was buried in the area, but later reinterred at Ohlsdorf, near Hamburg. The other four are commemorated on the Runnymede Memorial.
Captain: 40613 Flt Lt **Francis Edwin EUSTACE**, DFC, RAF - Age 25. 878hrs. 34th op.
Air Gunner: NZ401811 Sgt **Cecil Keith McKENZIE**, RNZAF - Age 23.
Air Gunner: NZ391862 Sgt **Peter Victor McLAREN**, DFM, RNZAF - Age 21. 271hrs. 33rd op.

Raid on Düsseldorf, Germany (by 44 aircraft - 1 lost)
150 Squadron, RAF (Newton, Nottinghamshire - 1 Group)
Wellington IC R1644/L - took off at 2305 and brought down by flak over the Netherlands at 0400, crashing at Grubbenvorst, on the west bank of the Maas, 6km NNW of Venlo. All six crew were buried in the Venlo Military Cemetery, but later reinterred in Jonkerbos War Cemetery, 4km SW of Nijmegen.
Captain: 79514 Plt Off **John Winston SIEVERS**, DFC, RAFVR - Age 23. 19th op.

Sun 29/Mon 30 Jun 1941

BOMBER COMMAND

Night cross country training
14 Operational Training Unit, RAF (Cottesmore, Rutland - 7 Group)
Hampden I P4300 - piloted by Plt Off J S Curl, RAF, and last heard from at 0103 when over the North Sea, its four crew being commemorated on the Runnymede Memorial.
WOpAG: NZ39288 Sgt **Albert Richard TOUT**, RNZAF - Age 22. 50hrs.

Raid on Hamburg, Germany (by 28 aircraft - 6 lost)
40 Squadron, RAF (Alconbury, Huntingdonshire - 3 Group)
Wellington II W5456/P - took off at 2245 and lost without trace, its five crew being commemorated on the Runnymede Memorial.
Captain: NZ391821 Flt Lt **Angus Bruce BAIRD**, RNZAF - Age 25. 564hrs. 34th op.
Baird's brother, Francis Stephen Baird, died on 17 January 1944 while flying with 2 Sqn, RNZAF.

Mon 30 Jun 1941

COASTAL COMMAND

Plt Off **D T HERRICK**, prisoner of war in France, died of wounds received when his aircraft was brought down on the 26th

Search for dinghy at approximate position 5325N:0230E
608 Squadron, RAF (Thornaby, Yorkshire - 18 Group)
Blenheim IV Z5982 - took off at 1139 piloted by Plt Off Sir I W MacRobert, RAF, but nothing further heard and presumed to have been lost at sea. Z5982 was carrying an additional crew member and all four are commemorated on the Runnymede Memorial.
Air Gunner: 76320 Fg Off **Rowland Keith KEATING**, mid, RAFVR - Age 31.

BOMBER COMMAND

Cloud cover raid on Kiel, Germany
35 Squadron, RAF (Linton-on-Ouse, Yorkshire - 4 Group)
Halifax I L9499/Q - took off at 1010 and seen to be attacked by fighters near the target area, and catch fire. An RAF Sgt survived to be taken prisoner, but his six comrades perished and are buried at Kiel.
Captain: 42768 Flt Lt **Thomas Douglas Inglis ROBISON**, RAF - Age 23. 550hrs. 36th op.

Tue 1/Wed 2 Jul 1941

BOMBER COMMAND

Raid on Brest, France (by 52 aircraft - 2 lost)

149 Squadron, RAF (Mildenhall, Suffolk - 3 Group)

Wellington IC R1408/J - took off at 2241 captained by Plt Off J E Horsfield, RAF, and brought down near Plouzane, 10km west of Brest. All six crew are buried at Brest.

2nd Pilot: NZ401768 Sgt **Brian Daniel James KENNEDY**, RNZAF - Age 25. 288hrs. 8th op.

Wed 2 Jul 1941

INDIA

Formation practice

20 Squadron, RAF (Begumpet, India)

Blenheim I L8404 - on completion of the exercise, began to carry out mock quarter and astern attacks on another Blenheim at 8000 feet. Shortly afterwards, when climbing to 9000 feet, the aircraft rolled over into an inverted spin. Failing to recover, it crashed and burst into flames at Ghizri, killing the three crew, all of whom are buried at Karachi.

Pilot: 40156 Flt Lt **Mark Walter Howchin UPHILL**, RAF - Age 26.

Uphill's brother, Paul Lailey Uphill, reportedly died in an aircraft accident in England in 1939, although this may have been a civil accident and not a military one.

Thu 3 Jul 1941

FIGHTER COMMAND

'Circus' - escorting 12 Blenheims on raid to Hazebrouck, Northern France

611 Squadron, RAF (Hornchurch, Essex - 11 Group)

Spitfire V R7293 (a presentation machine named *Sans Tache*) - took off at 1055 with eleven others. During the operation the Squadron dived to attack enemy aircraft. It was itself then bounced by another enemy formation and lost R7293 during the combats which followed, the pilot being buried at Longuenesse, on the western outskirts of St Omer.

Pilot: NZ401281 Sgt **Martin Kean McHUGH**, RNZAF - Age 23. 270hrs. 34th op.

Thu 3/Fri 4 Jul 1941

BOMBER COMMAND

Raid on Essen, Germany (by 90 aircraft - 6 lost)

75 (NZ) Squadron, RAF (Feltwell, Norfolk - 3 Group)

Wellington IC W5621 - took off at 2308 and lost without trace, all six crew being commemorated on the Runnymede Memorial.

Captain: NZ391846 Sgt **Ian Laurie REID**, RNZAF - Age 23. 434hrs. 27th op.
Observer: NZ40945 Sgt **John Ross NATION**, RNZAF - Age 22. 284hrs. 18th op.
Rear Gunner: NZ402111 Sgt **Roydon Horatio HARTSTONE**, RNZAF - Age 29. 155hrs. 18th op.

Reid's brother, Francis Graham Reid, died on 12 March 1944 while flying with 887 Sqn, FAA.

Sun 6/Mon 7 Jul 1941

BOMBER COMMAND

Raid on Brest, France (by 109 aircraft - 4 lost)

12 Squadron, RAF (Binbrook, Lincolnshire - 1 Group)

Wellington II W5360/U - took off at 2325 captained by Sqn Ldr A G G Baird, RAF, and brought down near Guilers, 7km NW of Brest. All six crew are buried at Brest.

2nd Pilot: NZ401290 Sgt **Charles Hepburn ROBSON**, RNZAF - Age 24. 263hrs. 5th op.

Raid on Munster, Germany (by 47 aircraft - 2 lost)

115 Squadron, RAF (Marham, Norfolk - 3 Group)

Wellington IC R1063/D - took off at 2326 and sent an emergency message at 0045 advising that it 'was going down in sea.' Although an extensive search was conducted no trace of R1063 was found. Two bodies were later washed up on the Dutch island of Schiermonnikoog, that of the captain on 5 August, both being buried there. A third body washed ashore elsewhere, but the other three airmen were lost without trace and are commemorated on the Runnymede Memorial.

Captain: NZ391842 Sgt **Oswald Arthur MATTHEWS**, RNZAF - Age 20. 433hrs. 25th op.

40 Squadron, RAF (Alconbury, Huntingdonshire - 3 Group)

Wellington IC N2843/L - took off at 2335 on same raid as the above and shot down by a night-fighter over the Dutch coast, crashing in flames into the sea off the island of Texel. The RAF Sgt wireless operator, although wounded, baled out and landed on the island, and was captured. The bodies of two crew later washed ashore, but of the three others, including the captain, no trace was found and they are all commemorated on the Runnymede Memorial.

Captain: NZ401417 Plt Off **John Edwin MacKenzie STEEDS**, RNZAF - Age 29. 288hrs. 7th op.

Mon 7 Jul 1941

MIDDLE EAST

Ground strafing of Sidi Azeiz and Gambut airfields in NE Libya

73 Squadron, RAF (Sidi Heneish, Egypt - 204 Group)

Hurricane I V7757 - took off from LG 02 (Sidi Barrani) at about 1730 as 'No 2' of 'Blue' section, accompanied by 'Red' and 'Yellow' sections. The attack was unwittingly opened on the incorrect airfield and, recognising their error, Blue and Yellow sections withdrew, losing sight of Red section as they did so, to carry out the intended raid on Gt Gambut. Strafing of enemy troops and transport followed, but of the six despatched only a single damaged Hurricane survived to return. All five fighters lost were believed to have been brought down by flak. The pilot of V7757 was lost without trace and is commemorated on the Alamein Memorial. Two of the others died of their injuries the following day.

Pilot: 65512 Plt Off **Rodney William Kinross WHITE**, RAFVR - Age 26.

73 Squadron, RAF (Sidi Heneish, Egypt - 204 Group)

Hurricane I W9197 - was 'No 2' of 'Yellow' section on the raid described above. The pilot died the day after being shot down and was buried at Bardia in a joint grave with the other pilot who also died that day. Both were later reinterred at Halfaya Sollum.

Pilot: 785002 Sgt **Gordon Archibald JUPP**, RAFVR - Age 28. 235hrs. 19th op.

Mon 7/Tue 8 Jul 1941

BOMBER COMMAND

Raid on Cologne, Germany (by 114 aircraft - 3 lost)

9 Squadron, RAF (Honington, Suffolk - 3 Group)

Wellington IC R1040 - took off at 2340 and collided with a night-fighter over Maastricht, Netherlands, and crashed. All six crew were buried at Venlo, but later reinterred in Jonkerbos War Cemetery, 4km SW of Nijmegen.

Captain: NZ401386 Plt Off **Douglas James JAMIESON**, RNZAF - Age 26. 318hrs. 14th op.
2nd Pilot: NZ401388 Sgt **Wilfred George JESSON**, RNZAF - Age 27.

Tue 8 Jul 1941

FIGHTER COMMAND

'Circus 39' - escorting three Stirling bombers to Mazingarbe or Lens, France

485 Squadron, RNZAF (Redhill, Surrey - 11 Group)

Spitfire IIA P8470/S (a presentation machine named *Foremost*) - took off at 0540 with 11 others and last seen at about 0700 off Gravelines, 19km WNW of Calais. The pilot is commemorated on the Runnymede Memorial.

Pilot: NZ39067 Sgt **William Nisbet HENDRY**, RNZAF - Age 23. 250hrs.
Hendry's brother, John Lawrence Hendry, died on 31 August 1943 while flying with 106 Sqn, RAF.

Escorting three Stirlings to Lille, France

611 Squadron, RAF (Hornchurch, Essex - 11 Group)

Spitfire V R7277 - took off at 1440 and lost without trace, the pilot being commemorated on the Runnymede Memorial.

Pilot: NZ39788 Sgt **Brian William FEELY**, RNZAF - Age 23. 273hrs. 53rd op.

Tue 8/Wed 9 Jul 1941

BOMBER COMMAND

Raid on Munster, Germany

9 Squadron, RAF (Honington, Suffolk - 3 Group)

Wellington IC T2973/G - took off at 2331 captained by Sgt B G Pitt, RAF, and brought down over Germany, crashing at Willich, 16km west of Düsseldorf. All six crew were buried in the prisoner of war cemetery at Neuss, about 10km SE of Willich, but later reinterred at Rheinberg, 12km south of Wesel.

2nd Pilot: NZ401268 Sgt **John Arthur GRADY**, RNZAF - Age 19. 273hrs. 9th op.

Fri 11 Jul 1941

FIGHTER COMMAND

Sweep over northern France

485 Squadron, RNZAF (Redhill, Surrey - 11 Group)

Spitfire IIA P7773/D - took off at 1400 with 12 others. The Squadron was attacked by enemy fighters at 23,000 feet over Gravelines. P7773 was lost without trace over the Straits of Dover, the pilot being commemorated on the Runnymede Memorial.

Pilot: 44231 Plt Off **Charles STEWART**, RAF - Age 24. 438hrs. 138th op.
Stewart's brother, Daniel Stewart, died on 14 March 1942 in a motor vehicle accident in Australia while returning from the Far East.

Sat 12 Jul 1941

NEW ZEALAND

RNZAF Station Whenuapai

Died of natural causes at Auckland Public Hospital and is buried at that city.

(Patrolman): NZ412050 LAC **Louis Frants Jocef PFEFFER**, RNZAF - Age 46.

The CWGC register gives Pfeffer's age as 43.

MIDDLE EAST

Delivery flight from England via Malta (to Egypt?)

15 Operational Training Unit, RAF (Harwell, Berkshire - 6 Group, Bomber Command)

Wellington IC Z8775 - piloted by Sgt E O Townsend, RAF. Crashed and caught fire immediately after take off from Safi at 2016 when flaps were prematurely retracted. All six crew are buried on Malta.

Air Gunner: NZ40957 Sgt **Ralph Winston ASKIN**, RNZAF - Age 23. 117hrs. 5th op(?)

Askin, previously with 15 OTU, had been posted from Bomber Command to the Middle East three days earlier and was making his way there by means of the delivery flight. The aircraft is noted in contemporary records as still being on the strength of 15 OTU at the time of its demise.

Tue 15/Wed 16 Jul 1941

BOMBER COMMAND

Raid on Duisburg, Germany (by 38 aircraft - 5 lost)

57 Squadron, RAF (Feltwell, Norfolk - 3 Group)

Wellington IC R1624 - took off at 2300 and later sent an unclear message, after which nothing further was heard. R1624 was brought down over the Belgian-Dutch border at 0131 and crashed alongside the Hamont to Bree road, killing the six crew. The captain, probably with the others, was buried in the Netherlands at Bocholt, 4km NW of Bree, but later all were reinterred at Heverlee, Belgium.

Captain: NZ40985 Sgt **Gordon David OSBORNE**, RNZAF - Age 20. 354hrs. 16th op.

75 (NZ) Squadron, RAF (Feltwell, Norfolk - 3 Group)

Wellington IC R3171 - took off at approximately 2245-2340 on same raid as the above and lost without trace. Two bodies later washed ashore onto the Dutch coast, one being that of the captain, who was buried at Petten Cemetery, 29km south of Den Helder, but later reinterred at Bergen-op-Zoom. Their four comrades are commemorated on the Runnymede Memorial.

Captain: NZ391833 Sgt **Robert Ewen Ernest FOTHERINGHAM**, RNZAF - Age 29. 411hrs. 29th op.

WOpAG: NZ40101 Sgt **Sydney Allan DYER**, RNZAF - Age 19. 122hrs. 14th op.

Front Gunner: NZ401227 Sgt **Phillip Edgar HARE**, RNZAF - Age 19. 169hrs. 21st op.

Rear Gunner: NZ40923 Sgt **Douglas Malcolm MacKINNON**, RNZAF - Age 20. 209hrs. 15th op.

Interestingly, this bomber crew did not include a single member of the RAF. In addition to the RNZAF personnel it included a member each of the RAAF and RCAF, a sign of the increasing influx of aircrew from the Dominions.

75 (NZ) Squadron, RAF (Feltwell, Norfolk - 3 Group)

Wellington IC W5663 - took off at 2305 captained by Plt Off W J Rees, RAF, on same raid as the above and hit by flak over the target area, then shot up by a night-fighter. Three of the crew were wounded in the latter attack, the 2nd pilot dying almost immediately and the front gunner 15 days later. The RCAF navigator fell out of the damaged aircraft and was taken prisoner on parachuting to earth. At 0335 the Wellington returned safely to base, the 2nd pilot being buried there a day or two later.

2nd Pilot: NZ401278 Sgt **David Campbell JOYCE**, RNZAF - Age 22. 217hrs.

Plt Off T B Robertson, RNZAF, also as a 2nd pilot, lost his life in this aircraft when 75 Sqn raided Cologne on the night of 15/16 October 1941.

218 Squadron, RAF (Marham, Norfolk - 3 Group)

Wellington IC R1536/G - took off at 2315 captained by Flt Lt J Stokes, RAF, on same raid as the above and shot down over the Netherlands by a night-fighter at 0055, crashing at Roggel, 10km NW of Roermond. All six crew were buried at Venlo, but later reinterred in Jonkerbos War Cemetery, 4km SW of Nijmegen.

Rear Gunner: NZ40758 Sgt **Henry David Patrick DYER**, RNZAF - Age 22.

Wed 16/Thu 17 Jul 1941

BOMBER COMMAND

Raid on Hamburg, Germany (by 107 aircraft - 5 lost)

150 Squadron, RAF (Snaith, Yorkshire - 1 Group)

Wellington IC R1495/B - took off at 2215 and shot down by flak over Germany at 0130, falling into the River Elbe. Only an RAF Plt Off survived and was taken prisoner. His five comrades are buried at Ohlsdorf, near Hamburg.

Captain: 43713 Plt Off **Edward Bethridge TOPP**, RAFVR - Age 24. 437hrs. 16th op.

2nd Pilot: NZ402183 Sgt **Alan Richard HARRIS**, RNZAF - Age 23. 223hrs. 2nd op(?)

Topp's surname is recorded as such in contemporary RNZAF and RAF documents but it appears in the CWGC register as Bethridge-Topp.The register dates the death of both airmen as the 16th.

40 Squadron, RAF (Alconbury, Huntingdonshire - 3 Group)
Wellington IC X9630/J - took off at 2320 on same raid as the above and during its return was shot down into the sea off the Dutch coast by a night-fighter. Only the RCAF observer survived and was taken prisoner. The bodies of three crew, including that of the captain, were later recovered and are buried at Den Burg on the island of Texel. Two others were lost without trace and are commemorated on the Runnymede Memorial.
Captain: NZ401369 Sgt **Alan William Porteous BIRD**, RNZAF - Age 22. 334hrs. 15th op.

Thu 17 Jul 1941

FIGHTER COMMAND

Offensive patrol
54 Squadron, RAF (Hornchurch, Essex - 11 Group)
Spitfire VA W3216 (a presentation machine named *Maidenhead & District*) - took off at 1605 with six others on the patrol, but crashed at Hawkinge, Kent, upon returning at 1705. The pilot is buried at Hornchurch.
Pilot: NZ401392 Sgt **Francis Mackay LAING**, RNZAF - Age 29. 59th op.

COASTAL COMMAND

Anti-shipping strike off St Nazaire, France
217 Squadron, RAF (St Eval, Cornwall - 19 Group)
Beaufort I AW189/X - took off at 2015 piloted by Flt Lt T F Kerr, RAF. Last seen making a climbing turn to port about 4 miles west of the target and believed to have been brought down by the intense flak barrage over the harbour. All four crew are buried at Escoublac-La-Baule, about 12km west of St Nazaire.
Observer: NZ401234 Sgt **Allan Douglas SCOTT**, RNZAF - Age 20. 89hrs. 17th op.

Fri 18 Jul 1941

FIGHTER COMMAND

Searchlight homing exercise under r/t control
54 Operational Training Unit, RAF (Church Fenton, Yorkshire - 81 Group)
Blenheim IF L8729 - was unwittingly vectored over the Cleveland hills, whilst circling at 1000-1500 feet to view the searchlight beams, and at 0435 flew into high ground near Cockayne ridge, 4 miles south of Chop Gate, about 12 miles east of Northallerton. The pilot is buried at Kirby Wharfe.
Pilot: NZ402908 Plt Off **Alexander Douglas McMURTRIE**, RNZAF - Age 21. 174hrs (97 solo)

BOMBER COMMAND

Cloud cover raid to Wesel, Germany
15 Squadron, RAF (Wyton, Huntingdonshire - 3 Group)
Stirling I N6030/P - took off captained by Fg Off S D Marshall, RAAF, but en route the cloud began to dissipate and the raid was abandoned. For reasons unknown, on the return journey N6030 crashed into the sea. Four bodies later washed up on the Dutch coast, but no trace of the other three, including the 2nd pilot, was found and they are commemorated on the Runnymede Memorial.
2nd Pilot: NZ39635 Sgt **Alfred James HIGGISON**, RNZAF - Age 24. 278hrs. 9th op.

Fri 18/Sat 19 Jul 1941

FIGHTER COMMAND

First night solo on type
54 Operational Training Unit, RAF (Church Fenton, Yorkshire - 81 Group)
Blenheim IF K7090 - while orbiting a beacon near the aircraft, entered a gradual diving turn and flew into the ground $1^1/_2$ miles north of the base. The pilot is buried at Kirby Wharfe. Due to the presence of enemy aircraft in the vicinity the airfield controller had had all lights extinguished, being unaware that the pilot of K7090 was inexperienced in night flying. Although death is recorded as being on the 18th, the actual time of this accident is unknown. If it occurred shortly after midnight, rather than before, then the event should appear under the night of 17/18 July.
Pilot: NZ402238 Sgt **Cyril John NEIGHBOUR**, RNZAF - Age 24. 232hrs solo.

Sat 19 Jul 1941

BOMBER COMMAND

Anti-shipping strike off The Hague, Netherlands
105 Squadron, RAF (Swanton Morley, Norfolk - 2 Group)
Blenheim IV V6039 - took off at 1122 and shot down into the sea by ship's flak during an attack on a convoy of 14 ships. The bodies of all three crew later washed up on Dutch shores, that of the pilot being buried at Noordwijk.
Pilot: NZ401790 Sgt **Ronald West TAYLOR**, RNZAF - Age 25. 346hrs.

Sun 20/Mon 21 Jul 1941

BOMBER COMMAND

Practising night landings
10 Operational Training Unit, RAF (Abingdon, Berkshire - 6 Group)
Whitley V N1527 - struck trees at 0130 and crashed 2 miles south of the base. All five crew died, both pilots being buried at Brookwood in Woking, Surrey.
Captain: NZ403444 Plt Off **Peter Maxwell GRAY**, RNZAF - Age 24. 164hrs.
2nd Pilot: NZ40966 Plt Off **Harry Gordon Compton FOOKS**, RNZAF - Age 23. 198hrs.
Fooks' brother, Ernest Berjeu Fooks, died on 7 January 1944 while flying with 180 Sqn, RAF.

Mon 21/Tue 22 Jul 1941

BOMBER COMMAND

Navigation exercise
17 Operational Training Unit, RAF (Upwood, Huntingdonshire - 7 Group)
Blenheim IV L9412 - presumed to have crashed into the sea, since the body of a crew member later washed ashore onto the dyke at Bensersiel, opposite the German island of Langeoog. The pilot and other crew member were not found, however, and they are commemorated on the Runnymede Memorial. The intended route was base-York-Helmsley-5410N:0200E-5320N:0200E-Market Rasen-base.
Pilot: NZ402155 Sgt **Charles William ANDREWS**, RNZAF - Age 25. 188hrs.

Tue 22 Jul 1941

FIGHTER COMMAND

Night flying test
96 Squadron, RAF (Cranage, Cheshire; detachment at Squires Gate, Lancashire - 9 Group)
Defiant I T4071 - took off from Squires Gate piloted by Plt Off R Smithson, RAF, and at 1859 flew into the side of Eddisbury Hill, a mile north of Delamere, about 12 miles ENE of Chester. The Defiant exploded on impact and caught fire, killing both crew, the gunner being buried at Byley.
Air Gunner: NZ40208 Sgt **Ivan Norton ROBINSON**, RNZAF - Age 20. 162hrs.

Tue 22/Wed 23 Jul 1941

FAR EAST

Flight from Port Victoria, Seychelles to Singapore
205 Squadron, RAF (Seletar, Singapore)
Catalina I W8426/U - took off at 0200 captained by Flt Lt E W Young, RAF, but about 10 minutes later, while heading back downwind at 1000 feet towards Port Victoria, crashed in flames into the sea 6 miles to the NE. All ten crew are commemorated on the Singapore Memorial. It is not known why the flying boat had reversed course towards its base so soon after departure.
3rd Pilot: 740053 Flt Sgt **Bruce Deans LYON**, RAFVR - Age 28. 132hrs.
Although other crew members are recorded in the CWGC register as having died on the 23rd, Lyon's date is incorrectly given as the 22nd.

Wed 23 Jul 1941

BOMBER COMMAND

Daylight raid on the battle cruiser *Scharnhorst* at La Pallice, France (by 6 aircraft - 1 lost)
15 Squadron, RAF (Wyton, Huntingdonshire - 3 Group)
Stirling I N6038/R - captained by Fg Off R B Campbell, RAAF, and came down into the North Sea, possibly through damage inflicted by flak or fighter, or a combination of both. Six of the crew are commemorated on the Runnymede Memorial. The observer's body was recovered on or near the Frisian island of Amrum, buried at Nebel, but later reinterred at Kiel, Germany. An official report originally stated N6038 had crashed off Milford Haven, Pembrokeshire, but this was later corrected.
Observer: NZ40193 Plt Off **Donald Alfred LEWIS**, RNZAF - Age 23. 309hrs. 26th op.

Thu 24 Jul 1941

BOMBER COMMAND

Raid against battle cruiser *Gneisenau* and heavy cruiser *Prinz Eugen* at Brest, France (by 100 aircraft - 12 lost)
40 Squadron, RAF (Alconbury, Huntingdonshire - 3 Group)
Wellington IC T2986/A - took off at 1030 and hit by flak during its bombing run, caught fire and blew up over St Pierre-Quilbignon, immediately SW of Brest. All six crew are buried at Brest.
Captain: NZ391832 Sgt **Mervyn EVANS**, DFM, RNZAF - Age 25. 492hrs. 30th op.

75 (NZ) Squadron, RAF (Feltwell, Norfolk - 3 Group)
Wellington IC N2854 - took off at about 1050 on same raid as the above and dropped out of formation just prior to reaching the target, after being hit by flak. An attack by fighters followed and N2854 was last seen diving vertically into the sea. All six crew are commemorated on the Runnymede Memorial.
Captain: NZ401033 Sgt **Donald Frederick STREETER**, RNZAF - Age 24. 361hrs. 20th op.
Front Gunner: NZ401817 Sgt **Graham Stuart WALKER**, RNZAF - Age 25. 218hrs. 20th op.

103 Squadron, RAF (Elsham Wolds, Lincolnshire - 1 Group)
Wellington IC N2770 - took off at 1100 captained by Sgt J S Bucknole, RAF, on same raid as the above. N2770 became separated from the formation during an attack by fighters at 15,000 feet and was last seen heading towards the sea, losing height with a failed port engine. The body of an RAF Sgt later washed ashore near Brest, but the other five crew were lost without trace and are commemorated on the Runnymede Memorial.
Air Gunner: NZ401439 Sgt **Edward Campbell McDONALD**, RNZAF - Age 21. 264hrs. 22nd op.

218 Squadron, RAF (Marham, Norfolk - 3 Group)
Wellington IC R1726/O - took off at 1126 on same raid as the above and brought down near the target. Two RAF Sgts who successfully baled out, and were taken prisoner, later stated that they were certain the aircraft had crashed into a small bay, 2 miles off shore NE of Brest. They were equally certain that they were the only ones to leave R1726. No trace was found of their four comrades, who are commemorated on the Runnymede Memorial.
Captain: NZ40648 Plt Off **Morrisson JOLLY**, RNZAF - Age 23. 456hrs. 22nd op.

101 Squadron, RAF (Oakington, Cambridgeshire - 3 Group)
Wellington IC R1702/F - took off at 1136 on same raid as the above and attacked by fighters in the target area. Two RAF Sgts succeeded in baling out and were later picked up by a German vessel. No trace was found of their four comrades, who are commemorated on the Runnymede Memorial.
Captain: 41832 Flt Lt **Frederick Henry CRAIG**, mid, RAF - Age 26. 631hrs. 31st op(?)

FIGHTER COMMAND

Escorting Blenheim bombers on raid to Cherbourg, France
485 Squadron, RNZAF (Redhill, Surrey - 11 Group)
Spitfire IIA P7831 (a presentation machine named *Tippera*) - took off from West Hampnett, Sussex, at 1350 with 10 others, being the only one not to return. Shortly after leaving the target area P7831 was bounced by enemy fighters and shot down. The pilot had radioed that he was baling out and his parachute was seen descending over the English Channel some 15 miles north of Cherbourg. Eight of the Squadron's Spitfires returned to the area later in the day to search but found no trace of him. He is commemorated on the Runnymede Memorial.
Pilot: NZ40240 Sgt **Jack MANEY**, RNZAF - Age 23. 355hrs. At least 14 ops.

Thu 24/Fri 25 Jul 1941

BOMBER COMMAND

Raid on Kiel, Germany (by 64 aircraft - 2 lost)
61 Squadron, RAF (North Luffenham, Rutland - 5 Group)
Hampden I AE189/G - took off at 2304 captained by Fg Off M Parry, RAF, and shot down over the Dutch coast by a night-fighter at 0105, crashing into the sea off Den Helder. The bodies of two crew members were later washed ashore, including that of a New Zealand air gunner, who is buried at Kiel. No trace was found of the other two and they are commemorated on the Runnymede Memorial.
Air Gunner: NZ401818 Sgt **Ivan Fenton WAUGH**, RNZAF - Age 20. 138hrs. 6th op.

Raid on Emden, Germany (by 49 aircraft - 3 lost)
103 Squadron, RAF (Elsham Wolds, Lincolnshire - 1 Group)
Wellington IC R1397 - took off at 2340 and believed to have fallen in flames over the Netherlands at 0354, exploding near the village of Bozum, between Leeuwarden and Sneek. All six crew died and all are commemorated on the Runnymede Memorial. However, it is now thought likely that they may in fact be the six unknown airmen buried at Leeuwarden Northern General Cemetery.
Captain: NZ40979 Plt Off **Mervyn Sydney LUND**, RNZAF - Age 23. 408hrs. At least 7th op.
Lund's brother, Clarence Patrick Lund, died just over a year later, on 29 July 1942, while flying with 7 Sqn, RAF.

Fri 25/Sat 26 Jul 1941

BOMBER COMMAND

Raid on Hannover, Germany (by 55 aircraft - 7 lost)
10 Squadron, RAF (Leeming, Yorkshire - 4 Group)
Whitley V Z6624/O - took off at 2221 captained by Plt Off W McN Spiers, mid, RAF, and brought down in the North Sea. The bodies of three crew members later washed ashore, two in Germany and that of the observer, on 15 or 25 August 1941, onto the Dutch island of Texel, where he is buried at Den Burg. No trace was found of the other three and they are commemorated on the Runnymede Memorial.

Observer: NZ391886 Sgt **George WELLS**, RNZAF - Age 23. 260hrs. 8th op.
Wells brother, Charles Easton Wells, died on 29 July 1942 while flying with 1651 Heavy Conversion Unit, RAF. By coincidence, he and Clarence Lund, footnoted in the previous entry, were to be lost on the same raid against Hamburg.

Sat 26 Jul 1941

TRAINING COMMAND

Training
1 Service Flying Training School, RAF (Netheravon, Wiltshire - 23 Group)
Hind L7205 - piloted by Leading Airman F Andrew. At about 0840, while engaged in aerobatics, made three or four stall turns before spinning into the ground $^1/_2$ mile east of Shrewton landing ground, 6 miles SW of Netheravon. Both pilots died, the New Zealander being buried at Netheravon.
Pilot: FX81925 Leading Airman **Grant RYALLS**, RN, of HMS *Daedalus* (FAA) - Age 22.

Tue 29 Jul 1941

NEW ZEALAND

Solo height test
2 Service Flying Training School (ATS), RNZAF (Woodbourne)
Harvard II NZ951 - climbed to 7000 feet, from which height it was seen to begin a dive. After recovery NZ951 began to lose height, finally crashing at 1430. The pilot had tried to leave the Harvard but pulled his ripcord too soon, the parachute fouling the machine. His body, with the badly torn parachute, was found at the Omaka riverhead, about a mile SW of Goulter Hill and 6 miles west of Woodbourne, and is buried at Marton. The aircraft was so completely wrecked that no explanation could be found as to why the pilot chose to bale out.
Pilot: NZ411906 LAC **Morris Desmond JURGENS**, RNZAF - Age 22. 93hrs (42 solo - 22 on Harvard)

Wed 30 Jul 1941

NEW ZEALAND

Dual instruction
4 Elementary Flying Training School, RNZAF (Whenuapai)
Tiger Moth II NZ673 - seen at low altitude recovering from a dive at 0825. As it lost height, NZ763 rolled over on its back and crashed almost vertically among trees in a gully on the farm of Mr J C H Lloyd at Riverhead, $2^1/_2$ miles NW of Whenuapai. Both pilots died and are buried at Purewa and Otahuhu respectively.
Pilot Instr: NZ40967 Plt Off **Athol FOSTER**, RNZAF - Age 25. 581hrs solo on Tiger Moth.
Pupil Pilot: NZ413544 LAC **Keith Millen McCULLOCH**, RNZAF - Age 18.
Foster's brother, Charles Grierson Foster, died on 31 March 1944 while flying with 550 Sqn, RAF.

FIGHTER COMMAND

Convoy patrol near Lowestoft
258 Squadron, RAF (Martlesham Heath, Suffolk - 11 Group)
Hurricane IIB Z3256/H - took off at 1350 with five others, the six operating in three sections of two. Shortly afterwards the pilot reported aileron control failure and Z3256 dived into the ground $1^1/_2$ miles west of Woodbridge, a mile or two north of the base. He is buried at Ipswich.
Pilot: NZ391841 Fg Off **George Montgomerie MARSHALL**, RNZAF - Age 23. 410hrs. 96th op.

Wed 30/Thu 31 Jul 1941

BOMBER COMMAND

Raid on Cologne, Germany (by 116 aircraft - 9 lost)
144 Squadron, RAF (North Luffenham, Rutland - 5 Group)
Hampden I AD784 - took off at 2348 and brought down into the North Sea. The body of an RAF Plt Off later washed ashore near Haamstede on the Dutch island of Schouwen, but no trace was found of the other three crew, who are commemorated on the Runnymede Memorial. Thunderstorms and heavy icing conditions were encountered during the raid and it appears unlikely any of the aircraft lost were brought down as the result of enemy action - six of the nine destroyed crashed in England.
Captain: NZ402241 Plt Off **Donald BROOK**, RNZAF - Age 26. 272hrs.

Sun 3 Aug 1941

FIGHTER COMMAND

Fighter attack exercises
52 Operational Training Unit, RAF (Debden, Essex - 81 Group)
Hurricane I W9149 - collided with another of the Unit's Hurricanes (R4183) while dogfighting. The two fighters crashed near Great Sampford, 7 miles SSE of Saffron Walden, killing both pilots, who are buried at Saffron Walden.
Pilot: NZ404335 Sgt **Ian McDonald CHEYNE**, RNZAF - Age 24. 215hrs.

52 Operational Training Unit, RAF (Debden, Essex - 81 Group)
Hurricane I R4183 - collided with Hurricane W9149 as detailed above.
Pilot: NZ403852 Sgt **William FLEMING**, RNZAF - Age 20. 181hrs.
Cheyne and Fleming were the first New Zealand graduates under the Empire Air Training Scheme (EATS) in Canada to lose their lives.

Mon 4 Aug 1941

FIGHTER COMMAND

Sweep - reconnaissance for E-boats
257 Squadron, RAF (Coltishall, Norfolk - 12 Group)
Hurricane IIB Z3164 - took off at 1615 with another as part of 'Yellow' section. During the reconnaissance an enemy bomber was intercepted and attacked several times. It seems that the Z3614 had been hit during the engagement, as at 1721 the pilot informed control he was having to bale out. Despite an intensive search nothing more was seen of him and he is commemorated on the Runnymede Memorial.
Pilot: 40441 Flt Lt **Kenneth William TAIT**, DFC mid, RAF - Age 22. 766hrs. At least 116th op.

Tue 5/Wed 6 Aug 1941

BOMBER COMMAND

Raid on Mannheim, Germany (by 98 aircraft - 3 lost)
106 Squadron, RAF (Coningsby, Lincolnshire - 5 Group)
Hampden I AE120 - took off at 2225 and shot down over the Netherlands by a night-fighter, crashing onto a building in Gendringen, 30km ESE of Arnhem near the German border. All four crew are buried at Gendringen. Within the building itself, two people were also killed and another two injured.
Captain: NZ402160 Sgt **William Leslie KNOWLES**, RNZAF - Age 21. 225hrs. 2nd op.

Wed 6/Thu 7 Aug 1941

BOMBER COMMAND

Raid on Karlsruhe, Germany (by 38 aircraft - 1 lost)
144 Squadron, RAF (North Luffenham, Rutland - 5 Group)
Hampden I AE140 - lost without trace, its four crew being commemorated on the Runnymede Memorial.
Captain: NZ402159 Plt Off **Stuart William Stott BEEDIE**, RNZAF - Age 23. 271hrs. 2nd op.

Raid on Frankfurt, Germany (by 53 aircraft - 6 lost)
150 Squadron, RAF (Snaith, Yorkshire - 1 Group)
Wellington IC W5721/Z - took off at 2235 captained by Plt Off C Landreth, RAF, and believed to have been brought down over Germany near Schermbeck, 15 miles NNW of Essen. The 2nd pilot, and most probably others of the crew, was buried at Schermbeck, but later all six were reinterred at Reichswald Forest, 5km SW of Kleve.
2nd Pilot: NZ403929 Plt Off **Alexander Cairns ANDERSON**, RNZAF - Age 27. 229hrs. 2nd op.

Raid on Mannheim, Germany (by 38 aircraft - 4 lost)
214 Squadron, RAF (Stradishall, Suffolk - 3 Group)
Wellington IC X9750 - lost without trace, its six crew being commemorated on the Runnymede Memorial.
Captain: NZ401259 Plt Off **Colin McDougall DIDSBURY**, RNZAF - Age 23. 337hrs. 19th op.

Fri 8/Sat 9 Aug 1941

BOMBER COMMAND

Raid on Kiel, Germany (by 54 aircraft - 5 lost)
61 Squadron, RAF (North Luffenham, Rutland; satellite airfield at Woolfox Lodge - 5 Group)
Hampden I X3127 - took off from Woolfox Lodge at 2218 captained by Plt Off J W W Graham, RAF, and brought down by flak after being coned by searchlights during its run in to the target. The air gunner was killed, either in the air or when the Hampden crashed landed with all on board at Lindau, 11km SW of Kappeln, to the NW of the target area. Buried in the Garrison Cemetery at Kiel, he was later reinterred in the Kiel War Cemetery. His three comrades survived to be taken prisoner.
Air Gunner: NZ401814 Sgt **Alfred Cecil NUTTALL**, RNZAF - Age 28. 169hrs. 7th op.

Tue 12 Aug 1941

FAR EAST

Formation practice
34 Squadron, RAF (Tengah, Singapore)
Blenheim IV V5877 - took off with others of the Squadron and during the practice collided with the formation leader over Johore and crashed. All three crew were buried at Bidadari Cemetery, but later reinterred at Kranji.
Pilot: NZ403548 Plt Off **John Francis HUNT**, RNZAF - Age 25.

BOMBER COMMAND

Raid on Knapsack & Quadrath power stations at Cologne, Germany (by 54 aircraft - 10 lost)

139 Squadron, RAF (Oulton, Norfolk - 2 Group)

Blenheim IV V6261/M - took off at 0949 for the raid on Knapsack. Already damaged by flak over the target area, V6261 was shot down at 1318 by fighters over the Netherlands on the return journey, crashing into the mouth of the River Scheldt, off Breskens. All three crew died, the bodies of two being recovered and buried at Bergen. The observer, however, was lost without trace and is commemorated on the Runnymede Memorial.

Pilot: NZ401763 Flt Lt **Alexander George HERBERT**, RNZAF - Age 23. 252hrs. 5th op(?)

FIGHTER COMMAND

'Circus' - escort to six Blenheims raiding shipyard at Le Trait, France

485 Squadron, RNZAF (Redhill, Surrey - 11 Group)

Spitfire IIA P7788/E - took off at 1725 with seven others. The formation was attacked by enemy fighters on the return journey, P7788 being lost without trace during the engagement. The pilot is commemorated on the Runnymede Memorial.

Pilot: 931283 Sgt **George Marchant PORTER**, RAFVR - Age 24. 220hrs. 9th op.

BOMBER COMMAND

Night flying practice

17 Operational Training Unit, RAF (Upwood, Huntingdonshire - 7 Group)

Blenheim I L1245 - piloted by Plt Off W E A Wand, RAF, and towards midnight was attacked three times by an enemy night-fighter intruding in the vicinity of the airfield. After the last attack, at about 2350, the Blenheim crashed and caught fire at Crow Farm, Wilburton, 5 miles SW of Ely, Cambridgeshire. All three crew died, the 2nd pilot being buried at Bury, 2 miles NE of Upwood.

2nd Pilot: NZ402841 Plt Off **Ronald Peter BELL**, RNZAF - Age 24. 196hrs.

Tue 12/Wed 13 Aug 1941

BOMBER COMMAND

Raid on Hannover, Germany (by 78 aircraft - 5 lost)

9 Squadron, RAF (Honington, Suffolk - 3 Group)

Wellington IC R1341/Z - took off at 2127 captained by Sgt E Lewin, RAF, and shot down over Germany, crashing near Rotenburg, 24km NE of Verden, to the east of Bremen. All six crew died and were buried at Rotenburg, but later reinterred at Becklingen, 17km south of Soltau.

Rear Gunner: NZ40206 Sgt **Ronald Ramsay PASSMORE**, RNZAF - Age 22. 235hrs. 26th op.

115 Squadron, RAF (Marham, Norfolk - 3 Group)

Wellington IC Z8835/U - took off at 2127 on same raid as the above. Brought down over Germany and believed to have crashed at Buxtehude, to the SW of Hamburg. Five of the crew, including the captain, were buried at Parnewinkel, but later reinterred at Becklingen, 17km south of Soltau. No trace of the sixth crew member could be found and he is commemorated on the Runnymede Memorial.

Captain: NZ40669 Sgt **John Trevor WALLACE**, RNZAF - Age 21. 411hrs. 23rd op.

Wallace's brother, Utrick Watson Wallace, died on 28 January 1943 while flying with 301 Ferry Training Unit, RAF.

Raid on Essen, Germany (by 35 aircraft - 1 lost)

115 Squadron, RAF (Marham, Norfolk - 3 Group)

Wellington IC T2563/D - took off at 2134 captained by Plt Off Wood, RAF, and on returning from the raid was attacked several times by an enemy night-fighter intruder. On catching fire, T2563 crash landed at 0220 on Smith's Farm at Scottow, 4 miles SSW of North Walsham, the pilot escaping unscathed while four others received minor injuries. The rear gunner, however, was killed during the attack and is buried at Scottow.

Rear Gunner: NZ401804 Sgt **Bernard George EVANS**, RNZAF - Age 25. 168hrs. 2nd op.

Sun 17 Aug 1941

MIDDLE EAST

Drowning accident

223 Squadron, RAF (LG 214 [Shandur], Egypt - 202 Group)

Accidentally drowned in a channel of running water near the Suez Canal and is buried at Suez.

(Observer): NZ401438 Sgt **Samuel John LUKE**, RNZAF - Age 22.

Tue 19 Aug 1941

NEW ZEALAND

RNZAF Station Ohakea

Died while undergoing an operation at the Palmerston North Public Hospital and is buried at Dannevirke.

(Aircrafthand): NZ391931 LAC **Robert John McQUILLAN**, RNZAF - Age 20.

Solo night flying training - circuits
2 Service Flying Training School (ITS), RNZAF (Woodbourne)
Harvard II NZ921 - took off normally shortly before 1930, but was then seen to gradually lose height from 800 feet until it struck a hill near Renwicktown and exploded. The pilot is buried at Martinborough.
Pilot: NZ412687 LAC **Arthur Ormond HATHAWAY**, RNZAF - Age 28. 75hrs (29 solo - 9 on Harvard)

FLYING TRAINING COMMAND

Training
1 Service Flying Training School, RAF (Netheravon, Wiltshire - 23 Group)
Hart Trainer K4979 - piloted by Leading Airman G R Watson, FAA. While low flying, hit high tension wires and crashed in flames at Claverton, 3 miles ESE of Bath, Somerset. Both pilots died, the New Zealander being buried at Luton, Bedfordshire.
Safety Pilot: FX84620 Leading Airman **Ewan Seymour INGLE**, RN, of HMS *Daedalus* (FAA) - Age 21.
In the CWGC register Ingle is incorrectly recorded as belonging to the RNZN, which was not formed until 1 October 1941.

FIGHTER COMMAND

'Circus 82'- escorting Blenheim bombers raiding power station at Gosnay, France
485 Squadron, RNZAF (Redhill, Surrey - 11 Group)
Spitfire IIA P7977/U - took off at 0955 with five others and lost without trace after combat with enemy fighters over the French coast. The pilot is commemorated on the Runnymede Memorial.
Pilot: NZ401761 Sgt **Keith Campbell Morison MILLER**, RNZAF - Age 24. 328hrs. 16th op.

'Circus 81'
452 Squadron, RAAF (Kenley, Surrey - 11 Group)
Spitfire VB P8717 - took off at 1005 and was one of two Squadron aircraft which failed to return. In an encounter with enemy fighters over the French coast P8717 was seen to lose a wing and spin down out of control. The pilot is buried at Dunkirk.
Pilot: Aus402232 Fg Off **William Davies ECCLETON**, RAAF - Age 25.

BOMBER COMMAND

'Beat 9'- anti shipping strike off the Dutch coast
114 Squadron, RAF (West Raynham, Norfolk - 2 Group)
Blenheim IV V6366/N - took off at 1715 and during an attack at about 1930 on a convoy 35km NW of the island of Vlieland was shot down into the sea by fighters, as were two accompanying Blenheims. V6366's air gunner survived to be taken prisoner, but his two comrades were lost without trace and are commemorated on the Runnymede Memorial.
Pilot: NZ402199 Plt Off **Roderick Urquhart McCRACKEN**, RNZAF - Age 22. 213hrs. At least 4th op.

Fri 22/Sat 23 Aug 1941

BOMBER COMMAND

Raid on Mannheim, Germany (by 97 aircraft - 1 lost)
106 Squadron, RAF (Coningsby, Lincolnshire - 5 Group)
Hampden I AE220 - took off at 2140 and shot down by flak, crashing between Plankstadt and Schwitzingen, about 13km SSE of the target. All four crew were buried at Mannheim, but later reinterred at Durnbach, 16km east of Bad Tölz.
Captain: NZ402201 Plt Off **Allan Frederick McGRUER**, RNZAF - Age 29. 460hrs. 10th op.

Sat 23 Aug 1941

COASTAL COMMAND

Night practice
413 Squadron, RCAF (Stranraer, Wigtownshire - 15 Group)
Catalina I AH556 - took off from Loch Ryan at 2255 captained by Pilot Ensign D A Eldred, United States Navy, but swung, left the water and crashed a few seconds later with full power on, and caught fire. Two of the crew survived, injured, but the other five died, including the 2nd pilot, whose body was later recovered and is buried at Stranraer. Investigation revealed that the elevator trim had been set in the fully nose down position, a recommendation being made that illumination of this control should be provided in future for night operations.
2nd Pilot: 42116 Fg Off **Alfred Lawrence FOWLER**, DFC, RAF - Age 23. 561hrs. 55 ops.

Sun 24/Mon 25 Aug 1941

BOMBER COMMAND

Raid on Düsseldorf, Germany (by 44 aircraft - 6 lost)
49 Squadron, RAF (Scampton, Lincolnshire - 5 Group)
Hampden I AD967/H - took off at about 2030 and while circling to land at 0250, upon return to England, collided with a Scampton-based 83 Sqn Hampden I (X3121), also returning from the same raid. Both bombers crashed

near Whale Jaws Farm, Hackthorn, 3 miles NE of the base, and all eight crew died, the captain of AD967 being buried at Scampton.
Captain: 968141 Sgt **Owen Bernard McMAHON**, RAF - Age 24. 414hrs. 18th op.

Mon 25 Aug 1941

FLEET AIR ARM - CARIBBEAN

Reconnaissance exercise
749 Squadron, FAA (Piarco, Trinidad - HMS *Goshawk*)
Walrus I L2239 - piloted by Sub Lt G L Stewart. Spun into the ground at Orange Grove Estate, Tacarigua. All three crew are buried at Port of Spain.
Observer: FX84602 Leading Airman **Brian Charles HAMILTON**, RN - Age 22.

Tue 26 Aug 1941

BOMBER COMMAND

Low level attack on Heligoland, Germany
82 Squadron, RAF (Watton, Norfolk - 2 Group)
Blenheim IV V6435/V - took off at 1053 from Bodney piloted by Flt Lt H R Shuttleworth, RAF, with four others, one of which lost contact and returned early to base. The other four were shot down by fighters near the target, with the loss of all 12 crew members. All three on board V6435 are commemorated on the Runnymede Memorial.
WOpAG: NZ402976 Sgt **Omer Wilbur John BISHOP**, RNZAF - Age 22. 70hrs.
Bishop's brother, Leon Eric Gordon Bishop, died on 11 June 1943 while flying with 207 Sqn, RAF

FIGHTER COMMAND

Training
61 Operational Training Unit, RAF (Heston, Middlesex - 81 Group)
Master W8583 - piloted by flying instructor Flt Sgt Whipps, RAF. While taxiing at 1550 was struck by one of the Unit's Spitfire Is (X4835) then in the process of taking off. The Spitfire pilot was admitted to hospital with head injuries, but both occupants of W8583 died, the pupil pilot being buried at Hounslow.
Pupil Pilot: NZ404876 Sgt **Ivan Leslie JONES**, RNZAF - Age 23.

Wed 27 Aug 1941

FIGHTER COMMAND

'Circus 85' - escort cover for six Blenheims on raid to St Omer, France
485 Squadron, RNZAF (Redhill, Surrey - 11 Group)
Spitfire VB W3527 - took off at 0620 with 10 others. The escorting fighter wing failed to rendezvous with the bombers and carried on without them. On encountering heavy fighter opposition over the French coast at 0832, W3527 was shot down, crashing 500 metres north of Audruicq. The pilot is buried at Dunkirk.
Pilot: NZ39928 Fg Off **William Arthur MIDDLETON**, RNZAF - Age 22. 370hrs. 74th op.

'Circus' - escort cover for nine Blenheims on raid against power station at Lille, France
222 Squadron, RAF (North Weald, Essex - 11 Group)
Spitfire VB W3723 - took off at 0705. Although the Squadron was attacked on the return journey, it was believed that W3723 might have been lost through a mid-air collision over the English Channel and not necessarily shot down. The pilot is commemorated on the Runnymede Memorial.
Pilot: 40313 Flt Lt **John Claverly MARTIN**, RAF - Age 27. 1069hrs. 131st op.

Training
256 Squadron, RAF (Squires Gate, Lancashire - 9 Group)
Defiant I N1745/O - while flying over the sea near Blackpool during the afternoon, collided with a Blackburn Botha of No 3 School of General Reconnaissance. N1745 crashed onto a Blackpool house in Reads Avenue, killing the pilot. The air gunner had attempted to bale out but was too low for his parachute to fully deploy. He fell into Regent Street and also died. Both airmen are buried at Lytham St Anne's. The tail separated from the Botha and fell into the sea near the city shore. The rest of the wreckage, however, crashed into the booking hall of the Blackpool Central railway station and besides the two crew, at least eight civilians were killed outright and six of the 27 injured later died also.
Pilot: NZ40760 Sgt **Lincoln John ELLMERS**, RNZAF - Age 24. 405hrs. 15 ops.
Air Gunner: NZ402166 Sgt **Noel Anthony John CLIFFORD**, RNZAF - Age 23. 149hrs.

Thu 28 Aug 1941

NEW ZEALAND

Air-to-air firing exercises
1 Service Flying Training School (ATS), RNZAF (Wigram)
Oxford I NZ252 - stalled at 100 feet while executing a cross wind approach at 1435 to the Lake Ellesmere landing ground. The nose and right wing dropped and NZ252 spun into the lake and overturned. The acting air gunner was

killed instantly and the pilot died of injuries later in the day, after admission to Christchurch Public Hospital. The pilot's remains were cremated at Wellington, while his comrade is buried at Springston.
Pilot: NZ411365 LAC **Ronald James BROWN**, RNZAF - Age 26. 124hrs (65 solo - 46 on Oxford)
(Pilot): NZ411921 LAC **Norman Joseph Graham MARSHALL**, RNZAF - Age 21. 130hrs.
Marshall was acting as air gunner for the exercise.

MIDDLE EAST

Pre-delivery test flight(?)
Fighter Defence Flight, RAF (Takoradi, Gold Coast)
Hurricane I Z4485 - executed a turn downwind while flying low over the Lagos Harbour or lagoon at 1155, struck the water and crashed. The pilot is buried at Lagos.
Pilot: 36273 Fg Off **Noel Carroll PETTIT**, RAF - Age 20. 315hrs.

Thu 28/Fri 29 Aug 1941

BOMBER COMMAND

Raid on Duisburg, Germany (by 118 aircraft - 9 lost)
78 Squadron, RAF (Middleton St George, Durham - 4 Group)
Whitley V Z6508 - took off at 0135 captained by Plt Off K W Davies, RAF, and presumed to have crashed into the sea, as the bodies of two New Zealand crew members were later washed ashore near Dunkirk and buried there. No trace of the other three crew was found and they are commemorated on the Runnymede Memorial. *Bomber Command Losses* details this loss under 27/28th August with the target as Mannheim, but is in error.
Observer: 916770 Sgt **Gerald Thomas HARPER**, RAFVR - Age 25.
WOpAG: NZ403041 Sgt **Peter GENNON**, RNZAF - Age 25. 87hrs. 1st op.
Harper's date and place of birth are unknown but he lived in New Zealand from at least 1930 to 1935.

Sat 30 Aug 1941

MIDDLE EAST

Raid on Tripoli, Libya
38 Squadron, RAF (Shallufa, Egypt; detachment at Luqa, Malta - 257 Wing)
Wellington IC X9737 - took off from Luqa at 1730 captained by Plt Off D P C F Mauchlen, RAF, and brought down near the target area, possibly being shot down by an Italian torpedo boat. The captain and observer both died and were buried in the Tripoli Catholic Cemetery, but later reinterred in Tripoli's War Cemetery. The other four crew survived to be taken prisoner.
Observer: NZ401217 Sgt **Selwyn Gibson SHIRLEY-THOMSON**, RNZAF - Age 30. 236hrs. At least 19th op.

COASTAL COMMAND

Training - low level attack on tower target
5 (Coastal) Operational Training Unit, RAF (Chivenor, Devon - 17 Group)
Beaufort I N1160 - captained by Fg Off K T Little, RAAF. Sideslipped after a turn at 1535 and, out of control, crashed into the sea about a half mile off Saunton Sands. All four crew died, including the captain and observer, who are buried at Heanton Punchardon. Investigators noted that it was an extremely bright day and difficult to judge height above the water when flying in certain directions.
Observer: NZ403833 Plt Off **David Giffin DOOLE**, RNZAF - Age 29.

Sun 31 Aug/Mon 1 Sep 1941

MIDDLE EAST

Low level attack on Benghazi Harbour, Libya
113 Squadron, RAF (Maaten Bagush, Egypt)
Blenheim IV T2066 - took off at 2000 piloted by Sgt G L R Sulman, RAF, with three others and was one of two which failed to return. T2066 is believed to have been brought down near the target area, the air gunner being the sole survivor. The captain and observer are commemorated on the Alamein Memorial.
Observer: NZ401730 Sgt **Douglas Lewin RHODES**, RNZAF - Age 21.

Tue 2/Wed 3 Sep 1941

BOMBER COMMAND

Raid on Berlin, Germany (by 49 aircraft - 10 lost)
61 Squadron, RAF (North Luffenham, Rutland - 5 Group)
Manchester I L7388 - took off at 2030 captained by Wg Cdr G E Valentine, DSO, RAF, and shot down by flak, crashing at Schönefeld, 16km SE of the target city's centre. All six crew and a 'passenger' were killed and are buried at Berlin. Valentine was the Squadron's commanding officer and the passenger the commanding officer of North Luffenham. Between them, the two COs and one other crew member held four DSOs and a DFC. Five of the aircraft lost on the raid were Hampdens which crashed in England after running out of fuel on the return journey.
Air Gunner: NZ401806 Sgt **Walter Duncan HAMER**, RNZAF - Age 21.

COASTAL COMMAND

'Trost' - night reconnaissance

235 Squadron, RAF (Dyce, Aberdeenshire; detachment at Sumburgh, Shetlands - 18 Group)

Blenheim IVF L9261/T - took off from Sumburgh at 2200 and signalled that it was lost, with unserviceable instruments and flying in low cloud and drizzle, before crashing into a hillside at 0305 and catching fire near Huxter, 25 miles north of Sumburgh. The three crew are buried at Lerwick.

Pilot: 37471 Flt Lt **Howard Moule HAMMOND**, RAF - Age 26. 874hrs solo.

Wed 3 Sep 1941

FLYING TRAINING COMMAND

Solo flying practice

10 Air Gunners School, RAF (Castle Kennedy, Wigtownshire - 25 Group)

Master III W8539 - took off at 1540, but 50 minutes later lost control in cloud at about 1000 feet, and crashed at Goldenloch, Glenluce, 9 miles WNW of Stranraer, and caught fire. The pilot is buried at Stranraer.

Pilot: NZ404358 Sgt **Owen Henry GLENGARRY**, RNZAF - Age 19. 148hrs.

FIGHTER COMMAND

Formation flying practice

136 Squadron, RAF (Kirton-in-Lindsey, Lincolnshire - 12 Group)

Hurricane IIB BD733 - took off at 1410 as 'No 3' in a section of three. While circling the airfield a few minutes later at 1500 feet, preparatory to joining up in squadron formation, overshot slightly and 'widened out.' In the process of regaining formation position, BD733 collided with the propeller and wing of one of the others and crashed a mile east of the airfield. The pilot is buried at Kirton-in-Lindsey. The other pilot involved in the collision is thought to have landed safely.

Pilot: NZ404364 Sgt **Theodore Stephen HEWLAND**, RNZAF - Age 22.

Thu 4 Sep 1941

NEW ZEALAND

Camera gun exercise against drogue target

3 Service Flying Training School (ATS), RNZAF (Ohakea)

Hind NZ1504 - turned too close while executing an attack at 0830 and struck the drogue with its port wing. NZ1504 began to spin, which at 1000 feet developed into a steep dive, and failed to recover before crashing into the sea off Paekakariki. The pilot is buried at Woodville.

Pilot: NZ412378 LAC **Arthur Louis SKINNER**, RNZAF - Age 26. 126hrs (65 solo - 45 on Hind)

Fri 5/Sat 6 Sep 1941

MIDDLE EAST

Raid on satellite airfield at Derna, Libya

37 Squadron, RAF (Shallufa, Egypt; detachment at Fuka - 257 Wing)

Wellington IC T2512 - took off at 2130 from Fuka and lost without trace, its six crew being commemorated on the Alamein Memorial.

Observer: NZ40628 Fg Off **Kenneth Stanley PETERSON**, DFC, RNZAF - Age 27. 451hrs. 19th op(?)

Sat 6 Sep 1941

BOMBER COMMAND

Air test

10 Squadron, RAF (Leeming, Yorkshire - 4 Group)

Whitley V Z6932 - piloted by Flt Sgt W Stuart, RCAF. Lost bearings owing to poor visibility and made a forced landing at Brunton, near Chathill, Northumberland, for a check of its position. Construction work was being undertaken on the airfield at the time and during take off in the afternoon, for return to base, Z6932 struck barrels on the runway, swung and hit a steam roller. It managed to become airborne, but in avoiding some high tension cables stalled and crashed. Only the rear gunner survived, albeit injured. His three comrades are buried at East Chevington.

2nd Pilot: NZ403785 Plt Off **Richard Scott AUSTIN**, RNZAF - Age 21. 235hrs. 1 op.

The CWGC register gives date of death for both pilots as the 7th. This may be an error, unless both airmen died of injuries on the following day.

Sat 6/Sun 7 Sep 1941

BOMBER COMMAND

Raid on Hüls, Germany (by 86 aircraft - 11 lost)

10 Squadron, RAF (Leeming, Yorkshire - 4 Group)

Whitley V Z6478/S - took off at 1949 captained by Sgt A Poupard, RAF, and shot down over the Netherlands by a night-fighter at 2300, crashing onto a German artillery practice range near Oldebroek, 15km SW of Zwolle. All five crew are buried at Oldebroek.

WOpAG: NZ39586 Sgt **Pine Tenga TAKARANGI**, RNZAF - Age 23. 80hrs. 1st op.

77 Squadron, RAF (Leeming, Yorkshire - 4 Group)
Whitley V Z6668 - took off at 2004 captained by Sgt D W Mercer, RAF, on same raid as the above and brought down over Germany near Quakenbrück. All five crew were buried in the parish cemetery there, but later reinterred at Rheinberg, 12km south of Wesel.
WOpAG: NZ403015 Plt Off **Richard John MINNIS**, RNZAF - Age 27. 175hrs. 7th op.

78 Squadron, RAF (Middleton St George, Durham - 4 Group)
Whitley V Z6881 - took off at 2037 captained by Plt Off F B Thorpe, RAF, on same raid as the above and shot down over the Netherlands by a night-fighter at 0125, crashing at Bergum, 12km ESE of Leeuwarden. All five crew are buried at Bergum.
WOpAG: NZ402979 Sgt **Geoffrey CARMAN**, RNZAF - Age 31. 19hrs(?). 1st op(?)

57 Squadron, RAF (Feltwell, Norfolk - 3 Group)
Wellington IC Z8794/H - took off at 2110 on same raid as the above and brought down over the Ruhr near Wanne-Eickel, where the captain was buried, probably with others of the crew. All six were later reinterred at Reichswald Forest, 5km SW of Kleve.
Captain: NZ401371 Sgt **Bernard Allison CLARK**, RNZAF - Age 28. 393hrs. 25th op.

Sun 7/Mon 8 Sep 1941

BOMBER COMMAND

Raid on Berlin, Germany (by 197 aircraft - 21 lost)
115 Squadron, RAF (Marham, Norfolk - 3 Group)
Wellington IC R1798/B - took off at 2023 captained by Sgt I P McH Gordon, RAF, and shot down over the Netherlands by a night-fighter at 0458, crashing 4km NE of Drachten, near Drachtstercompagnie, where all six crew are buried.
WOpAG: NZ401720 Sgt **Frederick Alphonsus BATY**, RNZAF - Age 30. 177hrs. 9th op.

104 Squadron, RAF (Driffield, Yorkshire - 4 Group)
Wellington II W5362/J - took off at 2034 captained by Sgt F Richardson, RAAF, on same raid as the above and lost without trace, its six crew being commemorated on the Runnymede Memorial.
Observer: 925336 Sgt **Russell Johnston MOUNTIER**, RAFVR - Age 20. 185hrs.

Wed 10 Sep 1941

FIGHTER COMMAND

Camera gun practice
61 Operational Training Unit, RAF (Heston, Middlesex - 81 Group)
Spitfire I X4552 - while attempting to follow the target aircraft at 0955, which was trying to break away in a climbing turn, stalled and spun into the ground at Burghfield, 4 miles SW of Reading, Berkshire. The pilot is buried at Hounslow.
Pilot: NZ404902 Sgt **Hector Lawrence CRAW**, RNZAF - Age 25. 165hrs.

Thu 11 Sep 1941

NEW ZEALAND

Dual instruction
3 Elementary Flying Training School, RNZAF (Harewood)
Tiger Moth II NZ883 - at 0750, when about to enter a third loop at 3000 feet, the port wing folded back to an angle of 45° with the fuselage, causing the machine to enter a steep left hand spin. The instructor, Plt Off **K S Blair**, closed the throttle and endeavoured to regain control, but to no avail and he ordered his pupil to abandon the aircraft before so doing himself, landing uninjured. The pupil evidently pressed his parachute's quick release box mechanism instead of that for the Sutton harness, leaving him with insufficient time to correct the error and bale out. He died when the trainer crashed a mile west of Kaiapoi and is buried at Onehunga. The wing collapse was caused through stripping of threads of the tie rod ends because of a maintenance error.
Pupil Pilot: NZ413860 LAC **Bryan Walter LAMBOURNE**, RNZAF - Age 18.
Blair died on 27 July 1942 while instructing at the same School.

Naval co-operation
2 Squadron, RNZAF (Nelson)
Vincent NZ327 - piloted by Plt Off **W B Pettet**, struck turbulence at 1445 while flying in the area of Flat Creek and Havelock, causing a crew member, who was not strapped in or wearing a parachute, to be thrown out. He is buried at Temuka. The pilot and the third crew member, AC2 **R B Anderson**, were uninjured and landed safely.
Wireless Op: NZ403239 AC1 **George Herbert William TAYLOR**, RNZAF - Age 20. 53hrs.

Fri 12/Sat 13 Sep 1941

BOMBER COMMAND

Night training

14 Operational Training Unit, RAF (Cottesmore, Rutland; satellite airfield at Saltby - 7 Group)

Hampden I P1301 - captained by Sgt A P Overall, DFM, RAF. Collided with the Hampden detailed below while in the circuit at Saltby, Leicestershire, 9 miles NNW of Cottesmore. All four crew died, three of whom, including a New Zealander, are buried at Cottesmore.

2nd Pilot: NZ34179 Sgt **Vincent Herbert GRIFFITHS**, RNZAF - Age 19. 246hrs.

Griffiths was probably acting as observer on this flight.

14 Operational Training Unit, RAF (Cottesmore, Rutland; satellite airfield at Saltby - 7 Group)

Hampden I P4303 - captained by Sgt D Bartlett, RAF. Collided with the Hampden detailed above and crashed at Saltby. All four crew died, including a New Zealander, who is buried at Cottesmore.

2nd Pilot: NZ401792 Sgt **Jack Edward WALL**, RNZAF - Age 19. 250hrs.

Wall was probably acting as observer on this flight.

Mon 15/Tue 16 Sep 1941

BOMBER COMMAND

Raid on Hamburg, Germany (by 169 aircraft - 12 lost)

75 (NZ) Squadron, RAF (Feltwell, Norfolk - 3 Group)

Wellington IC X3205 - took off at 1945 and while in the target area was hit by flak, and crashed. Four of the crew died and are buried at Ohlsdorf, near Hamburg. An RAF Sgt and RCAF Flt Sgt baled out safely and were captured.

Captain: NZ401793 Sgt **James Allen WARD**, VC, RNZAF - Age 22. 305hrs. 11th op.

Ward was the first New Zealand airman to be awarded the VC.

75 (NZ) Squadron, RAF (Feltwell, Norfolk - 3 Group)

Wellington IC X9759 - took off at approximately 1940-2115 on same raid as the above, and brought down over Germany, crashing near Hartenholm, about 40km north of the target. Two of the crew, including 2nd pilot Sgt **R B Blakeway**, RNZAF, survived to be taken prisoner. Their four comrades were buried at Neumünster, 6km NNW of Hartenholm, but later reinterred at Ohlsdorf, near Hamburg.

Captain: NZ40971 Sgt **Anthony Henry Ryder HAWKINS**, RNZAF - Age 20. 402hrs. 22nd op.

Tue 16 Sep 1941

FIGHTER COMMAND

Formation practice

61 Operational Training Unit, RAF (Heston, Middlesex - 81 Group)

Spitfire I R6611 - failed to recover from an aerobatic manoeuvre at 1100 and dived into the ground, and caught fire at Waterperry, Oxfordshire, 7 miles WSW of Haddenham, Buckinghamshire. The pilot is buried at Haddenham.

Pilot: NZ404347 Plt Off **James Edwards FALCONER**, RNZAF - Age 27. 181hrs.

CANADA

Motor vehicle accident

1 (Y) Depot, RCAF (Halifax, Nova Scotia - 3 Training Command)

In the evening an airman posted to the Depot for embarkation for Britain died, while on leave, in a motor vehicle accident on the highway at Blenheim, Ontario, 5 miles NW of Rondeau Harbour. He is buried at St Thomas, 53 miles NE of Blenheim.

(Observer): NZ404679 Sgt **James McIntosh FAWKNER**, RNZAF - Age 22.

Fawkner was the first member of the RNZAF to lose his life while undergoing training in Canada as part of the Empire Air Training Scheme.

Wed 17/Thu 18 Sep 1941

BOMBER COMMAND

Raid on Karlsruhe, Germany (by 38 aircraft - 1 lost)

75 (NZ) Squadron, RAF (Feltwell, Norfolk - 3 Group)

Wellington IC X9834 - took off at about 1930 captained by Sgt W B M Smyth, RAF, and shot down over Germany, crash landing on fire on the River Prüm at Holsthum, 12km SSW of Bitburg, Germany. Three crew members survived to be taken prisoner, although one was later shot while trying to escape the following December. The other three died, including the front gunner, who, probably with the others, was buried at Trier. All three were later reinterred at Rheinberg, 12km south of Wesel.

Front Gunner: NZ403003 Sgt **Howard Clive McLeish HASELDEN**, RNZAF - Age 22. 111hrs. 2nd op.

Thu 18 Sep 1941

BOMBER COMMAND

Attack on tanker off Belgian coast

88 Squadron, RAF (Attlebridge, Norfolk - 2 Group)

Blenheim IV Z7488/F - took off at 0940 with two others and one of two which failed to return. Under heavy fighter cover, a successful attack on the tanker was completed off Blankenberge in the face of fierce defensive fire from upwards of nine escorting flak ships. The tanker's stern was reported as being blown off. On turning for base, the Blenheims were attacked by some 20 enemy fighters and two were shot down into the sea. One crew was later sighted in a dinghy, but it is not known which this was. No further trace of Z7488's three crew was found and they are commemorated on the Runnymede Memorial.

Pilot: NZ39901 Plt Off **Tancred Eric COOPER**, RNZAF - Age 25. 444hrs.

Photography and wireless test

405 Squadron, RCAF (Pocklington, Yorkshire - 4 Group)

Wellington II W5492/K - captained by Sgt R G Chandos, RAF. Seen at 1105 to dive from 5000 feet and break up, portions of the tail unit breaking away at 1000 feet, followed by the port wing at 500 feet. The main wreckage fell and exploded on impact at Northfield Farm, 2 miles NNW of Pocklington. All eight on board died, the 2nd pilot and three others being buried at Barmby-on-the-Moor. The cause of the dive and subsequent break up was found to be due to the dinghy breaking loose from its stowage and lodging against the tailplane.

2nd Pilot: NZ402914 Sgt **Norman Watson THOMPSON**, RNZAF - Age 24.

FIGHTER COMMAND

'Circus 99' - escort to 11 Blenheims raiding a power station at Rouen, France

485 Squadron, RNZAF (Redhill, Surrey - 11 Group)

Spitfire VB AB903 - took off at 1405 with 10 others. A large number of enemy aircraft were engaged during the operation and AB903 was not seen again afterwards. The pilot is commemorated on the Runnymede Memorial.

Pilot: NZ402211 Flt Sgt **Arthur Ian PAGET**, RNZAF - Age 24. 308hrs. 48th op.

Fri 19 Sep 1941

FLYING TRAINING COMMAND

Practice instrument approaches

9 Service Flying Training School, RAF (Hullavington, Wiltshire - 23 Group)

Master I T8376 - took off with Fg Off Scragg as safety pilot and later seen entering a spin at about 1500 feet. Control appeared to be regained at about 500 feet, but T8376 was by then in a vertical dive from which it did not recover before crashing at Kingston St Michael, $2^1/_2$ miles south of Hullavington. Both pilots died, including the New Zealander, who is buried at Chippenham.

Pilot: NZ403439 Sgt **Joseph FYVIE**, RNZAF - Age 25. 208hrs.

Fri 19/Sat 20 Sep 1941

BOMBER COMMAND

Raid on Stettin, Germany (by 72 aircraft - 3 lost)

58 Squadron, RAF (Linton-on-Ouse, Yorkshire - 4 Group)

Whitley V Z6936/Q - took off at 2055 and brought down over Germany at Warnemünde, 11km NNW of Rostock. The captain, probably with his four comrades, was buried at Rostock, but all were later reinterred at Berlin.

Captain: 41313 Flt Lt **Russell Ernest ORCHARD**, DFC, RAF - Age 21. 677hrs.

Sat 20 Sep 1941

FAR EAST

Patrol off Chagos Archipelago, Indian Ocean

205 Squadron, RAF (Seletar, Singapore)

Catalina I W8411 - captained by Fg Off D Shaw, RAF. At about 1340, when it was two hours overdue, a signal was received advising that W8411 was about to land in the sea, but nothing further was heard. All nine crew are all commemorated on the Singapore Memorial.

2nd Pilot: NZ402224 Sgt **Lewis Llewelyn THOMAS**, RNZAF - Age 19.

FIGHTER COMMAND

'Circus 100B'- escort to bombers raiding Abbeville, France

602 Squadron, RAF (Kenley, Surrey - 11 Group)

Spitfire VB P8787 - took off at 1450 with 11 others and one of two lost. Several enemy aircraft were sighted during the raid, although few combats took place. Nothing was seen by Squadron members as to the fate of their two missing comrades. The body of P8787's pilot subsequently washed ashore onto the French coast during October and was buried in the cemetery at Chateau Ledquent, Marquise, 18km SW of Calais, but later reinterred in the Calais Canadian War Cemetery near Leubringhen, 6km NW of Marquise.

Pilot: 965683 Sgt **Ivor Morrison BROWN**, RAFVR - Age 27. 363hrs. 62nd op.

Brown's brother, Willis Stewart Brown, died on 3 October 1940 while flying with 20 Service Flying Training School, RAF.

Sat 20/Sun 21 Sep 1941

BOMBER COMMAND

Raid on Ostend, Belgium (by 28 aircraft - 1 lost)

101 Squadron, RAF (Oakington, Cambridgeshire - 3 Group)

Wellington IC X9922 - took off at 1912 and returned at 2300 in worsening weather with a failed wireless. In descending to locate its position X9922 collided with a tree on high ground, turned over and broke up at Preston Deanery, 4 miles SE of Northampton. Four of the crew died in the crash and the other two were injured, one seriously. The captain died next day in the Northampton General Hospital and is buried at Cambridge. Of 138 aircraft despatched by the Command to three enemy targets this night, 16 crashed in England or ditched near the coast, only three being lost over enemy territory.

Captain: NZ403434 Sgt **William Robert DIL**, RNZAF - Age 22. 308hrs (127 solo on Wellington) 6th op.

Sun 21 Sep 1941

FIGHTER COMMAND

Training exercises at 15000 feet

59 Operational Training Unit, RAF (Crosby-on-Eden, Cumberland - 81 Group)

Hurricane I P3623 - at 1540 dived vertically into the ground at high speed and caught fire at Westlington, 5 miles north of Carlisle. The pilot is buried at Carlisle. It was thought likely that he had either blacked out during a manoeuvre or was unconscious through an oxygen failure.

Pilot: NZ405227 Sgt **William BROWNLIE**, RNZAF - Age 24. 144hrs.

Wed 24 Sep 1941

MIDDLE EAST

Sea sweep

55 Squadron, RAF (LG 17 [Fuka Main], Egypt; detachment at LG 02 [Sidi Barrani] - 202 Group)

Blenheim IV Z6371 - took off at 1330 from Sidi Barrani piloted by Sgt W Bain, RAF. Made landfall at 1830 on returning from the sweep, but as Sidi Barrani's flare path could not be picked up made for Fuka. With an unserviceable wireless it was not possible to obtain a bearing. When what was believed to be Fuka's flarepath came into view the bomb load was jettisoned over the sea, pending a low level approach to inspect the ground at 2020. Moments later Z6371 struck a ridge and crashed, colliding with a tent on LG 68 ('Waterloo'), midway between Fuka and Bagush. Two of the crew were injured, but the observer died and was buried in the Maaten Bagush Cemetery, later being reinterred at El Alamein.

Observer: NZ401727 Sgt **Keith Allan HEPBURN**, RNZAF - Age 30. 104hrs. 5th op.

Thu 25 Sep 1941

FIGHTER COMMAND

Weather test(?)

245 Squadron, RAF (Chilbolton, Hampshire - 10 Group)

Hurricane IIB BD738 - took off at 1530 but during the flight collided with another of the Squadron's Hurricanes (AP524) when in the process of changing formation. Both aircraft crashed and burst into flames near Chilbolton, their pilots being buried at Brookwood in Woking, Surrey.

Pilot: 88246 Plt Off **Peter Gershom HERRICK**, RAFVR - Age 20. 450hrs. 74 ops.

Thu 25/Fri 26 Sep 1941

FLEET AIR ARM - ENGLAND

Night flying practice

761 Squadron, FAA (Yeovilton, Somerset - HMS *Heron*)

Fulmar II N4129 - spun in on turning to land and crashed near West Camel, 2 miles NE of the airfield. The pilot is buried at Yeovilton. Although death is recorded as being on the 25th, actual time of this accident is unknown. If it occurred shortly after midnight, rather than before, then the event should appear under the night of 24/25 September.

Pilot: Sub-Lt (A) **Douglas SIMMONDS**, RNVR - Age 23.

In the CWGC register Simmonds is incorrectly recorded as belonging to the RNZNVR, which was not formed until 1 October 1941.

Sat 27/Sun 28 Sep 1941

MIDDLE EAST

Raid on St Nicholas Harbour, Rhodes, Italy

108 Squadron, RAF (LG 221 [Fayid], Egypt - 202 Group)

Wellington IC T2729/V - captained by Sqn Ldr G F Irving, RAF. At 2145 a signal was received stating position as 3545N:2812E and that a course had been set for Turkey after an engine failure. This was followed by a message commencing 'LAN' but nothing further was heard. It was believed that T2729 ditched close to shore on the south side of the island of Rhodes. Its six crew are commemorated on the Alamein Memorial.

Air Gunner: 75697 Flt Lt **Malcolm James MILLER**, RAFVR - Age 31. 249hrs. 20th op.

Raid on Rhodes, Italy
37 Squadron, RAF (Shallufa, Egypt; detachment at LG 09 [Bir Koralyim] - 202 Group)
Wellington IC R1095/F - at 0130 five of the six crew were on board at a dispersal point on LG 09, awaiting their captain for take off, when a 250lb bomb became live and exploded. All five are buried at El Alamein. Another Wellington 15 yards away was severely damaged, two of its crew receiving head wounds.
Rear Gunner: 1066276 Sgt **Bernard George WATSON**, RAFVR - Age 32. 319hrs. 17th op.
CWGC register gives death date of this crew as the 27th, a contradiction possibly arising out of confusion between GMT and local time.

Sun 28 Sep 1941

MIDDLE EAST

Low level bombing attack on Bardia Harbour, Libya
55 Squadron, RAF (LG 17 [Fuka Main], Egypt - 202 Group)
Blenheim IV V5560 - took off at 0407 piloted by Sgt W A Ross, RAF, as one of three, two of which failed to return. After executing a low level attack on a merchant ship in the harbour V5560 was seen to continue inland at about 30 feet. Hit by flak, it crashed into a small house, killing the three crew and several German soldiers. The airmen were buried in the German Soldiers' churchyard near Bardia, but later reinterred at Halfaya Sollum in Egypt.
Observer: NZ401227 Sgt **Charles Ian JOHNSTON**, RNZAF - Age 28. 167hrs. 9th op.

COASTAL COMMAND

Bombing raid on St Nazaire, France
217 Squadron, RAF (St Eval, Cornwall - 19 Group)
Beaufort I W6501/D - took off at 1824 piloted by Plt Off Griffiths and brought down in vicinity of the target area. Two bodies later washed ashore but no trace could be found of the observer or fourth crew member, both of whom are commemorated on the Runnymede Memorial.
Observer: NZ401216 Sgt **Selwyn Clifford Edwin SAPWELL**, RNZAF - Age 23. 251hrs. 42nd op.

Sun 28/Mon 29 Sep 1941

BOMBER COMMAND

Raid on Frankfurt, Germany (by 44 aircraft - 6 lost)
99 Squadron, RAF (Waterbeach, Cambridgeshire - 3 Group)
Wellington IC Z8869 - took off at 1911 captained by Sgt J S Watt, RAF. Became lost on returning to England and at 0145 attempted a forced landing at Stone Farm, near Great Finborough, Suffolk. On touching down Z8869 tore through a hedge, struck a tree and haystack, then broke up and caught fire. Four of the crew died, including the observer, who is buried at Ringshall. The other two were seriously injured and died later.
2nd Pilot: NZ403937 Sgt **John Duncan BROADLEY**, RNZAF - Age 23. 213hrs.

Raid on Genoa, Italy (by 41 aircraft - 4 lost) (?)
99 Squadron, RAF (Waterbeach, Cambridgeshire - 3 Group)
Wellington IC T2879 (a presentation machine named *Ceylon I*) - took off at 1924 and apparently abandoned 10 or more hours later on the return journey. The Wellington crashed at 0740 near Landsmeer, a few kilometres north of Amsterdam. The bodies of three crew members later washed ashore onto the Dutch coast, including that of the front gunner, who is buried on the island of Texel at Den Burg. Their three comrades, including the captain, are commemorated on the Runnymede Memorial. Reflecting the increasing 'United Nations' composition of the bomber crews, these six airmen included members of the RAF, RAAF, RCAF and RNZAF. Some records give Genoa as the target while *Bomber Command Losses* states it to be Frankfurt - raids were made on both places this night. Time in the air suggests Genoa but the crash site fits either.
Captain: NZ401287 Flt Sgt **Joseph Stewart PARRY**, RNZAF - Age 26. 368hrs. 17th op.
Front Gunner: NZ402524 Sgt **Hugh McLennan MACDONALD**, RNZAF - Age 22. 148hrs. 8th op.

Mon 29 Sep 1941

FIGHTER COMMAND

Air-to-air camera gun exercise
59 Operational Training Unit, RAF (Crosby-on-Eden, Cumberland - 81 Group)
Master III W8594 - became lost during a heavy rainstorm with clouds and mist covering high ground. At 1530, while trying to locate its position, W8594 flew into a hillside at Tarn Beck Fell, 12 miles north of Gilsland. The pilot is buried at Carlisle.
Pilot: NZ402445 Sgt **George Frederick HILLIER**, RNZAF - Age 23. 184hrs.

Tue 30 Sep 1941

COASTAL COMMAND

Photographic reconnaissance of Kiel, Germany

1 Photographic Reconnaissance Unit, RAF (Benson, Oxfordshire - 16 Group)

Spitfire PR IV R7043 - lost without trace. The pilot's body later washed ashore and was buried at Wilhelmshaven, but later reinterred at Sage, 24km south of Oldenburg.

Pilot: 41080 Fg Off **James Frederick SWIFT**, DFC, RAF - Age 24. 986hrs. 45th op.

Unusually, Swift had completed an operational tour as a Wellington bomber pilot before converting to the Spitfire.

Wed 1 Oct 1941

NEW ZEALAND

Motor vehicle accident

RNZAF Station Harewood

Riding a motor cycle out of the aerodrome in the afternoon collided with a shingle lorry and was killed instantly, and is buried at Christchurch. The lorry driver suffered slight injuries.

(Flight Rigger): NZ414827 AC1 **Owen Bertie George SUTTON**, RNZAF - Age 25.

FLEET AIR ARM - SCOTLAND

Training exercise at Fort William

772 Squadron, FAA (Machrihanish, Argyllshire - HMS *Landrail*)

Roc L3073 - flew into a hill en route to Fort William in low visibility and burnt out at Airdo Farm, a mile NE of Cragnish Point, at the northern end of the Sound of Jura. Both crew died, the pilot being buried at Pennyfuir on the northern outskirts of Oban.

Pilot: Sub-Lt (A) **John William Terence COOPER**, RNZNVR - Age 21.

Coincidentally, 1 October 1941 was the date on which the New Zealand Naval Forces were officially reconstituted as the Royal New Zealand Navy. (Previously, naval airmen enlisting in New Zealand for service with the FAA joined as members of the New Zealand Division of the Royal Navy, although it seems few of them, if any, knew of this at the time.) Cooper was therefore almost certainly the first member of the new RNZN to lose his life.

Thu 2 Oct 1941

FIGHTER COMMAND

Tail chase training

58 Operational Training Unit, RAF (Grangemouth, Stirlingshire - 81 Group)

Spitfire I P9497 - at 1720, during an aerobatic manoeuvre at 5000 feet, entered a spin from which it did not recover, and crashed and caught fire at Cowie Colliery, Bannockburn, 2 miles SSE of Stirling. The pilot is buried at Grangemouth.

Pilot: NZ41318 Plt Off **Carlyle Gray EVERISS**, RNZAF - Age 26. 144hrs.

Fri 3 Oct 1941

FIGHTER COMMAND

Formation practice

610 Squadron, RAF (Leconfield, Yorkshire - 12 Group)

Spitfire IIA P7839 (a presentation machine named *City of Derry*) - lost height at 1145 while in a steep turn and dived into the ground a mile NNW of Bewholme, near Dunnington. The pilot is buried at Leconfield.

Pilot: NZ402911 Sgt **Andrew Albert TAYLOR**, RNZAF - Age 21. 243hrs. 3 ops.

Sat 4 Oct 1941

SOUTH PACIFIC

Instrument flying practice

Detached Flight, RNZAF (Nadi, Fiji)

DH60M Moth NZ522 - seen at 1145 flying on a steady course at 2000 feet over Nadi Bay, then suddenly to nose down and dive straight into the sea. Both airmen died, the pilot being commemorated on the Bourail Memorial. The safety pilot's body was recovered and buried at Lautoka, but later reinterred at Suva. No certain explanation could be found as to the cause of the accident, although a very remote possibility considered was that the pilot may have been struck by a stray bullet from Vincent NZ358, which was testing its front gun in the vicinity at the time.

Pilot: NZ404435 Plt Off **Ronald Leon Giles WILLIAMS**, RNZAF - Age 20. 92hrs solo (56 on Moth)

Safety Pilot: NZ404940 Plt Off **Francis Soltau PILLANS**, RNZAF - Age 26. 272hrs.

FIGHTER COMMAND

Formation practice

53 Operational Training Unit, RAF (Llandow, Glamorgan - 81 Group)

Spitfire I X4598 - thought to have been circling to establish its bearings at 0945, but stalled while turning and spun into a road, and caught fire near St Nicholas, 5 miles east of Cardiff. The pilot is buried at Llantwit Major.

Pilot: NZ41356 Sgt **Kenneth Charles PHILIP**, RNZAF - Age 22. 144hrs.

Thu 9 Oct 1941

MIDDLE EAST

Attack on railway line on east coast of Sicily

107 Squadron, RAF (Great Massingham, Norfolk; detachment at Luqa, Malta)

Blenheim IV Z7644 - took off with four others and one of two which failed to return. An Italian wireless report later stated that two Blenheims had collided and crashed on Sicily. Five of the six crew members died and are buried at Catania, including the pilot of Z7644.

Pilot: 41758 Fg Off **Neville Whitford WALDERS**, RAF - Age 24.

Fri 10 Oct 1941

MIDDLE EAST

Reconnaissance patrol, Crete-Libya

203 Squadron, RAF (LG 101 [7 miles SSE of Qasabab], Egypt - 201 Group)

Blenheim IV V6493/L - piloted by Fg Off J P Rowntree, RAF. At 1730 signalled estimated time of arrival but nothing further was heard. All three crew are commemorated on the Alamein Memorial.

Observer: NZ401435 Sgt **John Alfred EDWARDS**, RNZAF - Age 22. 282hrs. 34th op.

BOMBER COMMAND

Bombing practice

16 Operational Training Unit, RAF (Upper Heyford, Oxfordshire - 7 Group)

Hampden I P4319 - took off at 2115, piloted by Sgt C P A I Rose-Innes, RAF, climbed steeply, yawed to port and spun into the ground a mile SW of the airfield, and caught fire. All four crew died, three of whom, including the observer, are buried at Upper Heyford.

Observer: NZ401402 Plt Off **Brian William MILLER**, RNZAF - Age 20. 145hrs.

COASTAL COMMAND

'Nomad' - anti-shipping patrol between Texel and the Hook of Holland

500 Squadron, RAF (Bircham Newton, Norfolk - 16 Group)

Blenheim IV V6171/F - took off at 2125 and was later seen to crash into the sea at Overstrand, 2 miles SE of Cromer, pieces of wreckage later washing ashore. Two of the crew are commemorated on the Runnymede Memorial, but the pilot's body washed up further south along the coast and was buried at Caister north of Great Yarmouth on the 21st.

Pilot: 79738 Fg Off **Charles Cecil George WEBB**, RAFVR - Age 28. 547hrs. 59th op.

Fri 10/Sat 11 Oct 1941

BOMBER COMMAND

Raid on Cologne, Germany (by 69 aircraft - 6 lost)

75 (NZ) Squadron, RAF (Feltwell, Norfolk - 3 Group)

Wellington IC Z8909 - took off approximately 2350-0025 captained by Sgt R F Curlewis, RAAF, and brought down near the target area. All six crew were buried in the British 1914-18 War Cemetery in South Cologne, but later reinterred at Rheinberg, 12km south of Wesel.

2nd Pilot: NZ404427 Sgt **Colin Maurice THOMPSON**, RNZAF - Age 23. 219hrs. 3rd op.

Front Gunner: NZ404037 Sgt **Timothy Rowley MURPHY**, RNZAF - Age 20. 212hrs. 13th op.

Thompson's brother, William Alexander Thompson, died on 8 June 1944 while flying with 302 Ferry Training Unit, RAF.

Sun 12 Oct 1941

NEW ZEALAND

2 Elementary Flying Training School, RNZAF (Bell Block)

Received fatal injuries at 1545 at Kawaroa Park, New Plymouth, when one of the vertical iron bars holding a swing he was riding on gave way, causing him to fall head first on to the concrete and fracture his skull. Although rushed to the New Plymouth Hospital, he died at 1830 without regaining consciousness and is buried at New Plymouth.

(Pupil Pilot): NZ412650 LAC **Ronald Thomas BRANSGROVE**, RNZAF - Age 22.

Dual instruction

2 Elementary Flying Training School, RNZAF (Bell Block)

Tiger Moth II NZ659 - took off from Stratford airfield at about 1625 and carried out an aerobatic display with another of the School's Tiger Moths for about 20 minutes. While carrying out a downwind climbing turn to port at 500 feet NZ659 stalled and spun into the ground. The pilot instructor is buried at New Plymouth. The pupil, LAC **G F Forbes**, was admitted to the Stratford Hospital suffering only slight injuries.

Pilot Instr: NZ40819 Fg Off **Jack Rua COATH**, RNZAF - Age 31. 967hrs solo (713 on Tiger Moth)

Sun 12/Mon 13 Oct 1941

BOMBER COMMAND

Raid on Nuremberg, Germany (by 152 aircraft - 12 lost)

57 Squadron, RAF (Feltwell, Norfolk - 3 Group)

Wellington IC R1757/X - took off at 1900 captained by Sgt A W Jeffries, RAF, and brought down over the Netherlands, crashing near Blankenham, 14km WSW of Steenwijk. The sole survivor, an RCAF Plt Off, was taken prisoner. Four of his five comrades, including the observer, were buried in a grave dug by a Dutch family at the small fishing village of Vollenhove, 10km to the SSE of the crash area, but were later reinterred in a collective grave at Emmeloord. The fifth is buried at Kuinre, $4^1/_2$km NNW of Blankenham. Some records give R1757's target as Bremen, which was also raided this night.

Observer: NZ404531 Sgt **Percy Frederick Meadows COOKE**, RNZAF - Age 21. 243hrs. 4th op.

40 Squadron, RAF (Alconbury, Huntingdonshire - 3 Group)

Wellington IC X9619/M - took off at 1924 on same raid as the above, and brought down over Belgium near Dinant, where all six crew are buried.

Captain: NZ401757 Plt Off **Ian Murray Vass FIELD**, RNZAF - Age 26. 404hrs. 29th op.

Raid on Hüls, Germany (by 90 aircraft - 3 lost)

50 Squadron, RAF (Swinderby, Lincolnshire - 5 Group)

Hampden I AE367 - brought down 20km NNW of the target, crashing at Düsseldorf. All four crew died, the captain being buried at Dusseldorf, probably with the others. All four were later reinterred at Reichswald Forest, 5km SW of Kleve.

Captain: NZ402230 Plt Off **John Marshall WADDELL**, RNZAF - Age 22.

Tue 14 Oct 1941

COASTAL COMMAND

Navigation exercise

4 (Coastal) Operational Training Unit, RAF (Invergordon, Ross & Cromarty - 17 Group)

Lerwick I L7268 /Q - captained by Wt Off R G Bette, Free French Air Force. Shortly after setting course, at about 1100, the port engine began to 'pop'. The pilot put the nose down and throttled back, but got no response when pulling back on the control column. L7268 then dived steeply into the sea 6 miles ENE of Tarbat Ness. Nine of the 12 on board died, the other three surviving with injuries when thrown out from the top deck. The body of a Plt Off was later recovered but the others who died are commemorated on the Runnymede Memorial.

Observer: NZ404493 Sgt **Joseph Allan JOHNSTON**, RNZAF - Age 22. 224hrs.

The accident summary notes that two 'surplus' observers were being carried on the flight but it is not known if Johnston was one of these or part of the normal crew complement.

Wed 15 Oct 1941

NEW ZEALAND

Dual night flying practice

3 Service Flying Training School (ITS), RNZAF (Ohakea)

Oxford II NZ1238 - shortly after take off at 2145 from Bell Block was seen beginning a diving turn at low altitude. Failing to recover, NZ1238 struck the ground 2 miles NE of the airfield near Waitara and burst into flames. Both pilots were killed instantly and are buried at Auckland and Christchurch, respectively.

Pilot Instr: NZ1180 Flt Lt **Harold Cuthbert WILES**, RNZAF - Age 32. 1397hrs solo (222 on Oxford)
Pupil Pilot: NZ413529 LAC **Norman William WILSON**, RNZAF - Age 25.

FIGHTER COMMAND

Local flying practice

245 Squadron, RAF (Chilbolton, Hampshire - 10 Group)

Hurricane IIB Z3260 - took off at about 2035 and climbed to 1000 feet before starting a right hand turn, during which control was lost, Z3260 crashing into the ground 4 miles NW of Middle Wallop. The pilot is buried at Brookwood in Woking.

Pilot: NZ403544 Sgt **Allan Irving GATE**, RNZAF - Age 22. 253hrs.

BOMBER COMMAND

Night flying practice

11 Operational Training Unit, RAF (Bassingbourn, Cambridgeshire - 6 Group)

Wellington IC T2556 - lost flying speed after an overshoot at 2035, crashed and caught fire. One of the crew survived, injured, but his three comrades died, including the pilot, who is buried in the Bassingbourn cum Kneesworth Cemetery. Deteriorating weather conditions, which earlier in the day had curtailed commencement of the training programme, were thought to have been a factor in the accident.

Pilot: NZ391665 Sgt **Gordon James McGILL-NUTT**, RNZAF - Age 20. 159hrs (86 solo - 2 on Wellington)

Wed 15/Thu 16 Oct 1941

BOMBER COMMAND

Raid on Cologne, Germany (by 34 aircraft - 3 lost)

75 (NZ) Squadron, RAF (Feltwell, Norfolk - 3 Group)

Wellington IC W5663 - took off at approximately 1815-1855 captained by Sgt R C Barker, RAF, and brought down over Germany, crashing near Düsseldorf. Only an RAF Sgt and an RCAF Sgt survived to be taken prisoner. Their four comrades were buried in Düsseldorf's North Cemetery, but later reinterred at Reichswald.

2nd Pilot: NZ404948 Plt Off **Trevor Bernard ROBERTSON**, RNZAF - Age 26. 263hrs. 2nd op.

W5663 is the Wellington in which Sgt D C Joyce was killed during a raid on Duisburg on the night of 15/16 July 1941.

75 (NZ) Squadron, RAF (Feltwell, Norfolk - 3 Group)

Wellington IC X9916 - took off at approximately 1815-1855 captained by Sgt J A Matetich, RAF, on same raid as the above and brought down at Weiden, about 10km west of the target city's centre. The six crew were buried in Cologne's South Cemetery, but later reinterred at Rheinberg, 12km south of Wesel.

2nd Pilot: NZ404439 Sgt **Frederick Lionel Roy WOOD**, RNZAF - Age 23. 231hrs. 3rd op.

Observer: NZ391334 Flt Sgt **Neville Henry WELSH**, RNZAF - Age 20. 662hrs. At least 37th op.

Sun 19 Oct 1941

FIGHTER COMMAND

Forced landing approaches and aerobatics (slow rolls)

61 Operational Training Unit, RAF (Heston, Middlesex - 81 Group)

Spitfire I X4544 - after executing a loop at 2000-2500 feet at 1140, seen to bank steeply left, level out, then dive and crash at Slyfield Green, a mile north of Guildford, Surrey. The pilot is buried at Hounslow, Middlesex. An investigator's comment concluded that he may have been rendered unconscious through striking his head on the canopy during the aerobatic routine in the very turbulent conditions prevailing.

Pilot: NZ404382 Sgt **Donald Newsham LAW**, RNZAF - Age 23. 211hrs.

Mon 20 Oct 1941

MIDDLE EAST

High level formation bombing raid on Gambut landing ground, Libya

113 Squadron, RAF (Maaten Bagush, Egypt - 202 Group, Western Desert)

Blenheim IV V5641 - took off at 0710 piloted by 2nd Lt E H Burr, SAAF, with four others and failed to return. Accompanied by three Blenheims of 55 Sqn and a two-squadron escort of Hurricanes, as the formation turned left the target area enemy fighters attacked. During the engagement V5641 fell away and was not seen again. Its three crew are commemorated on the Alamein Memorial.

Observer: NZ40977 Sgt **Christopher Lewis JENKINSON**, RNZAF - Age 22. 177+hrs. 2nd op(?)

55 Squadron, RAF (LG 17 [Fuka Main], Egypt - 202 Group, Western Desert)

Blenheim IV V6228 - took off at 0720 with two others on same raid as the above and was attacked by four enemy fighters shortly after bombing at 0914. Took hits from the rear on the port engine and forward of the turret. Speed and height were gradually lost, V6228 ditching about 40 miles north of Gambut while two of the fighters circled overhead. The three crew died, however, and are commemorated on the Alamein Memorial.

Pilot: 40880 Sqn Ldr **Herbert George Percy BLACKMORE**, RAF - Age 27. 639hrs. 33rd op.

Observer: NZ401433 Sgt **William Dudley Linter COLE**, RNZAF - Age 21. 174hrs. 21st op.

FIGHTER COMMAND

Camera gun practice

61 Operational Training Unit, RAF (Heston, Middlesex - 81 Group)

Spitfire I R6765 - took off for practice with Spitfire I X4540 and during an attack or breakaway manoeuvre collided with it, causing both to crash near Heston. The pilot of R6765 is buried at Hounslow.

Pilot: NZ403977 Sgt **Charles Francis McCULLOUGH**, RNZAF - Age 21. 221hrs.

Mon 20/Tue 21 Oct 1941

BOMBER COMMAND

Raid on Bremen, Germany (by 153 aircraft - 10 lost)

408 Squadron, RCAF (Syerston, Nottinghamshire - 5 Group)

Hampden I P1212/T - took off at 1814, but on approaching Coningsby, Lincolnshire, for a landing upon return at 0045 both engines apparently cut and caused it to crash at Haltham, 4 miles NNW of the airfield. Three of the crew died and the fourth was seriously injured, although he too is thought to have died at a later date. The captain and RCAF observer are both buried at Newark-upon-Trent, Nottinghamshire.

Captain: NZ401801 Sgt **Desmond George BRADLEY**, RNZAF - Age 22. 322hrs. 10th op.

97 Squadron, RAF (Coningsby, Lincolnshire - 5 Group)
Manchester I L7462 - took off at 1830 captained by Plt Off W G Noble, RAF, on same raid as the above. A message received at 0030 indicated wireless difficulties were being experienced and L7462 is assumed to have crashed into the North sea. The bodies of two RAF Sgts washed ashore onto the East Anglian coast on 1 November. The 2nd pilot and three others are commemorated on the Runnymede Memorial.
2nd Pilot: NZ402184 Flt Sgt **Lester Louis HARRISON**, RNZAF - Age 26. 294hrs. 6th op.

106 Squadron, RAF (Coningsby, Lincolnshire - 5 Group)
Hampden I AD984 - took off at 1830 captained by Sgt J L Lockwood, RCAF, on the same raid as the above and brought down over Germany, crashing at Borby, to the north of Eckernförde, about 25km NW of Kiel. The observer was buried in the Kiel Garrison Cemetery, probably along with the other three crew members, but all were later reinterred in the Kiel War Cemetery.
Observer: NZ401296 Sgt **Geoffrey Fraser WILLIAMS**, RNZAF - Age 28.

Tue 21/Wed 22 Oct 1941

BOMBER COMMAND

Cross country
19 Operational Training Unit, RAF (Kinloss, Morayshire - 6 Group)
Whitley V P5103 - a severe glycol leak developed in the starboard engine, leading to engine failure and consequent loss of height. At 2359 P5103 dived from 1000 feet into the ground and caught fire at Ardestie, between Dundee and Abroath. One crew member was injured but the other four died, both pilots and the RCAF observer being buried at Kinloss.
Captain: NZ41390 Plt Off **Lewis Taylor YOUNIE**, RNZAF - Age 20. 163hrs.
2nd Pilot: NZ405322 Sgt **Harry Gervase ROUGHTON**, RNZAF - Age 24. 155hrs.

Wed 22 Oct 1941

NEW ZEALAND

Photography exercise, Wairau Valley
2 Service Flying Training School (ATS), RNZAF (Woodbourne)
Oxford I NZ1263 - took off at about 1515 and shortly afterwards stalled from 500 feet, dived and crashed into trees near Renwick, 2 miles west of the airfield. The pilots are buried at Waihi and Wellington respectively.
Pilot: NZ411161 LAC **Wylie MacDonald HOGG**, RNZAF - Age 19. 160hrs (48 solo - 29 on Oxford)
2nd Pilot: NZ413134 LAC **Barry Alfred SMITH**, RNZAF - Age 20.
Smith's brother, Maurice Victor Smith, RNZAF, died on 31 March 1943 while travelling as a passenger in a FAA aircraft in Scotland.

Wed 22/Thu 23 Oct 1941

BOMBER COMMAND

Raid on Mannheim, Germany (by 123 aircraft - 7 lost)
75 (NZ) Squadron, RAF (Feltwell, Norfolk - 3 Group)
Wellington IC X9914 - took off at approximately 1805-1815 captained by Sgt C Taylor, RAF, and brought down over Belgium near Werken, 8km east of Dixmude. All six crew are buried at Werken.
2nd Pilot: NZ401415 Sgt **Frederick Alexander SPARK**, RNZAF - Age 26. 282hrs. 1st op.

Thu 23 Oct 1941

COASTAL COMMAND

Convoy escort
236 Squadron, RAF (Carew Cheriton, Pembrokeshire - 19 Group)
Blenheim IVF V6013/V - returned to base at 2230 and while engaged in a very low circuit of the flarepath crashed into high ground at Hayes Farm gate on The Ridgeway Road, south of Carew Cheriton. One crew member was injured, but his two comrades died, the pilot being buried at Carew.
Pilot: NZ40651 Sgt **Francis Mervyn McCAFFRY**, RNZAF - Age 24. 432hrs. 54th op.

Thu 23/Fri 24 Oct 1941

BOMBER COMMAND

Reinforcement flight of eight aircraft from Alconbury to Luqa, Malta
40 Squadron, RAF (Alconbury, Huntingdonshire - 3 Group)
Wellington IC X9912 - took off at 2010 captained by Sgt J D Paine, RAF, and failed to arrive. Several of the Squadron's bombers only just made Malta after unexpectedly long flight times, because of unforecasted beam and head winds on the second half of the journey. Wellington X9912, off track and fuel exhausted, ditched into a rough running sea near Sicily. A Squadron ground crew member was the sole survivor of the ten on board. He drifted ashore onto the island some 30 hours later, unconscious, and was taken prisoner. The body of an RAF Sgt also washed up onto the island. No trace could be found of the other five aircrew or three ground crew, all of whom are commemorated on the Runnymede Memorial.

2nd Pilot: NZ403998 Sgt **William Douglas STUART**, RNZAF - Age 24.
Stuart, along with most of 40 Squadron's personnel and aircraft, was leaving Bomber Command on posting to Middle East Command. An echelon of the Squadron remained in the UK until February 1942, when it was renumbered 156 Sqn. Stuart's brother, Charles Alexander Stuart, died on 26 September 1942 while flying with 24 OTU, RAF.

Training(?)
10 Operational Training Unit, RAF (Abingdon, Berkshire - 6 Group)
Whitley V P5023 - took off at 0150 from Stanton Harcourt, Oxfordshire, captained by Flt Sgt C E Levitt, RAF. Shortly afterwards a mist began to settle over the airfield and P5023 was given a 'green' to land. While making a circuit the port wing struck the top of a 60 foot-high tree at Eaton, Berkshire, $2^1/_2$ miles SE of Stanton Harcourt. The impact caused the Whitley to roll left and crash inverted, and catch fire. Both pilots are buried at Brookwood in Woking, Surrey.
2nd Pilot(?): NZ41313 Sgt **William Royal CROWTHER**, RNZAF - Age 26. 166hrs.

Sat 25 Oct 1941

BOMBER COMMAND

Training
25 Operational Training Unit, RAF (Finningley, Yorkshire - 7 Group)
Hampden I P1234 - piloted by Plt Off J W Atkin, RCAF. Strayed off course in bad visibility and collided at 1245 with a barrage balloon cable over Weybridge, Surrey, crashed and caught fire. All four crew died, including the pilot and observer, who are buried at Brookwood in Woking.
Observer: NZ401398 Sgt **Lawrence Matthew George McDONALD**, RNZAF - Age 25.

FIGHTER COMMAND

Patrol
125 Squadron, RAF (Fairwood Common, Glamorgan - 10 Group)
Defiant I T3985 - took off at 2200 and swung away to starboard in error, either because of the pilot's inexperience in night flying or through his being distracted by radio traffic, or combination of both. Seven minutes later collided with a barrage balloon cable, both crew abandoning the machine by parachute. The air gunner landed safely, but the pilot died when he apparently baled out too low and is buried at Killay, 4 miles west of Swansea. The Defiant crashed into a hillside at Skewen, 4 miles NE of Swansea.
Pilot: NZ402170 Sgt **Clive Alan Gillions DALE**, RNZAF - Age 20. 303hrs.

Mon 27 Oct 1941

FIGHTER COMMAND

Aerobatic practice
610 Squadron, RAF (Leconfield, Yorkshire - 12 Group)
Spitfire IIA P7671 - stalled and spun into the ground at Winfield Farm, near Driffield, at 1450 after completing a slow roll. The pilot is buried at Leconfield. He was thought to have blacked out during one of the aerobatic manœuvres. A contributing factor may have been a spinal injury incurred four months earlier during a landing accident on return from an operational patrol.
Pilot: NZ402196 Plt Off **Thomas William McClure LECKIE**, RNZAF - Age 24. 266hrs. 10 ops.

Thu 30 Oct 1941

BOMBER COMMAND

Flight from Newmarket to Stradishall
138 Squadron, RAF (Newmarket, Suffolk - 3 Group)
Whitley V Z9223 - piloted by Flt Lt A J Oettle, DFC, RAF. Stalled on approach to Stradishall at 1159 and crashed just short of the runway. The three occupants, two crew and an LAC passenger, died, the observer being buried at Haverhill. Investigators noted that if the engines on the Whitley were opened full out with flaps down and the trimming wheel wound too far back, the nose would rise and the aircraft stall very quickly. On this occasion that tendency would have been aggravated by the fact that the passenger had seated himself in a position near the tail.
Observer: NZ40947 Sgt **Hugh Francis ROCHFORD**, DFM, RNZAF - Age 24.

Fri 31 Oct/Sat 1 Nov 1941

BOMBER COMMAND

Raid on Hamburg, Germany (by 123 aircraft - 4 lost)
51 Squadron, RAF (Dishforth, Yorkshire - 4 Group)
Whitley V Z9141/J - took off at 1739 captained by Flt Lt E A Barsby, RAF, and shot down over the Dutch coast by a night-fighter at 2350, crashing at De Cocksdorp, at the northern end of the island of Texel. All five crew are buried on the island at Den Burg.
2nd Pilot: NZ404424 Plt Off **Stanley David STEEL**, RNZAF - Age 21. 265hrs. 3rd op.

Raid on Dunkirk, France (by 25 aircraft - 1 lost)
76 Squadron, RAF (Middleton St George, Durham - 4 Group)
Halifax I L9602/N - took off at 1917 captained by Flt Sgt C S O'Brien, RCAF, and lost without trace, its seven crew being commemorated on the Runnymede Memorial.
2nd Pilot: NZ403876 Plt Off **Neil Frederick McLEAN**, RNZAF - Age 26. 307hrs. 4th op.

Tue 4/Wed 5 Nov 1941

COASTAL COMMAND

Convoy Strike off the West Frisian islands
500 Squadron, RAF (Bircham Newton, Norfolk - 16 Group)
Blenheim IV Z5959/X - took off at 2115 piloted by Fg Off W J Sipprell, RAF, and believed to have crashed into the sea off Terschelling Island. All three crew are commemorated on the Runnymede Memorial.
Observer: NZ402130 Sgt **Amos Thomas HALL**, RNZAF - Age 28. 2nd op.

Thu 6 Nov 1941

MIDDLE EAST

Acceptance test
148 Squadron, RAF (LG 213 [Kabrit], Egypt - 205 Group)
Wellington II Z8365 - took off captained by Plt Off H McMaster, RAAF, and later seen in vicinity of Kabrit at 800 feet with port engine stopped and propeller feathered. On reaching the NW corner of the airfield at 1240, began a turn to port as the starboard engine was opened up. The nose dropped as the turn tightened and Z8365 crashed out of control and burst into flames. All three crew are buried at Fayid.
2nd Pilot: 980442 Sgt **Donald Ferguson MUIR**, RAFVR - Age 26. 484hrs. 22 ops.
Muir's brother, Colin Bruce Muir, died on 4 December 1943 while flying with 487 Sqn, RNZAF.

Fri 7/Sat 8 Nov 1941

BOMBER COMMAND

Raid on Berlin, Germany (by 169 aircraft - 21 lost)
99 Squadron, RAF (Waterbeach, Cambridgeshire - 3 Group)
Wellington IC T2516 - took off at 1728 captained by Plt Off C G Gilmore, RAF, and brought down over the Netherlands, near Beilen, killing the six crew, who are buried there.
Observer: NZ403777 Flt Sgt **Ian Gordon HARROWBY**, RNZAF - Age 21. 247hrs. 5th op.

75 (NZ) Squadron, RAF (Feltwell, Norfolk - 3 Group)
Wellington IC X9951 - took off at approximately 1730-2054 captained by Plt Off W R Methven, RAF, on same raid as the above and brought down over Germany at 2300, crashing at Werdohl, about 65km east of Düsseldorf. All but one of the crew survived to be taken prisoner, including injured front gunner Sgt **T P Duffy**, RNZAF. The 2nd pilot died and was buried in Loh Military Cemetery at Lüdenscheid, 10km WSW of Werdohl, but later reinterred at Reichswald Forest, 5km SW of Kleve.
2nd Pilot: NZ404352 Sgt **John Cuthbert McKechnie GIBSON**, RNZAF - Age 29. 272hrs. 10th op.

75 (NZ) Squadron, RAF (Feltwell, Norfolk - 3 Group)
Wellington IC X9976 - took off at approximately 1730-2054 on same raid as the above and shot down over the Netherlands by a night-fighter at 0121, crashing near Oldeboorn, 15km SW of Drachten. All six crew were buried in the Utingeradeel Cemetery, but later reinterred at Bergen-op-Zoom.
Captain: NZ402843 Flt Sgt **John William BLACK**, RNZAF - Age 27. 316hrs. At least 7th op.
2nd Pilot: NZ404356 Sgt **Trevor Hedley GRAY**, RNZAF - Age 27. 135hrs. 7th op.
Observer: NZ402197 Plt Off **Eric LLOYD**, RNZAF - Age 28. 272hrs. 10th op(?)
Black's brother, Norman Scott Black, died on 1 March 1943 while flying with 76 Sqn, RAF.
Lloyd's brother, Raymond Edgar Lloyd, died on 19 April 1942 while flying with 37 Sqn, RAF.

'Rover patrol' to Munster, Germany (30 aircraft to various 'Rover' targets - 6 lost)
57 Squadron, RAF (Feltwell, Norfolk - 3 Group)
Wellington IC Z8985 - took off at 2010 and brought down over the Netherlands, crashing near Haastrecht, 4km SSE of Gouda. Three crew survived to be taken prisoner but their three comrades died. The captain was buried at Gouda, probably along with the others, but all were later reinterred at Bergen-op-Zoom.
Captain: Aus400580 Sgt **Andrew Harry Theodore COOK**, RAAF - Age 25. 288hrs. 10th op.

Sat 8 Nov 1941

FIGHTER COMMAND

'Ramrod' - escort to Hurricanes to bomb distillery at St Pol, France
72 Squadron, RAF (Gravesend, Kent - 11 Group)
Spitfire VB AB855 - took off at 1040. The escort was bounced by enemy fighters on the return journey, after which no trace of AB855 could be found. The pilot is commemorated on the Runnymede Memorial.
Pilot: NZ402232 Flt Sgt **David Richard WHITE**, RNZAF - Age 24. 344hrs. 53rd op.

Sat 8/Sun 9 Nov 1941

BOMBER COMMAND

Raid on Essen, Germany (by 62 aircraft - 7 lost)

75 (NZ) Squadron, RAF (Feltwell, Norfolk - 3 Group)

Wellington IC X9977 - took off at approximately 1730-1850 captained by Sgt G S Nunn, RAF, and shot down by over the Netherlands by a night-fighter at 2136, crashing at IJzevoorde, 3km east of Doetinchem. All six crew are buried at Doetinchem, 26km east of Arnhem.

2nd Pilot: NZ404011 Sgt **Thomas Young WYLLIE**, RNZAF - Age 25. 229hrs. 1st op.
Wireless Op: NZ403000 Flt Sgt **Alfred Hubert HARRISON**, RNZAF - Age 25. 188hrs. 11th op.

75 (NZ) Squadron, RAF (Feltwell, Norfolk - 3 Group)

Wellington IC Z8942 - took off at approximately 1730-1850 on same raid as above, and brought down over the Netherlands by flak at 2215, crashing at Zuidland, 17km SW of Rotterdam. Five of the crew are buried at Rotterdam. The sole survivor was the RNZAF rear gunner, Wt Off **L B H Hope**, who was taken prisoner. Tragically, he was to lose his life on 19 April 1945, less than three weeks prior to VE Day. While on a forced march from Stalag 357 at Fallingbostel to Lübeck with some 500 Allied prisoners, he was one of 30 who died at Gresse, 14km NE of Lauenburg, when RAF fighter-bombers attacked the column, mistaking them for enemy troops. The bodies were buried in the parish churchyard at Gresse, but in July 1947 all except that of an American were reinterred at Berlin.

Captain: NZ402530 Sgt **John Stephen WILSON**, RNZAF - Age 27. 305hrs. 13th or 14th op.
2nd Pilot: NZ402443 Plt Off **Ralph Owen FOSTER**, RNZAF - Age 29. 214hrs. 3rd op.
Rear Gunner: NZ40940 Wt Off **Lawrence Beresford Hamilton HOPE**, RNZAF - Age 28. 222hrs.

Sun 9/Mon 10 Nov 1941

BOMBER COMMAND

Raid on Ostend, Belgium (by 9 aircraft - 1 lost)

408 Squadron, RCAF (Syerston, Nottinghamshire - 5 Group)

Hampden I AE438/N - took off at 1755 captained by Plt Off J C Wilson, RCAF, and brought down over Belgium, crashing near Westkerke, 9km SE of Ostend. All four crew are buried at Westkerke.

Observer: NZ404556 Sgt **Evan Bertram Te Makahi ROBERTSON**, RNZAF - Age 29. 183hrs. 2nd op.

Tue 11 Nov 1941

FLEET AIR ARM - ENGLAND

(Duty?)

781 Squadron, FAA (Lee-on-Solent, Hampshire - HMS *Daedalus*)

Swordfish I W5847 - understood to have been operating with 55 Operational Training Unit, RAF, or from its base at Usworth, Co. Durham, 10 miles north of Durham. W5847 flew into high ground near Durham and caught fire, seriously injuring the pilot. He died later in the day at the Durham County Hospital and is buried at Hylton, near Sunderland, County Durham.

Pilot: Sub-Lt (A) **Gordon BLACK**, RNZNVR - Age 21.

FIGHTER COMMAND

Attack special objective south of Dieppe, France - ammunition wagons and dump

607 Squadron, RAF (Manston, Kent - 11 Group)

Hurricane IIB BE221 - took off at 1445 with seven others but went missing near St-Valery (whether St-Valery-sur-Somme, 46km to the NE of Dieppe or St-Valery-en-Caux, 27km to the west is not known). The pilot is buried at Dieppe. The formation had flown at low level to France with several other squadrons and found visibility so poor that some machines nearly hit cliffs, violent evasive action being required to clear them. One squadron turned back after becoming separated in 10/10ths cloud at low level.

Pilot: NZ404961 Plt Off **Linton William STEVENSON**, RNZAF - Age 24. 220hrs. 2nd op.

Escort to Hurricane fighter-bombers to St-Saëns, France

65 Squadron, RAF (Westhampnett, Sussex - 11 Group)

Spitfire VB W3629 - took off at 1455 on same raid as above, and failed to return. The circumstances of the Spitfire's final fate are unknown but the pilot is buried at Dieppe also. In addition to the poor weather, the Squadron's records note that accurate fire from two flak ships near the French coast was encountered en route to the target.

Pilot: NZ403470 Sgt **George Desmond MORCOM**, RNZAF - Age 25. 269hrs. 9th op.

Wed 12 Nov 1941

ARMY CO-OPERATION COMMAND

Practice(?)

41 Operational Training Unit, RAF (Old Sarum, Wiltshire - 70 Group)

Lysander II N1252 - overshot on making an approach at 1020 but was unable to apply sufficient power to go round again. N1252 made a brief climbing turn to port before losing height and crashing into some rifle butts. The air gunner was injured, but the pilot died and is buried at Salisbury. Subsequent investigation found that it was impossible to close or open the throttle fully due to the throttle nut being too tightly adjusted.

Pilot: NZ41325 Plt Off **Frederick Charles HACKETT**, RNZAF - Age 28. 177hrs.

Sat 15/Sun 16 Nov 1941

BOMBER COMMAND

Raid on Emden, Germany (by 49 aircraft - 7 lost)

99 Squadron, RAF (Waterbeach, Cambridgeshire - 3 Group)

Wellington IC L7873/J - took off at 1720 and lost without trace, its six crew being commemorated on the Runnymede Memorial.

Captain: NZ402891 Flt Sgt **Thomas Crawford Bentley PATTERSON**, RNZAF - Age 22. 278hrs. 7th op.

Patterson's brother, George Vernon Bentley Patterson, died on 8 August 1944 while flying with 630 Sqn, RAF.

Raid on Kiel, Germany (by 47 aircraft - 7 lost)

7 Squadron, RAF (Oakington, Cambridgeshire - 3 Group)

Stirling I W7445/V - took off captained by Sgt A C Bennett, RAAF, and crashed into a house immediately afterwards. The captain died and two of the crew survived unhurt. The four others were injured, including the 2nd pilot who was admitted to Addenbrooke's Hospital, Cambridge. He died the following day and is buried at Cambridge. A contributing factor to the crash was icing up of the windscreen during the take off run.

2nd Pilot: NZ403549 Sgt **Ian Hamish HUNTER**, RNZAF - Age 24. 274hrs. 5th op.

Wed 19 Nov 1941

FIGHTER COMMAND

Sector reconnaissance

153 Squadron, RAF (Ballyhalbert, Co. Down - 82 Group)

Defiant I N1645 - bounced on landing and went round again. On the second attempt at 2335 N1645 bounced again and collided with a mound of earth, and caught fire. Both air gunner and pilot were injured, but the latter died almost immediately and is buried at Ballyhalbert. The pilot, apparently concerned with the lack of suitable visibility for landing, had requested a floodlight to be switched on in order to make sure of the flarepath's position.

Pilot: NZ404880 Plt Off **Frederick Hugh ANDERSON**, RNZAF - Age 22. 237hrs.

Thu 20 Nov 1941

NEW ZEALAND

Solo night flying training

2 Service Flying Training School (ATS), RNZAF (Woodbourne)

Oxford I NZ1226 - took off shortly before 0300 and was seen to climb to 700 feet, then begin to lose height. The Oxford next approached the airfield at low level, passed over a floodlight, and crashed into the ground at Goulter Hill, 3 miles SW of the airfield, and caught fire. The pilot is buried at New Plymouth. Another of the School's Oxfords (NZ269) engaged in night flying also crashed at this time, its crew being distracted during approach by NZ1226's crash. They undershot and hit some tree tops, writing off the aircraft but escaping uninjured.

Pilot: NZ413547 LAC **Felix Patrick ULENBERG**, RNZAF - Age 26. 143hrs (102 solo - 21 on Oxford)

Navigation exercise - from base to Opunake and return

3 Service Flying Training School (ATS), RNZAF (Ohakea)

Oxford II NZ1245 - took off at 1030 but failed to return as scheduled at 1430. Despite an extensive aerial search and the offer of a £200 reward by Mr **W P Baker**, father of one of the pilots, the Oxford's fate remained a mystery until April 1943, when its remains were found in hill country near Maunganui Peak in the Akatarawas, 10 miles north of Upper Hutt. It appeared that NZ1245 became lost in bad weather and struck some trees. Rendered uncontrollable, it continued its course until, striking more trees, it overturned and crashed upside down on a hill, killing both pilots. Their bodies were recovered on 14 April and are buried at Matamata and Russell, respectively.

Pilot: NZ413814 LAC **Winston Stanley CANNELL**, RNZAF - Age 24. 137hrs (57 solo - 35 on Oxford)

(Pilot): NZ413804 LAC **Charles William BAKER**, RNZAF - Age 21. 100hrs.

Baker was probably acting as navigator for the exercise.

MIDDLE EAST

Offensive patrol El Adem area, south of Tobruk, Libya

250 Squadron, RAF (LG 123 [Fort Maddelena], Libya - 258 Wing, Western Desert)

Tomahawk IIB(?) (serial?) - took off at 1100 with 12 others and one of two which failed to return. Shortly after midday the Squadron intercepted a force of nearly 30 enemy aircraft and lost two of its number during the ensuing combats. The New Zealand pilot's body was found in the desert some weeks later by personnel of an Armoured Brigade. He was buried nearby, but later reinterred in Knightsbridge War Cemetery, a few miles from Acroma.

Pilot: NZ40983 Fg Off **Robert Sidney MASTERS**, RNZAF - Age 21. 518hrs.

FLYING TRAINING COMMAND

Dual training

8 Elementary Flying Training School, RAF (Woodley, Berkshire - 50 Group)

Magister I N3942 - while coming out of a loop at 1030 struck a tree and broke up after impact at Warren Row, 4 miles east of Maidenhead. The pupil pilot survived, injured, but the instructor died and was later cremated at Reading Crematorium, Caversham.

Pilot Instr: 108939 Plt Off **John Nesbitt SELLERS**, RAF - Age 28. 463hrs solo.

BOMBER COMMAND

11 Operational Training Unit, RAF (Bassingbourn, Cambridgeshire - 6 Group)
Died of natural causes on his fifth day at the RAF Hospital at Ely and is buried at Brookwood in Woking, Surrey.
(Pilot): NZ41471 Sgt **Francis Howard CREED**, RNZAF - Age 20.

Fri 21 Nov 1941

FIGHTER COMMAND

Practice
29 Squadron, RAF (West Malling, Kent - 11 Group)
Beaufighter IF R2183 - about to make a landing approach at 1430 but flew into high ground in poor visibility at Boxley Grange, 8 miles ENE of the airfield, killing both crew. The pilot is believed to have been first buried in the Boxley Grange Cemetery, but later reinterred at nearby Maidstone where he now lies.
Pilot: NZ404010 Sgt **Ralph Edmond William WILSON**, RNZAF - Age 21. 258hrs solo.

ENGLAND

(Duty?)
(Unit?) (Base?)
(Type?) (serial?) - a civilian scientist working for the Ministry of Aircraft Production lost his life in an aircraft accident in England on this date but further details have yet to be found.
(?): **George Gavin SAMSON**, Ministry of Aircraft Production - Age 34.
Because he was a civilian, Samson's name is not to be found in any CWGC register.

Sat 22 Nov 1941

MIDDLE EAST

Bomb tanks and mechanised transport on Acroma-El Adem road, Libya
45 Squadron, RAF (LG 75 [Maaten Bagush], Egypt - 270 Wing)
Blenheim IV Z6439 - took off at 1230 piloted by Wg Cdr J O Willis, DFC, RAF, with six others and failed to return. It had been intended that the Squadron be over the target at 1335, but avoiding storms en route delayed arrival by a crucial 30 minutes. Southeast of El Adem the formation and its escorting squadron of Tomahawks were intercepted by some 15 to 20 enemy fighters. Three Tomahawks and four of the Blenheims were brought down in the engagement. The three crew of Z6439 are buried in Knightsbridge War Cemetery, a few miles from Acroma.
Observer: NZ402122 Plt Off **Lawrence Philip BOURKE**, RNZAF - Age 31. 236hrs.

Sun 23 Nov 1941

FIGHTER COMMAND

Operational training
57 Operational Training Unit, RAF (Hawarden, Flintshire - 81 Group)
Spitfire I X4850 - circumstances unknown, but at about 1045 crashed near Moston Hill Farm, near Chester, Cheshire, 5 miles east of the airfield. The pilot is buried at Hawarden.
Pilot: NZ41973 Sgt **Douglas Anthony WITHER**, RNZAF - Age 22. 171hrs.

Tue 25 Nov 1941

MIDDLE EAST

Fighter sweep over Gambut area, Libya
112 Squadron, RAF (LG 122 [9 miles ENE of Fort Maddelena], Egypt - 262 Wing, Western Desert)
Tomahawk IIB AK461 - took off at 1525 with 11 others. Accompanied by seven Tomahawks from 3 Sqn, RAAF, the formation clashed with some 70 enemy aircraft attacking Allied troops and in the ensuing engagement AK461 was brought down. The pilot is commemorated on the Alamein Memorial.
Pilot: NZ402469 Sgt **Frederick Dunford GLASGOW**, RNZAF - Age 20. 123hrs. 5th or 7th op(?)

Operational
Sea Rescue Flight, RAF (Kabrit, Egypt; detachment at LG 39 [Burg el Arab South] - 201 Group)
Wellington IC T2551 - captained by Flt Sgt G L Swift, RAF. Stalled and side slipped into the ground while turning onto the flarepath on returning to land Burg el Arab at 1900. One crew member survived, injured, but his five comrades died. Their bodies were buried near the scene of the crash, but later reinterred at El Alamein. During the flight the front fuel filler cap on the port wing had come adrift, causing loss of lift and thereby raising the Wellington's stalling speed.
Observer: NZ401718 Fg Off **Walter John ALLATT**, RNZAF - Age 24.

Tue 25/Wed 26 Nov 1941

MIDDLE EAST

Bomb raid on Benghazi, Libya
37 Squadron, RAF (Shallufa, Egypt; detachment at LG 09 [Bir Koralyim] - 202 Group)
Wellington IC Z8798/L - took off at 1815 from LG 09 captained by Flt Lt **G E Guthrie**, RNZAF. At about 2130 an SOS was received but nothing further heard. It was later learned that Z8798 had been brought down about 50

miles short of the target, all five crew being taken prisoner. The 2nd pilot was subsequently liberated from a German camp by American forces in early April 1945, but then admitted to the 2nd Platoon, 308 Medical Bn Clearing Station on the 12th, suffering from an illness contracted while a prisoner. He died next day and was buried in the US Military Cemetery at Magraten, Netherlands, about 7km east of Maastricht, but later reinterred about 50km further north at Nederweert.
2nd Pilot: NZ402178 Wt Off **Gavin Matthew GOLDFINCH**, RNZAF - Age 23. 447hrs.

Sun 30 Nov 1941

FIGHTER COMMAND

Formation practice in cloud and Havoc night-fighter Co-operation
245 Squadron, RAF (Chilbolton, Hampshire - 10 Group)
Hurricane IIC BD953 - took off at 1500 but an hour later control was apparently lost while in cloud, the fighter diving into the ground at Greens Farm, Downend, a mile north of Chieveley, Berkshire. The pilot is buried at Brookwood in Woking.
Pilot: NZ405326 Sgt **Henry Lindsay SAUNDERCOCK**, RNZAF - Age 23. 219hrs.
Saundercock's brother, Charles Little Saundercock, died on 26 May 1943 while flying with 90 Sqn, RAF.

Sun 30/Mon 1 Dec 1941

BOMBER COMMAND

Raid on Hamburg, Germany (by 181 aircraft - 20 lost)
106 Squadron, RAF (Coningsby, Lincolnshire - 5 Group)
Hampden I P1228/L - took off at 1620 captained by Sgt A J Moore, RAF, and lost without trace, its four crew being commemorated on the Runnymede Memorial.
Observer: NZ404492 Sgt **Anthony Joseph William PARKER**, RNZAF - Age 26. 200hrs. 1st op.

Raid on Emden, Germany (by 50 aircraft - 3 lost)
10 Squadron, RAF (Leeming, Yorkshire - 4 Group)
Whitley V Z9166/O - took off at 1648 and lost sometime after a message sent at 1900 during the outward flight. An RAF Sgt's body was later recovered by a fishing vessel and buried at sea near Hartlepool, County Durham. No trace of his four comrades could be found, however, and all five crew are commemorated on the Runnymede Memorial.
Captain: NZ404400 Plt Off **Max Kiri NELSON**, RNZAF - Age 23. 364hrs. 5th op.

75 (NZ) Squadron, RAF (Feltwell, Norfolk - 3 Group)
Wellington IC Z1099 - took off at approximately 1650-1720 on same raid as the above and shot down over the Dutch coast by a night-fighter at 2305, crashing into the Waddenzee. All six crew died but only the body of an RAF Sgt was recovered, when it washed ashore on the island of Scheirmonnikoog. His five comrades are commemorated on the Runnymede Memorial.
Captain: NZ403959 Flt Sgt **Francis Charles HARRISON-SMITH**, RNZAF - Age 20. 326hrs. 19th op.

Wed 3 Dec 1941

COASTAL COMMAND

Escort patrol
201 Squadron, RAF (Lough Erne, County Fermanagh - 15 Group)
Sunderland II W3988/P - took off at 0630 captained by Flt Lt J G Fleming, RAF. Lost in poor visibility and failing light, at 1730 dropped flares in the vicinity of Carrowmore Point, Co. Clare, Eire. An hour later, after two landing attempts, W3988 set down near Doonbeg, 38 miles WNW of Limerick, losing a float in the process. All 11 crew, wearing their Mae Wests, safely disembarked into two dinghies. Unfortunately, in the rough seas, they were unable to keep together and paddle the few miles to shore. The only two survivors finally drifted ashore unconscious and were interned by the Irish authorities. Some of the dead later washed ashore, but the others, including the 2nd pilot, are commemorated on the Runnymede Memorial.
2nd Pilot: NZ401377 Plt Off **Wilfred Sefton EMMETT**, RNZAF - Age 26. 475hrs. 7th op.

Fri 5 Dec 1941

COASTAL COMMAND

Ferry flight from Oban to Pembroke Dock to collect the Squadron's new commanding officer
10 Squadron, RAAF (Oban, Argyllshire - 19 Group)
Sunderland I T9072/V - took off at 0925 captained by Sqn Ldr E B Courtney, RAAF, but on approaching Wales at 800 feet the port outer engine began backfiring, followed later by violent vibrations. T9072 began yawing to port and at 1121 become uncontrollable, diving into the sea at an angle of 45° to break up near the Skerries Lea Light, Anglesey. Eleven of the 15 on board, although injured, were rescued by the MV *Kilkenny*. The others, including two New Zealanders travelling as passengers, are commemorated on the Runnymede Memorial.
(Pilot): 36181 Flt Lt **William Samuel REA**, DFC, RAF - Age 26. 1628hrs. 77 ops.
(Pilot): 41499 Flt Lt **Richard John VAUGHAN**, DFC, RAF - Age 27. 1411hrs. 86 ops.
Rea and Vaughan, instructors with 4 (Coastal) Operational Training Unit, were travelling to Pembroke Dock in order to collect a Sunderland and ferry it to their base at Invergordon, Ross and Cromarty.

BOMBER COMMAND

Navigation exercise

20 Operational Training Unit, RAF (Lossiemouth, Morayshire - 6 Group)

Wellington I L4348 - took off captained by Flt Sgt H S D Goss, RAF, and later seen from another bomber on the same exercise to be burning on the ground about 14 miles from Strathy Point, Sutherland. L4348 had crashed at 1600 in conditions of clear weather and broken cloud, killing all five crew. Three, including a New Zealand wireless operator-air gunner, are buried at Wick, Caithness-shire.

WOpAG: NZ404111 Sgt **Allan James WILSON**, RNZAF - Age 20. 46hrs.

Sat 6 Dec 1941

MIDDLE EAST

Attack on Tripoli, Libya

89 Squadron, RAF (en route to Egypt; detachment at Malta)

Beaufighter IF T4784 - believed to have taken off from Luqa or Ta Kali, Malta, was brought down over Libya at Misurata, 175km SSE of Tripoli. Both pilot and navigator were captured, but the former was badly injured and died in a local hospital three days later. He was buried in the Misurata Cemetery, but later reinterred at Tripoli.

Pilot: NZ403769 Flt Sgt **Frederick Richard WALLIS**, RNZAF - Age 28. 333hrs. 4th op.

Sun 7 Dec 1941

MIDDLE EAST

Attack on a 'target in the Western Desert'

24 Squadron, SAAF (LG 130 [9 miles east of Bir Khamsa], Egypt)

Boston III (serial?) - believed to have taken off from LG 122 (9 miles ENE of Fort Maddelena) with a crew of four. Brought down behind enemy lines, with the observer thought to have been the sole survivor. He was taken prisoner but died in Germany from malnutrition and pleurisy on 6 March 1945, possibly at Bad Sulza, 15km SW of Naumburg. Post-war searches could find no trace of his grave and he is commemorated on the Alamein Memorial.

Observer: NZ401215 Wt Off **Cyril James RUDGE**, RNZAF - Age 23. 94hrs. 3rd op(?)

FLYING TRAINING COMMAND

Solo cross country exercise

2 Service Flying Training School, RAF (Brize Norton, Oxfordshire - 23 Group)

Oxford I P1930 - went out of control at 1030 in a snow shower and dived into the ground at Boughton Heath, a mile west of Bourton-on-the-Hill near Moreton-in-Marsh, Gloucestershire. The pilot is buried at Moreton-in-Marsh. An observation noted on the accident card is that this was the 'first accident [for 2 SFTS?] under adverse weather conditions, in which a Dominion trained pupil had been involved. [We] should warn New Zealand pupils in particular, to whom such conditions are novel, of what is involved in such flying.' The pilot had undergone training in New Zealand at New Plymouth and Ohakea before proceeding direct to England.

Pilot: NZ411387 Plt Off **Leslie Kitchener DUNCAN**, RNZAF - Age 25. 151hrs.

Duncan's brother, Stanley Logan Duncan, died on 30 August 1943 while flying with 16 Sqn, RNZAF.

FIGHTER COMMAND

Ferrying pilots from Litchfield, Hampshire to Hawarden

3 Delivery Flight, RAF (Hawarden, Flintshire - 9 Group)

Dominie I X7354 - weather conditions deteriorated as the flight progressed and in an unforecasted snowstorm was struck by lightning, rendering the blind flying instruments unserviceable. The pilot ordered his three passengers to bale out, but he and his wireless operator died when the Dominie crashed at 1720 at Hordley, 3 miles SSW of Ellesmere, Shropshire. Neither of the crew were able to leave X7354 quickly as they were not wearing parachute harnesses at the time of the lightning strike. The pilot's remains were cremated at the Birkenhead Crematorium, Cheshire.

Pilot: 60527 Plt Off **John Ernest MOODIE**, RAFVR - Age 24. 475hrs.

Moodie's brother, William Anderson Moodie, died two months later, on 9 February 1942 while flying with 232 Sqn, RAF.

Convoy patrol

3 Squadron, RAF (Hunsdon, Hertfordshire - 11 Group)

Hurricane IIC Z3916 - on returning in snow showers and thick cloud at 1810, struck a tree and dived into the ground a $^1/_4$ mile north of Thornwood village, in the neighbouring county of Essex, 6 miles SE of Hunsdon. The pilot is buried at Ware, Hertfordshire. Investigators highlighted the pilot's lack of night flying experience; he had been posted in from his operational training unit with a total of ten hours, of which only two had been on instruments.

Pilot: NZ404913 Plt Off **John Douglas McVAY**, RNZAF - Age 28. 245hrs. 3rd op.

BOMBER COMMAND

'Dark Night' training

21 Operational Training Unit, RAF (Moreton-in-Marsh, Gloucestershire - 6 Group)

Wellington IC Z8784 of 'E' Flight - lost height because of icing and at 2000 crashed into a ploughed field, colliding with trees before burning out. Five of the crew were injured but the captain died and is buried at Benson, Oxfordshire. Immediately prior to the crash the 2nd pilot had his feet up on the dashboard in a desperate attempt to gain leverage in pulling back the control column to overcome the locking effect of ice on the controls. Two more of the Unit's aircraft were lost on the same night, also due to icing up.

Captain: NZ405256 Plt Off **John Alexander GRIERSON**, RNZAF - Age 21. 202hrs (110 solo - 28 on Wellington)

COASTAL COMMAND

Anti-submarine sweep

269 Squadron, RAF (Kaldadarnes, Iceland)

Hudson III T9416 - took off at 1200 captained by Flt Sgt E R Budgell, RAF. Returning to base in unforecasted deteriorating weather at 1645, flew into a hillside at 650 feet and caught fire, 8 miles SW of Kaldadarnes. All four crew are buried at Reykjavik. An investigator's comment on the accident drew attention to the fact that the loss could have been avoided, had a landmark beacon been available on the coast to enable the pilot to locate the airfield.

2nd Pilot: NZ40244 Fg Off **Eric Edgar STEWART**, RNZAF - Age 22. 584hrs. 29th op.

Sun 7/Mon 8 Dec 1941

BOMBER COMMAND

Raid on Boulogne, France (by 19 aircraft - 1 lost)

61 Squadron, RAF (Woolfox Lodge, Rutland - 5 Group)

Manchester I L7494 - took off at 1742 captained by Sqn J L Riley, RAF, and exploded near the target before crashing into the sea. The body of an RAF Sgt later washed ashore near Boulogne, but his six comrades are commemorated on the Runnymede Memorial.

Air Gunner: NZ40942 Flt Sgt **James Benjamin LEIGH**, RNZAF - Age 23. 104hrs. 16th op.

Leigh's brother, Robert Eric Leigh, died on 10 February 1944 while flying with 109 Sqn, RAF.

Raid on Dunkirk, France (by 22 aircraft - 2 lost)

104 Squadron, RAF (Driffield, Yorkshire - 4 Group)

Wellington II Z8426/E - took off at 2005 captained by Sgt R K Anson, RAF, and lost without trace, its six crew being commemorated on the Runnymede Memorial.

2nd Pilot: NZ40421 Plt Off **Bernard Norris MURDOCH**, RNZAF - Age 27. 293hrs. 5th op(?)

Murdoch's brother, Graham Edward Murdoch, died on 9 June 1942 while flying with 75 (NZ) Sqn, RAF.

Mon 8 Dec 1941

MIDDLE EAST

Escort to 18 Blenheims bombing El Adem area, Libya

274 Squadron, RAF (LG 103 [10 miles south of Maaten Bagush], Egypt - 262 Wing - Western Desert)

Hurricane IIB BD885 - took off at 1125 with the rest of the Squadron as leader of a section of four. While over the target area a large force of enemy fighters was engaged. Heavily outnumbered, two Hurricanes were seen to go down, one in flames and another in a vertical dive, while another was obliged to force land at Tobruk. The pilot of BD885 was buried at the crash site, but later reinterred at Halfaya Sollum, Egypt.

Pilot: 42774 Flt Lt **Owen Vincent TRACEY**, DFC, RAF - Age 26.

Tue 9 Dec 1941

MIDDLE EAST

Flt Sgt **F R WALLIS** died at Misurata, Libya, of injuries received when his aircraft was brought down on the 6th.

MAINTENANCE COMMAND

(No.?) Ferry Pilots Pool(?), RAF (41 Group)

Died of natural causes.

(Pilot): 2nd Off **William Arundell STEWART**, Air Transport Auxiliary - Age 26.

Stewart's name, recorded in Brief Glory - The story of the A.T.A. *under a 'list of officially recognised ferry pilots and air crew who flew operational aircraft at A.T.A. Pools', and noted therein as 'killed while flying [sic] with A.T.A.', does not feature in any CWGC register.*

Wed 10 Dec 1941

FLEET AIR ARM - SCOTLAND

Solo training

767 Squadron, FAA (Arbroath, Angus - HMS *Condor*)

Albacore I N4184 - stalled immediately after take off and crashed a $^1/_2$ mile from the SW boundary of the airfield. The pilot is buried at Idle, a few miles north of Bradford, Yorkshire.

Pilot: Sub-Lt (A) **Lawrence Edmund MITCHELL**, RNZNVR - Age 24.

FIGHTER COMMAND

Air-to-sea firing practice

59 Operational Training Unit, RAF (Crosby-on-Eden, Cumberland - 81 Group)

Hurricane I W9342 - in rapidly deteriorating weather between 1500 and 1600, apparently entered a shallow dive through sea mist in order to establish position, but flew into the ground and caught fire, near Aspatria, 8 miles SW of Wigton. The pilot is buried at Carlisle.

Pilot: NZ411486 Plt Off **Alexander George WICKS**, RNZAF - Age 28. 152hrs.

Thu 11 Dec 1941

MIDDLE EAST

Attack on shipping in Argostóli Harbour, Cephalonia Island, Greece

18 Squadron, RAF (Luqa, Malta)

Blenheim IV Z7802 - took off with two others and brought down in or near the harbour during an attack on a ship. The bodies of its three crew were recovered from the sea, those of the pilot and observer being buried at Argostóli, possibly along with that of the air gunner. All were later reinterred at Athens.

Pilot: NZ403312 Flt Lt **Everest George EDMUNDS**, DFC, RNZAF - Age 26. 227hrs.

BOMBER COMMAND

Cloud cover raid to Brest, France

144 Squadron, RAF (North Luffenham, Rutland - 5 Group)

Hampden I AE353 - captained by Plt Off C B Payne, RAF. Returning from the raid crashed into the sea off the Sussex coast, killing the five crew. Four bodies were recovered, the observer and an RCAF Plt Off being buried nearby at Littlehampton. The fifth crew member is commemorated on the Runnymede Memorial. No communication was received from AE353 beforehand and it may have been damaged during the raid. A crew of five in a Hampden was unusual.

Observer: NZ402220 Plt Off **Archibald Locksley SMILEY**, RNZAF - Age 28. 188hrs. 1st op(?)

Thu 11/Fri 12 Dec 1941

BOMBER COMMAND

'Gardening' - mining in Kiel Bay, Germany (by 5 aircraft - 1 lost)

408 Squadron, RCAF (Balderton, Nottinghamshire - 5 Group)

Hampden I AE148/B(bar) - took off at 1715 piloted by Plt Off W F Hull, RCAF, and brought down over Denmark, crashing at Sanderum, 5km SW of Odense. All four crew are buried at Odense.

Observer: NZ404575 Flt Sgt **Duncan Luin TODD**, RNZAF - Age 29. 192hrs. 5th op.

Fri 12 Dec 1941

MIDDLE EAST

Fighter sweep of Derna area, Libya

112 Squadron, RAF (LG 122 [9 miles ENE of Fort Maddelena], Egypt - 258 Wing, Western Desert)

Tomahawk IIB AK457 - took off with the Squadron at 1525 and one of three which failed to return. Accompanied by 3 Sqn, RAAF, the formation engaged a large force of enemy aircraft over Tmimi, but lost three of its number during the ensuing combats. The pilot of AK457 is commemorated on the Alamein Memorial.

Pilot: NZ402473 Sgt **William Earl HOUSTON**, RNZAF - Age 19. 226hrs. 11th op.

BOMBER COMMAND

Cloud cover raid to Gelsenkirchen, Germany

106 Squadron, RAF (Coningsby, Lincolnshire - 5 Group)

Hampden I AE391 - took off at 1040 captained by Flt Lt H D Webber, RAF, and brought down over Germany, crashing between Oberhausen and Osterfeld. The 2nd pilot was buried at Düsseldorf, probably along with others of the crew, all four later being reinterred at Reichswald Forest, 5km SW of Kleve.

2nd Pilot: NZ402854 Flt Sgt **Samuel Victor CROWHURST**, RNZAF - Age 21. 319hrs. 6th op(?)

Crowhurst's brother, Jack Crowhurst, died on 14 January 1945 while serving at RNZAF Station Woodbourne.

Cloud cover raid to Cuxhaven, Germany

49 Squadron, RAF (Scampton, Lincolnshire - 5 Group)

Hampden I AD979 - took off at 1018 captained by Plt Off Robinson, and attacked an aerodrome at Cuxhaven from 100 feet. There was intense flak, the navigator being killed by a cannon shell and AD979 badly damaged. Returning to England at 1635, a crash landing was made at Bircham Newton, Norfolk. The dead crew member was buried at nearby at Great Bircham.

(Pilot): NZ402844 Flt Sgt **Stuart Corliss BLACK**, RNZAF - Age 27. 357hrs. 7th op(?)

Black was acting as navigator on this flight.

Fri 12/Sat 13 Dec 1941

MIDDLE EAST

Raid on enemy mechanised transport near Derna airfield, Libya

70 Squadron, RAF (LG 213 [Kabrit], Egypt - 205 Group)

Wellington IC X9686 - took off at 0130 captain by Fg Off K C Kitto, RAF, and brought down over Libya at about 0400, crashing near Martuba, to the south of Derna. All six crew were buried in a collective grave at the Martuba landing ground, but later reinterred in Knightsbridge War Cemetery, a few miles from Acroma.

Air Gunner: NZ40924 Flt Sgt **William Henry John MILNER**, RNZAF - Age 32. 532hrs. 51st op.

Sat 13 Dec 1941

FAR EAST

Squadron move from Sembawang to Ipoh, Malaya

453 Squadron, RAAF (Sembawang, Singapore)

Buffalo I W8158 - took off at about 1055 with two others. The formation became lost after being forced off course due to bad weather. Unwittingly the trio crossed over the Strait of Malacca and shortly after 1300, with fuel near exhausted, was obliged to carry out forced landings on Sumatra, probably to the SW of Pakanbaru. The first Buffalo, mistaking a paddy swamp for a hard field of corn or wheat, landed and instantly overturned several times. Its pilot, although injured, extricated himself safely from the inverted wreck and vainly tried to wave the others away. However, the second fighter continued to approach wheels up and hit a tree, overturned and caught fire, while W8158 dived or spun into the swamp during its approach and caught fire also. Both pilots were buried at Taloek. Postwar they were reinterred in the Dutch war cemetery at Medan, but at a later date were taken to Java for final reinternment at Djakarta.

Pilot: NZ41867 Plt Off **David Roderick Llewelyn BROWN**, RNZAF - Age 24.

BOMBER COMMAND

'Gardening' - cloud cover flight to lay mines off Brest, France (by 10 aircraft - 2 lost)

44 Squadron, RAF (Waddington, Lincolnshire - 5 Group)

Hampden I AE196 - took off at 1300 captained by the commanding officer, Wg Cdr S T Misselbrook, DSO, RAF, and lost without trace, its four crew being commemorated on the Runnymede Memorial.

(Pilot): NZ402865 Flt Sgt **Arthur Howard GUMBLEY**, RNZAF - Age 25. 305hrs. 8th op.

Although a pilot, Gumbley was acting as navigator on this flight.

Sat 13/Sun 14 Dec 1941

FLEET AIR ARM - MIDDLE EAST

Raid on Derna airfield, Libya

826 Squadron, FAA (Maaten Bagush, Egypt - HMS *Grebe*)

Albacore I T9151 - stopped at El Adem, Libya, at dusk for refuelling prior to undertaking the raid. T9151 was the first aircraft to depart, piloted by Sub Lt J S Kyffin, but took off downwind, failed to lift off, and collided with a Blenheim off the end of the runway. The Albacore's bomb load exploded and all three crew died. Although they were buried at the time at El Adem, their graves could not be located after hostilities and they are now commemorated on the Fleet Air Arm Memorial at Lee-on-Solent, Hampshire.

Observer: Lt **Thomas Ian HARRAY**, RN - Age 25.

Mon 15 Dec 1941

NEW ZEALAND

Operational perimeter patrol, area 'D'

2 Squadron, RNZAF (Nelson)

Hudson III NZ2007 - took off at 1650 and failed to return. Eight days passed before the bomber's fate was finally determined. In bad visibility at about 1900 NZ2007 had struck the side of a hill at 700 feet, 3 miles NW of Westhaven Inlet (also known as Whanganui Inlet). It exploded on impact, killing the four crew who are buried at Nelson, Stratford, Levin, and Wellington respectively.

Pilot: NZ401400 Plt Off **Alexander John MACFARLANE**, RNZAF - Age 25. 461hrs solo (92 on Hudson)
(Pilot): NZ404432 Plt Off **Louis George William VOLZKE**, RNZAF - Age 27. 342hrs.
WOp: NZ391703 Sgt **Keith Robertson JACK**, RNZAF - Age 20. 308hrs.
WOpAG: NZ401929 Sgt **Philip David HARRISON**, RNZAF - Age 20. 325hrs.

Volzke was acting as navigator for the exercise.
Jack had been promoted to Sgt the day of the crash.

BOMBER COMMAND

Training

11 Operational Training Unit, RAF (Bassingbourn, Cambridgeshire - 6 Group)

Wellington IC X3170 - while descending gradually at a high speed at 2300, flew into the ground at Whittlesey Mere, a few miles NW of Upwood, Huntingdonshire, and completely disintegrated. All six crew died, including the two pilots, who are buried in the Bassingbourn cum Kneesworth Cemetery.

Captain: NZ41363 Sgt **Albert Leonard SANDERSON**, RNZAF - Age 22. 214hrs (85 solo - 26 on Wellington)
2nd Pilot: NZ405240 Sgt **Dennis James DEMPSEY**, RNZAF - Age 22. 202hrs.

CANADA

Solo night circuits and landings
6 Service Flying Training School, RCAF (Dunnville, Ontario - 1 Training Command)
Harvard II 2762 - while turning to starboard on the downwind leg for landing at 2355, crashed and caught fire near Port Maitland, about 4 miles SE of the airfield. The pilot is buried at Dunnville.
Pilot: NZ411516 LAC **Alexander Collard WILLIS**, RNZAF - Age 25. 113hrs (35 on Harvard)
Willis was the first RNZAF airman to lose his life in a Canadian flying accident under the Empire Air Training Scheme.

Mon 15/Tue 16 Dec 1941

NEW ZEALAND

Motor vehicle accident
RNZAF Station Levin
Accidentally knocked down by a fire tender shortly after 2145 while in the vicinity of the flarepath at Ohakea. Admitted to Palmerston North Pubic Hospital with serious injuries, he made a partial recovery and returned to his home town of Ormondville. In May the following year he was admitted to the Dannevirke Public Hospital for an operation, and for another on June 10. The last proved unsuccessful, however, and he died next morning at 1000, and is buried at Ormondville.
(Aircrafthand): NZ414047 AC2 **Douglas Lionel FORWARD**, RNZAF - Age 21.

Tue 16 Dec 1941

NEW ZEALAND

Transit from Birdlings Flat to Wigram
1 Service Flying Training School (ATS), RNZAF (Wigram)
Oxford I NZ280 - took off at 1355, but strayed off course in heavy rain and low cloud and ten minutes later flew into a hillside at 900 feet in Gebbies Valley, about 2 miles NE of Motukarara, near the crest of a ridge, and caught fire. The staff pilot and his three pupil pilot passengers died and are buried at Onehunga, Waimairi, Christchurch, and Wellington respectively.
Pilot: NZ412249 Plt Off **Alan Leslie MARSHALL**, RNZAF - Age 21. 200hrs solo (103 on Oxford)
(Pilot): NZ413810 LAC **Ronald Stephen BROOKER**, RNZAF - Age 26.
(Pilot): NZ413986 LAC **Brian Patrick FITZGERALD**, RNZAF - Age 19.
(Pilot): NZ413435 LAC **William Ian LINDSAY**, RNZAF - Age 29.
All three pupil pilots were due for graduation in four days time.

Tue 16/Wed 17 Dec 1941

FIGHTER COMMAND

Transport flight from England to Gibraltar (or vice versa?)
24 Squadron, RAF (Hendon, Middlesex - 11 Group)
Hudson III V9115 - piloted by Fg Off J H M Davies, DFC, RAF, crashed into the sea off the coast of Spain near Corunna. The four crew and their passengers - two Navy personnel and a civilian - were lost without trace, the former being commemorated on the Runnymede Memorial.
(Pilot): 36240 Fg Off **Douglas Veale GILMOUR**, mid, RAF - Age 26. 810hrs. At least 36th op.
Gilmour was acting as navigator on this flight. Prior to joining 24 Sqn he had completed a tour of 35 ops with Bomber Command. The number undertaken with 24 Sqn is unknown.

Thu 18 Dec 1941

FAR EAST

Formation practice
488 Squadron, RNZAF (Kallang, Singapore)
Buffalo I W8175 - one of three on approach to land back at the airfield at 1537. Coming in low over the eastern boundary, the formation cleared a line of trees, but the starboard wing of W8175 collided with trees of the next row, causing the fighter to crash and burst into flames. The pilot was buried in Singapore's Bidadari Civil Cemetery, but later reinterred at Kranji. The approach was being made towards the sun, which may have obscured the pilot's vision.
Pilot: NZ41882 Sgt **Alexander Gray CRAIG**, RNZAF - Age 23. 115hrs solo.

BOMBER COMMAND

Raid on warships at Brest, France (by 47 aircraft - 8 lost)
7 Squadron, RAF (Oakington, Cambridgeshire - 3 Group)
Stirling I N3680/Y - took off at 0940 captained by Flt Lt B Parnell, mid, RAF, and shot down into the sea by fighters off the French coast. All seven crew are commemorated on the Runnymede Memorial.
2nd Pilot: NZ404425 Sgt **Kenneth Roy TAYLOR**, RNZAF - Age 26. 324hrs. 9th op.

15 Squadron, RAF (Wyton, Huntingdonshire - 3 Group)
Stirling I N3665/S - took off at 0950 captained by Flt Lt G G Heathcote, mid, RAF, on same raid as the above. Last seen in the target area with an engine streaming smoke and being engaged by fighters, N3665 was believed to have crashed into the sea. The body of an RAF Plt Off later washed ashore, while his six comrades are commemorated on the Runnymede Memorial.
Mid-Upper Gunner: NZ401267 Flt Sgt **Clarence Phillip GOULDING**, RNZAF - Age 29. 103hrs. 1stop.

Fri 19 Dec 1941

NEW ZEALAND

Solo flying practice
1 Service Flying Training School (ITS), RNZAF (Wigram)
Oxford I NZ1314 - apparently caught in a severe hailstorm which critically damaged its port wing. At 1445 was seen flying at 500 feet with pieces of fabric disintegrating and falling away. NZ1314's nose dropped and the throttles were opened to level up, but the machine continued to lose height, veering to port before striking the ground at high speed at Yaldhurst, about 4 miles NW of the airfield. The pilot is buried at Christchurch. Two other Oxfords from Wigram were damaged by hail in flight during the day.
Pilot: NZ415003 LAC **Clifford Henry John MILLAR**, RNZAF - Age 21. 33hrs solo (10 on Oxford)

Sun 21 Dec 1941

NEW ZEALAND

Circuits and landings training
1 Squadron, RNZAF (Whenuapai)
Hudson III NZ2020 - seen shortly before 1500 flying over hilly country in extremely turbulent conditions and apparently experiencing engine difficulties. NZ2020 stalled a few minutes later at about 1500 feet and began a steep diving turn to starboard, from which it failed to recover, crashing 5 miles west of Helensville and exploding. All three on board died, two being buried at Auckland and Christchurch, respectively, while the last named was cremated at Auckland. All RNZAF Hudsons in New Zealand and Fiji were grounded for several days afterwards.
Captain: NZ1281 Fg Off **Eric Francis HOLDAWAY**, RNZAF - Age 34. 745hrs.
(Pilot): NZ413393 Sgt **George Bevan EATHORNE**, RNZAF - Age 27. 152hrs.
(Pilot): NZ413396 Sgt **Eric ELDERSHAW**, RNZAF - Age 21. 150hrs.
Eathorne and Eldershaw were probably on board for type familiarisation and instruction.

COASTAL COMMAND

Bombing oil refinery at Donges, near St Nazaire, France
206 Squadron, RAF (Aldergrove, County Antrim; detachment at Chivenor, Devon - 15 Group)
Hudson V AM837/N - took off from Chivenor at 1530, with five others, and brought down over France at La Breteche, St Viaud, 7km SE of Donges. All four crew all buried at Nantes.
Pilot: NZ40666 Flt Lt **Ian Wingate TERRY**, RNZAF - Age 25. 589hrs. 32nd op.

Mon 22 Dec 1941

BOMBER COMMAND

Ferry flight to the Handley Page factory at Radlett, Hertfordshire
28 Conversion Flight, RAF (Leconfield, Yorkshire - 4 Group)
Halifax I L9522 - took off at 1025 captained by Flt Lt Owen, RAF, and seen about 40 minutes later flying low over Lincolnshire, in the vicinity of Knipton, heading towards Barkston Woods in poor visibility. L9522 crashed a few minutes later into a wooded area on the top of a high ridge known as Terrace Hills. All eight on board died, two New Zealanders being buried at Grantham, Lincolnshire, and St Mary Church, Devonshire, respectively.
(Pilot): NZ41464 Plt Off **Richard Percival William BARKER**, RNZAF - Age 25. 206hrs. 1 op.
(Pilot): NZ41301 Plt Off **William Stuart BEATTIE**, RNZAF - Age 24. 184hrs. At least 3 ops.
Barker's brother, Desmond Wallace Ferguson Barker, died on 12 February 1941 while flying with 83 Sqn, RAF. Barker and Beattie were on the strength of Leeming-based 10 Sqn, while the other six airmen belonged to the Conversion Flight. Although listed as 'crew' it seems likely that one or both were actually travelling as passengers to Radlett.

Tue 23 Dec 1941

BOMBER COMMAND

Training
21 Operational Training Unit, RAF (Moreton-in-Marsh, Gloucestershire; satellite at Edgehill, Oxfordshire - 6 Group)
Wellington IC DV422 - took off from Edgehill captained by Act Sqn Ldr R G Williams, DFC AFM, RAF. While demonstrating performance in the near stalled condition at 1420, lost control and spun into the ground, and caught fire at Wild Spring Farm, Little Tew, 4½ miles ENE of Chipping Norton. All six crew died, two New Zealand pupil pilots being buried at Moreton-in-Marsh.
Pilot: NZ411978 Plt Off **Sidney Arnold CRUMP**, RNZAF - Age 26. 138hrs.
Pilot: NZ41898 Plt Off **Ian Curtis GRANT**, RNZAF - Age 24. 140hrs.

Tue 23 Dec/Christmas Eve 1941

BOMBER COMMAND

Raid on Brest, France (by 47 aircraft - 1 lost)

75 (NZ) Squadron, RAF (Feltwell, Norfolk - 3 Group)

Wellington IC Z8834 - took off at approximately 1623-1720 and circled the base in poor weather on returning from the raid. At 2347, while awaiting permission to land, Z8834 flew into the ground 2 miles east of Berner's Heath, about 5 miles SW of Thetford. Five crew were injured, but the captain died and is buried at Feltwell.

Captain: NZ403936 Flt Sgt **Loch Lomond BENTLEY**, RNZAF - Age 28. 324hrs. 7th op(?)

CANADA

Solo night training exercise

12 Service Flying Training School, RCAF (Brandon, Manitoba - 2 Training command)

Crane I 7773 - took off at 0050, but almost immediately lost height during a right hand turn and crashed into the ground a $^1/_2$ mile SE of the airfield. The pilot is buried at Brandon.

Pilot: NZ413387 LAC **Henry George DEERY**, RNZAF - Age 27. 122hrs (59 solo - 36 on Crane)

Wed, Christmas Eve 1941

COASTAL COMMAND

Flight to Wick, Caithness-shire, for detached operations

254 Squadron, RAF (Dyce, Aberdeenshire - 18 Group)

Blenheim IVF L9313 - on climbing to 150 feet after take off at 1415, the starboard engine began to misfire and flames appeared round the cowling. These were extinguished but the engine eventually cut. Continued flying straight and level on the port engine, but stalled and crashed at Pitmedden Churchyard, 9 miles north of the airfield. One of those on board survived injured, but the other four died, the pilot being buried at Dyce.

Pilot: NZ39933 Flt Lt **Dick Middleton POYNTER**, RNZAF - Age 26. 556hrs. 61 ops.

Thu, Christmas Day 1941

FAR EAST

Interception

67 Squadron, RAF (Mingaladon, Burma - 221 Group)

Buffalo I W8248 of 'B' Flight - took off in the late afternoon as one of three in 'Green' Section to intercept two incoming raids of about 60 aircraft each headed for the airfield and the nearby city of Rangoon. During the ensuring combats all three Green Section pilots were shot down and killed. The body of the pilot of W8248 was recovered and buried in the Mingaladon Cantonment Cemetery. However, as it was not possible for the authorities to move or maintain the grave post-war he is also commemorated on a memorial in the Taukkyan War Cemetery, 21 miles north of Rangoon.

Pilot: NZ405269 Sgt **Edward Bertram HEWITT**, RNZAF - Age 19.

67 Squadron, RAF (Mingaladon, Burma - 221 Group)

Buffalo I W8206 of 'B' Flight - took off as a member of 'Green' Section to intercept enemy raids as described above and shot down. The pilot is commemorated on the Singapore Memorial.

Pilot: NZ404393 Sgt **Ronald Percy McNABB**, RNZAF - Age 19.

67 Squadron, RAF (Mingaladon, Burma - 221 Group)

Buffalo I AN216 of 'A' Flight - took off as member of 'Red' Section to intercept the same raid as described above and shot down. The pilot's body was recovered and buried in the Mingaladon Cantonment Cemetery. However, as it was not possible for the authorities to move or maintain the grave post-war he is also commemorated on a memorial in the Taukkyan War Cemetery, 21 miles north of Rangoon.

Pilot: NZ41486 Sgt **John MACPHERSON**, RNZAF - Age 25.

Sat 27 Dec 1941

BOMBER COMMAND

Anti-shipping diversionary strike off Norwegian coast in support of *Operation ARCHERY*

110 Squadron, RAF (Wattisham, Suffolk; detachment at Lossiemouth, Morayshire - 2 Group)

Blenheim IV Z7317 - took off from Lossiemouth during the morning with five others, four of which failed to return. The four were shot down by flak or fighters while attacking a convoy, all their crews being killed. The body of Z7317's pilot was later recovered and is buried at Sola, near Stavanger. His two comrades were lost without trace and are commemorated on the Runnymede Memorial. *Archery* was the first combined Army-Navy-Air Force operation of the war against German-held territory. The commando attack on the Vaagsö and Maalöy islands in Vaags Ford was a complete success but cost Bomber Command eight of the 29 aircraft despatched.

Pilot: NZ403118 Flt Lt **Richard Trevor BLEWETT**, RNZAF - Age 24. 285hrs.

110 Squadron, RAF (Wattisham, Suffolk; detachment at Lossiemouth, Morayshire - 2 Group)

Blenheim IV Z7442 - piloted by Sgt N Kaby, RAF, took off on same raid as the above and lost without trace. The pilot's body was later recovered and is buried at Trondheim. His two comrades are commemorated on the Runnymede Memorial.

Observer: NZ404480 Plt Off **Robert Charles DAVIS**, RNZAF - Age 25. 209hrs.

Sat 27/Sun 28 Dec 1941

BOMBER COMMAND

Raid on Düsseldorf, Germany (by 132 aircraft - 8 lost)

58 Squadron, RAF (Linton-on-Ouse, Yorkshire - 4 Group)

Whitley V Z6841/F - took off at 1656 and lost without trace, its five crew being commemorated on the Runnymede Memorial.

Captain: NZ404464 Plt Off **Melville John Hyland HUNT**, RNZAF - Age 21. 372hrs. 9th op.

57 Squadron, RAF (Feltwell, Norfolk - 3 Group)

Wellington IC Z1097 - took off at 1700 captained by Wt Off T Purdy, DFM, RAF, on same raid as the above and shot down over the Netherlands by a night-fighter at 2030, crashing near Someren, 30km west of Venlo. An RAF Sgt survived to be captured, but his five comrades died and were buried at Venlo, later being reinterred at Eindhoven.

2nd Pilot: NZ404996 Sgt **Maxwell Joseph CRONIN**, RNZAF - Age 21. 243hrs. 6th op.

Sun 28 Dec 1941

BOMBER COMMAND

Raid on Emden, Germany (by 40 aircraft - 2 lost)

405 Squadron, RCAF (Pocklington, Yorkshire - 4 Group)

Wellington II W5561/J - took off at 1705. May have been brought down near Aurich, 22km NE of Emden, as the captain and at least two others of the crew were at first buried there in the Engerhave Cemetery. Their three comrades also died and the bodies of all six were later reinterred at Sage, 24km south of Oldenburg.

Captain: NZ39298 Sgt **Edwin John WILLIAMS**, RNZAF - Age 28. 411hrs. 18th op.

Williams brother, Leslie Williams, died on 17 February 1944 while flying with 43 Sqn, RAF.

Raid on Hüls, Germany (by 81 aircraft - 4 lost)

408 Squadron, RCAF (Balderton, Nottinghamshire - 5 Group)

Hampden I P1165/B - took off at 1855 captained by Plt Off S B K Brackenbury, RCAF, and shot down over the Netherlands by a night-fighter, crashing 2km east of Winterswijk, 55km east of Arnhem. The captain was captured, but his three comrades died and are buried at Winterswijk.

Observer: NZ404557 Flt Sgt **Desmond William THANE**, RNZAF - Age 21. 231hrs. 8th op.

144 Squadron, RAF (North Luffenham, Rutland - 5 Group)

Hampden I P1295/L - captained by Sgt J W Woodall, RAF, on same raid as the above and lost without trace. The bodies of two crew, including a New Zealander, were recovered and are buried at Cherbourg, France. The other two crew are commemorated on the Runnymede Memorial.

(Pilot): NZ402917 Sgt **Edward Boyce WHITAKER**, RNZAF - Age 21. 305hrs.

Whitaker was probably acting as observer on this flight.

Mon 29 Dec 1941

FAR EAST

Training

Malayan Volunteer Air Force (Katong, Singapore)

Gipsy Moth '6' of 'A' Flight - collided over Singapore with a 488 Sqn, RNZAF, Buffalo I (W8200) flown by Sgt **W R De Maus**. A wing was torn from the Moth and the aircraft crashed into Lavender Street. The pilot and his passenger, Petty Officer W Smith, RN, of HMS *Repulse*, are buried at Kranji. The Buffalo, suffering only minor damage, made a safe landing at Kallang (adjacent to Katong) immediately afterwards.

Pilot: NZ412668 Plt Off **William Walter Garfield FELTON**, RNZAF - Age 23.

De Maus died on 16 May 1944 while flying with 4 OTU, RNZAF.

Wed, New Year's Eve 1941

NEW ZEALAND

Instrument flying training

2 Service Flying Training School (ATS), RNZAF (Woodbourne)

Oxford II NZ1243 - while engaged in an instrument take off at 1015, collided with another of the School's Oxfords (NZ1257) circling the field at 800 feet preparatory to landing. The latter entered a spin from which it failed to recover, while NZ1243 began a diving turn to port and also crashed. Both machines came down about a mile from the airfield, near Renwick. All four pilots died, and are buried at Wellington.

Pilot Instr: NZ401309 Sgt **Matthew William COXON**, RNZAF - Age 34. 229hrs solo (137 on Oxford)
Pilot: NZ414725 LAC **James Duff HEWETT**, RNZAF - Age 28.

Hewett's father, also named James Duff Hewett, was a pilot in the First World War, a prominent aviator in the interwar years, and a serving officer in the RNZAF at the time of his son's death.

Camera gun exercise

2 Service Flying Training School (ATS), RNZAF (Woodbourne)

Oxford I NZ1257 - returning from the exercise when it collided with NZ1243 as described above.

Pilot: NZ414291 LAC **Charles William HILDRETH**, RNZAF - Age 27. 83hrs solo (62 on Oxford)
Pilot: NZ414250 LAC **Gordon Keith CLARIS**, RNZAF - Age 29. 122hrs.

1942

WITHIN SIX MONTHS the Americans, British, and Dutch had been swept out of South-East Asia and most of the western Pacific by the Japanese, who were then working across New Guinea and down the Solomon Islands. Singapore fell on 15 February to become the worst British military defeat in history, while the retreat from Burma would be recorded as the longest. Four days later northern Australia withstood the first of 50 raids flown against it in 1942. The very day Java capitulated, on 8 March, a reconnaissance aircraft from a Japanese submarine was engaged on a sortie over Wellington and again five days later over Auckland. Of nearly 50 New Zealand airmen who died in the Far East during the year, 37 were lost during the first 65 days.

Anticipating possible invasion, emergency measures in January saw the RNZAF organise FAFAI Squadrons (Forces Available For Anti-Invasion) within all of its flying training schools. Even the humble Tiger Moth was modified to carry bombs. With most of its fighting men serving in the Middle East and Europe, New Zealand was essentially left to 'make do' with a combination of Home Guard and Territorial personnel, about 150,000 in all, equipped with insufficient arms and little ammunition, complemented by regular forces personnel primarily engaged in training and coastal patrol work. Meanwhile the population at large focused on tighter 'Blackout' restrictions and energetically applied itself to digging slit trenches as far south as Invercargill for protection, in anticipation of attack from the air. Ten Kittyhawk fighters for the RNZAF arrived in early April, providing for the first time the basis of a credible local air defence.

But Japan's rapid run of success disguised the fact that she had overreached herself, militarily and industrially. A sea-borne invasion bound for Port Moresby in early May was thwarted by the Americans in the Battle of the Coral Sea, and withdrew. A month later another Japanese strike force was decisively defeated off Midway Island, with the destruction of its four fleet carriers and most of its naval aircrew elite. Plans to advance on Fiji, New Caledonia and Samoa were now postponed - later cancelled - thus removing any temptation to attack New Zealand. In June US Marines arrived in New Zealand and by August were engaged in combat on Guadalcanal, where they would eventually halt the enemy's advance and force him to retreat. RNZAF Hudsons operating in the Pacific, initially from Fiji, and New Caledonia from July 1942, moved forward to Guadalcanal late in November. The Air Force's fatal Pacific casualties for 1942 were limited to 20, but 15 of them were accounted for in the loss of three of the Hudsons. A further 85 died in New Zealand, a reflection of the growing volume of flying training, although the deaths of 16 were actually unrelated to flying. During the year a further 15,000 airmen and women enlisted in the RNZAF, a total as great as that for the years 1939-41 combined.

Flying casualties in Canada too, with over 20 killed, highlighted the growing numbers of RNZAF personnel receiving training there. A similar number died in the RAF's Flying Training Command, mainly while undertaking specialised flying courses. Numbers entering the FAA were also building, 27 losing their lives in various parts of the world but primarily through accidents in Britain. Another 16 Air Force personnel died while engaged on ferrying duties, often undertaken through hostile skies off the coast of France and over the Mediterranean. A few were aircrew who had just completed their training, only to be lost on new aircraft delivery flights from Canada across the Atlantic.

In North Africa a final Axis offensive launched from Libya in late June drove deep into Egypt, being halted at El Alamein, only a few hundred kilometres west of Cairo. Fighting continued into September before the line stabilised and enabled the British to prepare a counter offensive for 23 October - the Second Battle of El Alamein. Now hopelessly outnumbered, the enemy retreated, fighting a stubborn rearguard action across Libya and on into Tunisia. Further west, on 8 November, an Allied invasion force came ashore at French North Africa to attack him from the rear. The majority of nearly 140 New Zealand airmen who died in the Middle East-Mediterranean theatre in 1942 lost their lives in support of the desert campaigns, half of them while serving with the heavy bomber squadrons. Malta's survival was also ensured by year's end. It had narrowly avoided capitulation after fending off daily, sometimes almost hourly, raids originating from Sicily, less than 100km away, and surviving on marginal supplies brought through at great cost by battered convoys. These difficulties notwithstanding, unrelenting offensive operations against Axis ports and convoys continued throughout the siege.

Around Britain and over the Atlantic, Coastal Command continued with its long and often unrewarding task of patrolling for elusive U-boats, aiding in their destruction and fending them off convoys. For Fighter Command it was 1941 revisited, with the continuation of regular sweeps over the Low Countries and Northern France. Superior equipment in the shape of an improved Spitfire IX and the new Typhoon became operational during the year. The Typhoon became a useful deterrent to low-level hit and run raiders and developed into an excellent fighter-bomber. Some 50 and 70 New Zealanders, respectively, died in each Command, with a disconcerting 50 percent accounted for by flying accidents.

Three days before Singapore fell, another humiliation befell the British as the German capital ships *Scharnhorst, Gneisenau* and *Prinz Eugen*, broke out from Brest and slid through the English Channel to reach German ports almost unscathed. Bomber, Coastal and Fighter Commands, and the FAA, were belatedly brought into action but achieved little at great cost. A consolation was that two of the ships were damaged on striking mines (or *Vegetables* in RAF parlance) laid earlier by Bomber Command. This hazardous activity would grow greatly in importance during the year, the number of mines sown rising nearly six-fold from a wartime total of less than 2000 up to February. On 19 August a predominately Canadian force was landed at the French port of Dieppe in an action designed to test German defences. An unmitigated disaster - it cost the RAF over 100 aircraft and the Luftwaffe only half as many - it nonetheless provided painful lessons, which were to prove crucial to the success of future Allied amphibious operations.

Having reached its nadir the previous November Bomber Command survived a searching review and was about to be transformed. In February the Air Ministry, finally acknowledging unrealised expectations of precision bombing, authorised the targeting of the densely built up parts of large German cities - 'area bombing' as it came to be known. Eight days later Air Chief Marshal Sir Arthur Harris - 'Bomber' Harris to the public but 'Butch' to his crews - took over the Command and gradually set about rebuilding it into a force to be reckoned with. In his favour was the imminent operational debut of the Lancaster heavy bomber and the first of many electronic aids to come - *Gee*. Bombing now became increasingly concentrated in terms of bombers and time over the target.

On the night of 30/31 May, by drawing on bombers from the operational training and conversion units, a thousand-bomber raid, two and a half times greater than any previously undertaken, was flown against Cologne. This figure was also made possible through a recent policy decision which, among other things, resulted in dispensing with 2nd pilots. Just under 1000 bombers were despatched two nights later to Essen and again to Bremen in late June, although the results were disappointing when compared with Cologne. While the '1000' concept would not be repeated in the near future the experience confirmed that the new methods were sound. On 15 August would come a further revolution with the creation of a Path Finder Force (PFF) whose task it was to locate and illuminate targets for the Main Force. Two days later 12 B-17 Fortresses carried out the United States Army Air Force's first daytime bombing raid from Britain. It was the very small beginnings of 'round the clock' bombing and a sign of the increasing pressure Germany would have to face from the air in the west.

That the much improved Command was still up against an enemy no less resourceful in defence, however, is clearly demonstrated by the dreadful cost in casualties. On the night of 28/29 July, for instance, more New Zealand airmen died during a raid on Hamburg alone than on any other day or night of the war, anywhere. No 75 (NZ) Sqn in particular was badly mauled, with 21 dead. New Zealand casualties in the Command exceeded 360 killed for the six months July to December - their highest of the war - greater in total than all wartime RNZAF deaths in the Pacific. Although greater numbers died in RAF Bomber Command overall in the later years, 1942, with a record 4% average loss rate (i.e. one in every 25 aircraft despatched), was by far the grimmest for bomber crew survival: those fortunate enough to complete their first tour of 30 ops, let alone a double one of 50, were few.

This year was the hardest for the New Zealanders. Exceeding 1060, the total number of dead was 135% greater than that for 1941. But, while 3 September 1942 marked the halfway point of the war, it is sobering to note that only one third of the ultimate number of deaths had yet occurred. While the Allies would suffer no more major defeats, victory would not come cheaply. Casualty figures for each of the next two years would be little better than those of 1942, although they would be spread more thinly over the greater numbers then serving. Almost all of these would come from the ranks of those who had enlisted during the first half of the war.

Fri 2 Jan 1942

FLEET AIR ARM - ENGLAND

Formation practice

888 Squadron, FAA (Lee-on-Solent, Hampshire - HMS *Daedalus*)

Martlet II AM995 - on peeling away from formation at 1440 dived into the ground and caught fire near Cheesefoot Head (Hill?), 3 miles ESE of Winchester. The pilot is buried at Gosport.

Pilot: Sub-Lt (A) **William Clive HOOSON**, RNZNVR - Age 21.

Sat 3 Jan 1942

FAR EAST

Attack on convoy

34 Squadron, RAF (Tengah, Singapore)

Blenheim IV V5388 - took off with two others on a patrol along the western coast of Malaya. At least two of the Blenheims located and bombed two vessels off Kuala Selangor, and a number of small craft were machine gunned ten miles south of Penang. V5388 was brought down further north, some 200 yards off the coast, about 3 miles from Yen. All three crew are buried at Taiping.

Pilot: NZ405289 Plt Off **James Garfield LOMAX**, RNZAF - Age 29.

FIGHTER COMMAND

Training

53 Operational Training Unit, RAF (Llandow, Glamorgan - 81 Group)

Spitfire I P9491 - found on the following day to have crashed into a hillside $2^1/_4$ miles SSW of Ton Pentre, 12 miles north of Llandow. The pilot is buried at Llantwit Major.

Pilot: NZ405330 Sgt **Phillip Neville SHAW**, RNZAF - Age 19. 160hrs.

Sun 4/Mon 5 Jan 1942

MIDDLE EAST

Raid on airfield at Castelvetrano, Sicily

40 Squadron, RAF (Luqa, Malta)

Wellington IC Z9036 - took off with three others and brought down by flak during a low level attack on the airfield. All six crew are buried at Catania.

Captain: NZ403463 Flt Sgt **James Frederick LEWTHWAITE**, RNZAF - Age 23. 282hrs. 18th op.

Mon 5 Jan 1942

FAR EAST

Scramble to meet incoming raid

243 Squadron, RAF (Kallang, Singapore)

Buffalo I W8199 - took off with others at about 0700 but shortly afterwards broke formation and returned to the airfield. Ground visibility was limited due to smoke haze and while taxiing along the runway Buffalo W8179 (see below) collided with it while taking off to take its place. Both fighters caught fire and were destroyed, the pilots being buried at Kranji.

Pilot: NZ402467 Sgt **Paul Lester ELLIOTT**, RNZAF - Age 19. 230hrs solo (84 on Buffalo)

243 Squadron, RAF (Kallang, Singapore)

Buffalo I W8179 - destroyed in the collision described above while taking off at 0715.

Pilot: NZ404954 Plt Off **Ronald Spencer SHIELD**, RNZAF - Age 23. 240hrs solo (94 on Buffalo)

Tue 6 Jan 1942

BOMBER COMMAND

Cross country

19 Operational Training Unit, RAF (Kinloss, Morayshire - 6 Group)

Whitley V N1498 - while descending through cloud at 2030 on the last leg of the flight, flew into snow covered high ground and caught fire at Tomatin House, 15 miles SE of Inverness. Three crew survived but the other four died, including two New Zealanders, who are buried at Kinloss.

(Pilot): NZ41875 Plt Off **John Graham CASTLING**, RNZAF - Age 25. 164hrs.

(Pilot): NZ41987 Plt Off **John George IRVINE**, RNZAF - Age 22. 190hrs.

Castling and Irvine had graduated together in New Zealand on course 12B at Blenheim. Although both were pilots their crew roles for this flight are unknown. Neither was acting as captain, however.

Fri 9 Jan 1942

MIDDLE EAST

Bomber escort

229 Squadron, RAF (Mechili or Antelat, Libya - 262 Wing, Western Desert)

Hurricane IIB Z5588 - took off at 1400 with nine others. During the escort four of the Hurricanes were shot down when the Squadron was intercepted by six enemy fighters. One pilot died and is commemorated on the Alamein Memorial.

Pilot: NZ392011 Sgt **Arnold George BROWNE**, RNZAF - Age 27. 256hrs.

The CWGC register gives Browne's age as 30.

BOMBER COMMAND

Raid on Cherbourg, France (by 31 aircraft - 4 lost)

458 Squadron, RAAF (Holme-on-Spalding Moor, Yorkshire - 1 Group)

Wellington IV R1785 - took off at 0435 captained by Fg Off B P Hickey, RAAF, and brought down by flak at 0723, crashing at Gonneville, 12km east of the target. All six crew are buried at Cherbourg.

Observer: NZ403773 Plt Off **Robert BIRNIE**, RNZAF - Age 24. 301hrs. 5th op.

12 Squadron, RAF (Binbrook, Lincolnshire - 1 Group)

Wellington II W5356 - took off at 0438 captained by Sgt D Butterworth, RAF, on same raid as the above. Returning on one engine after a glycol leak had developed in the port engine, W5356 crashed a half-mile NW of Wythall, 5 miles NNW of Redditch, Worcestershire, while attempting an emergency landing at about 0900. As it struck the ground the machine swung round, broke its back, and caught fire. Three of the crew died and the other three were injured, the observer so badly that he died two days later at the Hollymoor Hospital in Birmingham. He is buried at Sunderland, County Durham.

Observer: NZ404478 Plt Off **Wilfred Henry BURNHAM**, RNZAF - Age 21. 234hrs. 1st op.

Fri 9/Sat 10 Jan 1942

FAR EAST

Raid on Sungei Patani airfield, Malaya

34 Squadron, RAF (Tengah, Singapore)

Blenheim IV V5597 - took off with five others, only two of which reached the target in the poor weather prevailing. All returned except V5597, which was thought to have been brought down at about 0330 on encountering severe turbulence in a tropical storm. All three crew are commemorated on the Singapore Memorial.

Pilot: NZ402905 Sgt **James Ellis SMITH**, RNZAF - Age 25.

Sat 10 Jan 1942

FIGHTER COMMAND

'Jim Crow' - Shipping reconnaissance

152 Squadron, RAF (Coltishall, Norfolk - 12 Group)

Spitfire IIA P8266 - took off at 1145 and lost without trace, the pilot being commemorated on the Runnymede Memorial.

Pilot: NZ404963 Sgt **Henry Twite STURT**, RNZAF - Age 21.

Sat 10/Sun 11 Jan 1942

BOMBER COMMAND

Intruder raid on Soesterberg airfield, Netherlands

82 Squadron, RAF (Watton, Norfolk - 2 Group)

Blenheim IV V6440/X - took off at 1728 from Bodney piloted by Flt Lt K H Nash, RAF, and brought down by flak at 2050, crashing at Soesterberg. All three crew are buried at Oud Leusden on the southern outskirts of Amersfoort.

Observer: NZ39965 Sgt **Harry Chapman DOWNS**, RNZAF - Age 22. 190hrs. 2nd op.

Raid on Wilhelmshaven, Germany (by 124 aircraft - 6 lost)

12 Squadron, RAF (Binbrook, Lincolnshire - 1 Group)

Wellington II W5611 - took off at 1803, but on returning to Lincolnshire towards 2330, the port engine cut and height was rapidly lost. The engine picked up again for a few minutes but this was followed by several loud explosions and a sheet of flame. Height continued to be lost, the Wellington finally crashing a mile east of Brumby, near Scunthorpe. Two of the crew were injured, but their four comrades died, including the captain, who is buried at Binbrook.

Captain: NZ402210 Flt Sgt **Carroll Frederick O'CONNELL**, RNZAF - Age 23. 282hrs (199 solo - 105 on Wellington) 6th op.

Sun 11 Jan 1942

BOMBER COMMAND

Plt Off **W H BURNHAM** died of injuries received in a crash landing on the 9th.

Mon 12 Jan 1942

FAR EAST

Interception patrol
243 Squadron, RAF (Kallang, Singapore)
Buffalo I W8234 - took off at about midday and was lost during an engagement with some 70 enemy aircraft SE of Changi. The pilot's parachute was later retrieved from the sea, but no trace of him could be found and he is commemorated on the Singapore Memorial.
Pilot: NZ404942 Sgt **Noel Bain RANKIN**, RNZAF - Age 28.

Interception patrol
243 Squadron, RAF (Kallang, Singapore)
Buffalo I W8137 - took off during the afternoon to intercept a second enemy raid by a large number of aircraft and shot down. The pilot was buried at Johore Bahru, but reinterred in the Bidadari Civil Cemetery at Singapore in 1947, before final internment in the island's Kranji War Cemetery.
Pilot: NZ41995 Sgt **Reginald James NEWMAN**, RNZAF - Age 22.

ATLANTIC

In transit on convoy HX168 from Canada to Britain
On board SS *Yngaren* (having embarked from 1 [Y] Depot, RCAF - 3 Training Command)
The *Yngaren*, a 5246-ton Swedish motor ship, had sailed in convoy from Halifax on the 2nd, bound for Hull with six passengers aboard. The convoy became scattered in bad weather and the *Yngaren*, the only ship in HX168 to be lost to submarine attack, was torpedoed by U-43 at 0400 on the 12th, sinking 600 miles west of Ireland. The only two survivors, a Dane and a Swede, were rescued from their raft a month later, while the airmen who died are commemorated on the Ottawa Memorial. German records time the sinking at 0802 at position 'AL1938'. U-43's Wolfgang Lüth, the second most successful U-boat commander of the war, was credited with sinking over 222,000 tons of shipping during the war.
(Observer): NZ411357 Plt Off **Michael Patrick ANGLAND**, RNZAF - Age 23.
(Observer): NZ405230 Plt Off **Maurice Melton CAMPBELL**, RNZAF - Age 27.
(Observer): NZ404363 Plt Off **William Upton HEWETT**, RNZAF - Age 29.

Tue 13 Jan 1942

FAR EAST

Interception
243 Squadron, RAF (Kallang, Singapore)
Buffalo I W8238 - took off towards midday and was lost south of Singapore during an engagement with an incoming light bomber force under fighter escort. The pilot is commemorated on the Singapore Memorial.
Pilot: NZ405819 Sgt **Arthur Russell REYNOLDS**, RNZAF - Age 19.

COASTAL COMMAND

Navigation exercise and air-to-sea cannon firing test
235 Squadron, RAF (Dyce, Aberdeenshire - 18 Group)
Beaufighter IC T4764/T - took off at 1440 and crashed into the sea sometime after 1500, killing all three occupants. The following day one body was retrieved from the sea near Belhelvie, 7 miles north of Aberdeen, but no trace could be found of the pilot and the other crew member, and they are commemorated on the Runnymede Memorial. The Beaufighter's dinghy was also recovered from the sea, close to a balloon of a type towed by ships. It was thought that in poor visibility T4764 might have collided with the balloon which had broken free.
Pilot: NZ404881 Plt Off **Desmond Thomas AUSTEN**, RNZAF - Age 22. 288hrs.

Wed 14 Jan 1942

NEW ZEALAND

Wind speed and direction finding exercise
3 Service Flying Training School (ATS), RNZAF (Ohakea)
Oxford II NZ1237 - took off shortly after 1500 and failed to return. Despite an intensive aerial search no trace of the aircraft or its occupants could be found. Nearly four months later, during the first week of May, the wreckage and crew remains were stumbled upon at Mitre Peak in the Tararua ranges, 13 miles WSW of Otaki, by a Mr **M Trass**. The Oxford had apparently become lost while flying in cloud and at 1550 struck the mountain slope. The bodies were buried at the scene, both airmen also being commemorated on the Provincial Memorial at Auckland.
Pilot: NZ414963 LAC **James Edgar Norman CORIN**, RNZAF - Age 29. 80hrs (49 solo - 25 on Oxford)
Pilot: NZ391792 LAC **Ronald Alfred Walter ABRAHAMS**, RNZAF - Age 22.
Abrahams father, Walter Alfred Abrahams, died of natural causes on 10 July 1944 in Auckland while serving with the RNZAF.

MIDDLE EAST

Anti-shipping strike

21 Squadron, RAF (Luqa, Malta)

Blenheim IV Z7431/F - took off at 1145 with three others, being one of two which failed to return. An enemy merchant ship and destroyer were attacked at 1320 after being located SE of the Kerkenna Islands, off the Tunisian coast at position 3450N:1200E. One of the Blenheims struck the merchant ship's mast with its starboard wing and plunged into the sea. Blenheim Z7431, following close behind, was hit by flak while pulling up over the vessel. It then climbed almost vertically, exploded and crashed into the sea in flames. There were no survivors from either. Five of the six crew are commemorated on the Malta Memorial but, rather oddly, the sixth, an airman from the other Blenheim, is, according to the CWGC register, commemorated on the Alamein Memorial.

Pilot: NZ402852 Plt Off **Kenneth James COAKLEY**, RNZAF - Age 22.

Thu 15 Jan 1942

FAR EAST

Interception

488 Squadron, RNZAF (Kallang, Singapore)

Buffalo I W8183 - took off at 0950 to intercept an incoming raid and was at once attacked and badly damaged. Its mortally wounded pilot crash landed on a narrow strip of land by the oil tanks alongside the Alexandra Hospital, but was dead by the time ground personnel reached the wrecked machine. At first buried in the military cemetery at Bukit Timah, he was later reinterred in the Bidadari civil cemetery, before final reinternment at Kranji.

Pilot: NZ402868 Plt Off **Greville Lloyd HESKETH**, RNZAF - Age 26.

Interception

243 Squadron, RAF (Kallang, Singapore)

Buffalo I W8178/P - took off during the morning on an interception flight and was lost without trace, the pilot being commemorated on the Singapore Memorial.

Pilot: NZ402888 Sgt **John Benjamin OLIVER**, RNZAF - Age 23.

Thu 15/Fri 16 Jan 1942

BOMBER COMMAND

Raid on Hamburg, Germany (by 96 aircraft - 11 lost)

150 Squadron, RAF (Snaith, Yorkshire - 1 Group)

Wellington IC Z1078 - took off at 1723 captained by Plt Off B A Macdonald, RCAF. On return to England, encountered a snow storm and at 2355 crashed onto West Hill in the Cheviot hills, a mile south of Dunsdale, Northumberland, and caught fire. The observer was killed, while the 2nd pilot died of injuries on the way to hospital in the early hours of the 16th. A further crew member also died of injuries on the 16th. The other three suffered injuries also, but are believed to have survived. Both New Zealanders are buried at East Chevington. The Wellington was one of seven aircraft to crash in England on return, most of them because of the bad weather.

2nd Pilot: NZ405278 Sgt **Laurence Warren HUNT**, RNZAF - Age 20. 223hrs. 3rd op.

Observer: NZ405378 Sgt **Thomas Walter IRVING** RNZAF - Age 23.

214 Squadron, RAF (Stradishall, Suffolk - 3 Group)

Wellington IC R1759 - took off at 1740 captained by Flt Sgt E E Hale, RCAF, on same raid as the above, and presumed to have crashed into the North Sea off the Yorkshire coast. The captain's body and that of an RCAF crew member later washed ashore, their four comrades being commemorated on the Runnymede Memorial.

Observer: NZ403487 Plt Off **Henry Joseph WOODROW**, RNZAF - Age 27. 244hrs. 6th op.

10 Squadron, RAF (Leeming, Yorkshire - 4 Group)

Halifax II L9622/G - took off at 1812 captained by Sgt M S Schneider, RCAF, on same raid as the above. On its return L9622 passed over the base preparatory to landing, but a few minutes later, at 2330, crashed and caught fire a mile north of Northallerton. Sgt Schneider survived but was too severely injured at the time to give an account of the accident. The other six crew died, the observer and two RCAF airmen being buried at Leeming.

Observer: NZ403615 Plt Off **Michael VON DADELSZEN**, RNZAF - Age 25. 248hrs. 8th op.

Raid on Emden, Germany (by 50 aircraft - 3 lost)

104 Squadron, RAF (Driffield, Yorkshire - 4 Group)

Wellington II W5493/D - took off at 1835 captained by Sgt J Wilmot, RAF. While returning to Yorkshire on a south-easterly heading at 0005, struck the side of a hill about five feet below its summit at Bumper Castle, Arden Great Moor, at the northern end of the Hambleton Hills. Two of the crew were injured but their four comrades died, including the observer and 2nd pilot, who are buried at Topcliffe.

Observer: NZ401302 Sgt **Douglas Reginald BRADLEY** RNZAF - Age 28. 163hrs.

104 Squadron, RAF (Driffield, Yorkshire - 4 Group)

Wellington II W5417/F - took off at 1838 captained by Sgt B A Adams, RAF, on same raid as the above and lost without trace. All six crew are commemorated on the Runnymede Memorial.

Observer: NZ401396 Plt Off **Royal George LINDSAY**, RNZAF - Age 30. 253hrs. 8th op.

Fri 16 Jan 1942

CANADA

1 Air Navigation School, RCAF (Rivers, Manitoba - 1 Training Command)
Died of a cerebral haemorrhage at the RCAF Rivers station hospital and is buried at Rivers.
(Observer): NZ402518 **Sgt John HARDYMENT**, RNZAF - Age 26. 98hrs.
The CWGC register incorrectly gives Hardyment's age as 31.

Sat 17 Jan 1942

FAR EAST

Enemy air raid on Singapore
205 Squadron, RAF (Seletar, Singapore)
Catalina I Z2151/Y - at about 1100 two enemy fighters strafed the Squadron's Catalinas moored at Seletar. Catalina 'Y' and another burst into flames and sank, while two others were badly damaged. Two airmen aboard 'Y' were killed during the attack, including a New Zealander whose body was later recovered and buried in the Bidadari cemetery, but later reinterred at Kranji. Also on board was LAC **F Gluyas**, RNZAF, who, although wounded, succeeded in swimming away from the sinking Catalina.
(Flight Rigger): NZ391096 LAC **Frank Frederick EDWARDS**, RNZAF - Age 26.

BOMBER COMMAND

Landing practice
12 Operational Training Unit, RAF (Chipping Warden, Northamptonshire - 6 Group)
Wellington IC R1340 - took off at 1115 but swung to the right when the starboard engine cut, stalled and struck a tree, then crashed. All three crew died, the captain and 2nd pilot being buried at Brookwood in Woking, Surrey.
Captain: NZ411963 Sgt **Arthur Roy WILSON**, RNZAF - Age 24. 190hrs (103 solo - 3 on Wellington)
2nd Pilot NZ411939 Sgt **Kenneth Stanley ROGERS**, RNZAF - Age 26. 185hrs.
Wilson's brother, John Colin Wilson, died on 21 May 1941 while flying with 107 Sqn, RAF.

Sat 17/Sun 18 Jan 1942

FERRY COMMAND

Delivery of aircraft to Far East (for 62 Sqn)
Overseas Aircraft Despatch Unit, RAF (Portreath, Cornwall - 44 Group)
Hudson III V9126 - flew from Portreath, Cornwall, to Gibraltar on the 10th, captained by Sgt A W Story, RAF. When completing the second leg to Malta at 0600, flew into the ground near Siggiewi on approach to Luqa. Only one crew member survived. His three comrades are buried on Malta.
Observer: NZ402234 Sgt **Noel Francis WOULDES**, RNZAF - Age 20.
Wouldes had been posted to Coastal Command's 59 Sqn only three weeks prior to commencement of the flight. It is not known if he was scheduled to stay in the Far East upon completion of the delivery or make his way back to Britain and 59 Sqn.

BOMBER COMMAND

Raid on Bremen, Germany (by 83 aircraft - 3 lost)
142 Squadron, RAF (Waltham, Lincolnshire - 1 Group)
Wellington IV Z1330/B - took off at 1740 captained by Plt Off A C Pickett, RAF. Last heard from when 75 miles ENE of base and assumed to have crashed into the North Sea. All six crew are commemorated on the Runnymede Memorial.
Observer: NZ401753 Plt Off **Daniel Joseph CLIFFORD**, RNZAF - Age 23. 215hrs. 2nd op.

Sun 18 Jan 1942

FAR EAST

Scramble to intercept raid
488 Squadron, RNZAF (Kallang, Singapore)
Buffalo I W8141 - took off at 1145 with others, but on climbing through 16,000 feet the formation was attacked from above by enemy fighters. In the ensuing engagement W8141 was last seen at about 1230 diving into a cloud over the Strait of Singapore. The pilot is commemorated on the Singapore Memorial.
Pilot: NZ41881 Plt Off **Ernest Wilkin COX**, RNZAF - Age 22.

Mon 19 Jan 1942

FAR EAST

Offensive patrol
488 Squadron, RNZAF (Kallang, Singapore)
Buffalo I W8171 - took off at about 1340 with three others and headed north along the western Malayan coast to strafe ground targets a mile north of Muar township. Some 50 minutes later, en route near Batu Pahat, enemy fighters were sighted below. However, as the Buffaloes dived to intercept they were themselves attacked from above, two being shot down. Although W8171 was seen to crash, no trace of its pilot could be found and he is commemorated on the Singapore Memorial.
Pilot: NZ411421 Plt Off **Kenneth Jack McANENY**, RNZAF - Age 19.

488 Squadron, RNZAF (Kallang, Singapore)
Buffalo I AN189 - the other fighter shot down during the patrol detailed above. The pilot baled out, only to be captured by the Japanese. Later transported to Thailand, he died there of dysentery on Christmas Day 1943 and was buried at Tenbaya, but later reinterred at Thanbyuzayat in Burma.
Pilot: NZ41876 Wt Off **Cecil Deryck CHARTERS**, RNZAF - Age 29.

BOMBER COMMAND

Navigation exercise
9 Squadron, RAF (Honington, Suffolk - 3 Group)
Wellington III X3370/D - captained by Sgt R D Telling, RAF. At 1050, while flying at 250 feet, the outer half of the starboard wing detached and X3370 crashed at Holly Farm, just north of Thetford, Norfolk, about 7 miles from the base. All seven crew died, including the New Zealand observer, who is buried at Honington. Subsequent investigation revealed a material defect in the wing. Pilots who had flown X3370 previously stated that its starboard wing tended to drop suddenly and unexpectedly.
Observer: NZ40618 Flt Lt **Philip Henry CRESWELL**, mid, RNZAF - Age 25. 418hrs. 33 ops.
Creswell was on board to instruct others in navigation.

Training flight over sea
20 Operational Training Unit, RAF (Lossiemouth, Morayshire - 6 Group)
Wellington IC R1646 - at first presumed to have been lost at sea, the wrecked bomber was later found on 22 February by personnel of 56 Maintenance Unit, RAF. It had crashed at 1250hrs and burnt out in inaccessible snow covered mountainous country near Braemar, Aberdeenshire. All eight crew died, including the captain, who is buried at Dyce. Contemporary records give the site as 'at Cluny, near Braemar,' but it is believed that this refers to an area in the vicinity of Clunie Water or Clunie Glen, both of which lie a few miles to the south of Braemar.
Captain: NZ40667 Fg Off **James Williamson THOMSON**, DFC, RNZAF - Age 25. 659hrs. 31 ops.

CANADA

Solo interception exercise
12 Service Flying Training School, RCAF (Brandon, Manitoba - 2 Training Command)
Crane I 7747 - returned in formation with other solo pilots on the exercise at 1500, but on breaking away collided with Crane 7756. Both crashed about 3½ miles NE of Knox Station, about 5 miles NE of Forrest. Both pilots are buried at Brandon.
Pilot: NZ412238 LAC **Ross Francis KITTO**, RNZAF - Age 23. 146hrs (76 solo - 54 on Crane)

12 Service Flying Training School, RCAF (Brandon, Manitoba - 2 Training Command)
Crane I 7756 - destroyed in the collision with Crane I 7747 as detailed above.
Pilot: NZ413063 LAC **Geoffrey Herbert Weston HAMILTON**, RNZAF - Age 22. 127hrs (72 solo - 48 on Crane)

Tue 20 Jan 1942

INDIA COMMAND

Scramble during attack on airfield
67 Squadron, RAF (Mingaladon, Burma - 221 Group)
Buffalo I W8229 - landed at Moulmein with another Buffalo to refuel after completing a morning reconnaissance to Raheng, Thailand. When about to depart, the airfield came under attack by strafing fighters. Rather than have their aircraft destroyed on the ground, both pilots took off immediately but were at once attacked, shot down and killed. They were buried at Moulmein, but the graves could not be located after the war and their names are now commemorated on the Singapore Memorial.
Pilot: NZ402565 Plt Off **Paul Middleton BREWER**, RNZAF - Age 22.

67 Squadron, RAF (Mingaladon, Burma - 221 Group)
Buffalo I W8240 - the other fighter lost in the above engagement. The pilot managed to evade his attackers when taking off but was shot down on turning back to aid his outnumbered companion.
Pilot: NZ404859 Sgt **John Gilbert FINN**, RNZAF - Age 22.

Tue 20/Wed 21 Jan 1942

BOMBER COMMAND

Raid on Emden, Germany (by 20 aircraft - 4 lost)
142 Squadron, RAF (Waltham, Lincolnshire - 1 Group)
Wellington IV Z1207/U - took off at 1910 captained by Plt Off J G Scott, RCAF, and believed to have been shot down by a night-fighter after 2100, to crash into the sea off the West Frisian Islands. The bodies of two RAF crew members later washed ashore onto the Dutch islands of Terschelling and Ameland respectively. Their four comrades are commemorated on the Runnymede Memorial.
Air Gunner: NZ401499 Flt Sgt **Albert Williams KERRISK**, RNZAF - Age 23. 289hrs. 22nd op.

Wed 21 Jan 1942

NEW ZEALAND

Cross country

2 Service Flying Training School (ATS), RNZAF (Woodbourne)

Harvard II NZ961 - took off at 0700 for Lake Grassmere and was last seen at 1500 feet flying away from the landing ground. At about 0800 a witness at Cape Campbell heard an engine roar followed by an explosion in the direction of the sea. The following day the pilot's father, Mr **J R McKenzie**, offered a reward of £3000 for the discovery alive of either of the two airmen, or £1000 for discovery of the wreckage. Apart from a Harvard seat cushion washed up on a beach about ten days later, no other sign of the aircraft or its occupants was found. The two airmen are commemorated on the Provincial Memorials at Christchurch and Wellington, respectively.

Pilot: NZ414650 LAC **Donald Hugh McKENZIE**, RNZAF - Age 20. 119hrs (80 solo - 59 on Harvard)

Pilot: NZ412412 LAC **Jack Trevor VILLERS**, RNZAF - Age 25.

Thu 22 Jan 1942

FAR EAST

Interception of incoming raid

243 Squadron, RAF (Kallang, Singapore)

Buffalo I W8147/O - took off late morning with others and one of two which failed to return. Enemy fighters escorting the raiding bombers shot W8147 down into the sea NE of Horsburgh Light, located off the SE tip of Malaya. The pilot is commemorated on the Singapore Memorial.

Pilot: NZ405519 Sgt **Vincent ARTHUR**, RNZAF - Age 19.

Enemy raid on airfield

488 Squadron, RNZAF (Kallang, Singapore)

The Squadron was scrambled at 1115 to intercept a bombing raid on its aerodrome. Two ground crew, one RAF and the other RNZAF, had just started up one of the Buffaloes when they were killed. The New Zealander is buried at Kranji.

Flight Rigger: NZ391391 AC1 **Ivan Victor Strachan ANDERSON**, RNZAF - Age 20.

Scramble to intercept enemy raid on airfield

488 Squadron, RNZAF (Kallang, Singapore)

Buffalo I W8191 - took off at 1120 with three others in response to the scramble described above. Bombs were raining down just as the fighter became airborne and the blast from one blew W8191 into a petrol dump adjacent to the airfield boundary. The pilot suffered serious shrapnel wounds and died in hospital at 1150 on the 25th. He was buried at 0800 the following morning, but as the grave could not be located post-war he is commemorated on the Singapore Memorial.

Pilot: NZ411390 Plt Off **Leonard Raymond FARR**, RNZAF - Age 24.

Fri 23 Jan 1942

NEW ZEALAND

Air gunnery exercise

2 Service Flying Training School (ATS), RNZAF (Woodbourne)

Harvard II NZ957 - arrived at Lake Grassmere at 1525, just after the exercise, but swung on landing and was momentarily out of position on the airfield. One of the School's Gordons (NZ638), piloted by Plt Off **P S Rishworth**, accompanied by LAC **A J Griffiths**, had been given an all clear signal by an airman at its wingtip and took off with a drogue streamed behind. The Gordon pilot was unable to see the Harvard directly ahead and collided with it. Rishworth was slightly injured but his drogue operator escaped unscathed. The Harvard pilot, however, died and is buried at Napier.

Pilot: NZ414663 LAC **John William Thomas NELSON**, RNZAF - Age 20. 74hrs solo (54 on Harvard)

Sat 24/Sun 25 Jan 1942

INDIA COMMAND

Raid on Bangkok docks

113 Squadron, RAF (Mingaladon, Burma - 221 Group)

Blenheim IV Z7582/L - took off at 1916 captained by Flt Sgt P N Keeley, RAF, with seven others and brought down by flak from a Japanese gunboat, crashing about 1km SW of Talat Plu Railway Station, Bangkok. All three crew were buried nearby in a collective grave by local villagers, but in 1948 the remains were reinterred at Kanchanaburi.

Observer: NZ402125 Sgt **Alvin Messines DINGLE,** RNZAF **-** Age 24.

Sun 25 Jan 1942

FAR EAST

Plt Off **L R FARR**, RNZAF, died of wounds received during an enemy air raid on the 22nd.

FIGHTER COMMAND

Fighter sweep

485 Squadron, RNZAF (Kenley, Surrey - 11 Group)

Spitfire VB AB788/T - took off at 1130 with 11 others and when over the English Channel inexplicably stalled, spun then dived vertically into the sea. A high wind and heavy sea were running at the time and although the others of AB788's section circled the crash area no sign of life could be found. The pilot is commemorated on the Runnymede Memorial. As the Squadron had not seen action with the enemy the most likely cause of the accident appears have been oxygen failure.

Pilot: 79515 Flt Lt **Robert Harold STRANG**, RAFVR - Age 23. 499hrs. 104th op.

Mon 26 Jan 1942

FAR EAST

Attack on invasion force approaching Endau, Malaya

100 Squadron, RAF (Seletar, Singapore - 225 Group)

Vildebeest III K6386 - took off at about 1350 as part of the first of two raids directed against the convoy. When the ships were sighted and dived upon at about 1500 the bomber twice came under fighter attack, the pilot and air gunner being killed during the second pass. The third crew member, wounded RAAF Sgt G W McCabe, managed to crawl into the cockpit and steer K6386 away to the south before it finally fell into the sea a quarter mile south of Lalang Island. McCabe survived the war as a PoW, but the bodies of his companions were never found and they are commemorated on the Singapore Memorial.

Pilot: NZ401769 Flt Sgt **David Bruce Smaill LEE**, RNZAF - Age 29.

36 Squadron, RAF (Seletar, Singapore - 225 Group)

Albacore I T9135 - took off at about 1615 as part of the second of the above raids. On approaching the target at about 1730 the aircraft was shot down, crashed and caught fire on a beach near Endau. All three crew died, the pilot's body being thrown clear and later recognised by a downed survivor from the first raid. Post-war no graves could be found for any of the three airmen and they are all commemorated on the Singapore Memorial.

Pilot: NZ40965 Sgt **Andrew Martin Hearn FLEMING**, RNZAF - Age 30.

Fleming's brother, John Stuart Selby Fleming, died on 19 November 1943 while flying with 90 Sqn, RAF.

36 Squadron, RAF (Seletar, Singapore - 225 Group)

Vildebeest III K6392 - took off at about 1615 as part of the second of the above raids. K6392 was shot down at about 1730 and crashed into a jungle swamp $6^1/_2$ miles SE of Endau at positon 02.35.30N:103.41.30E. All three crew died, their remains being recovered from the crash site post-war and interred in a collective grave at Kranji.

Pilot: NZ41503 Sgt **Thomas Steele TANNER**, RNZAF - Age 23.

Mon 26/Tue 27 Jan 1942

BOMBER COMMAND

Raid on Emden, Germany (by 33 aircraft - 3 lost)

51 Squadron, RAF (Dishforth, Yorkshire - 4 Group)

Whitley V Z9315/V - took off at 1703 and brought down over Germany near Meppen, 45km south of the target. Only an RCAF Sgt survived and was captured. His four comrades were buried in the Evangelical Cemetery at Meppen, but later reinterred at Reichswald Forest, 5km SW of Kleve. The Whitley was one of two aircraft on the raid detailed to carry leaflets rather than bombs, the first leaflet-only (or 'Nickel') raid made on Germany since April 1940.

Captain: NZ404866 Sgt **Harvey Jennings GRUT**, RNZAF - Age 29. 285hrs. 9th op.

2nd Pilot: NZ41328 Sgt **Robert Alfred HAMILTON**, RNZAF - Age 21. 238hrs. 5th op.

51 Squadron, RAF (Dishforth, Yorkshire - 4 Group)

Whitley V Z9423/S - took off at 1712 captained by Plt Off J F Wright, RAF, on same raid as the above and lost without trace, its five crew being commemorated on the Runnymede Memorial.

Observer: NZ404527 Sgt **Hugh Stainger BROWN**, RNZAF - Age 24.

Raid on Hannover, Germany (by 71 aircraft - 2 lost)

12 Squadron, RAF (Binbrook, Lincolnshire - 1 Group)

Wellington II W5585/D - took off at 1815 captained by Sgt J O McKnight, RCAF, but bombs fell out and exploded as it left the runway. The port engine then lost power and the nose was pushed down in an unsuccessful bid to maintain flying speed. Height was lost and during the crash landing which followed W5585 collided with a pill box and caught fire. The captain and one other crew member survived, injured, but the other four died, including the observer, who is buried at Binbrook. Four other aircraft were damaged by the bomb blasts.

Observer: NZ402454 Flt Sgt **James Alexander FLEMING**, RNZAF - Age 31. 199hrs. 6th op.

Fri 30 Jan 1942

AUSTRALIA

RAAF (Sydney (?), New South Wales)
Died of illness at No 113 Australian General Hospital, Concord, and was cremated in Sydney.
(Administrative Officer): 3072 Plt Off **Leslie Clarence ADAMS**, RAAF - Age 41.
Adams served at RAAF Station Headquarters, Darwin, until falling ill, and left there on 4 January for hospital in NSW. The CWGC register gives his date death as the 30th, but RAAF records conflict as to whether he died on the 29th or 30th.

FLYING TRAINING COMMAND

Cross country dual
2 Service Flying Training School, RAF (Brize Norton, Oxfordshire - 23 Group)
Oxford II N6431 - while flying in low cloud and heavy rain at 1520, collided with 24 Squadron Leopard Moth AW156. The Oxford came down at Buscot Park, a few miles WNW of Faringdon, injuring one crew member and killing the other two, the pilot instructor being buried at Black Bourton.
Pilot Instr: NZ401247 Plt Off **Ralph Norman ALLEN**, DFM, RNZAF - Age 23. 569hrs (444 solo - 129 on Oxford) Approx. 30 ops.

Sat 31 Jan 1942

FAR EAST

Scramble to meet incoming enemy raid
258 Squadron, RAF (Tengah, Singapore - 226 Group)
Hurricane I (serial?) - took off in the morning with seven others to intercept a large formation of enemy bombers over Johore and shot down during the engagement. The pilot is commemorated on the Singapore Memorial.
Pilot: NZ40236 Plt Off **Bruce Alexander McALISTER**, RNZAF - Age 24
The CWGC register date of 3 February 1942 and the footnoted death date of 4 April 1942 on page 258 of Vol I of New Zealanders with the Royal Air Force *are both in error. The above date is to be found in fellow-Squadron member Terence Kelly's autobiographical accounts,* Hurricane over the Jungle *and* Battle for Palembang. *The scramble was the first operational sortie by the Squadron since its arrival in the Far East.*

MIDDLE EAST

Calibration test
1 General Reconnaissance Unit, RAF (Ismailia, Egypt - 250 Wing)
Wellington I L4374 - captained by Flt Sgt E R Penhale, RAF, took off for the test and later reported engine failure 40 miles NNW of Port Said, but was lost without trace. The four crew are commemorated on the Alamein Memorial. The GRU operated DWI (Directional Wireless Installation) equipped Wellingtons which operated as magnetic mine sweepers.
Observer: NZ402120 Sgt **Arthur Raymond BATLEY**, RNZAF - Age 26. 252hrs.

Mon 2 Feb 1942

CANADA

Formation flying practice
4 Service Flying Training School, RCAF (Saskatoon, Saskatchewan - 2 Training Command)
Crane I 7703 - while being flown solo, collided at 1530 with another Crane (7831) in the formation and crashed 18 miles south of Moon Lake. The pilot is buried at Saskatoon. The instructor in the other Crane and his pupil pilot, LAC **F L Perrers**, RNZAF, were uninjured and landed their damaged machine safely.
Pilot: NZ414641 LAC **Mervyn David KNIGHT**, RNZAF - Age 26.
Perrers died on 23 November 1943 while flying with 630 Sqn, RAF.

Tue 3 Feb 1942

FAR EAST

Loading cargo aboard SS *Talthybius* in Singapore Harbour
1 Aerodrome Construction Squadron, RNZAF (Singapore)
Owing to the disappearance of local labour, a working party from the Squadron was engaged in straightening and loading additional cargo in the 10,254-ton troopship's holds. Berthed at Go-Down 31/32, the *Talthybius* had escaped with some near misses during a mid-morning bombing, but received two direct hits in the holds on a second raid about midday. The NCO in charge was killed outright and seven airmen suffered moderate to severe burns, one dying later in the day and another on the 8th. The five other wounded were admitted to hospital for treatment and were captured by the Japanese when Singapore fell on the 15th. One of the prisoners died in 1944 when the ship transporting him to Japan was torpedoed and sunk by an American submarine. The other four, who survived the war, were LACs **E W Dense, N E Docherty, S O Meredith** and **S E Sargeant.**
Killed outright and buried at Kranji:-
(Master Mech.): NZ414099 Flt Sgt **Emil Algernon GIFFORD**, RNZAF - Age 33.

Died at the Queen Alexandra Military Hospital later the same day. Buried in the Bidadari Cemetery, Singapore, but his grave could not be located post-war and he is now commemorated on the Singapore Memorial:-
(Driver): NZ414145 LAC **Colin McFarlane KAYSER**, RNZAF - Age 24.

Died from wounds at the Queen Alexandra Military Hospital on the 8th. Buried in the grounds of the hospital, but his grave could not be located post-war and he is now commemorated on the Singapore Memorial:-
(Driver): NZ414133 LAC **Raymond George CARTER**, RNZAF - Age 25.

Suffered severe burns and was admitted to the Queen Alexandra Military Hospital. Taken prisoner as the city fell on the 15th and transported to Thailand. Later moved to Moulmein in Burma and thence to Singapore for transportation by sea to Japan aboard the 9,419-ton Rakuyo Maru. The ship was intercepted on 12 September 1944 by the American submarine Sealion II and torpedoed about 100 miles NE of the Paracel Islands in the South China Sea. Lost without trace when the ship sank and is commemorated on the Singapore Memorial:-
(Driver): NZ414144 LAC **Francis Laurence SCHDROSKI**, RNZAF - Age 24.

FIGHTER COMMAND

Delivery flight(?)
3 Delivery Flight, RAF (High Ercall, Shropshire - 9 Group)
Spitfire (serial?) - took off on a flight to Leconfield, Yorkshire, but crashed en route at Weston Coyney, 4 miles ESE of Stoke-on-Trent, Staffordshire. The pilot is buried at Atcham, Shropshire. The Delivery Flight apparently operated as a detached flight of 131 Sqn, which was based at Atcham.
Pilot: NZ405241 Sgt **Leslie George Raymond DONALDSON**, RNZAF - Age 23. 221hrs.

Wed 4 Feb 1942

FAR EAST

Raid on airfield at Kluang, Malaya
62 Squadron, RAF (P II [Parabumilih], Sumatra - 225 Group)
Hudson III AE485 - took off with five others to bomb Kluang airfield in Johore. Escorting fighters failed to rendezvous with the formation, as did Blenheim bombers which were also to attack the airfield. After bombing alone the Squadron came under fighter attack for 20 minutes. However, it was not noticed if the missing machine was lost through flak over the target or to the fighters, or a combination of both. The Hudson crashed at about 0600 in the jungle near Kampong Frampil in the Batu Pahat area, about 25 miles WSW of Kluang, killing the four crew. Local natives, acting on Japanese instructions, buried the airmen later the same day in a collective grave. In 1950 the bodies were reinterred at Kranji in Singapore. Some sources incorrectly quote AE495 for the Hudson serial - in fact AE495 was a Hudson operated by the RNZAF and never flew in the Far East. This is believed to be a typographical error for AE485, a Hudson which is known to have been delivered to the Far East and lost during the campaign.
Pilot: NZ39634 Wt Off **Douglas Stanley HUNTER**, RNZAF - Age 24.
(Pilot): NZ401783 Wt Off **Denis Allan SCRIMGEOUR**, RNZAF - Age 21.
Scrimgeour is likely to have been acting as observer on this flight.

MIDDLE EAST

Low level attack on docks at Palermo, Sicily
21 Squadron, RAF (Luqa, Malta)
Blenheim IV Z9824 - took off mid-morning with five others and one of four which failed to return. One aircraft was lost on the northerly approach over the sea as an unplanned landfall was made west of Palermo a little after 1100. The others carried on and attacked a train near Carini and a bridge at Scoppe as alternative targets of opportunity. Shortly afterwards the formation entered an area of cloud shrouded hills, only two of these Blenheims finally returning to base. The three missing machines all crashed on the island, either through striking high ground or being brought down by flak. Two crew of Z9824 died and are buried at Catania. The third airman, an RAF Sgt, survived to be taken prisoner.
Pilot: NZ402189 Flt Sgt **Jack Kendrick IBBOTSON**, RNZAF - Age 21. 363hrs. 5th op(?)

Fri 6 Feb 1942

MIDDLE EAST

Strafing enemy on the road between Barce and Cyrene, Libya
272 Squadron, RAF (LG 10 [Gerawala], Egypt - Western Desert)
Beaufighter IC T4875/H - took off at 0745 with three others and was hit by flak in the port engine while strafing near Barce, the starboard engine also failing soon afterwards. The observer survived the crash landing which followed, but owing to his wounds, the fire and exploding ammunition, he was unable to save his pilot who lay either dead or unconscious in the wreckage. At first buried in a military cemetery near Barce, the body was reinterred at Benghazi after the cessation of hostilities.
Pilot: NZ39904 Fg Off **Hector Hugh CRAWFORD**, RNZAF - Age 25. 704hrs. Approx. 112th op.

Anti-shipping strike on Buerat el Hsun, Libya
21 Squadron, RAF (Luqa, Malta)
Blenheim IV Z7308 - took off at 1415 leading three others and one of three which failed to return. One Blenheim was obliged to abort the mission and return early. The others apparently failed to find any ships in the target area and headed back to base, but on nearing Malta were intercepted by enemy fighters and all promptly shot down into the sea near the island of Filfla. There were no survivors, all three crew of Z7308 being commemorated on the Malta Memorial.
Pilot: NZ39946 Sqn Ldr **Russell Redway STEWART**, RNZAF - Age 26. 397hrs. 25th op.

21 Squadron, RAF (Luqa, Malta)
Blenheim IV Z9725 - one of the three shot down into the sea during the above raid. The bodies of two crew members were later recovered and buried on Malta, but the pilot was lost without trace and is commemorated on the Malta Memorial.
Pilot: NZ403940 Flt Sgt **Gordon Russell CAMERON**, RNZAF - Age 22. 273hrs.

Fri 6/Sat 7 Feb 1942

BOMBER COMMAND

Raid on Brest, France (by 57 aircraft - 3 lost)
142 Squadron, RAF (Waltham, Lincolnshire - 1 Group)
Wellington IV Z1247/K - took off at 1735 captained by Flt Lt S St C McNeilly, DFC, RAF, and lost without trace, its six crew being commemorated on the Runnymede Memorial.
Observer: NZ403605 Plt Off **Arthur Gordon HUGGETT**, RNZAF - Age 32. 260hrs. 8th op.

Sat 7 Feb 1942

FIGHTER COMMAND

Low level practice attack on coastal battery
256 Squadron, RAF (Squires Gate, Lancashire - 9 Group)
Defiant I V1116/H - while executing a turn 15 to 20 feet above the water near Lytham Pier at 1630 its port wingtip struck the water, the wing tore away and V1116 crashed into the sea. Both crew died, but no trace of the pilot was found and he is commemorated on the Runnymede Memorial.
Pilot: 112509 Plt Off **William John Paterson OLNEY**, RAFVR - Age 21.

Sun 8 Feb 1942

FAR EAST

LAC **C M KAYSER** died of injuries received in an air raid on the 3rd.

MIDDLE EAST

Escort to Blenheim bombers on raid to Derna, Libya
112 Squadron, RAF (Gambut, Libya - 258 Wing)
Kittyhawk I AK585 - took off at 1250 and one of three which failed to return. After the attack the formation was intercepted by enemy fighters over the Gulf of Bomba and three Kittyhawks were shot down while defending the Blenheims. Two of the three pilots who died, including that of AK585, are commemorated on the Alamein Memorial.
Pilot: NZ404365 Flt Sgt **Brian Patrick HOARE**, RNZAF - Age 20. 269hrs. 24th op.

FIGHTER COMMAND

Formation flight
60 Operational Training Unit, RAF (East Fortune, East Lothian - 81 Group)
Defiant I N1705 - took off piloted by Sgt T Lawrence, RAF, but shortly afterwards the weather deteriorated. Later the pilot reported over the r/t that he had 'hit something' while flying in cloud. Although a wing was badly damaged he indicated he could still maintain some control over the aircraft. However, on approach to base at 1300 control was lost and the Defiant rolled at low altitude, crashed and caught fire near North Berwick, $3^1/_2$ miles north of East Fortune. The two crew are buried at Haddington.
Air Gunner: NZ403607 Sgt **Andrew LAWRIE**, RNZAF - Age 21. 53hrs.

Mon 9 Feb 1942

FAR EAST

Interception
232 Squadron, RAF (P I [Palembang], Sumatra; detachment at Kallang, Singapore - 226 Group)
Hurricane IIB BG797(?) - took off from Kallang late morning, as 'Blue 2', with others to intercept an incoming air raid, but was one of two almost immediately redirected to deal with low flying aircraft reported to be harrassing Allied troops near Johore Bharu. Shortly after arriving over the area the section sighted an enemy fighter below and dived to attack. As 'Blue 1' was about to fire his attention was diverted by a second enemy machine which he attacked instead. On breaking off combat Blue 1 then saw Blue 2 dive steeply into a field and explode. The pilot's

body could not be found post-war and he is commemorated on the Singapore Memorial. Enemy flak was active during the assault and it could have been this or a fighter which brought him down.
Pilot: NZ404923 Sgt **William Adamson MOODIE**, RNZAF - Age 22.
The CWGC register incorrectly gives 5 February as Moodie's date of death. His brother, John Ernest Moodie, died two months earlier, on 7 December 1941 while flying with 3 Delivery Flight, RAF.

FIGHTER COMMAND

Search for aircraft
504 Squadron, RAF (Kirkistown, Co. Down - 82 Group)
Spitfire IIB P8576 (a presentation machine named *City of Bundaberg & District*) - took off at about 1220 and failed to return. While engaged on the search over the Irish Sea the fighter developed a glycol leak west of the Isle of Man. The pilot apparently died of exposure after baling out over the sea, his body later being found in a dinghy washed ashore 6 miles north of the Mull of Galloway, Wigtownshire. He is buried at Stoneykirk. One or more other aircraft are believed to have been lost also while searching for him.
Pilot: NZ403424 Flt Sgt **Edward Dennis CANNON**, RNZAF - Age 25. 249hrs.

Tue 10 Feb 1942

FLEET AIR ARM - SCOTLAND

R/T exercise
888 Squadron, FAA (HMS *Formidable* - Fleet Aircraft Carrier)
Martlet II AM982 - in the morning, stalled on approach to land and spun into the Firth of Clyde, sinking immediately. The pilot is commemorated on the RNZN Memorial at Devonport, Auckland.
Pilot: Sub-Lt (A) **Jock BENNETT**, RNZNVR - Age 22.

Tue 10/Wed 11 Feb 1942

CANADA

Night flying exercise
4 Service Flying Training School, RCAF (Saskatoon, Saskatchewan - 2 Training Command)
Crane I 7868 - took off from Vanscoy, 17 miles SW of Saskatoon, at 0130 captained by Plt Off F G Hovey, RCAF, and flew into the ground about 400 yards after lifting off. 7868 bounced and struck the ground a second time before finally coming to rest and bursting into flames. The instructor and one of his pupil pilots died, the latter being buried at Saskatoon. The other pupil, LAC **D M Hunter**, RNZAF, escaped slightly injured.
Pupil Pilot: NZ414712 LAC **Leslie George Hamilton WILSON**, RNZAF - Age 26. 117hrs.
Hunter was shot down on 21 July 1943 while flying with 39 Sqn, RAF, and became a German prisoner of war.

Wed 11 Feb 1942

FAR EAST

Air raid
1 Airfield Construction Squadron, RNZAF (in transit)
The SS *Darvel*, evacuating the second of two parties of the Squadron and other troops from Singapore to Java, was caught en route in a bombing raid at 1130 whilst sheltering during daylight at anchor in Banka Strait. No direct hits were received but the ship was riddled with bomb splinters, although she did manage to safely limp into Batavia the following afternoon. Some 20 men of the Unit were wounded during the attack, several seriously, one of whom died within the hour. His body was buried at sea shortly afterwards and he is commemorated on the Singapore Memorial.
(Driver): NZ414146 LAC **Albert Desmond STANFORD**, RNZAF - Age 31.

FLYING TRAINING COMMAND

Dual instruction
6 Flying Instructors School, RAF (Staverton, Gloucestershire - 50 Group)
Tiger Moth II T7850 - piloted by Fg Off G G W Pledger, RAF, collided at 1617 with a 9 Service Flying Training Unit Master I (N2766) and crashed near Pardward Mansions, Cheltenham. Both two-man crews died, the New Zealander in T7850 and an airman from the Master being buried at Gloucester.
Pilot: NZ413085 Plt Off **Stanley Charles JONES**, RNZAF - Age 24.

Thu 12 Feb 1942

NEW ZEALAND

Flying practice
1 Squadron, RNZAF (Whenuapai; detachment at Waipapakauri)
Vildebeest IV NZ125 - took off from Waipapakauri at 1030 piloted by Sgt **J M Worden**, but stalled from 20 feet almost immediately and crashed, bursting into flames. The pilot and one of his passengers were seriously injured, the second passenger, AC2 **K I Curtis**, only slightly so. All three were admitted to the Kaitaia hospital but one of the airmen died later in the day and is buried at Auckland.
(Flt Mechanic): NZ392075 AC1 **David Joseph SUTTON**, RNZAF - Age 27.

COASTAL COMMAND

'The Channel Dash' - breakout from the French port of Brest of the German battle-cruisers *Scharnhorst* and *Gneisenau* and the cruiser *Prinz Eugen* to Germany; Allied code name *Operation FULLER*

217 Squadron, RAF (Thorney Island, Sussex - 16 Group)

Beaufort II AW278/F - took off at 1420 captained by Flt Lt A J H Finch, DFC, RAF, and was lost during an attack on the ships off the Dutch coast. All four crew died, the observer being commemorated on the Runnymede Memorial.

Observer: NZ403778 Flt Sgt **Albert Henry Leeds JACKSON**, RNZAF - Age 31. 182hrs. 3rd op.

FIGHTER COMMAND

118 Squadron, RAF (Ibsley, Hampshire - 10 Group)

Spitfire VB AD239 (a presentation machine named *Hoeloesoengai*) - took off at 1435 from West Malling, Kent, with the rest of the Squadron, which was tasked with providing fighter cover for the attacks being made during *Operation FULLER*. Spitfire AD239 and another became engaged with enemy fighters off the coast of Ostend after they had become separated from the Squadron in the poor weather. One returned safely to base, but AD239 was not seen again and its pilot is commemorated on the Runnymede Memorial.

Pilot: NZ41501 Plt Off **Ivan Stuart STONE**, RNZAF - Age 25. 221hrs. 7th op.

BOMBER COMMAND

Many attacks were made by the Command throughout the day (by 242 aircraft in total -16 lost)

420 Squadron, RCAF (Waddington, Lincolnshire - 5 Group)

Hampden I AT134 - took off at 1445 during *Operation FULLER* to attack the warships off the Hook of Holland and was lost. One crew member is commemorated on the Runnymede Memorial, but the bodies of his three comrades later washed ashore along the Dutch coast, including that of the pilot on 16 June at Zandvoort, immediately SW of Haarlem. He is buried at Amsterdam.

Pilot: 41406 Sqn Ldr **George Lawlor Bernays HARRIS**, RAF - Age 28. 773hrs. 41st op.

455 Squadron, RAAF (Wigsley, Nottinghamshire - 5 Group)

Hampden I P1156 - took off from Swinderby, Lincolnshire, at 1522 during *Operation FULLER* captained by Flt Lt W E Perrin, RAF, and lost without trace from an attack off the Dutch coast. Its four crew are commemorated on the Runnymede Memorial.

Observer: NZ404522 Plt Off **Alexander Richard ABBOTT**, RNZAF - Age 23. 266hrs. 9th op.

Thu 12/Fri 13 Feb 1942

BOMBER COMMAND

'Gardening' - mining in the Frisian islands (by 21 aircraft - 1 lost)

455 Squadron, RAAF (Wigsley, Nottinghamshire - 5 Group)

Hampden I AD783/U - took off at 2345 to drop mines off Emden, Germany, but returned early due to engine trouble. At 0255, while circuiting the base at 800 feet, the Hampden dived into the ground at Eagle, 2 miles SW of Wigsley, and caught fire. All four crew died, the pilot being buried at Newark-upon-Trent.

Pilot: NZ402874 Flt Sgt **John Bernard KENNEDY**, RNZAF - Age 26. 353hrs (140 solo - 65 on Hampden) 13th op.

Sat 14 Feb 1942

FAR EAST

In transit from Singapore to Batavia, Java

RAF (in transit)

The 707-ton armed auxiliary *Li Wo*, escaping from Singapore and evading several air attacks en route, in the morning stumbled into a Japanese invasion convoy escorted by several destroyers and a cruiser. Despite hopelessly uneven odds, the former passenger steamship attacked several of the transports, ramming one after expending all her ammunition, all the while coming under fire from the cruiser. The burning *Li Wo* was finally abandoned as she heeled over to port and sank, only a handful of those on board surviving to be taken prisoner. Among those lost was a New Zealand pilot who is commemorated on the Singapore Memorial. After the war the *Li Wo's* captain, Lt T S Wilkinson, RNR, was posthumously awarded the VC for the action.

(Pilot): NZ412243 Sgt **Ian Allan Rogers McDONALD**, RNZAF - Age 30.

McDonald arrived in Singapore about six weeks prior to the Japanese attack and was posted to Seletar. It is not known which unit, if any, he may have served with prior to being hospitalised. According to a surviving sailor, McDonald had travelled with him from Hospital to the docks to escape Singapore on the Li Wo.

Alexandra Military Hospital, RAF (Singapore)

A few days prior to the evacuation of Singapore an RNZAF pilot was transferred to the hospital with a suspected kidney ailment. Nothing further was heard of him and his death in Japanese hands is officially presumed to have occurred on this date (although it seems likely that he may have been among those massacred on the 15th - see entry for Griffiths below). No trace of his body could be found after the war and he is commemorated on the Singapore Memorial.

(Pilot): NZ412722 Sgt **Ernest William MOTTRAM**, RNZAF - Age 23.

Mottram arrived in Singapore on 8 December 1941 and was posted to Seletar. It is not known which unit, if any, he may have served with prior to being hospitalised.

FLYING TRAINING COMMAND

Aerobatics

6 Service Flying Training School, RAF (Staverton, Gloucestershire - 50 Group)

Magister I N3829 - while engaged in a mock attack at 1415 on a Tiger Moth at 500 feet, flicked over following a vertical bank and dived into the ground at Lake Cover, Croome Court estate, to the NW of Upton-on-Severn, Worcestershire. Both Magister pilots are buried at Gloucester.

Pilot: NZ413130 Sgt **Charles Cecil SHAW**, RNZAF - Age 20.

Pilot: NZ413161 Sgt **Vernon Alfred WALTERS**, RNZAF - Age 20.

Shaw and Walters graduated together in New Zealand on Course 17A at Wigram.

BOMBER COMMAND

Cross country training, practice bombing and photography

12 Operational Training Unit, RAF (Chipping Warden, Northamptonshire - 6 Group)

Wellington IC R1027 - after an approach to land upon return at 2325, opened up to go round again but the starboard engine cut. Control was lost and the starboard wingtip collided with a tree as the Wellington flew into the ground and caught fire. All five crew died, both pilots being buried at Brookwood in Woking, Surrey.

Captain: NZ404944 Plt Off **John Reginald REENBERG**, RNZAF - Age 26. 244hrs (91 solo - 21 on Wellington)

2nd Pilot: NZ41997 Plt Off **Frederick Burns RICHARDSON**, RNZAF - Age 21. 261hrs.

Sun 15 Feb 1942

FAR EAST

Alexandra Military Hospital, RAF (Singapore)

One of a number of hospital staff and patients murdered by Japanese troops as the island surrendered. No trace of his body could be found post-war and he is commemorated on the Singapore Memorial.

Surgeon: 23103 Sqn Ldr **Eustatius William Barton GRIFFITHS**, MB ChB FRCS(E) DOMS, RAF - Age 38.

MIDDLE EAST

Return of detachment from Malta to Egypt after convoy escort duties

252 Squadron, RAF (LG 229 [Idku], Egypt - detachment at Luqa, Malta - 201 Group)

Beaufighter IC T4651/C - took off from Luqa at 0840 with four others, but recalled almost immediately as an incoming air raid was identified. Some of the formation returned safely but several failed to receive the message and flew on, as insufficient time had been available on Malta to have their sets adjusted for the correct frequency. Fifteen to 20 enemy fighters soon intercepted them and shot T4651 down into the sea a few miles east of the island. Its two crew and a Squadron fitter travelling as a passenger are commemorated on the Alamein Memorial.

Pilot: NZ403972 Flt Sgt **Clive Herbert MULHOLLAND**, RNZAF - Age 26. 317hrs.

It will be noted that Mulholland is commemorated on the Alamein Memorial rather than that on Malta, although his last flight was from the Island.

Sun 15/Mon 16 Feb 1942

BOMBER COMMAND

Raid on St Nazaire, France (by 26 aircraft - 3 lost)

77 Squadron, RAF (Leeming, Yorkshire - 4 Group)

Whitley V Z9231/U - took off at about 1800 captained by Flt Sgt G R Millar, RCAF, and on return to England at 0020 crashed with a failed engine while attempting an emergency landing at Colerne airfield, Wiltshire. All five crew died, the captain and observer being buried at Colerne.

Observer: NZ404494 Sgt **Reginald Andrew KEMP**, RNZAF - Age 21. 295hrs. 9th op.

Mon 16 Feb 1942

NEW ZEALAND

Dual instruction

1 Elementary Flying Training School, RNZAF (Taieri)

Tiger Moth II NZ766 - seen at noon spinning from 4000 feet, completing eleven turns before disappearing behind some trees to crash near Bush Road, Mosgiel. The pilot is buried at Wellington, but pupil LAC **G Rogers** survived with serious injuries. It was thought that one of the pilots feet may have become jammed and prevented free rudder movement.

Pilot Instr: NZ404387 Plt Off **Hugh Maxwell McGILL**, RNZAF - Age 23. 661hrs solo (570 on Tiger Moth)

Mon 16/Tue 17 Feb 1942

FERRY COMMAND

Ferry flight from Gibraltar to Malta

Overseas Aircraft Despatch Unit, RAF (Portreath, Cornwall - 44 Group)

Wellington IC AD635 - took off for Malta but flew off course and crashed into the mountains near Kef Oum Teboul, about 8 miles east of La Calle, Algeria, near the Tunisian border. All six crew were buried at the scene of the crash, but later reinterred at Bone.

Captain: NZ404564 Plt Off **Allan O'BRIEN**, RNZAF - Age 31. 2nd op.
2nd Pilot: NZ41499 Sgt **Walter Brian STEHR**, RNZAF - Age 21. 220hrs. 2nd op.

O'Brien and Stehr had previously trained at 21 OTU, Bomber Command. They were then briefly posted to 15 OTU, prior to commencing the ferry flight on 9 February on being posted to the Middle East.

BOMBER COMMAND

'Gardening' - mining in the Frisian islands (by 49 aircraft - 2 lost)

61 Squadron, RAF (Woolfox Lodge, Rutland - 5 Group)

Manchester I L7433 - took off at 1910 captained by Flt Sgt P H G Webster, RAF, and was hit by flak, crashing into the sea off the Dutch island of Terschelling. All seven crew died, three being buried on the island at West-Terschelling when their bodies washed ashore, including that of the 2nd pilot which was recovered on 21 June. The others are commemorated on the Runnymede Memorial.

2nd Pilot: NZ41490 Sgt **Frederick Stanley PARTRIDGE**, RNZAF - Age 28. 271hrs. 4th op.

Tue 17 Feb 1942

NEW ZEALAND

Dual instruction

2 Service Flying Training School (ITS), RNZAF (Woodbourne)

Harvard II NZ967 - seen at 1315 to begin a steep diving turn to starboard with engine at high power. At under a 100 feet, while attempting to flatten out, the starboard wing folded back and detached, causing NZ967 to dive into the ground in an almost inverted position in the Omaka Valley, 8 miles SW of the airfield. The pilots are buried at Taradale and Wanganui, respectively.

Pilot Instr: NZ402176 Fg Off **John Angus FERGUSON**, RNZAF - Age 22. 796hrs (700 solo - 348 on Harvard)
Pilot: NZ414618 LAC **Allen Russell HANSEN**, RNZAF - Age 23.

FLYING TRAINING COMMAND

Training

5 Service Flying Training School, RAF (Tern Hill, Shropshire - 21 Group)

Master III W8571 - with instructor Wt Off N Arnold, RAF, spun off a steep climbing turn at low altitude at 1100, almost recovered, but struck trees and crashed at Chetwynd Heath, Gloucestershire. The pilot is buried near Tern Hill at Stoke-Upon-Tern.

Pilot: 126992 Lt **Thomas FOSTER-BARHAM**, 1st Bn, The Black Watch (Royal Highland Regiment), attached RAF - Age 22.

Foster-Barham's date and place of birth are unknown, but he lived in New Zealand from at least 1930 to 1934.

BOMBER COMMAND

Training

11 Operational Training Unit, RAF (Bassingbourn, Cambridgeshire - 6 Group)

Wellington IC T2710 - took off from Steeple Morden at 2010 captained by Sgt A Livingstone, RAF, but immediately lost flying speed, crashed and caught fire. Four crew members died, including two New Zealanders, who are buried in the Bassingbourn cum Kneesworth Cemetery.

WOpAG: NZ401265 Sgt **Hylton Carlyle GEORGE**, RNZAF - Age 28. 30hrs.
WOpAG: NZ437013 Sgt **Percy John Havelock HAMLIN**, RNZAF - Age 26.

Sat 21 Feb 1942

COASTAL COMMAND

'Shadow' - off the Danish coast

500 Squadron, RAF (Bircham Newton, Norfolk - 16 Group)

Hudson VAM697/K - took off at 1528 and lost without trace, the observer being commemorated on the Runnymede Memorial.

Observer: NZ401641 Plt Off **Robert CROZIER**, RNZAF - Age 24. 203hrs. 4th op.

Crozier's brother, Alexander Crozier, died on 2 March 1944 while flying with 23 Sqn, RAF.

Sun 22/Mon 23 Feb 1942

FLEET AIR ARM - SHETLANDS

Night search for enemy fleet

832 Squadron, FAA (HMS *Victorious* - Fleet Aircraft Carrier)

Albacore I X8962 - piloted by Lt R C G de Longueuil, RN, and last seen 30 miles east of the Shetlands in snowstorm conditions. The three crew are commemorated on the Lee-on-the-Solent Fleet Air Arm Memorial.

Observer: Lt Cdr **Arthur James Philson PLUGGE**, mid, RN - Age 30.

Plugge was 832 Squadron's commanding officer. His brother, John Aicken Plugge, died on 1 May 1934 while flying at the RAF College, Cranwell.

Mon 23 Feb 1942

SOUTH PACIFIC

Familiarisation

70th Pursuit Squadron, USAAF (Nawera, Fiji)

P-39D Airacobra 41-7107 - while carrying out a second familiarisation sortie the pilot engaged in a series of mock attacks on an RNZAF Vincent of 4 Squadron engaged in a ferry flight. At 1545, after carrying out several climbing and diving turns, the Airacobra flicked over during a steep climbing turn and entered a spin from which it failed to recover. The machine crashed through a native hut at Nainulomulo village, $5^1/_2$ miles ESE of Nadi. Three occupants in the hut were injured and the pilot died. He was buried the following day at Lautoka, but later reinterred at Suva.

Pilot: NZ1182 Flt Lt **Eric Neville GRIFFITHS**, RNZAF - Age 28. 1668hrs solo.

A member of 4 Squadron, RNZAF, Griffiths had been attached to the American unit a few days prior to the accident. In addition to an unknown number of operational patrols and convoy escort flights around Fiji he had also flown operationally on the Republican side during the Spanish Civil War of 1936-39.

ARMY CO-OPERATION COMMAND

Training

41 Operational Training Unit, RAF (Old Sarum, Wiltshire - 70 Group)

Magister I T9884 - dived towards the ground at 1045, pulled up steeply, stalled and spun into the ground near Wincanton, Somerset. Both pilots are buried at Salisbury.

Pilot: NZ412651 Plt Off **Ernest William BRIGHT**, RNZAF - Age 26. 192hrs.

2nd Pilot: NZ412772 Plt Off **David George WEST**, RNZAF - Age 22. 399hrs.

West's brother, Peter Wynn Mason West, died on 4 July 1942 while flying with 61 Sqn, RAF.

Mon 23/Tue 24 Feb 1942

MIDDLE EAST

Raid on Martuba airfield, Libya

108 Squadron, RAF (LG 221 [Fayid], Egypt - 242 Wing, 205 Group)

Wellington IC AD599 - captained by Fg Off B A Willis, RAAF. Other crews on the raid reported seeing a bomber catch fire and explode on hitting the ground. The six crew were buried close to the wreckage north of Martuba at Sahel-el-Emeilet, Derna, but later reinterred in Knightsbridge War Cemetery, a few miles from Acroma, Libya.

Rear Gunner: NZ401436 Flt Sgt **Cecil Montrose GRAY**, RNZAF - Age 28. 330hrs. 19th op.

Gray's brother, Murray Llewellyn Gray, died on 18 June 1943 while flying with 20 Sqn, RNZAF.

Tue 24 Feb 1942

CANADA

Formation flying practice

4 Service Flying Training School, RCAF (Saskatoon, Saskatchewan - 2 Training Command)

Crane I 7717 - while in formation at 1620, momentarily disappeared under the leader, who noticed 7717's wing shudder just before the trainer entered a dive. It then appeared to recover, climb again, but entered another dive, striking the ground and catching fire near Osler, 17 miles NNW of the airfield. The pilot is buried at Saskatoon.

Pilot: NZ414274 LAC **Leslie William GILLAN**, RNZAF - Age 20. 135hrs (71 solo - 48 on Crane)

Wed 25 Feb 1942

NEW ZEALAND

Air to ground firing exercise

1 Squadron, RNZAF (Whenuapai)

Vildebeest III NZ114 - seen at 0730 to descend towards Muriwai Beach and carry out a low pass over a truck. Climbing away afterwards in a steep left-hand turn, NZ114 stalled at 250 feet and dived steeply into the sea. Two of the crew died, while the third, wireless operator AC1 **G H Tate**, suffered serious injuries. The pilot is buried at Auckland and wireless operator **Mannell** at Masterton. Three airmen - Act Cpl **G F G Brown**, LAC **G S D'Anvers** and Sgt **J A McNie** - were later commended for their attempts to rescue the crew from the sea.

Pilot: NZ404969 Sgt **Alfred William TODMAN**, RNZAF - Age 22. 349hrs (267 solo - 192 on Vildebeest)

(WOp): NZ412455 AC1 **Edward James MANNELL**, RNZAF - Age 20. 203hrs.

FAR EAST

Interception scramble

605 Squadron, RAF (Tjilitjan, Java)

Hurricane I (BG821?) - took off with four others to intercept a morning raid on nearby Batavia or Semplak, being one of two which failed to return. During an engagement with enemy fighters was shot down, crashing in a swamp near Krawang, about 30 miles NE of Batavia. The pilot was buried at the scene of the crash, but post-war searches could not locate his grave and he is now commemorated on the Singapore Memorial.

Pilot: NZ391831 Flt Lt **Henry Augustine DOBBYN**, RNZAF - Age 22.

FIGHTER COMMAND

Local flying practice and instruction

485 Squadron, RNZAF (Kenley, Surrey - 11 Group)

Magister I T9947 - while low flying at 1545, stalled above some trees, crashed and caught fire near Three Bridges. Both pilots are buried near the base at Whyteleafe.

Pilot: NZ41484 Sgt **Arnold George McNEIL**, RNZAF - Age 22. 207hrs.

2nd Pilot: NZ41307 Sgt **James Joseph BURKE**, RNZAF - Age 23. 228hrs.

FLYING TRAINING COMMAND

Formation practice

1 Service Flying Training School, RAF (Netheravon, Wiltshire - 23 Group)

Master III W8798 - collided with trees while low flying at 1845 and crashed near Everleigh, Hampshire. The pilot is buried at Brookwood in Woking.

Pilot: NZ2517 Leading Airman **Charles Anthony GALLAGHER**, RNZN - Age 23.

Gallagher was from HMS Daedalus*, of the Fleet Air Arm, but undergoing his basic training with the RAF.*

Wed 25/Thu 26 Feb 1942

BOMBER COMMAND

Navigation exercise over the Irish Sea

27 Operational Training Unit, RAF (Lichfield, Staffordshire - 6 Group)

Wellington IC X9682 - captained by Sgt D H Dyson, RAAF. Sent an SOS message at 2256 but nothing further was heard, its six crew being commemorated on the Runnymede Memorial.

WOpAG: NZ403788 Sgt **Roy Gordon HARKINS**, RNZAF - Age 22. 40hrs.

Thu 26 Feb 1942

COASTAL COMMAND

Meteorological flight(?)

1401 Meteorological Flight, RAF (Bircham Newton, Norfolk - 16 Group)

Spitfire IIA P7546 - at 1110 dived out of a 500 foot cloud base and into the ground at East Wretham airfield. The pilot is buried at Honington, Suffolk.

Pilot: NZ391363 Flt Sgt **William David Angus McLEOD**, RNZAF - Age 25. 423hrs.

FIGHTER COMMAND

Circuits and landings

256 Squadron, RAF (Squires Gate, Lancashire - 9 Group)

Defiant I T3995/U - at 2005, under difficult weather conditions, stalled on approach to the airfield, crashed and caught fire. Both crew are buried at Lytham St Anne's.

Pilot: NZ404376 Sgt **Francis William JOYCE**, RNZAF - Age 22. 196hrs (150 solo - 48 on Defiant)

Air Gunner: NZ402117 Flt Sgt **Gordon Fitzgerald WALDEN**, RNZAF - Age 26. 222hrs. 16 ops.

FLEET AIR ARM - CARIBBEAN

Air to sea firing exercise

750 Squadron, FAA (Piarco, Trinidad - HMS *Goshawk*)

Albacore I N4177 - piloted by Lt (A) J L Cribb, RNVR. Rolled over while executing a turn at 1055 and dived straight into the sea, sinking immediately at Cangregos Point, off Waterloo. The pilot and the two trainee observers on board died. The New Zealand observer is commemorated on the RNZN Memorial at Devonport, Auckland, and his two comrades on the FAA Memorial at Lee-on-Solent.

Observer: NZ2524 Leading Airman **Robert Murray PATERSON**, RNZN - Age 24.

Fri 27 Feb 1942

FAR EAST

Evacuation from Tjilatjap, Java, to Broome, Western Australia

QANTAS Empire Airways

S.23 Empire Flying Boat G-AETZ (named *Circe*) - captained by Capt W B Purton and last reported in when 150 miles outbound from Tjilatjap, only to be shot down by enemy fighters shortly afterwards at about midday. The Japanese chanced upon *Circe* while returning from a successful attack on the elderly American aircraft carrier

Langley which was en route to Tjilatjap. Searchers failed to locate any wreckage and no trace could be found of the airliner's four crew or sixteen passengers. The First Officer, as a member of the RAAF Reserve, is commemorated on the Northern Territory Memorial at Adelaide River, 72 miles south of Darwin.
First Officer: Aus267574 Fg Off **Mervyn William BATEMAN**, RAAF Reserve - Age 30.
The CWGC register records Bateman's date of death as the 28th, a date also quoted in some sources for when Circe *was shot down. The Australian Official History volume,* Royal Australian Air Force 1939-1942, *however, in a detailed account of events of the day, confirms that the loss occurred on the 27th.*

Fri 27/Sat 28 Feb 1942

FERRY COMMAND

Ferry flight from Gibraltar to Malta
1 Overseas Aircraft Despatch Unit, RAF (Portreath, Cornwall - 44 Group)
Wellington IC Z9097 - flew off course and crashed in Tunisia, killing the six crew, who are buried at Enfidaville.
Captain: NZ404945 Plt Off **Stewart Walton REID**, RNZAF - Age 24.
2nd Pilot: NZ41380 Sgt **John Noel VINCENT**, RNZAF - Age 22.
Reid and Vincent had trained at 11 OTU, Bomber Command. They were then briefly posted to 15 OTU, prior to beginning the ferry flight from England on being posted to the Middle East.

BOMBER COMMAND

Raid against the battle-cruiser *Scharnhorst* at Wilhelmshaven, Germany (by 33 aircraft - 3 lost)
77 Squadron, RAF (Leeming, Yorkshire - 4 Group)
Whitley V Z9148/L - brought down over Germany and crashed near Esens, 35km WNW of Wilhelmshaven. All five crew died, the two New Zealanders being buried in the Esens Cemetery, probably with some or all of the others. All were later reinterred at Sage, 24km south of Oldenburg.
Captain: NZ403976 Plt Off **Morris Donald Darwin McCARTHY**, RNZAF - Age 19. 419hrs. 14th op.
Observer: NZ404029 Flt Sgt **Bert Gordon HIBELL**, RNZAF - Age 21. 266hrs. 9th op.
McCarthy's brother, Raymond William McCarthy died over southern Germany on 15 April 1943 while flying with 7 Sqn, RAF. After the war Raymond's body was exhumed from Stuttgart and buried alongside his brother's at Sage.

Sat 28 Feb 1942

BOMBER COMMAND

Air test
75 (NZ) Squadron, RAF (Feltwell, Norfolk - 3 Group)
Wellington III X3355/Y - took off shortly before 1600 captained by Sgt **R A Colville**, RNZAF. Immediately afterwards the starboard engine cut and on attempting a return to the airfield the port engine began to splutter. On approaching for a forced landing, X3355 stalled, crashed and caught fire near Lime Kiln Farm at Brandon Fields, $3^1/_2$ miles ESE of Brandon, Suffolk. The two pilots were injured; the 2nd pilot died the same day in the RAF Hospital at Ely, Cambridgeshire, and is buried at Feltwell. Four ground crew were also on board, two of whom were killed, the others being injured.
2nd Pilot: NZ402449 Sgt **Henry William WOODHAM**, RNZAF - Age 27. 297hrs. 7ops.

Sun 1 Mar 1942

FAR EAST

Strafing enemy invasion force
605 Squadron, RAF (Tjilitjan, Java)
Hurricane I (serial?) - took off at dawn to attack enemy troops landing at Indramaju, about 90 miles east of Batavia. Damaged by flak during a low level strafing run along the beach and belly-landed in a paddy field a mile or two inland. The pilot was seen to safely get out and set his fighter on fire. He waved to his No 2 circling overhead, but nothing further was heard of him and he is commemorated on the Singapore Memorial.
Pilot: NZ411455 Fg Off **Noel Callan SHARP**, DFC, RNZAF - Age 20.
The CWGC register incorrectly gives Sharp's date of death as 20 February 1942.

Wed 4 Mar 1942

FAR EAST

Raid on Kyaikto village and transport on road between Bilin and Thaton, east side of Sittang River, Burma
113 Squadron, RAF (Magwe, Burma - 221 Group)
Blenheim IV Z7592 - took off at 1500 piloted by Flt Lt S F Lee, RAAF, accompanied by three Blenheims of 45 Sqn. While flying line astern at 1500 feet the formation was attacked and Z7592 dived to low level in attempt to escape, but was set upon by four fighters which sent it down to crash. The air gunner was the sole survivor and reached friendly territory a few days later. The remains of the observer could not be found post-war and he is commemorated on the Singapore Memorial, as is the pilot. An official source places the crash site as 60 miles east of Pegu and west of the Sittang River but as this is geographically impossible it is probable that west of the Salween River was intended.
Observer: NZ402123 Sgt **Kenneth William BRETT**, RNZAF - Age 26.
The CWGC register incorrectly records Brett's pilot as having died on 15 March 1943.

Wed 4/Thu 5 Mar 1942

MIDDLE EAST

Raid on shipping in Benghazi Harbour, Libya

70 Squadron, RAF (LG 104 [Qotafiya II/El Daba], Egypt - 231 Wing, 205 Group)

Wellington IC Z1042/S - took off at 2300 captained by Flt Lt P J Fisher, RAF, and last heard from when signalling identification to base at 0429. The wreckage of Z1042 was found about 10 days later in the No 1 Armoured Division's operational area, approximately 40-50 miles south of Bomba. Two bodies were discovered at the site, but not those of the other four crew, including the 2nd pilot. After hostilities the graves of the former could not be located either and all six airmen are commemorated on the Alamein Memorial.

2nd Pilot: NZ41308 Plt Off **Geoffrey Pender CHAMBERS**, RNZAF - Age 21. 286hrs. 10th op(?)

Fri 6/Sat 7 Mar 1942

FAR EAST

Raid on airfield at Kalidjati, Java

36 Squadron, RAF (Tasikmalaja, Java)

Vildebeest III (serial?) - took off from Andir at about 2200 with two others and was hit by flak or the blast of its own bombs during a third bombing dive on the airfield, the observer being slightly wounded. About midnight, while struggling to clear mountains on the return journey, the engine failed. The observer and gunner were ordered to bale out, but the pilot died shortly afterwards when the Vildebeest crashed into a paddy field near the Estate Pasir Padang, about 20 miles from his base. His body was recovered and buried on the estate, but post-war the grave could not be found and he is commemorated on the Singapore Memorial.

Pilot: NZ40955 Flt Sgt **Bruce APPLEBY**, mid, RNZAF - Age 22.

In transit from Tjilatjap to Tasikmalaja airfield, Java

RAF Works Branch, RAF (in transit)

A large party of RAF airmen, evacuated earlier by sea from Singapore to Tjilatjap, boarded two trains at Poerworkerto in the evening for transfer to Tasikmalaja. En route both trains were ambushed by enemy troops, a number of airmen being killed and many others wounded, including a New Zealand civilian in RAF uniform who was slightly wounded in the arm. Shortly after midnight, or within a day or two, he was captured and taken to Batavia. He later escaped with three others, but the party was betrayed by natives and recaptured about 40km away. Returned to Batavia, they were given a perfunctory trial, found guilty, and shot by firing squad in front of other prisoners. The exact date of their execution is unknown, being quoted post-war by an eyewitness as simply 'sometime in May [1942].' Although not in full agreement with the above, an account by airman James Home on pages 38-39 of his book, *Their Last Tenko*, appears to relate to the this event.

(Stores Superintendent): Honorary Wt Off **Eion FIELD**, RAF - Age 24.

At the time of the Japanese invasion Field was employed under the Air Ministry Directorate of Works in a civilian capacity. As Singapore fell he and fellow workers were put into RAF uniform as a precautionary measure to avoid their being treated as spies if captured. It appears that his remains were not located post-war and no record of his name can be found in the CWGC registers, presumably because of his 'civilian' status.

Sat 7 Mar 1942

NEW ZEALAND

Solo training

2 Elementary Flying Training School, RNZAF (Bell Block)

Tiger Moth II NZ844 - became lost over featureless country in low cloud and poor weather. On finding a break in the clouds shortly before noon, descended and attempted a forced landing in a football field at Tokirima, 14 miles WSW of Taumarunui. However, while negotiating the difficult and obstructed approach to the field NZ844 stalled, crashed and caught fire. The pilot is buried at Palmerston North.

Pilot: NZ415764 LAC **Ronald Frederick Charles JONES**, RNZAF - Age 20. 11hrs solo.

Sun 8 Mar 1942

FAR EAST

Surrender and capture on Java

605 Squadron, RAF (Tasikmalaja, Java)

All Allied forces in Java officially surrendered to the Japanese this day. Surviving aircrew from the Squadron were taken prisoner on the 8th or during the next few weeks. For a time some of the airmen endeavoured to remain hidden in the hills and survive off the land. However, this existence soon proved to be unsustainable and they eventually had to give themselves up. It is not known exactly when the New Zealander mentioned below surrendered, but he was later transported with others to help build an airfield on the island of Ambon, in the Moluccas group some 250 miles south of the equator. On 29 November 1943, he was one of some 750 prisoners being redrafted back to Java when their 4646-ton steamship *Suez Maru* was torpedoed and sunk by the US submarine *Bonefish* NE of Kangean island in the Java sea. Almost all of those on board died and he is commemorated on the Singapore Memorial.

(Pilot): NZ411484 Flt Lt **Grahame Prebble WHITE**, RNZAF - Age 22.

COASTAL COMMAND

Squadron move to Thorney Island, Sussex

489 Squadron, RNZAF (Leuchars, Fifeshire - 16 Group)

Blenheim IVF Z6342 - after a brief en route stop, took off from Worthy Down, Hampshire, at 1750 piloted by Flt Sgt Wilkinson. Z6342 immediately suffered an engine failure, stalled from about 100-150 feet while turning to starboard, crashed and burnt out. All three crew died, the wireless operator-air gunner being buried at West Thorney.

WOpAG: NZ405506 Plt Off **Cedric Hubert Owen PRIEST**, RNZAF - Age 28.

Mon 9 Mar 1942

FLEET AIR ARM - ENGLAND

Local flying

825 Squadron, FAA (Lee-on-Solent, Hampshire - HMS *Daedalus*)

Swordfish II DK687 - struck the sea while low flying over the Solent, bounced, stalled from 30 feet and crashed into the water, sinking immediately. Both crew died, the pilot being commemorated on the RNZN Memorial at Devonport, Auckland, and his comrade on the FAA Memorial at Lee-on-Solent.

Pilot: Sub-Lt (A) **David Winton ADAMS**, RNZNVR - Age 20.

Mon 9/Tue 10 Mar 1942

BOMBER COMMAND

Raid on Essen, Germany (by 187 aircraft - 6 lost)

12 Squadron, RAF (Binbrook, Lincolnshire - 1 Group)

Wellington II W5442/B - took off at 2039 captained by Plt Off R H Buchanan, RAF, and brought down by flak, crashing at 2204 by a Dutch road called Kagerweg, near Beverwijk, 20km south of Bergen. Three of the crew survived to be taken prisoner, but their three comrades died and are buried at Bergen.

Observer: NZ401413 Flt Sgt **Reuben Acton SCRAGG**, RNZAF - Age 24. 288hrs. 14th op.

150 Squadron, RAF (Snaith, Yorkshire - 1 Group)

Wellington IC DV447 - took off at approximately 1915-2045 captained by Sqn Ldr J P V Nicholls, mid, RAF, on same raid as the above, and brought down near Gescher, 17km north of Essen. All six crew were buried in Gescher Cemetery, but later reinterred at Reichswald Forest, 5km SW of Kleve.

Observer: NZ404567 Plt Off **Oswald William SUCKLING**, RNZAF - Age 26. 293hrs. 10th op.

Tue 10 Mar 1942

FIGHTER COMMAND

Low level demonstration attacks during Army manoeuvres near Loch Watten

54 Squadron, RAF (Castletown, Caithness-shire - 14 Group)

Spitfire VB BL785 - 54 Sqn and another were charged with 'beating up' gun positions, etc, during the day. BL785 came in too low during a dummy attack on some lorries along the Wick-Thurso road, 2 miles west of Watten, and struck one of them. The driver and a passenger were killed, as was the Spitfire pilot, who is buried at Olrig, adjacent to Castletown.

Pilot: NZ401378 Plt Off **Graeme Holt FENTON**, RNZAF - Age 20. 408hrs.

Tue 10/Wed 11 Mar 1942

BOMBER COMMAND

Raid on Essen, Germany (by 126 aircraft - 9 lost)

455 Squadron, RAAF (Wigsley, Nottinghamshire - 5 Group)

Hampden I AT192/R - took off at 1917 captained by Plt Off O Gandy, RAF, and brought down over Germany between Repelen and Baeri, about 10km NW of Duisburg. All four crew were buried nearby in Lohmannseide Cemetery, but later reinterred at Reichswald Forest, 5km SW of Kleve.

Observer: NZ401800 Plt Off **Lewis Alfred WRIGHT**, RNZAF - Age 26. 295hrs. 16th op.

149 Squadron, RAF (Mildenhall, Suffolk - 3 Group)

Stirling I R9295/G (a presentation machine named *East India III*) - took off at 2000 captained by Fg Off C L Pilkington, RAF, on same raid as the above. Damaged by flak during the raid, on returning to base at 0130 R9295 overshot and crashed into some trees at Holywell Row, 2 miles NW of Mildenhall. One crew member survived, injured, but the other seven died, including the 2nd pilot, who is buried at Beck Row, 2 miles NW of Mildenhall.

2nd Pilot: 72585 Sqn Ldr **Lloyd Watt COLEMAN**, DFC*, RAFVR - Age 24. 734hrs. 53rd op.

Wed 11 Mar 1942

NORTHERN IRELAND

Low flying reconnaissance during Army manoeuvres
231 Squadron, RAF (Maghaberry, Co. Antrim)
Lysander III T1577 - struck a farmhouse at 0825 while low flying near Dungannon, Co. Tyrone, crashed and burst into flames at Aktinsallagh, Glebe, near Coalisland. Both crew are buried at Belfast.
Pilot: NZ41938 Plt Off **William Russell RATTRAY**, RNZAF - Age 22. 226hrs (153 solo - 73 on Lysander)

Thu 12/Fri 13 Mar 1942

BOMBER COMMAND

Raid on Emden, Germany (by 40 aircraft - 3 lost)
58 Squadron, RAF (Linton-on-Ouse, Yorkshire - 4 Group)
Whitley V Z9227/R - took off at 1915 and shot down over the Netherlands by a night-fighter, crashing at 2304 into the sea between the island of Ameland and the mainland. All five crew died, the bodies of two later being recovered - one on Ameland and the other on the mainland. The others, including the captain, are commemorated on the Runnymede Memorial.
Captain: NZ41371 Sgt **Charles Ronald SPENCE**, RNZAF - Age 25. 237hrs. 6th op.

Raid on Kiel, Germany (by 68 aircraft - 6 lost)
75 (NZ) Squadron, RAF (Feltwell, Norfolk - 3 Group)
Wellington III X3282 - took off at approximately 1940-2035 captained by Sgt J F M Parnham, RAF, and brought down near Schleswig, 40km NW of the target. All six crew died, at least five, including two New Zealanders, being buried in the Garrison Cemetery at Schleswig, but all six were later reinterred at Kiel.
2nd Pilot: NZ402534 Flt Sgt **John Lukies BROWN**, RNZAF - Age 22. 282hrs. 9th op.
Front Gunner: NZ402974 Flt Sgt **Campbell Ewen Justin AITCHISON**, RNZAF - Age 22. 291hrs. 24th op.

75 (NZ) Squadron, RAF (Feltwell, Norfolk - 3 Group)
Wellington III X3585 - took off at approximately 1940-2035 captained by Fg Off J F K Sandys, RCAF, on the same raid as the above and lost without trace, its six crew being commemorated on the Runnymede Memorial.
2nd Pilot: NZ404985 Sgt **Roy Joffre Desmond WOODCOCK**, RNZAF - Age 26. 259hrs. 3rd op.
Observer: NZ401756 Plt Off **John EARLE,** RNZAF - Age 29. 255hrs. 3rd op.
WOp: NZ404095 Flt Sgt **Henry John PRICE,** RNZAF - Age 25. 135hrs. 2nd op.
Rear Gunner: NZ39857 Flt Lt **Thomas James Desmond BABER**, mid Czech War Medal, RNZAF - Age 23. 435hrs. 39th op.

75 (NZ) Squadron, RAF (Feltwell, Norfolk - 3 Group)
Wellington III X3588 - took off at approximately 1940-2035 captained by Sqn Ldr P J R Kitchin, DFC, RAF, on the same raid as the above and lost without trace, its six crew being commemorated on the Runnymede Memorial.
2nd Pilot: NZ41932 Plt Off **William James PARTON**, RNZAF - Age 20. 201hrs. 1st op(?)
Rear Gunner: NZ40914 Flt Sgt **Lloyd Montgomery CHAMBERLAIN**, RNZAF - Age 28. 318hrs. 8th op.

Fri 13 Mar 1942

NEW ZEALAND

1 Service Flying Training School (ITS), RNZAF (Wigram)
Died of natural causes at the Christchurch Public Hospital and is buried at Cust.
(Aircrafthand): W1515 Woman Auxiliary 1st Class **Leone Beatrice WOODFIELD**, WAAF - Age 24.
Woodfield was the first woman in the New Zealand air service to die. The CWGC register incorrectly records her rank as ACW1 (Aircraftwoman 1st Class).

FLEET AIR ARM - SCOTLAND

Reconnaissance exercise
819 Squadron, FAA (Hatston, Orkneys - HMS *Sparrowhawk/Tern*)
Swordfish I W5923/C - took off from Twatt during the afternoon for the exercise over Scapa Flow and lost without trace. The pilot is commemorated on the RNZN Memorial at Devonport, Auckland, and his two comrades on the FAA memorial at Lee-on-Solent.
Pilot: Sub-Lt (A) **David Anthony HOWDEN**, RNZNVR - Age 21.

ARMY CO-OPERATION COMMAND

Air firing exercise
239 Squadron, RAF (Gatwick, Surrey - 35 Wing)
Hurricane IIC BN864 - flew into cloud-covered high ground at 1755 in the Quantock Hills near Crowcombe, 8 miles NW of Taunton, Somerset. The pilot is buried at Bridgwater, 8 miles NNW of Taunton.
Pilot: NZ402456 Plt Off **James Joseph Jack GREVATT**, RNZAF - Age 24. 291hrs.

BOMBER COMMAND

Circuits and landings

16 Operational Training Unit, RAF (Upper Heyford, Oxfordshire - 7 Group)

Hampden I X2906 - entered a flat spin at 1445 while flying in hazy weather and dived into the ground 3 miles east of Croughton airfield, near Mixbury. The two crew are buried at Upper Heyford.

Pilot: NZ411440 Plt Off **Reginald Alfred Stewart PARKER**, RNZAF - Age 25. 187hrs (94 solo - 5 on Hampden)

Fri 13/Sat 14 Mar 1942

AUSTRALIA

Motor vehicle accident

1 Aerodrome Construction Squadron, RNZAF (in transit)

Severely injured in a motor vehicle accident near Heathcote, about 55 miles WNW of Melbourne, and died almost immediately at the scene in the early hours of the morning. Buried at Heathcote. The ACS had reached Melbourne, Victoria, only three days earlier, after its evacuation from Java on board the SS *Marella*.

(Driver): NZ403073 LAC **Daniel STEWART**, RNZAF - Age 27.

Stewart's brother, Charles Stewart, died on 11 July 1941 while flying with 485 Sqn, RNZAF.

BOMBER COMMAND

Raid on Dunkirk, France (by 19 aircraft - 2 lost)

460 Squadron, RAAF (Breighton, Yorkshire - 1 Group)

Wellington IV Z1251/X - took off at 1934 and brought down over Belgium, crashing near the canal at St Joris-lez-Nieuport. All six crew are buried in the village cemetery.

Captain: NZ403951 Flt Sgt **Patrick John Denis COONEY**, RNZAF - Age 22. 336hrs. 8th op.

12 Squadron, RAF (Binbrook, Lincolnshire - 1 Group)

Wellington II Z8578 - took off at 2015 on same raid as the above and lost without trace, its six crew being commemorated on the Runnymede Memorial.

Captain: NZ401291 Flt Sgt **Leslie McKenzie SCOTT**, RNZAF - Age 25. 410hrs. 14th op.

Sat 14 Mar 1942

FLYING TRAINING COMMAND

Solo night flying training

17 (Pilots) Advanced Flying Unit, RAF (Watton, Norfolk - 21 Group)

Master II DK873 - crashed shortly after take off at 0410 at Red Barn Farm, near Carbrooke, 3 miles ENE of the airfield. The pilot is buried at Watton.

Pilot: NZ413528 Sgt **Bottrell Stanley WILDER**, RNZAF - Age 26. 165hrs.

Sun 15/Mon 16 Mar 1942

FLYING TRAINING COMMAND

Solo practice - circuits and bumps

17 (Pilots) Advanced Flying Unit, RAF (Watton, Norfolk - 21 Group)

Master II DK865 - took off at about 0020, climbed to 1000 feet then dived. A recovery was made at about 100 feet and a steep climb followed. DK865 then appeared to stall and again recover at about 100 feet. Afterwards it flew slowly down wind, losing height until lost to view, before crashing at Little Ellingham. The pilot is buried at Watton.

Pilot: NZ41172 Sgt **Andre HOLMES,** RNZAF - Age 21. 185hrs (93 solo - 19 on Master)

Mon 16 Mar 1942

BOMBER COMMAND

Cross country from Feltwell to Aldergrove, Co. Antrim, Northern Ireland

57 Squadron, RAF (Feltwell, Norfolk - 3 Group)

Wellington III X3599 - weather and visibility was poor on reaching Northern Ireland, by which time it had drifted off course. At 1230 X3599 crashed into a hill near Thomas Mount, 2 miles south of Newcastle, Co. Down, and caught fire. Six of those on board died, including a WAAF. Three, including the captain, are buried at Belfast. At least one airman survived, injured.

Captain: NZ402060 Fg Off **Harold Eric HUNTER**, RNZAF - Age 25. 405hrs (272 solo - 190 on Wellington) 17 ops.

Wed 18 Mar 1942

NEW ZEALAND

RNZAF Station Harewood
Granted leave until 2359 on the 18th, but failed to report for duty the following day. The airman's body was not found until 8 May 1944 and at the inquest it was presumed he had taken his own life at Springfield on or about 18 March 1942. He is buried at Christchurch.
(Clerk): NZ403834 Cpl **Leslie Errington FORD**, RNZAF - Age 48.

Sun 22 Mar 1942

NEW ZEALAND

Solo flying practice
2 Service Flying Training School (ATS), RNZAF (Woodbourne)
Harvard II NZ989 - while 'tail chasing' another of the School's Harvards (NZ977) at 1515 up the Dee River valley, a tributary of the Ure River, was unable to maintain height with rising terrain. While attempting a steep right hand turn to leave the valley, NZ989 stalled and crashed into a hillside, catching fire near the headwaters of the Dee, about 11 miles SW of Lake Grassmere. The pilot died, but owing to the dense bush and transportation difficulties he was buried at the scene, being also commemorated on the Provincial Memorial at Christchurch. The other Harvard, flown by LAC **J C Voss**, was unable to climb out of the valley either and forced landed into some trees, the uninjured pilot being rescued three days later by some musterers.
Pilot: NZ415413 LAC **Brian Cecil HEAPS**, RNZAF - Age 21. 109hrs solo (79 on Harvard)

Tue 24 Mar 1942

FAR EAST

Surrender and capture on Java
84 Squadron, RAF (Kalidjati, Java)
Although hostilities in Java had officially ceased on 8 March, five officers of the Squadron, with a number of other RAF personnel, were not taken prisoner for another 16 days. One of them, a New Zealand equipment officer, in September led a draft of about 40 prisoners being shipped to Japan via Singapore. He was later transferred to a camp at Hakodate on 1 December, but died there of pneumonia on 3 January 1943. His body was cremated and the ashes placed in a Japanese monastery at Hakodate. After the war the ashes were transferred to the British Commonwealth War Cemetery at Yokohama.
(Equipment Off): 74591 Flt Lt **Harper Mervyn BONGARD**, RAFVR - Age 24.

COASTAL COMMAND

Convoy escort
612 Squadron, RAF (Reykjavik, Iceland)
Whitley VII Z6637/K - took off at 0735 but at 0943 sent an SOS, followed by a long dash, after which nothing further was heard. As a good bearing had been obtained another aircraft diverted to search the area at position 6227N:2400W-6305N:2330W. Cloudy conditions with sleet and snow were encountered, however, and no sign of the aircraft could be found. All six crew died, the captain being commemorated on the Runnymede Memorial.
Captain: NZ401279 Flt Sgt **Reginald John Brydges LINDESAY**, RNZAF - Age 30. 411hrs. 20th op.

BOMBER COMMAND

Training
1515 Beam Approach Training Flight, RAF (Swanton Morley, Norfolk - 2 Group)
Oxford I V4063 - piloted by Act Fg Off D M Craik, RAF. Collided at 1600 with another of the Flight's Oxford Is as noted below. Both aircraft crashed at Foxley Wood, a mile NE of Bawdeswell. The two crew in each Oxford died, the New Zealanders in each being buried at Swanton Morley.
Pilot: NZ411891 Sgt **Everard Howard HAWKEN**, RNZAF - Age 20. 228hrs.
Hawken was on attachment from 15 (Pilots) Advanced Flying Unit at the time.

Training
1515 Beam Approach Training Flight, RAF (Swanton Morley, Norfolk - 2 Group)
Oxford I V4137 - piloted by Flt Sgt B Johnston, DFM, RAF. Collided with Oxford I V4063 as described above.
Pilot: NZ413545 Sgt **Stephen Harry ROBINSON**, RNZAF - Age 27. 194hrs.
Robinson was on attachment from 15 (Pilots) Advanced Flying Unit at the time.

Wed 25 Mar 1942

FERRY COMMAND

Ferry flight from England to Middle East via Gibraltar
Overseas Aircraft Despatch Unit, RAF (Portreath, Cornwall - 44 Group)
Wellington IC AD640 - took off from Gibraltar at 1255 and immediately suffered a starboard engine failure, crashing into the sea 1000 yards east of North Front. Five of the crew died, including the captain, who is buried at

Gibraltar with two of his comrades. The others are commemorated on the Runnymede Memorial. The sole survivor, 2nd pilot Sgt **J E Clarke**, RNZAF, was rescued slightly injured after two hours in the water.
Captain: NZ405324 Sgt **Henry John McNEILL**, RNZAF - Age 25. 120hrs solo (60 on Wellington)
McNeill, previously with 15 Operational Training Unit, Bomber Command, had been posted to the Middle East via the OADU and was making his way there by means of the delivery fight.
Clarke died on 30 May 1943 while flying with 10 Sqn, RAF.

BOMBER COMMAND

Night training
14 Operational Training Unit, RAF (Cottesmore, Rutland - 7 Group)
Hampden I P5398 - piloted by Sgt E J Biddulph, RAF. Drifted 8 miles off track and at 2110, apparently out of control, crashed and caught fire at Whittle Farm, Brockhampton, 2 miles north of Cheltenham, Gloucestershire. All four crew are buried at Little Rissington. According to the CWGC register's next of kin addresses, in addition to the New Zealander the crew comprised a Rhodesian and a Newfoundlander - both members of the RAF - and an airman of the RAAF.
WOpAG: NZ403630 Sgt **Arthur William WESTGATE**, RNZAF - Age 25. 101hrs.

Thu 26 Mar 1942

NEW ZEALAND

Navigation exercise - 'Route 6C'; base-Jack's Point (Timaru)-Position 'X'-base
1 Service Flying Training School (ATS), RNZAF (Wigram)
Oxford I NZ278 - took off at 0800 in suitable weather and last communicated with at 0914. Believed to have encountered low cloud and poor visibility while flying from Timaru out to sea to position 'X' (4430S:17320E). Thirty-five Oxford sorties totalling nearly 90 hours up to 1 April failed to find any trace of NZ278. The two pilots are commemorated on the Provincial Memorial at Wellington and the wireless operator on the Provincial Memorial at Auckland.
Pilot: NZ415187 LAC **Verdun Arthur GODFREY**, RNZAF - Age 25. 74hrs solo (43 on Oxford)
(Pilot): NZ415685 LAC **Cecil Harold COOK**, RNZAF - Age 24.
WOp: NZ405626 AC1 **Donald Alexander PATERSON**, RNZAF - Age 33.
Cook was acting as navigator on the flight.

FIGHTER COMMAND

'Ramrod 17' - escort to 24 Boston bombers on raid to Le Havre, France
485 Squadron, RNZAF (Kenley, Surrey - 11 Group)
Spitfire VB W3577/P - took off at 1510 with 12 others of the Squadron. Engagements with enemy fighters took place during the operation and W3577 was afterwards found to be missing. The pilot is commemorated on the Runnymede Memorial.
Pilot: NZ402194 Flt Sgt **William Max KREBS**, RNZAF - Age 24. 381hrs.

457 Squadron, RAAF (Redhill, Surrey - 11 Group)
Spitfire VC AB495 - took off at 1535 with 'B' Flight on the same operation as the above. During an engagement with enemy fighters a few miles off the French coast AB495 was last seen at about 2000 feet in a steep dive and trailing smoke. The pilot is commemorated on the Runnymede Memorial.
Pilot: Aus402236 Fg Off **Brian Joseph HALSE**, RAAF - Age 29.

CANADA

31 Air Navigation School, RAF (Port Albert, Ontario - 1 Training Command)
Died of natural causes at the Alexandria Hospital, London, Ontario, and is buried at London.
(Pilot): 40336 Fg Off **Trevor Ernest BRAMMALL**, RAF - Age 23.

Thu 26/Fri 27 Mar 1942

BOMBER COMMAND

Raid on Essen, Germany (by 115 aircraft - 11 lost)
57 Squadron, RAF (Feltwell, Norfolk - 3 Group)
Wellington III X3665 - took off at 1940 and brought down near the target area in the southern Duisburg suburb of Wedau. The six crew died, at least five of whom, including the captain, were buried at Düsseldorf, but all six were later reinterred at Reichswald Forest, 5km SW of Kleve.
Captain: NZ404960 Flt Sgt **Ronald Vernon SNOOK**, RNZAF - Age 20. 342hrs. 14th op.

Fri 27/Sat 28 Mar 1942

BOMBER COMMAND

'Gardening' - mine laying in the 'Yams' area - Weser Estuary, Germany

408 Squadron, RCAF (Balderton, Nottinghamshire - 5 Group)

Hampden I L4140/B(bar) - took off at 1919 and believed to have been shot down by a night-fighter. All four crew are commemorated on the Runnymede Memorial.

Pilot: NZ402237 Plt Off **Gordon Findlay BEAVER**, RNZAF - Age 26. 572hrs. 37th op.

Sat 28 Mar 1942

AIR TRAINING CORPS

164 Air Training Corps Squadron (Wimbledon, Surrey)

Died of natural causes, the remains being cremated at Wandsworth, London.

(Administration Officer?): 63692 Fg Off **Robert James KIRKWOOD**, RAFVR - Age 64.

Sat 28/Sun 29 Mar 1942

BOMBER COMMAND

Raid on Lübeck, Germany (by 234 aircraft - 13 lost)

75 (NZ) Squadron, RAF (Feltwell, Norfolk - 3 Group)

Wellington III X3462/N - took off at approximately 1930-2205 and brought down over Germany, crashing near Kiel. All six crew were buried in the German Naval Garrison Cemetery, but later reinterred in the adjacent Kiel War Cemetery.

Captain: NZ404882 Plt Off **Maurice Perrott BELL**, RNZAF - Age 26. 371hrs. At least 10th op.
2nd Pilot: NZ405237 Sgt **Franklyn Bertram CRAN**, RNZAF - Age 21. 289hrs. 5th op.
Observer: NZ404028 Sgt **Claude Joseph HARRIS**, RNZAF - Age 31. 316hrs. 8th op.

Sun 29/Mon 30 Mar 1942

BOMBER COMMAND

'Gardening' - minelaying in the Frisian Islands and off Denmark (by 26 aircraft - 2 lost)

106 Squadron, RAF (Coningsby, Lincolnshire - 5 Group)

Manchester I L7394 - took off at 1930 captained by Flt Sgt E R Dimond, RAF, and lost without trace, its seven crew being commemorated on the Runnymede Memorial.

2nd Pilot: NZ405228 Flt Sgt **Henry Chaytor BUSSELL**, RNZAF - Age 21. 207hrs. At least 3rd op.

Mon 30 Mar 1942

ARMY CO-OPERATION COMMAND

Practice landings at satellite airfield

42 Operational Training Unit, RAF (Andover, Hampshire - 70 Group)

Blenheim IV V5572 - piloted by Plt Off A F McTavish, RCAF. After a landing attempt at 1605 the port engine failed to pick up and V5572 spun off a turn from 400 feet, crashing and catching fire a mile and half west of Andover airfield. All three crew died, the pilot and wireless operator-air gunner being buried at Andover.

WOpAG: NZ403006 Sgt **Ronald Crawford HUNTER**, RNZAF - Age 22. 11hrs(?)

Mon 30/Tue 31 Mar 1942

BOMBER COMMAND

Raid on the German battleship *Tirpitz* at anchor near Trondheim, Norway (by 34 aircraft - 6 lost)

10 Squadron, RAF (Leeming, Yorkshire - 4 Group)

Halifax II W1044/D - took off from Lossiemouth, Morayshire, at 1851 and brought down in Asen Fjord, crashing close to the battleship. All seven crew died, only three bodies being recovered. The others, including the captain, are commemorated on the Runnymede Memorial.

Captain: NZ403419 Fg Off **Neil Ralph BLUNDEN**, RNZAF - Age 26. 436hrs.

Wed 1/Thu 2 Apr 1942

BOMBER COMMAND

Low level attacks on railway targets at Hanau & Lohr, Germany (by 49 aircraft - 13 lost)

57 Squadron, RAF (Feltwell, Norfolk - 3 Group)

Wellington III Z1565 - took off at 2055 to raid Hanau and brought down near the target area, an RAF Sgt being the sole survivor and taken prisoner. His five comrades were buried at Hanau, but later reinterred at Durnbach, 16km east of Bad Tölz.

Captain: NZ404936 Sgt **William John PATERSON**, RNZAF - Age 22. 336hrs. 10th op.

57 Squadron, RAF (Feltwell, Norfolk - 3 Group)
Wellington III X3410 - took off at 2120 on same raid as the above and brought down in Germany. All six crew are buried in a collective grave at Rheinberg, 12km south of Wesel.
Captain: 36112 Sqn Ldr **Guy de Laval HARVIE**, RAF - Age 28. 680hrs. 6th op.

57 Squadron, RAF (Feltwell, Norfolk - 3 Group)
Wellington III X3748 - took off at 2125 on same raid as the above and brought down near Sprendlingen, 9km south of the centre of Frankfurt. Four RAF Sgts survived to be taken prisoner. The captain and RCAF 2nd pilot were buried at Sprendlingen, but later reinterred at Reichswald Forest, 5km SW of Kleve
Captain: NZ403457 Flt Sgt **Reginald Eric KNOBLOCH**, RNZAF - Age 29. 346hrs. 12th op.

214 Squadron, RAF (Stradishall, Suffolk - 3 Group)
Wellington IC X9979 - captained by Plt Off J H A Baker, RAF, on same raid as the above, and brought down at Bad Siedlung, near Russelsheim, about 20km SW of Frankfurt. All six crew were buried in Wald Cemetery at Russelsheim, but later reinterred at Durnbach, 16km east of Bad Tölz.
2nd Pilot: NZ403547 Flt Sgt **Donald Albert Stanley HAMILTON**, RNZAF - Age 22. 359hrs. 10th op.

Night flying training
20 Operational Training Unit, RAF (Lossiemouth, Morayshire - 6 Group)
Wellington IC R1168 - captained by Fg Off D H Jeffery, RAF. When heading back to base at 0113 flew into a hillside at Sinsharnie, 3 miles NW of Huntly, Aberdeenshire. All six crew died, including a New Zealander, who is buried at Lossiemouth.
WOpAG: NZ404088 Flt Sgt **Berry Arthur NEILL**, RNZAF - Age 24. 154hrs.

Thu 2 Apr 1942

NEW ZEALAND

Training - course and speed judging of launch
School of General Reconnaissance , RNZAF (Omaka)
Vincent NZ315 - took off at 0840, being one of seven Vincents engaged in an exercise with the 'target' launch *Lassie*. Thirty-five minutes into the flight NZ315 was seen to descend over Queen Charlotte Sound towards another launch, a half mile away from the target and en route from Tory Channel to Picton. Diving on the launch from 400 feet, NZ315 flew over at mast height and 200 yards further on pulled up into a stall turn at 200 feet. The nose dropped as the Vincent swung round 180° to starboard and dived straight into the sea some 50 yards forward of the launch, about a half mile west of Deiffenbach Light and 300 yards from shore. All three crew died, their bodies being found in the wreckage when it was retrieved two days later. The pilot's remains were cremated at Christchurch and the others are buried at Wellington and Auckland, respectively.
Pilot: NZ414665 Plt Off **Nigel Bracken Hirst NEWTON**, RNZAF - Age 32. 112hrs (25 on Vincent)
Observer: NZ402119 Fg Off **Patrick Howard McCAWE**, RNZAF - Age 25.
Wireless Op: NZ411307 LAC **Cyril Vincent WILLIAMS**, RNZAF - Age 22.

Good Friday 3 Apr 1942

MIDDLE EAST

Diversionary sweep over Gazala area, Libya
274 Squadron, RAF (LG 139 [Gambut Main, No 1], Libya - 243 Wing, Western Desert)
Hurricane IIB Z5086 - took off late morning with seven others. At 1200 the formation was bounced from above by enemy fighters and Z5086 shot down, its pilot being commemorated on the Alamein Memorial.
Pilot: NZ401275 Flt Sgt **Alexander Edwin HOWELL**, RNZAF - Age 26. 344hrs. 38th op.

Easter Saturday 4 Apr 1942

FIGHTER COMMAND

'Circus 119' - escorting 12 Boston bombers on raid to St Omer railway yards, France
485 Squadron, RNZAF (Kenley, Surrey - 11 Group)
Spitfire VB BM231 - took off at 0940 with 11 others and one of two which failed to return. On crossing the French coast south of Boulogne after the Bostons had bombed, the raiding force was attacked by about 40 enemy fighters. Two Spitfires were lost, one of which was seen smoking and going down in a gradual dive. No trace of the pilot of BM231 was found, however, and he is commemorated on the Runnymede Memorial.
Pilot: NZ402543 Plt Off **Edward Fred CHANDLER**, RNZAF - Age 24. 273hrs.
Chandler's brother, John Lincoln Chandler, died on 30 June 1943 while flying with 131 Sqn, RAF.

485 Squadron, RNZAF (Kenley, Surrey - 11 Group)
Spitfire VB P8724 - lost on same operation as the above. Flying as 'Red 4', P8724 crashed near the target area, the pilot being buried in the Longuenesse Souvenir Cemetery near St Omer. However, after the war his grave could not be identified and in addition to a special memorial within the cemetery he is also commemorated on the Runnymede Memorial.
Pilot: NZ411392 Plt Off **Thomas Tristram FOX**, RNZAF - Age 21. 259hrs. 5th op.

Easter Sunday 5 Apr 1942

FERRY COMMAND

Ferry flight from England to Middle East

Overseas Aircraft Despatch Unit, RAF (Portreath, Cornwall - 44 Group)

Beaufort I AW298 - took off from Gibraltar captained by Plt Off H P McKee, RCAF, on second leg of the delivery flight bound for Malta. On approach to Luqa at 2300 in a slight ground mist, AW298 crashed into a dispersal area. All four crew are buried on Malta.

WOpAG: NZ403591 Plt Off **Edmund Patrick SMITH**, RNZAF - Age 26.

Smith, previously with 5 (Coastal) Operational Training Unit, Coastal Command, may have been posted to the Middle East and been making his way there by means of the delivery flight.

Easter Sunday 5/Easter Monday 6 Apr 1942

BOMBER COMMAND

Raid on Cologne, Germany (by 263 aircraft - 7 lost)

75 (NZ) Squadron, RAF (Feltwell, Norfolk - 3 Group)

Wellington III X3489 - took off at approximately 2330-0125 captained by the commanding officer, Wg Cdr R Sawrey-Cookson, DSO DFC, RAF, and brought down over the target, crashing into the city centre. All six crew were buried in the English Military Cemetery in South Cologne, but later reinterred at Rheinberg, 12km south of Wesel. Sixteen spectators on the ground watching the burning wreckage were themselves killed when the bomb load exploded, a further 30 being injured - thus more than doubling the toll attributed to the actual bombing.

2nd Pilot: NZ41977 Plt Off **William Finlay BUDGE**, RNZAF - Age 24. 203hrs. 1st op(?)

Easter Monday 6/Tue 7 Apr 1942

BOMBER COMMAND

Raid on Essen, Germany (by 157 aircraft -6 lost)

61 Squadron, RAF (Woolfox Lodge, Rutland - 5 Group)

Manchester I L7470 - took off at 0035 captained by Flt Sgt E W Noble, RAF, and shot down over Belgium by a night-fighter, crashing at Piringen, 16km ESE of St Trond. All seven crew were buried the following day in the cemetery for Allied airmen at the German air base at Brustem, near St Trond, but later reinterred at Heverlee.

2nd Pilot: NZ41341 Sgt **Lloyd John Holmes LINCOLN,** RNZAF - Age 26. 240hrs. 3rd op.

Lincoln's brother, Stanley Holmes Lincoln, was killed almost two months later while flying a Lancaster with the same squadron on 3 June. They enlisted on the same date, held consecutive service numbers and served several months together with 61 Sqn. By extraordinary coincidence each was shot down by a night-fighter over Belgium (within 80km of one another) while engaged on a raid on the same target city, both as 2nd Pilot. They now lie buried some 20km apart.

CANADA

Night navigation

10 Air Observer School, RCAF (Chatham, New Brunswick - 3 Training Command)

Anson I 6631 - took off at 2155 piloted by L Berriman and failed to return. An extensive search was conducted but no trace of 6631 or its three crew could be found. Two months later, on 18 June, the body of observer **McKinnon** was found washed up on the shore of the Northumberland Strait near Shediac, where he is now buried. No trace of his fellow observer was found, however, and he is commemorated on the Ottawa Memorial. Berriman is believed to have been a civilian pilot flying under contract as part of the Empire Air Training Scheme (EATS); his name is not therefore recorded by the Commonwealth War Graves Commission. Upon the finding of McKinnon's body it was assumed that the Anson had either crashed in the sea or into ice floating in the Strait.

Observer: NZ413946 LAC **Cyril Robert Hudson McFARLANE**, RNZAF - Age 30. 47hrs.

Observer: NZ415340 Cpl **Harold Stuart McKINNON**, RNZAF - Age 31.

Tue 7/Wed 8 Apr 1942

BOMBER COMMAND

Cross country exercise

22 Operational Training Unit, RAF (Wellesbourne Mountford, Warwickshire - 6 Group)

Wellington IC R1773 - at 0120 the starboard engine failed while the bomber was at 2000 feet over Berkshire. Unable to maintain height on the one engine, it crashed into a tree $^3/_4$ mile SW of Great Coxwell on attempting a forced landing. One crew member survived, injured, but the other five died, the pilot instructor and two RCAF pupil pilots being buried at Black Bourton, Oxfordshire. Some of the height loss was attributed to the pilots changing seats in the turbulent air conditions prevailing at the time of the emergency.

Pilot Instr: NZ402231 Plt Off **John Mervyn WARNOCK**, RNZAF - Age 26. 510hrs. 19 ops.

Wed 8 Apr 1942

FLEET AIR ARM - ENGLAND

Practising section attacks

759 Squadron, FAA (Yeovilton -HMS *Heron*)

Sea Hurricane IB P3398 - returning from the practice at 1715, collided head on with another Sea Hurricane IB (Z7152) at 1000 feet. The two crashed at West Camel, just over a mile NE of Yeovilton, the former in Bower's Field and the latter in Bindle's Field. Both pilots died, the New Zealander being buried at Yeovilton.

Pilot: Midshipman (A) **George Denys COOK**, RNZNVR - Age 19.

Wed 8/Thu 9 Apr 1942

BOMBER COMMAND

Raid on Hamburg, Germany (by 272 aircraft - 5 lost)

57 Squadron, RAF (Feltwell, Norfolk - 3 Group)

Wellington III X3757 - took off at 2205 and lost without trace, its six crew being commemorated on the Runnymede Memorial.

Captain: NZ404397 Plt Off **Noel Percy MORSE**, RNZAF - Age 27. 356hrs. 17th op.

Observer: NZ404577 Flt Sgt **George Hillary VOGAN**, RNZAF - Age 21. 304hrs. 18th op.

The CWGC register incorrectly gives Vogan's age as 23.

83 Squadron, RAF (Scampton, Lincolnshire - 5 Group)

Manchester I L7427/Q - took off at 2215 captained by Plt Off J H Morphett, RAF, on same raid as the above. Last heard from at 0010, when over Germany west of Lastrup, L7427 was apparently brought down soon afterwards as it crashed a short distance away, to the NE of Cloppenburg. An RAF Plt Off survived, to be taken prisoner, but died in November. His six comrades are buried in a collective grave at Sage, 24km south of Oldenburg.

Observer: NZ402188 Flt Sgt **Geoffrey Douglas HUTCHINSON**, RNZAF - Age 27. 266hrs. 8th op.

Thu 9 Apr 1942

NEW ZEALAND

RNZAF Station Nelson

Died of illness in the Nelson Public Hospital and is buried at that city.

(Patrolman): NZ411568 LAC **Otto BAY**, RNZAF - Age 50.

CANADA

Instrument flying training

10 Service Flying Training School, RCAF (Dauphin, Manitoba- 2 Training Command)

Crane I 7957 - piloted by Sgt D D Dean, RCAF. Executed a turn and fell out of control at 1500 while flying low in icy and gusty conditions. Both pilots died when 7957 crashed and caught fire some miles SW of Makinak, the New Zealander being buried at Dauphin.

Pupil Pilot: NZ405796 LAC **Selwyn Fitzgerald BRADLEY**, RNZAF - Age 26. 124hrs (53 solo - 16 on Crane)

Fri 10/Sat 11 Apr 1942

BOMBER COMMAND

Raid on Essen, Germany (by 254 aircraft - 15 lost)

144 Squadron, RAF (North Luffenham, Rutland - 5 Group)

Hampden I AT218 - captained by Sgt H C Weeks, RAAF, and brought down over Germany near Bad Berleburg, Wittgenstein. Only one crew member survived to be taken prisoner. His three comrades were buried at Bad Berleburg, but later reinterred at Hannover

WOpAG: NZ404070 Flt Sgt **Maurice Young KENNEDY**, RNZAF - Age 26. 150hrs. 1st op.

Sun 12 Apr 1942

BOMBER COMMAND

Night flying test

144 Squadron, RAF (North Luffenham, Rutland - 5 Group)

Hampden I AT155 - at 1540, crashed and caught fire at Ridlington, 5 miles west of the base. Both two crew died, the captain being buried at North Luffenham.

Captain: NZ402877 Flt Sgt **David Lloyd Carnegy LIDDELL**, RNZAF - Age 21. 366hrs. 6 ops.

Sun 12/Mon 13 Apr 1942

BOMBER COMMAND

Raid on Essen, Germany (by 251 aircraft - 16 lost)

115 Squadron, RAF (Marham, Norfolk - 3 Group)

Wellington III X3596/B - took off at 2257 captained by Sgt A E Holder, RAF, and shot down over the Netherlands by a night-fighter at 0100, crashing into the Ijsselmeer, east of Urk. All six crew are buried at Amsterdam.

2nd Pilot: NZ404994 Sgt **Noel Hensley BLAIR**, RNZAF - Age 24. 314hrs.

Mon 13 Apr 1942

FIGHTER COMMAND

Practice(?)

616 Squadron, RAF (Kings Cliffe, Northamptonshire - 12 Group)

Spitfire VB AD543 - took off at about 1440 and began a roll shortly afterwards at 100 feet, but failed to recover before crashing into a field adjoining the base. The pilot is buried at Wittering.

Pilot: NZ404904 Sgt **George Leslie DAVIDSON**, RNZAF - Age 26. 407hrs. 26 ops.

COASTAL COMMAND

Flight to St Eval, Cornwall

172 Squadron, RAF (Chivenor, Devonshire - 19 Group)

Wellington VIII Z8721 - took off at 1400 captained by Plt Off H W Russ, RCAF. Encountered bad weather and poor visibility en route and flew directly into a Beckland Bay cliff face, 3 miles east of Hartland Point, Devonshire. All five crew are buried at Heanton Punchardon, near Chivenor.

2nd Pilot: NZ401248 Fg Off **Ernest Anthony BLAIR**, RNZAF - Age 28. 423hrs.

Mon 13/Tue 14 Apr 1942

BOMBER COMMAND

'Gardening' - mine laying off Frisian islands and the German Bight (by 47 aircraft - 1 lost)

15 Squadron, RAF (Wyton, Huntingdonshire - 3 Group)

Stirling I W7519 - took off from Alconbury at 2130 to lay mines off the island of Wangerooge and shot down over the North Sea by a night-fighter. All seven crew and an Army captain on board died. The airmen are commemorated on the Runnymede Memorial and their Army comrade on the Brookwood Memorial.

Captain: NZ40922 Plt Off **Ronald Victor Neal McKAY**, RNZAF - Age 22. 1889hrs. 108th op.

2nd Pilot: NZ404030 Flt Sgt **Robert Henry HIGGINS**, RNZAF - Age 31. 379hrs. 10th op.

McKay served as an air gunner with Coastal Command, completing 102 ops before remustering in mid-1941 for pilot training.

Tue 14 Apr 1942

MIDDLE EAST

Torpedo strike on convoy

39 Squadron, RAF (LG 86 [Amriya], Libya - detachment at LG 147 [Bu Amud] - 201 Group)

Beaufort I N1166 - took off at from LG 147 at 1200 piloted by Plt Off B W Way, RAF, with seven others (including a 22 Sqn Beaufort) and one of five which failed to reach Malta after the attack. An escort of four Beaufighters, when at the limit of its endurance towards 1700, was obliged to return to base immediately prior to the convoy being located and attacked at 3459N:1559E, about 80 miles SW of Malta. On their run in the Beauforts were intercepted by a large fighter force and three, including N1166, were quickly shot down into the sea. All of their four-man crews died and are commemorated on the Malta Memorial. Two other aircraft were badly shot up and later ditched, one within 6 miles of the island base.

Observer: NZ403595 Flt Sgt **Keith Marius DIXON**, RNZAF - Age 24. 222hrs. 4th op.

BOMBER COMMAND

Cross country training

21 Operational Training Unit, RAF (Moreton-in-Marsh, Gloucestershire - 6 Group)

Wellington IC R1085 - captained by Sgt J M S Millar, RAF. At 1531 a propeller flew off and in attempting an emergency landing R1085 stalled, crashed and caught fire at Wollerton, a mile or two SSW of Tern Hill, Shropshire. All six crew died, five, including a New Zealander, who is buried at Stoke-upon-Tern.

Air Gunner: NZ41626 Sgt **Trevor Victor YDE**, RNZAF - Age 21. 122hrs.

Wed 15/Thu 16 Apr 1942

BOMBER COMMAND

Raid on Dortmund, Germany (by 152 aircraft - 4 lost)

149 Squadron, RAF (Lakenheath, Suffolk - 3 Group)

Stirling I N6068/T - took off at 0020 captained by Plt Off M L Field, RAF, and brought down over Belgium at 0450, crashing at Stene, in the south of Ostende. All eight crew are buried at Ostende.

2nd Pilot: NZ41887 Plt Off **Alexander Leslie ELLIS**, RNZAF - Age 25. 223hrs. 2nd op.

Fri 17 Apr 1942

BOMBER COMMAND

Low level raid on Augsburg, Germany (by 12 aircraft - 7 lost)

44 Squadron, RAF (Waddington, Lincolnshire - 5 Group)

Lancaster I L7536/H - took off at 1512 captained by Sgt G T Rhodes, RAF, and shot down over France en route to the target at 1655 by fighters, crashing near le Vieil-Evreux, 6km ESE of Overact. All seven crew are buried at Overact.

2nd Pilot: NZ41300 Sgt **Lawrence Heaton BAXTER**, RNZAF - Age 23. 231hrs. 2nd op(?)

CANADA

Radio navigation exercise No 1

6 Service Flying Training School, RCAF (Dunnville, Ontario - 1 Training Command)

Harvard II 2839 - about 50 minutes after take off, at 1400, was seen to be in difficulty at a 'fair' height. Losing altitude, 2839 flew erratically and parts of the structure were seen to fall away as it dived out of control to explode on the ground a mile west of Fisherville, about 12 miles west of Dunnville. Both pilots died, the New Zealander being buried at Dunnville.

Pilot: NZ412526 LAC **Alastair Logie RANKIN**, RNZAF - Age 22. 197hrs (97 solo - 51 on Harvard)

Fri 17/Sat 18 Apr 1942

BOMBER COMMAND

Raid on Hamburg, Germany (by 173 aircraft - 9 lost)

158 Squadron, RAF (Driffield, Yorkshire - 4 Group)

Wellington II W5575/V - took off at 2341 and last heard of at 0556 over the North Sea at position 5431N:0309E, after which it was presumed to have ditched or crashed. All five crew are commemorated on the Runnymede Memorial.

Captain: NZ404934 Flt Sgt **Edward Alan OLLIVER**, RNZAF - Age 22. 357hrs. 12th op(?)

57 Squadron, RAF (Feltwell, Norfolk - 3 Group)

Wellington III X3542 - took off at 2351 on same raid as the above, and shot down from 10,000 feet over Germany by a night fighter at 0300, crashing at Rickert, 3km north of Rendsburg. Three RAF crew members survived, to be taken prisoner, but their three comrades died. The captain's body was retrieved, but the wreckage sank quickly in soft ground and prevented recovery of the other two. The captain was buried at Rendsburg, but later reinterred at Kiel, while his two companions are commemorated on the Runnymede Memorial.

Captain: NZ404438 Flt Sgt **Warwick St George Ruxton WILSON**, RNZAF - Age 25. 382hrs. 21st op.

Wilson's brother, Ian Lewis Ruxton Wilson, died on 1 May 1944 while flying with 838 Sqn, FAA.

Sat 18 Apr 1942

FLEET AIR ARM - ENGLAND

Fighter training

759 Squadron, FAA (Yeovilton, Somerset - HMS *Heron*)

Sea Hurricane IB Z7068 - suffered an engine failure and while attempting a forced landing at 1015 crashed at Pylle, 3 miles SSW of Shepton Mallet. The pilot was seriously injured and admitted to Bath Military Hospital, where he died at 1750 on 1 May. He is buried at Englishcombe, just SW of Bath.

Pilot: Sub-Lt (A) **Jack Radford PEPPERELL**, RNZNVR - Age 23

Sat 18/Sun 19 Apr 1942

MIDDLE EAST

Raid on Benghazi Harbour, Libya

37 Squadron, RAF (Shallufa, Egypt - detachment at LG 09 [Bir Koralyim] - 205 Group)

Wellington IC AD642 - took off from LG 09 at 2236 captained by Sgt B R Steward, RAF, but signalled at 0138, 'operations abandoned'. On returning to base in bad weather AD642 flew into the ground and caught fire 3 miles SE of the landing ground. All six crew were buried in the El Daba Military Cemetery, but later reinterred at El Alamein.

Observer: NZ403608 Flt Sgt **Raymond Edgar LLOYD**, RNZAF - Age 20. 456hrs. 39th op.

Lloyd's brother, Eric Lloyd, died on 8 November 1941 while flying with 75 (NZ) Sqn, RAF.

Sun 19/Mon 20 Apr 1942

BOMBER COMMAND

'Gardening' - minelaying off the Frisian Islands (by 51 aircraft - 3 lost)

156 Squadron, RAF (Alconbury, Huntingdonshire - 3 Group)

Wellington III X3485 - took off at 2107 captained by Plt Off R W Fox, RAF, and presumed to have been brought down into the sea, killing the six crew. The body of an RAF Flt Sgt was later recovered near the Dutch island of Ameland and is buried there. His comrades are commemorated on the Runnymede Memorial.

2nd Pilot: NZ405215 Sgt **Thomas Brian AINGER**, RNZAF - Age 26. 280hrs. 6th op(?)

Mon 20 Apr 1942

COASTAL COMMAND

Navigation training exercise

58 Squadron, RAF (St Eval, Cornwall - 19 Group)

Whitley V Z9388/W - while over the Atlantic a signal was sent advising of return to base because of engine failure. As the information provided was corrupted its position could not be plotted and a signal on a different frequency was sent seeking clarification. Z9388's operator was then told to stand by while another aircraft was

cleared. However, no further contact could be made and despite a 10-aircraft search the next day no trace of the Whitley could be found. The body of one crew member was later recovered off the coast of County Cork, Ireland, but his five comrades were lost without trace and are commemorated on the Runnymede Memorial.
Captain: NZ41485 Sgt **Kenmir Loudon McADAM**, RNZAF - Age 24. 267hrs (124 on Whitley) 9 ops.

Tue 21 Apr 1942

SOUTH PACIFIC

2 Aerodrome Construction Squadron, RNZAF (Nadi, Fiji)
Died of peritonitis at Suva and was buried in the Lautoka Cemetery, but later reinterred at Suva.
(Shotfirer): NZ40398 Act Cpl **Harold Gordon Cyril INWOOD**, RNZAF - Age 24.

Wed 22 Apr 1942

FLEET AIR ARM - ENGLAND

Reconnaissance practice
759 Squadron, FAA (Yeovilton, Somerset - HMS *Heron*)
Sea Hurricane IB Z7080 - an emergency forced landing was attempted at 1005 because of a suspected coolant leak. On approach Z7080 spun in and crashed at Bower Hinton Farm, $1^1/_2$ miles NE of South Petherton. The pilot is buried at Yeovilton.
Pilot: Sub-Lt (A) **Rex McMILLAN**, RNZNVR - Age 20.

Reconnaissance practice
760 Squadron, FAA (Yeovilton, Somerset - HMS *Heron*)
Sea Hurricane IB R4214 - crashed at 1520 at Milton, near Gillingham, Dorset. The pilot is buried at Yeovilton.
Pilot: Sub-Lt (A) **Alexander Neilson THOMSON**, RNZNVR - Age 21.

Wed 22/Thu 23 Apr 1942

BOMBER COMMAND

'Gardening' - minelaying off the Frisian Islands (by 63 aircraft - 1 lost)
408 Squadron, RCAF (Balderton, Nottinghamshire - 5 Group)
Hampden I AE237/X - took off at 1946 and believed to have been shot down off the Dutch coast by a night-fighter at 0007, crashing into the sea NW of Ameland Island. The four crew are commemorated on the Runnymede Memorial.
Captain: NZ41920 Sgt **Allan James MACGREGOR**, RNZAF - Age 22. 254hrs. 5th op.

Raid on Le Havre, France (by 23 aircraft - 2 lost)
75 (NZ) Squadron, RAF (Feltwell, Norfolk - 3 Group)
Wellington III X3667 - took off at approximately 2120-2215 and believed to have been brought down into the sea near the target area. The body of the RAF Sgt rear gunner later washed ashore near Le Havre, but his five RNZAF comrades were lost without trace and are commemorated on the Runnymede Memorial.
Captain: NZ404916 Flt Sgt **Thomas Stanley MAHOOD**, RNZAF - Age 25. 344hrs. 12th op.
2nd Pilot: NZ411998 Sgt **Cyril Charles WRIGHTSON**, RNZAF - Age 22. 279hrs. 4th op.
Observer: NZ402172 Flt Sgt **Sydney Houston DROMGOOLE**, RNZAF - Age 28. 418hrs. 26th op.
Wireless Op: NZ403580 Sgt **Reginald Joseph Stephen KELLY**, RNZAF - Age 24.
Front Gunner: NZ403019 Flt Sgt **Henry Thomas Owen McMAHON**, RNZAF - Age 27.
Dromgoole's brother, Ian Dromgoole, died on 19 May 1940 while flying with 16 Sqn, RAF.

Raid on Cologne, Germany (by 69 aircraft - 5 lost)
75 (NZ) Squadron, RAF (Feltwell, Norfolk - 3 Group)
Wellington III X3705 - took off at 2205 captained by Flt Sgt **I J McLachlan**, RNZAF. Shortly after midnight on the return journey X3705 was attacked by a night-fighter, the 2nd pilot being killed and the rear gunner, Sgt **D S Tutty**, RNZAF, seriously wounded. The bomber was coaxed back to base where a successful wheels-up landing was accomplished without further injury to the crew. The 2nd pilot is buried at Feltwell. Other RNZAF crew members who made up the six crew were wireless operator Flt Sgt **J W Walters** and front gunner Sgt **A G E Pugh**.
2nd Pilot: NZ41981 Plt Off **Cedric Niel FOUNTAIN**, RNZAF - Age 23. 289hrs. 11th op.
Walters died on 20 May 1944 while flying with 7 Sqn, RAF.

75 (NZ) Squadron, RAF (Feltwell, Norfolk - 3 Group)
Wellington III X3487 - took off at 2215 captained by Plt Off E G D Jarman, RAF, on same raid as the above. On the return journey an attack by a night-fighter killed the rear gunner and wounded the navigator, wireless operator and 2nd pilot. The captain and front gunner escaped injury and a successful crash landing was accomplished back at base at 0440. However, the 2nd pilot died later that morning and he and the air gunner are buried at Feltwell.
2nd Pilot: NZ411929 Plt Off **Trafford McRae NICOL**, RNZAF - Age 21. 328hrs. 6th op.
Rear Gunner: NZ402999 Sgt **Richard James HARRIS**, RNZAF - Age 24. 129hrs. 9th op.

9 Squadron, RAF (Honington, Suffolk - 3 Group)
Wellington III X3759 - took off at 2220 captained by Sgt W T Ramey, RCAF, on same raid as the above. On final approach to Honington when returning at 0440, X3759 flew into the ground during a right hand turn, bounced, overturned and caught fire. Only the RAF Sgt rear gunner survived, injured. His five comrades are buried at Honington.
Observer: NZ401765 Flt Sgt **John Whiteside JASPER**, RNZAF - Age 27.

Thu 23 Apr 1942

MIDDLE EAST

Almaza Transit Centre, RAF (Egypt)
Died of illness at No 3 New Zealand General Hospital, Heliopolis, and is buried in that city.
(Observer): NZ401304 Flt Sgt **Alan STEVENS-JORDAN**, RNZAF - Age 25.

Fri 24 Apr/Anzac Day 1942

COASTAL COMMAND

Night search for missing aircraft
502 Squadron, RAF (St Eval, Cornwall - 19 Group)
Whitley V Z6733/N - took off at 2112 captained by Plt Off J G Brown, RAF, to search for the Squadron's Whitley V, Z6552/Z, which had suffered an engine failure on patrol earlier in the day and had ditched at 4900N:0650W, south of Land's End. At 0505 a signal was received that the searching aircraft also was experiencing engine trouble and heading back to base. Last plotted on approach towards Land's End at 0519, for a time Z6733 appeared to remain stationary at position 4948N:0540W, but nothing further was seen or heard of it. Three bodies were later recovered but the other three, including the 2nd pilot, were lost without trace and are commemorated on the Runnymede Memorial. It was thought the engine problems were probably caused by wood chippings from the camouflaged runways entering the oil cooler and caused overheating. Further post-accident comment suggested that provision of fully feathering propellers and a fuel jettison system might have prevented the losses.
2nd Pilot: NZ41473 Plt Off **Geoffrey MacLennan DENSEM**, RNZAF - Age 28. 235hrs (150 solo - 99 on Whitley) 1st op(?)

Sat, Anzac Day 1942

FIGHTER COMMAND

'Circus'- escort to six Bostons on raid against Abbeville marshalling yards, France
72 Squadron, RAF (Biggin Hill, Kent - 11 Group)
Spitfire VB W3406 - took off at 1545 with 12 others. The Squadron was bounced by enemy fighters SE of Le Touquet, W3406 last being seen in a controlled dive before crashing near Le Touquet. The pilot was buried at there, but later reinterred at Etaples.
Pilot: NZ403978 Flt Sgt **John Grant McCUTCHAN**, RNZAF - Age 27. 323hrs. 30th op.

Sat, Anzac Day/Sun 26 Apr 1942

MIDDLE EAST

Raid on Benghazi Harbour, Libya
70 Squadron, RAF (LG 104 [Qotafiya II], Egypt - 231 Wing, 205 Group)
Wellington IC Z8984/D - took off at about 2000 captained by Sgt P R Darby, RAF, and while over the target area was hit by flak which killed the 2nd pilot. The other five crew baled out before the crippled bomber crashed some 15 miles ESE of Benghazi, near Berka, still with a full bomb load aboard. No trace of the dead airman was found and he is commemorated on the Alamein Memorial. Three of the survivors were taken prisoner, while the others evaded capture.
2nd Pilot: NZ41375 Sgt **George Lockhart TAYLOR**, RNZAF - Age 22. 303hrs. 4th op(?)

Sun 26 Apr 1942

FIGHTER COMMAND

High cover escort to Hurribombers on raid against Calais, France
485 Squadron, RNZAF (Kenley, Surrey - 11 Group)
Spitfire VB BM139/E - took off at 1710 with others of the Squadron and one of three which failed to return. Enemy fighters in strength intercepted the escort over coastal France and inflicted heavy loss. Flt Sgt **T C Goodlet**, RNZAF, was brought down over France, baling out to become a prisoner. Fg Off **J C Pattison**, RNZAF, baled out into the sea near Dungeness after gliding across the English Channel with a shot up engine and was rescued uninjured from his dinghy within about 90 minutes. The pilot of BM139, wounded in the fighting and his machine critically damaged, also baled out near the English coast, at 1810. Although soon rescued by launch, he died in hospital that evening and is buried near the base at Whyteleafe.
Pilot: NZ404385 Flt Sgt **John Raby LIKEN**, RNZAF - Age 28. 345hrs. 38th op.

Sun 26/Mon 27 Apr 1942

MIDDLE EAST

Raid on Martuba, Libya

104 Squadron, RAF (LG 213 [Kabrit], Egypt - 236 Wing, 205 Group)

Wellington II Z8500/B - captained by Sgt H J Bennett, RAF, and lost without trace. Two of the six crew were reported to have been buried by the Germans at Martuba, but after hostilities had ceased the graves, apparently effaced by the elements, were unable to be located. Nothing further was heard of the other four, including the 2nd pilot, and all six are commemorated on the Alamein Memorial.

2nd Pilot(?): NZ404329 Sgt **William John BRANDON**, RNZAF - Age 20. 296hrs. 4th op(?)

Mon 27/Tue 28 Apr 1942

BOMBER COMMAND

Raid on Cologne, Germany (by 97 aircraft - 7 lost)

150 Squadron, RAF (Snaith, Yorkshire - 1 Group)

Wellington III X3288/H - took off at 2145 captained by Sgt A G Hutchinson, RAF, and brought down over the target area. the six crew were buried in a collective grave in south Cologne, but later reinterred at Rheinberg, 12km south of Wesel.

WOpAG: 1186829 Sgt **Phillip Goldham FUSSELL**, RAF - Age 29.

'Nickel' - leaflet raid on Lille-Bethune area, France (by 5 aircraft - 2 lost)

27 Operational Training Unit, RAF (Lichfield, Staffordshire - 6 Group)

Wellington IC Z8901/V - took off at 2241 captained by Sgt G A Dale, RAF, and brought down over Beldium at Sautour, 27km SSE of Charleroi, some 90km ESE of Lille. All five crew died and are buried at Charleroi, including two New Zealanders in a joint grave. An accident [sic] card raised for this loss was highly critical of how the raid had been organised - the route for OTU aircraft coincided with that for the Command's main raid on Cologne and no anti-glow paint had been applied to the exhaust stubs of the OTU machines.

WOpAG: NZ403766 Sgt **Wilfred George MUTTON**, RNZAF - Age 21. 1st op(?).

WOpAG: NZ403008 Sgt **Eric Charles INDER**, RNZAF - Age 29. 144hrs. 1st op(?).

The CWGC register incorrectly gives Inder's age as 31.

Tue 28 Apr 1942

BOMBER COMMAND

Circuit and landing practice

22 Operational Training Unit, RAF (Wellesbourne Mountford, Warwickshire - 6 Group)

Wellington IC R3198 - captained by Sgt J A Schweitzer, RCAF. While executing a turn at about 0550 crashed and caught fire at Low Farm, a half mile from the base. All three crew are buried at nearby Stratford-on-Avon.

WOpAG: NZ41638 Sgt **Herbert Charles CROUCH**, RNZAF - Age 24. 155hrs.

Tue 28/Wed 29 Apr 1942

BOMBER COMMAND

Raid on the battleship *Tirpitz* in Trondheim Fjord, Norway (by 34 aircraft - 2 lost)

35 Squadron, RAF (Linton-on-Ouse, Yorkshire - 4 Group)

Halifax II W7656/P - took off at 2037 from Kinloss captained by Flt Lt D Petley, RAF, and was hit by flak and set afire at 4000 feet over the target area. Rapidly descending into the fjord, W7656 hit the water at high speed opposite Vilkhamer, its nose breaking up on impact and the Halifax sinking immediately. Four RAF members of the crew survived to be taken prisoner, although it is not clear if they had baled out or were survivors of the crash. Two of their comrades died and are commemorated on the Runnymede Memorial.

Wireless Op: NZ402989 Flt Sgt **Arthur William Smith EVANS**, RNZAF - Age 23. 231hrs. 14th op.

Wed 29/Thu 30 Apr 1942

BOMBER COMMAND

Raid on Gnome-Rhône aero-engine factory at Gennevilliers, Paris, France (by 88 aircraft - 3 lost)

57 Squadron, RAF (Feltwell, Norfolk - 3 Group)

Wellington III X3640 - took off at 2125 and over Paris collided with another aircraft on the raid, crashing on the north side of Maisons-Laffitte in Le Parc De Maisons-Laffitte. All five crew are buried at Bievres, 25km SSE of the crash site.

Captain: NZ404359 Fg Off **Kenneth Fenton HEALD**, RNZAF - Age 23. 508hrs. 26th op.

May 1942

FAR EAST

Wt Off **E FIELD**, RAF, was executed by a Japanese by firing squad about this time after escape and recapture. It will be recalled that he was originally captured on or shortly after the night of 6/7 March 1942 when his train was ambushed in Java.

Fri 1 May 1942

FLEET AIR ARM - ENGLAND

Sub-Lt (A) **J R PEPPERELL** of 759 Sqn, FAA, died of injuries received in a flying accident on 18 April.

FIGHTER COMMAND

High cover escort to Hurribombers raiding Shell factory at Marquise, France
457 Squadron, RAAF (Redhill, Surrey - 11 Group)
Spitfire VB BM188 - took off at 1830 leading 11 others of the Squadron and one of two which failed to return. Enemy fighters intercepted the formation just prior to its crossing the French coast on the return journey. The leading Spitfire and its No 2 were last seen diving on an enemy aircraft, with some ten of the enemy diving on them in turn. Both pilots are commemorated on the Runnymede Memorial.
Pilot: 41608 Flt Lt **Harold Leslie NORTH**, DFC, RAF - Age 22.

Sat 2 May 1942

BOMBER COMMAND

Cross country exercise
13 Operational Training Unit, RAF (Bicester, Oxfordshire - 7 Group)
Blenheim IV L9206 - piloted by Sgt J R Young, RCAF. At 1135 the bomber dived vertically into the ground and caught fire 3 miles east of Billesdon, Leicestershire. All three crew died, the pilot and navigator being buried at North Luffenham, Rutland, 11 miles to the east of the crash site.
Navigator: NZ404987 Sgt **Hugh MORRISON**, RNZAF - Age 22. 213hrs.

Sun 3/Mon 4 May 1942

MIDDLE EAST

Raid on shipping in Tripoli Harbour, Libya
108 Squadron, RAF (LG 221 [Fayid], Egypt - 231 Wing, 205 Group)
Liberator II AL511 - took off at 1830 and shot down over Libya by a night-fighter, crashing near Soluch, about 35 miles SSE of Benghazi. Four of the crew died, including two New Zealanders, while the other three succeeded in baling out and were captured. One of the prisoners, Gp Capt F M V May, died in hospital on 22 May and was buried at Barce, about 60 miles NE of Benghazi, but later reinterred at Benghazi. Sandstorms are believed to have obliterated the grave sites of the others, for post-war they could not be located, and these airmen are now commemorated on the Alamein Memorial.
Captain: 36243 Flt Lt **Duncan Harold McARTHUR**, DFC, RAF - Age 21. 1036hrs. 58th+ op(?)
Air Gunner: NZ402104 Flt Sgt **Denis Frederick CLIFFORD**, RNZAF - Age 26. 296hrs. 22nd op(?)
The CWGC register incorrectly gives McArthur's age as 23.

Mon 4 May 1942

FIGHTER COMMAND

'Rodeo 18' - to Hardelot-Sangatte, France
485 Squadron, RNZAF (Kenley, Surrey - 11 Group)
Spitfire VB BM196 (a presentation machine named *Borough of Brentford and Chiswick*) - took off at 1905 with others of the Squadron and one of two later brought down into the English Channel. While still climbing, the Squadron was bounced by four enemy fighters, and by about 20 more while in the Fort Du Boulogne area. Their aircraft seriously damaged during the engagements, two pilots were obliged to bale out over the Channel and await rescue from their dinghies. One, Plt Off **J M Checketts**, RNZAF, was picked up by launch about an hour later. The other was less lucky, not being pulled from the sea until several hours afterwards by a friendly trawler. He died of exposure that evening and is buried near the base at Whyteleafe.
Pilot: NZ403553 Wt Off **David Malyon RUSSELL**, RNZAF - Age 25. 245hrs.
Russell's brother, John Wilfred Russell, died on 5 June 1943 while travelling as a passenger with 6 Sqn, RNZAF.

Tue 5 May 1942

ARMY CO-OPERATION COMMAND

'Silent sortie' - Anti-aircraft co-operation
Anti-Aircraft Co-operation Unit, RAF (Towyn, Merionethshire - 70 Group)
Henley I L3435 of 'C' Flight - took off at 0905, but 10 minutes later stalled off a steep right-hand turn at 100 feet and crashed into the sea 300 yards from shore. One crew member survived, injured, but the pilot died and is buried at Towyn.
Pilot: NZ411489 Sgt **James Herbert WILSON,** RNZAF - Age 24. 263hrs (172 solo - 102 on Henley)

Tue 5/Wed 6 May 1942

BOMBER COMMAND

Raid on Stuttgart, Germany (by 77 aircraft - 6 lost)

7 Squadron, RAF (Oakington, Cambridgeshire - 3 Group)

Stirling I N3710/M - took off at 2133 and brought down while returning over France, crashing at Wailers (Wallers?). All seven crew are buried in a collective grave at Avesnes-sur-Helpe, 14km south of Maubeuge, together with an RAAF airman who was killed during another raid to Stuttgart the following night.

Captain: NZ404384 Flt **Sgt Jeffrey LEWIS**, RNZAF - Age 21. 388hrs. 20th op.

Wed 6 May 1942

FLEET AIR ARM - MADAGASCAR

Operation IRONCLAD - bombing French batteries at Antsirane

829 Squadron, FAA (HMS *Illustrious* - Fleet Aircraft Carrier)

Swordfish II DK788 - captained by Sub-Lt (A) G H Ervin, RNVR, and shot down by flak while bombing the batteries opposite Diégo Suarez, crashing near the French military barracks. All three crew were buried at Antsirane, but later reinterred at Diégo Suarez. *Operation IRONCLAD* was the code name for the plan to capture Diégo Suarez.

Observer: Sub-Lt (A) **Arnott Roy MEIKLEJOHN**, RNZNVR - Age 25.

Meiklejohn was actually a member of 810 Sqn, also embarked on Illustrious. *It is not known why he was crewing a Swordfish belonging to 829 Sqn. He is incorrectly shown in the CWGC register as being of the RNVR.*

BOMBER COMMAND

Cross country and sea navigation exercise

13 Operational Training Unit, RAF (Bicester, Oxfordshire - 7 Group)

Blenheim IV L8755 - took off at 1105 piloted by Flt Sgt T A Crawford, RCAF, as one of twelve despatched on the exercise and presumed to have ditched or crashed into the North Sea. The pilot's body later washed ashore onto the Dutch island of Ameland, but no trace his two comrades was found and they are commemorated on the Runnymede Memorial.

Navigator: NZ404388 Sgt **Donald John McKENZIE**, RNZAF - Age 27.

Wed 6/Thu 7 May 1942

BOMBER COMMAND

Raid on Stuttgart, Germany (by 97 aircraft - 8 lost)

115 Squadron, RAF (Marham, Norfolk - 3 Group)

Wellington III X3466/N - took off at 2130 and brought down over France, crashing at Marle, 21km NNW of Laon. The navigator is buried at Laon and the captain commemorated on the Runnymede Memorial. Three RAF Sgts survived to be taken prisoner, although one died in captivity in March 1944.

Captain: 36261 Flt Lt **Nathaniel PATERSON**, RAF - Age 25. 897hrs. 33rd op.

Navigator: NZ403965 Plt Off **Patrick Henry Francis William LELAND**, RNZAF - Age 27. 316hrs. 6th op(?)

Thu 7/Fri 8 May 1942

BOMBER COMMAND

'Gardening' - minelaying off Denmark and Germany (by 81 aircraft - 2 lost)

57 Squadron, RAF (Feltwell, Norfolk - 3 Group)

Wellington III Z1564 - took off at 2210 and brought down over Denmark, crashing opposite Esbjerg, near Nordby, Fanø Island, probably just off shore. All five crew are buried at Fourfelt, to the NW of Esbjerg, the captain's recovered body being laid to rest on 17 October.

Captain: NZ41946 Plt Off **John Adam SLIGO**, RNZAF - Age 24. 406hrs. 17th op.

Fri 8 May 1942

FIGHTER COMMAND

Patrol to guard special train between Ashburton and Dartmouth

154 Squadron, RAF (Church Stanton, Devonshire - 10 Group)

Spitfire VB BL438 - took off at 0905 with three others and crashed in unknown circumstances on Dartmoor. The seriously injured pilot was taken to the military hospital at Moreton Hampstead, 10 miles north of Ashburton, but died later in the day and is buried at Exeter. The train was taking the King and Queen to Exeter.

Pilot: NZ405314 Sgt **John OLIVER**, RNZAF - Age 27. 315hrs. 8th op.

Fri 8/Sat 9 May 1942

BOMBER COMMAND

Raid on Warnemünde, Germany (by 193 aircraft - 20 lost)

158 Squadron, RAF (Driffield, Yorkshire - 4 Group)

Wellington II W5562/H - took off at 2230 and brought down into the water near the target area, killing the crew of five. The bodies were recovered and buried a few kilometres south of Warnemünde, at Rostock, but later reinterred at Berlin.

Captain: NZ404852 Flt Sgt **Ian Wynn DAVIES**, RNZAF - Age 25. 357hrs. 12th op.

Davies brother, Evan Bertram Davies, died on 20 August 1944 while flying with 150 Sqn, RAF.

44 Squadron, RAF (Waddington, Lincolnshire - 5 Group)

Lancaster I R5555/P - captained by Wg Cdr P W Lynch-Blosse, DFC, RAF, on same raid as the above and brought down near Lambrechtshagen, 9km SSE of the target. All eight crew were buried nearby, but later reinterred at Berlin.

Navigator: NZ404477 Plt Off **Rex Vivian BEATSON**, RNZAF - Age 31. 253hrs. 8th op.

Sat 9 May 1942

FIGHTER COMMAND

Wing sweep over Northern France

72 Squadron, RAF (Biggin Hill, Kent - 11 Group)

Spitfire VB BM384 - took off at 1200 with others of the Squadron as 'Red 3'. No enemy opposition was encountered during the sortie, but on the return over the English Channel 'Red' Flight dropped down to sea level to investigate an object which turned out to be a raft. During this inspection Red 3 was seen to hit the water and disappear. The pilot is commemorated on the Runnymede Memorial.

Pilot: NZ405350 Plt Off **David Owen WATERS**, RNZAF - Age 22. 257hrs. 31st op.

Sat 9/Sun 10 May 1942

FERRY COMMAND

Delivery flight from Canada to Britain via Reykjavik, Iceland

Ferry Command, RAF (Canadian HQ, Dorval, Quebec)

Hudson IIIA FH269 - took off from Gander, Newfoundland, captained by Wg Cdr C F Herrington, RAF. At 0145 GMT FH269 signalled it was returning to Gander with engine trouble, but nothing further was heard. The pilot and navigator are both commemorated on the Runnymede Memorial, while the RCAF wireless operator-air gunner is remembered on the Ottawa Memorial.

Navigator: NZ413717 Plt Off **Richard John SOTHAM**, RNZAF - Age 25

Sotham is thought to have been posted to Britain and travelling there by means of the delivery flight.

Tue 12 May 1942

INDIA

Test flight

34 Squadron, RAF (Allahabad, India - 221 Group)

Blenheim IV V6254 - took off during the evening and made a forced landing 23 miles SSW of Bamrauli after an engine failure. All three crew were injured, the pilot dying at 0900 next day in the British Military hospital at Allahabad. He was buried in the Cantonment Cemetery at Allahabad, but later reinterred at Delhi.

Pilot: NZ404377 Flt Sgt **Harold KAY**, RNZAF - Age 21. 295hrs (215 solo - 128 on Blenheim)

Official records are divided over whether Kay died on the 12th or 13th. The flight may have taken place on the 11th but the weight of evidence suggests it was the 12th.

MIDDLE EAST

Interception of incoming air raid

603 Squadron, RAF (Ta Kali, Malta)

Spitfire VC BR127 - took off at 1110 with seven others of the Squadron and shot down during combat with enemy fighters, diving into the sea at high speed. The pilot is commemorated on the Malta Memorial

Pilot: NZ391843 Plt Off **Herbert Robert MITCHELL**, RNZAF - Age 25. 566hrs. 128th op(?)

Wed 13 May 1942

FLEET AIR ARM - SCOTLAND

Training

753 Squadron, FAA (Arbroath, Angus - HMS *Condor*)

Swordfish I L2838 - piloted by Lt C H Dupere, RN, and collided at 2000 feet with a 767 Sqn Swordfish I (K8434) at 1535. Both machines crashed at Shandford Farm, Finavon, $4^1/_2$ miles NNW of Forfar, killing the three crew in each. The observer of L2838 is buried at Arbroath, as are the crew of K8434.

Observer: NZ2825 Leading Airman **Brian Arthur PATTERSON**, RNZN - Age 24.

Wed 13/Thu 14 May 1942

MIDDLE EAST

Raid on Benghazi, Libya

104 Squadron, RAF (LG 106 [13 miles east of Daba], Egypt - 236 Wing, 205 Group)

Wellington II Z8574/L - captained by Plt Off A F O'Donnell, RAF, and lost without trace, its six crew being commemorated on the Alamein Memorial.

2nd Pilot: NZ405307 Sgt **Stanley Harold MORTIMORE**, RNZAF - Age 21. 314hrs. 6th op.

Fri 15/Sat 16 May 1942

BOMBER COMMAND

'Gardening' - mine laying in the Western Baltic (by 50 aircraft - 4 lost)

75 (NZ) Squadron, RAF (Feltwell, Norfolk - 3 Group)

Wellington III X3482/J - took off at approximately 2245-2305 to lay mines in Kiel Bay and brought down on or near the North Frisian island of Sylt. All five crew were buried in the island's main town of Westerland, but later reinterred at Kiel.

Captain: NZ403437 Flt Sgt **Myles Frederick Gordon FRASER**, RNZAF - Age 22. 555hrs. 35th op.
Navigator: NZ402221 Flt Sgt **Albert Ivan SMITH**, RNZAF - Age 27. 345hrs. 21st op.
Rear Gunner: NZ404107 Sgt **Norman Edward WHITING**, RNZAF - Age 27. 190hrs. 19th op.

Sun 17 May 1942

COASTAL COMMAND

Torpedo strike against the heavy cruiser *Prinz Eugen*

42 Squadron, RAF (Leuchars, Fifeshire - 18 Group)

Beaufort II AW383/V - took off at 1755 piloted by Fg Off R B Archer, DFC, RAAF, with 11 others and one of three which failed to return. At 2020 the Squadron, escorted by four Beaufighters and six Blenheims, sighted and attacked the cruiser off Lista Fjord, about 80km west of Kristiansand, Norway. About 20 enemy fighters intercepted the strike force and after the engagement three aircraft were seen burning on the water. Three of AW383's crew were lost without trace and are commemorated on the Runnymede Memorial, while the body of a wireless operator-air gunner was later recovered and buried on the island of Blome, NW of Bergen.

Navigator: NZ404033 Flt Sgt **David North KEELING**, RNZAF - Age 23. 331hrs. 21st op.

Sun 17/Mon 18 May 1942

BOMBER COMMAND

'Gardening' - mine laying off the Frisians, Heligoland and Denmark (by 60 aircraft - 7 lost)

15 Squadron, RAF (Wyton, Huntingdonshire - 3 Group)

Stirling I W7531/F (a presentation machine named *MacRobert's Reply*) - took off at 2140 captained by Sqn Ldr J C Hall, DFC mid, RAF, to mine Danish waters and was hit twice by flak, crashing at 0210 at Galsklint, 2km west of Middelfart. Only an RAF Sgt survived severely injured and was taken prisoner. His eight comrades died, all but one (an RAF Sgt who is commemorated on the Runnymede Memorial) being buried on the 21st at Odense, about 45km to the east of the crash site.

Navigator: J/4758 Fg Off **John Patrick RYAN**, RCAF - Age 30.

Tue 19 May 1942

CANADA

Gunnery exercise

9 Bombing & Gunnery School, RCAF (Mont Joli, Quebec - 3 Training Command)

Battle I 2022 - piloted by Plt Off A F Halamka, RCAF. At about 1000 black and white smoke was seen streaming from the Battle immediately before it crashed and burned in a field near Mont Joli's No 2 Central Tower, between Ste Flavie and Ste Luce. During 2022's descent the pilot was seen to fall into the waters of the adjacent Gulf of St Lawrence and he is remembered on the Ottawa Memorial. An RAAF and RNZAF wireless operator-air gunner and their RCAF instructor died also, the former two being buried at Metis Beach.

WOpAG: NZ413287 LAC **Kenneth George WEAL**, RNZAF - Age 19.

Tue 19/Wed 20 May 1942

BOMBER COMMAND

Raid on Mannheim, Germany (by 197 aircraft - 11 lost)

156 Squadron, RAF (Alconbury, Huntingdonshire - 3 Group)

Wellington III X3671 - took off at 2217 and shot down over the Franco-Belgian border, exploding on impact at Moulins de Thilay, north of Willerzie. All six crew are buried in a collective grave at Hargnies, 6km ENE of Haybes, France.

Captain: 41858 Sqn Ldr **Thomas Campbell McGILLIVRAY**, DFC, RAF - Age 26. 955hrs. 46th op.

McGillivray was a prominent figure in pre-war boxing circles in New Zealand and won many fights under his professional name of Don Stirling.

35 Squadron, RAF (Linton-on-Ouse, Yorkshire - 4 Group)
Halifax II W1101/S - took off at 2230 on same raid as the above, and brought down at Heuchelheim, 15km NW of the target city's centre. All six crew were buried in the main cemetery at Mannheim, but later reinterred at Durnbach, 16km east of Bad Tölz.
Captain: NZ403987 Flt Sgt **Dudley Herman REED**, RNZAF - Age 29. 430hrs. 10th op.
Rear Gunner: NZ40209 Flt Sgt **Leslie Plimmer RUSSELL**, RNZAF - Age 25. 351hrs. 82nd op(?)
Seventy-three of Russell's ops were undertaken while flying as air gunner in Defiants with 264 Sqn.

12 Squadron, RAF (Binbrook, Lincolnshire - 1 Group)
Wellington II W5458 - took off at 2240 on same raid as the above and presumed to have been brought down into the North Sea off the Dutch coast, the bodies of two RAF Sgt crew members later being washed ashore onto the island of Vlieland. Their four comrades are commemorated on the Runnymede Memorial.
Captain: NZ402177 Plt Off **William John FULTON**, RNZAF - Age 26. 445hrs. 22nd op.
Navigator: NZ404014 Flt Sgt **William Alan BAIRD**, RNZAF - Age 25. 276hrs.

7 Squadron, RAF (Oakington, Cambridgeshire - 3 Group)
Stirling I W7520/S - took off at 2258 on same raid as the above and collided with an enemy night-fighter, both crashing near Brustem, 3km SE of St Trond. Both crews died, the seven from the Stirling being buried on the airfield at St Trond, but later reinterred at Heverlee, near Louvain.
Captain: 65540 Flt Lt **Hector Garrick PILLING**, DFC, RAF - Age 29.
Pilling is believed to have been born in Fiji but he lived in New Zealand during at least 1926-1927 and 1929-1930.

Thu 21/Fri 22 May 1942

BOMBER COMMAND

Cross country training
75 (NZ) Squadron, RAF (Feltwell, Norfolk - 3 Group)
Wellington III Z1566 - after take off the weather unexpectedly deteriorated. Z1566 flew above cloud for a considerable time before being obliged to break through in the early morning hours to establish its position, but at 0340 flew into a hillside at Grindon Farm, 6 miles ESE of Leek, Staffordshire. The rear gunner, Sgt **S J Chappell**, RNZAF, was the sole survivor, albeit injured. His five comrades died and all except the RAF Sgt wireless operator are buried at Buxton, Derbyshire, 11 miles north of Grindon.
Captain: NZ411919 Plt Off **Andrew Donald MACKAY**, RNZAF - Age 22. 309hrs (239 solo - 102 on Wellington) 3 ops.
Navigator: NZ405331 Sgt **Wilfred Hubert SMEATON**, RNZAF - Age 28.
Air Bomber: NZ403566 Sgt **James Wilfred BEAVEN**, RNZAF - Age 31. 163hrs.

Sat 23 May 1942

FLYING TRAINING COMMAND

Solo flarepath landing practice
11 (Pilots) Advanced Flying Unit, RAF (Shawbury, Shropshire - satellite at Wheaton Aston, Staffordshire - 21 Group)
Oxford II P1862 - dived into the ground and caught fire on a second approach to Wheaton Aston at 0310. The pilot is buried at Shawbury.
Pilot: NZ413823 Sgt **Edward Charles Hibbs DOHERTY**, RNZAF - Age 24. 180hrs (107 solo - 84 on Oxford)

Sun 24/Mon 25 May 1942

BOMBER COMMAND

Cross country training
22 Operational Training Unit, RAF (Wellesbourne Mountford, Warwickshire - satellite at Stratford - 91 Group)
Wellington IC DV840 - took off from Stratford captained by Sgt C R Whitworth, RAF, but at the conclusion of the exercise at 0240, overshot and swung badly while making an approach in a strong 30° crosswind. Control was lost and DV840 crashed onto another of the OTU's Wellingtons (AD625), which was stationary at its taxiing post, and caught fire. Only one of the five crew survived, albeit injured. The navigator is buried at Stratford-on-Avon.
Navigator: NZ411721 Sgt **Ronald Ernest HERBERT**, RNZAF - Age 23. 228hrs.

Training
22 Operational Training Unit, RAF (Wellesbourne Mountford, Warwickshire - satellite at Stratford - 91 Group)
Wellington IC AD625 - captained by Fg Off J J Gale, RAF, was awaiting clearance for take off when struck by Wellington DV840 as described above. At least three crew members died, all of whom are buried at Stratford-on-Avon.
Navigator: NZ404977 Sgt **John Syddall WILLIAMS**, RNZAF - Age 26. 133hrs.
WOpAG: NZ412313 Sgt **Douglas CALLAGHAN**, RNZAF - Age 23. 98hrs.
WOpAG: NZ405532 Sgt **Mervyn WILKINSON**, RNZAF - Age 21.

Mon 25 May 1942

MIDDLE EAST

Interception of incoming raid

601 Squadron, RAF (Luqa, Malta - Ta Kali, Malta)

Spitfire VC BR354 - took off at about 1500 with 15 others of 126 and 601 Squadrons and shot down, crashing into the sea near St Paul's Bay. An air and sea search proved unsuccessful in finding any trace of BR354, or its pilot, who is commemorated on the Malta Memorial.

Pilot: NZ405229 Plt Off **Walter Alexander CALDWELL**, RNZAF - Age 19. 295hrs.

Tue 26 May 1942

NEW ZEALAND

Dual instruction - instrument training

2 Service Flying Training School (ITS), RNZAF (Woodbourne)

Harvard II NZ920 - approaching to land at 0945, the pilot of another of the School's Harvards (NZ923), LAC **R C C Nairn**, saw NZ920 approaching him rapidly from the port quarter. He was unable to take avoiding action, and it collided with him and crashed, catching fire near the Delta landing ground, 10 miles west of Blenheim. Its two crew are buried at Wanganui and Wellington, respectively. Nairn was commended for the excellent handling and safe landing of his seriously damaged machine. The crashed Harvard had been flying towards the sun at the time and it was thought that its glare may have obscured the crew's view of Nairn's camouflaged aircraft.

Pilot Instr: NZ39903 Fg Off **Robert William Douglas COWPER**, RNZAF - Age 23.
Pilot: NZ42421 LAC **Derek Thomas LIPPITT**, RNZAF - Age 20.

FIGHTER COMMAND

Practice interception with 303 Sqn

616 Squadron, RAF (Kings Cliffe, Northamptonshire - 12 Group)

Spitfire VI BR172 - seen to dive at great speed before the wings broke off, causing it to crash and burn out at Dunholme, 5 miles east of RAF base Scampton, Lincolnshire. The pilot is buried at Scampton.

Pilot: NZ404007 Plt Off **Leicester Bond WARE**, RNZAF - Age 24. 476hrs. 48 ops.

The CWGC register incorrectly gives Ware's age as 26.

Thu 28 May 1942

INDIA

Practice(?)

67 Squadron, RAF (Alipore, India - 224 Group)

Hurricane IIB BG875 - stalled at 1030 while pulling up from low level inverted flight and crashed close to the gasworks at Alipore, on the outskirts of Calcutta. The pilot is buried at Calcutta. The accident card records the time as 0500 but this is believed to refer to GMT and not local time.

Pilot: NZ41514 Sgt **Charles Thomas KRONK**, RNZAF - Age 23. 305hrs (226 solo - 35 on Hurricane) 60 ops.

Sat 30 May 1942

FLEET AIR ARM - USA

Training

Naval Air Station Miami, United States Navy (Florida - HMS *Saker II*)

Buffalo F2A-3 BuNo01579 - spun into the ground 5 miles west of the airfield at 0830. The pilot died of injuries later in the day and is buried at Miami.

Pilot: Sub-Lt (A) **Francis Patrick Joseph FLANNAGAN**, RNZNVR - Age 22.

Sat 30/Sun 31 May 1942

BOMBER COMMAND

Intruder raid on night-fighter airfield at Vechta, Germany

114 Squadron, RAF (West Raynham, Norfolk - 2 Group)

Blenheim IV V5645/R - took off at 2124 and shot down over Germany, two of the crew being killed and the pilot seriously wounded. The latter was taken to the Marine Hospital at Sonderbusch where he died on 1 June. He was buried at Wilhelmshaven, but later reinterred with the other members of his crew at Sage, 24km south of Oldenburg. Fifty-six intruders operated during the night against enemy airfields in support of the first 1000-bomber raid, V5645 being one of two which failed to return.

Pilot: NZ41320 Plt Off **John Joseph FOX**, RNZAF - Age 24. 324hrs. 3rd op.

It will be recalled that on 15 May 1941 Fox survived the mid-air collision at 2SFTS which resulted in the death of Alec Haigh.

Raid on Cologne, Germany (by 1047 aircraft - 52 lost): The first 1000-bomber raid

22 Operational Training Unit, RAF (Wellesbourne Mountford, Warwickshire - 91 Group)

Wellington IC DV701/P - took off at 2248 from Elsham Wolds, Lincolnshire, and shot down over Germany, crashing at Zons a/m Rhine, 11km south of Düsseldorf. The bodies of the captain and at least two others were recovered at the time, but post-war their graves could not be located and all five airmen are commemorated on the Runnymede Memorial.

Captain: NZ401164 Fg Off **Harold Roger BLAKE**, DFC, RNZAF - Age 24.

Blake had completed an operational tour in 1941 with 103 and 150 Sqns and was instructing with 20 OTU at Lossiemouth, Morayshire, when called upon to captain the 22 OTU machine for this raid.

22 Operational Training Unit, RAF (Wellesbourne Mountford, Warwickshire - 91 Group)

Wellington IC R1235/O - took off at 2255 from Elsham Wolds, Lincolnshire, captained by Flt Sgt C J Matthews, RAF, on same raid as the above, and brought down over Germany, crashing at Kalenberg, 4km NE of Kall. All five crew are buried at Rheinberg, 12km south of Wesel.

Air Bomber: NZ405515 Sgt **Denis Henry EDWARDS**, RNZAF - Age 22. 1st op(?)

Edwards brother, Norman John Edwards, died four days later while flying with 101 Sqn, RAF.

1502 Beam Approach Training Flight, RAF (Driffield, Yorkshire - 4 Group)

Whitley V Z9307/H - took off at 2310 on same raid as the above captained by Sqn Ldr J A S Russell, DFC, RAF, and shot down by a night-fighter over Belgium, crashing at 0114 in the SW suburbs of Antwerp near Hoboken. The captain and a fellow RAF Fg Off survived to be taken prisoner, but their three comrades died and are buried at Antwerp.

2nd Pilot: NZ404884 Plt Off **Denis Grant BOX**, RNZAF - Age 23. 364hrs. 1st op.

Box, serving with 9 Air Observers School, was attached to the BAT Flight for this raid.

158 Squadron, RAF (Driffield, Yorkshire - 4 Group)

Wellington II Z8577/T - took off at 2319 on same raid as the above and en route to the target was shot down by a night-fighter over the Netherlands, crashing in the Oosterschelde near Schelphoek. The captain's body washed ashore next day, as did those of his four comrades over the next six months. All are buried at Vlissingen

Captain: 41694 Sqn Ldr **Donald Joseph HARKNESS**, DFC, RAF - Age 25. 34th+ op.

Harkness had served a year in the RNZAF as a clerk prior to embarking for Britain in November 1938 for a short service commission in the RAF.

26 Operational Training Unit, RAF (Wing, Buckinghamshire - 92 Group)

Wellington IC W5704/S - took off at 2345 from Graveley, Huntingdonshire, captained by Fg Off W R H Whiting, RAF, on same raid as the above, and shot down by a night-fighter over the Netherlands at about 0130, crashing near Middelbeers, 15km west of Eindhoven. All five crew are buried at Eindhoven.

Air Bomber: NZ404073 Sgt **James Milton HALL**, RNZAF - Age 26. 1st op(?)

Wireless Op: NZ404629 Sgt **Allen Norman YOUNG**, RNZAF - Age 21. 1st op(?)

Both Hall and Young were undergoing training with 27 OTU when called upon to crew the 26 OTU Wellington for this raid.

405 Squadron, RCAF (Pocklington, Yorkshire - 4 Group)

Halifax II W7707/K - took off at 0022 captained by Sgt L A Wadman, RCAF, on same raid as the above and lost without trace, its seven crew being commemorated on the Runnymede Memorial.

Air Bomber: NZ404616 Sgt **Andrew Fraser McLEAN**, RNZAF - Age 31. 107hrs. 1st op(?)

McLean had been undergoing Halifax training with 1652 Conversion Unit since 1 May, prior to being called upon to crew the 405 Sqn Halifax for this raid. The CWGC register records his unit as 78 Sqn but it appears unlikely he ever served with it.

61 Squadron, RAF (Syerston, Nottinghamshire - 5 Group)

Lancaster I R5561 - took off at 0040 captained by Plt Off J B Underwood, RAF, on same raid as above, and brought down over Germany at Niederaussem, 21km WNW of the target city's centre. All seven crew were buried in the Military Cemetery in South Cologne, but later reinterred at Rheinberg, 12km south of Wesel.

2nd Pilot: NZ41637 Plt Off **Benson USHER**, RNZAF - Age 25. 325hrs. 10th op.

Sun 31 May 1942

MIDDLE EAST

Strafing and bombing attacks

112 Squadron, RAF (LG 139 [Gambut No 1 or Main, Libya - 239 Wing, 211 Group)

Kittyhawk I AK999 - took off at 1845 and failed to return. During the Army's advance later into Libya the pilot was found buried alongside his wrecked fighter near Derna. Unfortunately, a post-war search failed to again locate the grave, as it had been effaced by sand storms, and the pilot is now commemorated on the Alamein Memorial.

Pilot: NZ41348 Plt Off **Stuart Stewart MITCHELL**, RNZAF - Age 21. 234hrs.

Mitchell's brother, John Stewart Mitchell, died on 2 March 1940 while flying with 1 Sqn, RAF.

Mon 1 Jun 1942

BOMBER COMMAND

Plt Off **J J FOX**, RNZAF, died of injuries received when he was shot down on the night of 30/31 May.

Mon 1/Tue 2 Jun 1942

BOMBER COMMAND

Raid on Essen, Germany (by 956 aircraft - 34 lost)

115 Squadron, RAF (Marham, Norfolk - 3 Group)

Wellington III X3721/F - took off at 2255 captained by Fg Off L C Williams, RAF, and lost without trace, its five crew being commemorated on the Runnymede Memorial.

Navigator: NZ404378 Plt Off **Douglas George KING**, RNZAF - Age 23. 261hrs. 5th op.

83 Squadron, RAF (Scampton, Lincolnshire - 5 Group)

Lancaster I R5564/P - took off at 2310 captained by Plt Off F R Busby, RAF, on same raid as the above, and presumed to have been brought down into the North Sea. The captain's body washed ashore onto the Dutch coast on 23 July and was buried at Amsterdam. Of his six comrades no trace was found and they are commemorated on the Runnymede Memorial.

Navigator: NZ404526 Plt Off **Kelvin Cholwill BILLING**, RNZAF - Age 24. 302hrs. 11th op.

10 Squadron, RAF (Leeming, Yorkshire - 4 Group)

Halifax II W1143/F - on same raid as the above and brought down over the Netherlands, crashing into the Haringvliet off Hellevoetsluis, 25km SW of Rotterdam. An RAF Sgt survived, being the only one to successfully bale out before the Halifax hit the water. The captain's body was later recovered and is buried at Rotterdam, but the five others who died are commemorated on the Runnymede Memorial. Halifax W1143 belonged to Croft-based 78 Sqn but had been loaned to 10 Sqn for this raid.

Captain: NZ404442 Plt Off **Henry Garfield CLOTHIER**, RNZAF - Age 26. 374hrs. 8th op.

Navigator: NZ403599 Fg Off **John Raymond Francis GANDERTON**, RNZAF - Age 26. 328hrs. 8th op.

20 Operational Training Unit, RAF (Lossiemouth, Morayshire - 91 Group)

Wellington IC X9975 - took off from Stanton Harcourt, Oxfordshire, captained by Plt Off S S Martin, DFC, RCAF, on same raid as the above, and brought down into the North Sea. The body of an RAF crew member later washed ashore on the Netherlands coast near Bergen aan Zee, but no trace of his five comrades was found and they are commemorated on the Runnymede Memorial.

2nd Pilot: NZ402438 Flt Sgt **Bernard Joseph DALEY**, RNZAF - Age 25. 205hrs. 2nd op.

Tue 2 Jun 1942

MIDDLE EAST

Air-sea-rescue flight for missing Wellington

80 Squadron, RAF (LG 121 [17 miles E of Sidi Barrani], Egypt - 243 Wing, 211 Group, Western Desert)

Hurricane IIC BM977 - took off at 1315 as one of a pair to search for a missing Wellington and was last seen 4 miles out to sea flying SE in the Sollum area, near the Egyptian border with Libya. The pilot is commemorated on the Alamein Memorial. Some records state that the Hurricane was simply engaged on a sea patrol over a convoy and make no mention of searching for a Wellington.

Pilot: NZ411394 Sgt **James Adam GLYNN**, RNZAF - Age 28. 234hrs. 14th op.

ARMY CO-OPERATION COMMAND

Photographic sweep over the Cherbourg Peninsula, France

140 Squadron, RAF (Mount Farm, Oxfordshire - 34 Wing)

Spitfire PR IV X4502 - took off at 1120 and lost without trace, its pilot being commemorated on the Runnymede Memorial. Later it was reported that an intercepted German r/t message claimed a British fighter shot down 4 miles west of the Channel Island of Alderney at the time of this sortie.

Pilot: 46821 Plt Off **Francis John BLACKWOOD**, OBE, RAF - Age 28.

Tue 2/Wed 3 Jun 1942

BOMBER COMMAND

Raid on Essen, Germany (by 195 aircraft - 15 lost)

15 Squadron, RAF (Wyton, Huntingdonshire - 3 Group)

Stirling I N3728/T - took off at 2348 and shot down over the Netherlands by a night-fighter, crashing at 0310 between Melick and Herkenbosch, three to 4km SE of Roermond. All seven crew were buried at Venlo, 25km to the NE, but later reinterred in Jonkerbos War Cemetery, 4km SW of Nijmegen.

Captain: NZ39902 Wt Off **Andrew Jack COWLRICK**, RNZAF - Age 26. 555hrs. 28th op.

57 Squadron, RAF (Feltwell, Norfolk - 3 Group)
Wellington IC X9787 - took off at 2350 on same raid as the above, and shot down over Germany by a night-fighter, crashing at 0159 about 6km north of Gildehaus. All five crew died, the captain, possibly with others of the crew, being buried in the Konigseschstrasse Cemetery at Rheine. All five were later reinterred at Reichswald Forest, 5km SW of Kleve.
Captain: NZ41574 Flt Sgt **Vincent Patrick CUMMOCK**, RNZAF - Age 25. 452hrs. 6th op.

61 Squadron, RAF (Syerston, Nottinghamshire - 5 Group)
Lancaster I R5613 - took off at 0001 captained by Plt Off R E Clark, DFM, RAF, on the same raid as the above and shot down over Belgium by a night-fighter, crashing at 0226 at La Hulpe, 14km SE of Brussels. An RAF Sgt was the sole survivor and succeeded in evading capture to return safely Britain. The body of an RAF Flt Sgt was not recovered and he is commemorated on the Runnymede Memorial, while the five others who died are buried at Brussels.
2nd Pilot: NZ41342 Sgt **Stanley Holmes LINCOLN**, RNZAF - Age 24. 295hrs. 10th op.
Lincoln's brother, Lloyd John Holmes Lincoln, was killed almost two months earlier, on 7 April, while flying a Manchester with the same squadron. They enlisted on the same date, held consecutive service numbers and served several months together with 61 Sqn. By extraordinary coincidence, both were shot down by a night-fighter over Belgium (within 80km of each other) while engaged on a raid to the same target city as 2nd Pilot. They now lie buried some 20km apart.

Wed 3/Thu 4 Jun 1942

BOMBER COMMAND

Raid on Bremen, Germany (by 170 aircraft - 14 lost)
115 Squadron, RAF (Marham, Norfolk - 3 Group)
Wellington III X3635/J - took off at 2300 captained by Fg Off T R R Wood, RAF, and brought down over Germany, only an RAF Plt Off surviving to be taken prisoner. The captain's body was recovered and is buried at Becklingen, 17km south of Soltau, but the other four crew were lost without trace and are commemorated on the Runnymede Memorial.
2nd Pilot: NZ41978 Flt Sgt **John Wynne CHAPMAN**, RNZAF - Age 27. 231hrs. 1st op(?)
The CWGC register incorrectly gives Chapman's age as 29.

207 Squadron, RAF (Bottesford, Leicestershire - 5 Group)
Lancaster I R5847/Y - took off at 2337 captained by Wt Off C Wathey, RAF, on same raid as the above, and shot down over the Netherlands by a night-fighter, crashing at 0114 at Zwiggelte, 3km NW of Westerbork. All seven crew are buried in a collective grave at Westerbork.
Navigator: NZ404540 Sgt **Russell Royl Alwyn JONES**, RNZAF - Age 21. 304hrs. 13th op.
The CWGC register give Jones's second name as Roy, but all other documents sighted show Royl.

101 Squadron, RAF (Bourn, Cambridgeshire - 3 Group)
Wellington III X3473/S - took off on same raid as the above and lost over Germany on spinning out of control while taking evasive action, crashing near Delmenhorst, about 10km WSW of Bremen. Two RAF Pilot Officers survived to be taken prisoner. Their three comrades were buried in a collective grave at Delmenhorst, but later reinterred at Becklingen, 17km south of Soltau.
Captain: 79510 Flt Lt **Norman John EDWARDS**, RAFVR - Age 28. 832hrs. 25th op.
Edwards brother, Denis Henry Edwards, died four days earlier while flying with 22 OTU, RAF.

Fri 5/Sat 6 Jun 1942

BOMBER COMMAND

Raid on Essen, Germany (by 180 aircraft - 15 lost)
10 Squadron, RAF (Leeming, Yorkshire - 4 Group)
Halifax II W7696/H - took off at 2259 captained by Flt Sgt P G Rochford, DFM, RAF, and brought down over Germany between Liblar and Knapsack, 14km SW of Cologne. Four of the crew survived to be taken prisoner. Their three comrades were buried in the South Cologne Cemetery, but later reinterred at Rheinberg, 12km south of Wesel.
Navigator: NZ403610 Fg Off **Edward Basil Hamilton OPENSHAW**, RNZAF - Age 33. 351hrs. 14th op.
The CWGC register incorrectly gives Openshaw's age as 31.

150 Squadron, RAF (Snaith, Yorkshire - 1 Group)
Wellington III X3674/H - took off at 2320 on same raid as the above, and brought down over the Netherlands, crashing onto land owned by Mr Van Os at Benschop, 14km SW of Utrecht. All six crew are buried at Oud Leusden on the southern outskirts of Amersfoort, 20km ENE of Utrecht.
Captain: NZ39557 Flt Sgt **Thomas KAY**, RNZAF - Age 23.

57 Squadron, RAF (Feltwell, Norfolk - 3 Group)
Wellington IC HF915 - took off at 2325 on same raid as the above, and lost without trace, its five crew being commemorated on the Runnymede Memorial.
Captain: NZ41975 Flt Sgt **William James BARNES**, RNZAF - Age 21. 269hrs. 4th op.

Sat 6 Jun 1942

NEW ZEALAND

Solo flying practice

3 Elementary Flying Training School, RNZAF (Harewood)

Tiger Moth II NZ801 - seen at about 0915 in a left hand gliding turn at 1000 feet, but crashed shortly afterwards on the north side of the Waimakariri River near West Eyreton, apparently as the result of a spin. The pilot was seriously injured and admitted to the Christchurch Public Hospital, but died there the following morning and was later cremated at Auckland.

Pilot: NZ421080 LAC **Douglas MOIR**, RNZAF - Age 20. 15hrs solo (all on Tiger Moth)

MIDDLE EAST

Practice formation flying

238 Squadron, RAF (Gambut West, Libya - 243 Wing, 211 Group, Western Desert)

Hurricane IIB BG902 - appeared to lag behind but indicated no sign of trouble. The formation climbed through a small cloud near Sidi Azeiz, to the ESE of Gambut, but afterwards BG902 was nowhere to be seen. Wreckage of the burnt out Hurricane was later located, and the pilot buried at Sidi Omar, south of Sidi Azeiz, but later reinterred at Halfaya Sollum, Egypt.

Pilot: NZ41863 Plt Off **Alan William ASMUSS**, RNZAF - Age 21. 298hrs solo (100 on Hurricane)

Mon 8/Tue 9 Jun 1942

BOMBER COMMAND

Raid on Essen, Germany (by 170 aircraft - 20 lost)

75 (NZ) Squadron, RAF (Feltwell, Norfolk - 3 Group)

Wellington III X3587 - took off at approximately 2325-2345 and brought down over Germany, crashing near Düsseldorf. All five crew were buried in the North Cemetery at Düsseldorf, but later reinterred at Reichswald Forest, 5km SW of Kleve.

Captain: NZ41950 Plt Off **Rupert John SMITH**, RNZAF - Age 26. 399hrs. 12th op.

Air Bomber: NZ405508 Sgt **William Robertson RAMSAY**, RNZAF - Age 25. 59hrs. 8th op.

75 (NZ) Squadron, RAF (Feltwell, Norfolk - 3 Group)

Wellington III Z1573 - took off at approximately 2325-2345 on same raid as above, and brought down over the Netherlands, crashing in the Krabbenkreek between St Philipsland and the Anna Jacobapolder, about 20km NW of Bergen-op-Zoom. All five crew died, four of whom are buried at Bergen-op-Zoom. The air bomber is commemorated on the Runnymede Memorial.

Captain: NZ411927 Plt Off **Graham Edward MURDOCH**, RNZAF - Age 26. 397hrs. 18th op.

Navigator: NZ41544 Sgt **Albert William O'DOWD**, RNZAF - Age 25.

Air Bomber: NZ405494 Sgt **Leon Gaston KNIGHT**, RNZAF - Age 22. 119hrs. 5th op.

Murdoch's brother, Bernard Norris Murdoch, died on 7 December 1941 while flying with 104 Sqn, RAF.

35 Squadron, RAF (Linton-on-Ouse, Yorkshire - 4 Group)

Halifax II W7699/F - took off at 2329 on same raid as the above, and shot down into the North Sea by a night-fighter. The bodies of three crew washed up onto the Dutch coast over the next two months. Of the other three, including the captain, no trace was found and they are commemorated on the Runnymede Memorial.

Captain: NZ404491 Plt Off **Angus Carr MACKENZIE**, RNZAF - Age 29. 494hrs.

Mackenzie's brother, Donald Carr Mackenzie, died a year and three days later, on 12 June 1943 while flying with 467 Sqn, RAAF. His fiancée, Jane Winstone, died on 10 February 1944 while flying with the Air Transport Auxiliary.

Wed 10/Thu 11 Jun 1942

WEST AFRICA COMMAND

Escort to convoy WS19Q

95 Squadron, RAF (Jui, Sierra Leone)

Sunderland I L5805/B - took off at 2255 and rendezvoused with the convoy off Freetown at 0030 at position 0902N:1529W, then flew off to the NE. At 0400 the convoy sighted what resembled a white Very light at 0830N:1422W at a low height. Despite searches over the next few days no trace of L5805 or its crew could be found, although oil and what may have been petrol were sighted on the 13th at position 'JKCR 3450'. Eight of the nine crew are commemorated on the Malta Memorial, while the ninth, an RAF Fg Off, is commemorated on the Alamein Memorial.

Captain: NZ401778 Fg Off **Jack PYBUS**, RNZAF - Age 28. 1101hrs.

Thu 11 Jun 1942

NEW ZEALAND

AC2 **D L FORWARD** died after an operation following complications after being injured in a motor vehicle accident on 15/16 December 1941.

Emergency Precautions Scheme co-operation
1 Elementary Flying Training School, RNZAF (Taieri)
Tiger Moth II NZ769 - while carrying out a dummy air raid attack at 1040 over Gore, collided with cables across the main street, crashed onto the road, bounced, then skidded along to strike a service car driven by Mr **B N Smith**. The wreckage finally came to rest against a parked truck on the west side at the intersection with Mersey Street. Although pulled clear of the wreckage by Constable **E W Fibbes**, the pilot died before reaching hospital and is buried at Onehunga. Five civilians were also injured as the Tiger Moth careered along the street - Miss **June McKenzie** and Messrs **John Leslie Anderson, Ronald Hill, Leonard Frank Shave** and **Leslie Osborne Wallis.**
Pilot: NZ412657 Plt Off **Geoffrey Thomas COLDHAM**, RNZAF - Age 20. 550hrs solo (400 on Tiger Moth)

Fri 12 Jun 1942

FIGHTER COMMAND

Formation practice
59 Operational Training Unit, RAF (Crosby-on-Eden, Cumberland - 81 Group)
Hurricane I N2428 - at 1350 the formation entered cloud during which N2428 lost contact with the others. The Hurricane was later found to have crashed and caught fire 3 miles south of Shankend, Roxburghshire. The pilot is buried at Carlisle, Cumberland.
Pilot: NZ413834 Sgt **Donald FRASER**, RNZAF - Age 21. 206hrs (112 solo - 24 on Hurricane)

Fri 12/Sat 13 Jun 1942

MIDDLE EAST

Raid on Benghazi, Libya
104 Squadron, RAF (LG 106 [13 miles east of Daba], Egypt - 236 Wing, 205 Group)
Wellington II Z8592/Z - took off at 2145 captained by Plt Off F H Shackleton, RAF, and at 0140 was seen to be coned by searchlights at about 7000 feet over the target. Heavily engaged by flak, it lost height rapidly, with tracer and the lights following it down to almost the horizontal until it crashed into the sea and exploded a mile or two SW of Benghazi Harbour. All six crew are commemorated on the Alamein Memorial.
2nd Pilot: NZ402550 Flt Sgt **Alfred William CORBY**, RNZAF - Age 28.

Mon 15 Jun 1942

MIDDLE EAST

Convoy patrol
601 Squadron, RAF (Luqa, Malta)
Spitfire VC BR306 - took off at 1303 with three others. The Allied convoy was being attacked by a force of fighter-escorted dive bombers as the Flight arrived on the scene. In the ensuing engagement BR306 was shot down into the sea, its pilot being commemorated on the Malta Memorial.
Pilot: NZ405293 Sgt **Jack Nock McCONNELL**, RNZAF - Age 21.
See also footnote to entry under 27 May 1941 for Fg Off Murray MacKenzie regarding erroneous reference in the Official History as to when and where McConnell died.

Tue 16 Jun 1942

INDIA

Solo ferrying fight from Allahabad to Asansol
60 Squadron, RAF (Asansol, India - 221 Group)
Blenheim IV Z7903 - about five minutes after a morning take off and a single circuit, was seen to lose height and disappear behind some trees, where it crashed 2 miles SW of the airfield. The pilot was buried at Allahabad, but later reinterred at Delhi. A reconstruction of events established that there had been a long delay in starting up on what was an extremely hot day, the pilot spending a long time in the cockpit prior to take off. It was thought that he fainted from heat exhaustion once airborne and lost control. His death resulted not from injury caused by the crash, but as a result of his body jerking down between the safety straps and being strangled upon Z7903 impacting with the ground.
Pilot: NZ405328 Sgt **William Alexander SHANNON**, RNZAF - Age 25. 187hrs. 1 op.

COASTAL COMMAND

Experimental torpedo test
Torpedo Development Unit, RAF (Gosport, Hampshire - 16 Group)
Beaufighter VIF X8065 - piloted by Wg Cdr R G Shaw, RAF, and crashed at 1515 at the junction of Angelsey and Fort Roads, Alverstoke, SW of Gosport, after port engine failure. Carriage of the torpedo had exacerbated the handling difficulties created by engine failure, causing X8065 to flick into a spin and dive into the ground. The two crew are buried at Rowner, 2 miles NW of Gosport.
(Pilot): 82655 Flt Lt **Michael Orme DAVENPORT**, RAFVR - Age 22.
Although a pilot, Davenport was apparently acting as observer on this flight. His date and place of birth are unknown, but he lived in New Zealand from at least 1931 to 1933.

BOMBER COMMAND

Cross country training

1651 Heavy conversion Unit, RAF (Waterbeach, Cambridgeshire - 3 Group)

Stirling I N6088 - seen at about 1608 in a right hand spiral dive with the port wing on fire. The outer port mainplane broke away, causing N6088 to crash onto the Great North Road, $2^1/_2$ miles north of East Retford and 9 miles south of Finningley, Nottinghamshire. All eight crew died, most of whom, including the captain, are buried at Finningley.

Captain: NZ411491 Plt Off **Milan Roy SCANSIE**, RNZAF - Age 24. 306hrs (111 solo - 5 on Stirling)

Wed 17 Jun 1942

MIDDLE EAST

Cover patrol over Boston bombers

73 Squadron, RAF (Gambut, Libya - 243 Wing, 211 Group)

Hurricane IIC BN277 - took off at 1230 leading the Squadron patrol. As the formation approached to land at the end of the sortie it was bounced by enemy fighters which promptly shot down four of the Hurricanes, killing two of the pilots. BN277 was brought down 2 miles west of Gambut, the pilot's body being buried nearby, but later reinterred at Halfaya Sollum.

Pilot: 40786 Sqn Ldr **Derek Harland WARD**, DFC*, RAF - Age 24. 1400hrs. 262nd op(?)

Fri 19 Jun 1942

INDIA

1 Operational Training Unit, RAF (Risalpur, India - 1 [Indian] Group)

Died from typhoid fever at 1245 at the British Military Hospital at Murree. Buried at Lower Topa, but later reinterred at Karachi.

(Pilot): NZ411976 Plt Off **James Gordon BOOKLASS**, RNZAF - Age 25. 148hrs.

FIGHTER COMMAND

En route to Lossiemouth, Morayshire, for Army Co-operation exercise.

289 Squadron, RAF (Turnhouse, Midlothian - 13 Group)

Defiant I N1569 - took off from Inverness at about 1417 piloted by Sgt M Munro, RAF. A few minutes later, on climbing to about 500 feet, there was a loud explosion from the engine. The pilot closed and reopened the throttle, but there was another explosion and N1569 began to vibrate violently. Both crew baled out over the Moray Firth, the pilot successfully and without injury, but the air gunner died and is buried at Inverness. It was believed he struck some part of the structure when abandoning the aircraft.

Air Gunner: NZ41572 Sgt **Henry Lisle CHRISTIAN**, RNZAF - Age 26. 123hrs.

Fri 19/Sat 20 Jun 1942

MIDDLE EAST

Raid on Benghazi, Libya

37 Squadron, RAF (LG 09 [Bir Koralyim], Egypt - 231 Wing, 205 Group)

Wellington IC DV669/W - a stricken aircraft thought to be DV669 was last seen going down steadily near the target just after midnight. All six crew died and are buried at Benghazi.

Captain: NZ405281 Fg Off **William Lynn KAUTER**, RNZAF - Age 26. 321hrs. 11th op.
2nd Pilot: NZ41376 Plt Off **William George Duncan THURSTON**, RNZAF - Age 27. 340hrs. 11th op.
WOpAG: NZ403572 Flt Sgt **Lorenzo Richard FEASEY**, RNZAF - Age 20. 6th op.

BOMBER COMMAND

Raid on Emden, Germany (by 194 aircraft - 10 lost)

15 Squadron, RAF (Wyton, Huntingdonshire - 3 Group)

Stirling I R9352/T 2315 - took off at 2315 captained by Plt Off G J Hains, RAF, and brought down over Germany, crashing at Siemolten. The seven crew died, at least six of the whom, including a New Zealander, were buried at Lingen-Ems, but all seven were later reinterred at Reichswald Forest, 5km SW of Kleve

Air Gunner: NZ403576 Flt Sgt **John Theodore HEURTLEY**, RNZAF - Age 23. 228hrs. 9th op.

Sat 20 Jun 1942

FIGHTER COMMAND

Air test and weather check

616 Squadron, RAF (Kings Cliffe, Northamptonshire - 12 Group)

Spitfire VB AD559 - took off for the test after repair to a glycol pipe. Later seen carrying out aerobatics, in bad visibility at a low height, before crashing at 1815 into the sea 2 miles off Sheringham, Norfolk. The pilot is buried at Scottow.

Pilot: NZ402452 Sgt **Warwick Ashling CLOUSTON**, RNZAF - Age 21. 231hrs (155 solo - 17 on Spitfire)

Sat 20/Sun 21 Jun 1942

BOMBER COMMAND

Raid on Emden, Germany (by 185 aircraft - 11 lost)

75 (NZ) Squadron, RAF (Feltwell, Norfolk - 3 Group)

Wellington III X3760 - took off at approximately 2329-2359 and believed to have been shot down by a night-fighter, crashing into the sea off the Frisian islands, killing the five crew. The air bomber was lost without trace and is commemorated on the Runnymede Memorial. The bodies of his four comrades later washed ashore, those of the captain (on 12 July) and wireless operator onto the Dutch island of Schiermonnikoog, where they are buried. The navigator and rear gunner came ashore further north onto the German coast - the former on 15 July on the dyke near Neuharlingersiel - and were buried at Wittmund, 15km inland, directly south of the island of Spiekeroog, but were later reinterred at Sage, 24km south of Oldenburg.

Captain: NZ405030 Fg Off **Allen Armistice FRASER**, RNZAF - Age 23. 460hrs. 33rd op.
Navigator: NZ391379 Plt Off **Wallace Edward BUCKLEY**, RNZAF - Age 28.
Wireless Op: NZ402982 Flt Sgt **Arthur Stafford CHRISTIE**, RNZAF - Age 21. 542hrs. 35th op.
Air Bomber: NZ404587 Sgt **Clifton Robert BRAILEY**, RNZAF - Age 23. 128hrs. 2nd op.
Rear Gunner: NZ40927 Plt Off **Raymond Wickliffe John TRENGROVE**, RNZAF - Age 20.
This was the first all-New Zealand heavy bomber crew loss.

Sun 21/Mon 22 Jun 1942

NEW ZEALAND

Boating accident

3 Elementary Flying Training School, RNZAF (Harewood)

Drowned when the auxiliary yacht *Pera* he was travelling on from Lyttelton to Redcliffs broke up on the Sumner bar. His body was recovered about midnight and is buried at Christchurch. The *Pera*, with four men aboard, had safely negotiated the bar about noon under power, then ran aground on a sand spit and became stranded after the engine failed at the critical moment. Because of the receding tide and the *Pera's* very large keel it was decided to attempt to refloat her at next high tide, scheduled for 2220. The four men remained aboard in anticipation of the refloating, but shortly before high tide the weather suddenly deteriorated with a strong NW gale across the estuary. Only one survived when the yacht was swamped.

(Aircrafthand): NZ422913 AC1 **Ernest William DOBBIE**, RNZAF - Age 23.

MIDDLE EAST

Raid on Benghazi, Libya

148 Squadron, RAF (LG 106 [13 miles east of Daba], Egypt - 236 Wing, 205 Group)

Wellington IC ES990 - took off at 0005 captained by Wg Cdr D A Kerr, DSO, RAF, and blew up over the target area at 0210 when hit by flak. The 2nd pilot was the sole survivor. Blown out of the aircraft, he woke up on the ground about 5 miles east of Benghazi sometime later with an open parachute beside him and was captured. His five comrades are commemorated on the Alamein Memorial.

Navigator: NZ405369 Flt Sgt **Donald William CROSS**, RNZAF - Age 24. 332hrs. 11th op.
WOpAG: NZ404052 Flt Sgt **George BOOTH**, RNZAF - Age 30.

Mon 22 Jun 1942

MIDDLE EAST

Accidental explosion

203 Squadron, RAF (LG 39 [Burgh el Arab South], Egypt - 235 Wing, 201 Group)

An organised bathing party from the Squadron found a metal float on the beach and loaded it on the truck taking them back to base. The 'float', later identified as a German explosive sweep cutter, detonated as the vehicle entered the camp. Five airmen died and are buried in a collective grave at El Alamein.

WOpAG: 47580 Plt Off **Robert Douglas LEWIS**, DFM, RAF - Age 25.
English-born Lewis came to New Zealand at an early age. Upon completion of his education he served as a cadet with the Shaw Savill and Albion Line and enlisted with the RAF while in England during August 1939. His brother, William Radcliffe Lewis, died on 25 August 1944 while flying with 180 Sqn, RAF.

Mon 22/Tue 23 Jun 1942

BOMBER COMMAND

Raid on Emden, Germany (by 227 aircraft - 6 lost)

57 Squadron, RAF (Feltwell, Norfolk - 3 Group)

Wellington III X3758 - took off at 2330 and believed to have been brought down over the North Sea off the Frisian Islands, killing the five crew. The captain and navigator are commemorated on the Runnymede Memorial, while the others were buried at Wittmund, Germany, but later reinterred at Sage, 24km south of Oldenburg.

Captain: NZ41338 Sgt **James Beattie LARKINS**, RNZAF - Age 20. 253hrs. 4th op.

61 Squadron, RAF (Syerston, Nottinghamshire - 5 Group)

Lancaster I R5517 - took off at 2340 captained by Sqn Ldr I G A McNaughton, RCAF, on same raid as the above, and brought down over Germany or into the sea offshore, killing the seven crew. An RAF Sgt is commemorated

on the Runnymede Memorial. A New Zealand air gunner, possibly with some or all of the five others, was buried at Wittmund, but all six were later reinterred at Sage, 24km south of Oldenburg.
Air Gunner: NZ401440 Plt Off **James Day McKEOWN**, RNZAF - Age 22. 210hrs.

Tue 23 Jun 1942

MIDDLE EAST

Fighter sweep over Fort Maddelena-Sidi Omar, Libya
238 Squadron, RAF (LG 76 & LG 07 [Matruh West], Egypt - 243 Wing, 211 Group)
Hurricane IIB BH233 - took off from LG07 at 1020 with three others and one of two which failed to return, both believed to have been brought down by flak. The pilot of BH233 is buried at Tobruk. The Squadron was moving from LG 76 to LG 07 this day, the former being located at 3110N:2605E, about 30 miles SSE of Sidi Barrani, and LG 07 about 23 miles west of Mersa Matruh Square.
Pilot: NZ405273 Fg Off **Norman Robert HOLMES**, RNZAF - Age 22. 230hrs.

ARMY CO-OPERATION COMMAND

Navigation practice
1 Coastal Artillery Co-operation Unit (Detling, Kent - 35 Wing)
Spitfire IIB P8531 (a presentation machine named *Who's a-fear'd*) - took off at about 1140 and a few minutes later stalled off a turn and spun into the ground at Camer Farm NE of Meopham. The pilot is buried at Maidstone.
Pilot: NZ413101 Plt Off **David Lindsay MALLOCH**, RNZAF - Age 21. 268hrs (55 on Spitfire)

Tue 23/Wed 24 Jun 1942

BOMBER COMMAND

'Gardening' - minelaying off coast from Frisian Islands to Lorient, France (by 52 aircraft - 3 lost)
103 Squadron, RAF (Elsham Wolds, Lincolnshire - 1 Group)
Wellington IC T2921/T - took off at 2331 captained by Sgt T W B Emmott, RAF, and believed to have been shot down by a night-fighter near Terschelling Island at either 0112 or 0146, crashing into the sea. Two crew members were lost without trace and are commemorated on the Runnymede Memorial, while the body of another came ashore on the German coast. The other two, including a New Zealander, washed ashore onto the Dutch island of Ameland and are buried at Nes.
WOpAG: NZ404051 Sgt **Albert Frank BISSELL**, RNZAF - Age 27. 48hrs.

Wed 24/Thu 25 Jun 1942

MIDDLE EAST

Raid on Benghazi, Libya
37 Squadron, RAF (LG 09 [Bir Koralyim], Egypt - 231 Wing, 205 Group)
Wellington IC DV643 - captained by Flt Lt C P T Halliwell, DFC, RAF, and lost without trace, its six crew being commemorated on the Alamein Memorial.
2nd Pilot: NZ405239 Plt Off **Mervyn Charles DE CLIFFORD**, RNZAF - Age 21. 392hrs. 14th op.

Thu 25 Jun 1942

MIDDLE EAST

Aerodrome and road strafing
272 Squadron, RAF (LG 229 [Idku], Egypt - detachment at LG 104 [Qotafiya II/El Daba] - Western Desert)
Beaufighter IC T4981/C - believed to have taken off from LG 104 at 1815 with eight others and was hit by flak at 1905 while strafing in the Mersa Matruh area. With its starboard engine on fire, T4981 apparently tried to maintain height in order to reach the sea and ditch away from enemy forces. However, it hit dunes above the beach (at grid reference 633341) and blew up. The two crew were buried beside the wreckage, but later reinterred at Halfaya Sollum in a joint grave.
Pilot: 87339 Fg Off **James Edward MAKGILL**, RAFVR - Age 28. 483hrs.

FIGHTER COMMAND

Training - climb to 18,000 feet
55 Operational Training Unit, RAF (Annan, Dumfriesshire - 81 Group)
Hurricane I N2666 - took off at about 0910, but towards the end of the exercise at 1030 struck trees and crashed at Easton, to the SE of Drumburgh, Cumberland, about 9 miles SE of the base. The pilot died is buried at Annan.
Pilot: NZ39979 Sgt **Sydney ALCOCK**, RNZAF - Age 23. 190hrs (105 solo - 8 on Hurricane)

Thu 25/Fri 26 Jun 1942

BOMBER COMMAND

Raid on Bremen, Germany (by 960 aircraft - 50 lost)
24 Operational Training Unit, RAF (Honeybourne, Gloucestershire - 92 Group)
Whitley V BD379 - took off at 2220 on the raid and was last heard from at 0411 signalling that it was going to ditch. BD379's position at the time was calculated to be NNW of the Dutch island of Terschelling. An RAAF crew

member is commemorated on the Runnymede Memorial, while the bodies of his four comrades later washed ashore onto Terschelling and were buried at West-Terschelling.
Captain: NZ403970 Fg Off **James Brian MONRO**, RNZAF - Age 21. 622hrs.
Monro had completed an operational tour with 51 Sqn, RAF, and was serving as an instructor with the OTU.

26 Operational Training Unit, RAF (Wing, Buckinghamshire - 92 Group)
Wellington IC DV721/N - took off at 2240 on same raid as the above, and brought down over Germany, crashing near Brunsbüttelkoog, to the east of Brunsbüttel. All five crew were buried in the Brunsbüttelkoog Cemetery, but later reinterred at Kiel.
Captain: NZ403997 Wt Off **Colin John STIRLING**, DFM, RNZAF - Age 24. 414hrs. 33rd op(?)
Stirling had completed an operational tour with 150 Sqn, RAF, and was serving as an instructor with the OTU.

11 Operational Training Unit, RAF (Bassingbourn, Cambridgeshire - 91 Group)
Wellington IC X3213/L - took off at 2312 captained by Flt Sgt R Holden, RAF, on same raid as above, and brought down by flak over Germany, crashing near Aschhauserfeld, to the NE of Bad Zwischenahn. All five crew were all buried in the Oldenburg New Cemetery on 1 July, but later reinterred at Sage, 24km south of Oldenburg.
WOpAG: NZ41601 Sgt **Leonard Allen MITCHELL**, RNZAF - Age 20.
Rear Gunner: NZ403571 Sgt **Victor Arthur Bullecourt FALK**, RNZAF - Age 25. 158hrs. 1st op.

11 Operational Training Unit, RAF (Bassingbourn, Cambridgeshire - 91 Group)
Wellington IC DV778/A - took off at 2314 captained by Sgt J C Cubitt, RAF, on same raid as the above, and brought down into the sea in or near the Heligoland Bight. The captain and navigator **Flower** are commemorated on the Runnymede Memorial. The body of navigator **Parish** later washed ashore near Cuxhaven and was buried there, but later reinterred at Becklingen, 17km south of Soltau. The New Zealand wireless operator-air gunner was recovered from the harbour entrance at Westeraccumersiel, opposite the island of Langeoog, and buried at Wittmund alongside the body of an RAF crew member, but both were later reinterred at Sage, 24km south of Oldenburg
Navigator(?): NZ412472 Sgt **William Harvey FLOWER**, RNZAF - Age 22. 75hrs. 1st op.
Navigator(?): NZ41546 Sgt **Richard John PARISH**, RNZAF - Age 23. 230hrs. 1st op.
WOpAG: NZ403631 Sgt **Stanley Lyell WILTON**, RNZAF - Age 32. 107hrs. 1st op.
Two navigators in the one crew was unusual but probably came about through trying to make up the numbers for this '1000 bomber' raid. One of them almost certainly would have acted as air bomber.

11 Operational Training Unit, RAF (Bassingbourn, Cambridgeshire - 91 Group)
Wellington IC R1078/Q - took off at 2316 captained by Plt Off S G King, RCAF, on same raid as the above, and shot down over Germany by a night-fighter, crashing near Schale, 21km NE of Rheine. The captain and air gunner Sgt **W N Hollands**, RNZAF, successfully baled out and became prisoners. Their three comrades were buried in the Konigseschstrasse Cemetery at Rheine, but later reinterred at Reichswald Forest, 5km SW of Kleve
WOpAG: NZ41564 Sgt **Frederick George BARBER**, RNZAF - Age 23.
WOpAG: NZ411100 Sgt **Ivan Harry REEVES**, RNZAF - Age 23. 99hrs. 2nd op.

50 Squadron, RAF (Swinderby, Lincolnshire - 5 Group)
Manchester I L7289 - took off at 2320 on same raid as the above, and hit by flak over the target area, crashing into a field at Grambke, to the NE of Bremen. All seven crew were buried in Bremen's Waller Cemetery, but later reinterred at Becklingen, 17km south of Soltau.
Captain: NZ41361 Sgt **John Charles ROY**, RNZAF - Age 21. 416hrs. 15th op.

83 Squadron, RAF (Scampton, Lincolnshire - 5 Group)
Lancaster I R5620/H - took off at 2348 on same raid as the above, and brought down over Germany, crashing by a large ammunition dump near Winkelsett, about 30km SW of Bremen. The captain, and possibly four or more of the others were buried at Vechta, 22km SW of Winkelsett, but all seven crew were later reinterred at Sage, 24km south of Oldenburg.
Captain: NZ403119 Plt Off **James Ralph FARROW**, RNZAF - Age 20. 495hrs. 22nd op.

12 Operational Training Unit, RAF (Chipping Warden, Northamptonshire - 91 Group)
Wellington IC R1410/M - captained by Sgt J T Shapcott, RAF, on same raid as the above, and brought down into the North Sea, killing the five crew. The bodies of two RAF members were later washed ashore onto the Dutch islands of Ameland and Texel, respectively. Their three comrades are commemorated on the Runnymede Memorial.
WOpAG: NZ411797 Sgt **Frederick Robert WEST**, RNZAF - Age 26. 109hrs. 1st op(?)

ARMY CO-OPERATION COMMAND

Intruder operations in support of the '1000 bomber' raid on Bremen (by 56 aircraft - 2 lost)
13 Squadron, RAF (Odiham, Hampshire - 36 Wing)
Blenheim IV T2254/A - took off at 2350 from Wattisham, Suffolk, piloted by Fg Off P H Looker, RAF, to attack the Dutch airfield of Venlo and shot down over Belgium by a night-fighter, crashing at Aartselaar, a few kilometres south of Antwerp. All three crew are buried at Antwerp. The aircraft was one of 15 which had been detached from Army Co-operation Command to operate under the control of Bomber Command's 2 Group in support of the night's raid on Bremen.
Navigator: NZ404496 Flt Sgt **William George O'NEILL**, RNZAF - Age 21. 365hrs. 2nd op.

Fri 26 Jun 1942

FLEET AIR ARM - INDIAN OCEAN

Ferrying from RN Air Station Cochin to RN Aircraft Repair Yard at Coimbatore, India

Cochin Storage Section, FAA (India - HMS *Lanka*)

Albacore I T9169 - lost without trace en route, its pilot is commemorated on the RNZN Memorial at Devonport, Auckland. In the poor weather prevailing the only possible route was said to be via a two mile-wide gap at Palghat, between the Nilgiris and Annamallie mountains.

Pilot: Sub-Lt (A) **John Allnutt PICKARD**, RNZNVR - Age 24.

Pickard was on the strength of the FAA Pilot Pool at China Bay (Trincomalee), Ceylon, at the time of his death. It will be recalled that he was injured during an air raid on London on the night of 16/17 April 1941.

Fri 26/Sat 27 Jun 1942

MIDDLE EAST

Bombing over the Battle Area

40 Squadron, RAF (Shallufa, Egypt - 205 Group)

Wellington IC Z9028(?)/V - brought down over Egypt, crashing just SE of the Siwa Road near the railway line 36 miles WSW of Mersa Matruh (at map ref 713335 Egypt & Cyrenaica 1/250,000 Sheet 5). All six crew were buried in a collective grave near the wreckage, but later reinterred at Halfaya Sollum.

Captain: NZ412680 Sgt **Stanley Hearne GUNN**, RNZAF - Age 29. 330hrs. 5th op.

Bombing convoys between Mersa Matruh, Egypt, and Benghazi, Libya

70 Squadron, RAF (LG 224 [Cairo West, Egypt] - 231 Wing, 205 Group)

Wellington IC Z9102/A - captained by Sgt F V Clayton, RAF, and shot down by a night-fighter shortly before 0045, crashing near El Daba, Egypt. Four crew members successfully baled out and were returned to their base after being picked up at about 0700 by an Army truck. The front gunner and captain both died, a report later being received that the former's body had been found and been buried some 2 miles south of El Daba. However, after hostilities no grave for either airman could be located and both are commemorated on the Alamein Memorial.

Front Gunner: NZ404048 Flt Sgt **Maurice Lloyd BERG**, RNZAF - Age 23. 450hrs. 15th op.

Sat 27 Jun 1942

ARMY CO-OPERATION COMMAND

Training or practice

41 Operational Training Unit, RAF (Old Sarum, Wiltshire - satellite at Oatlands Hill - 70 Group)

Tomahawk IIB AK159 - at Oatlands Hill one of the Unit's Mustang Is (AG396), its pilot undertaking his first solo on the type, swung on take off at 1340 and collided with the stationary AK159 which was waiting to begin its sortie. The Mustang pilot suffered slight injuries and a rigger on the ground was more seriously hurt. The Tomahawk pilot, however, died and is buried at Salisbury.

Pilot: NZ414251 Plt Off **Alfred Graham CLARK**, RNZAF - Age 32. 233hrs.

Sun 28 Jun 1942

BOMBER COMMAND

Night flying test

11 Operational Training Unit, RAF (Bassingbourn, Cambridgeshire, satellite at Steeple Morden, Hertfordshire - 91 Group)

Wellington IC R1445 - took off from Steeple Morden at 1750 and immediately suffered an engine failure. While carrying out a circuit preparatory to making an emergency landing R1445 lost speed in a turn, stalled and crashed into houses near Ashwell, 2 miles SW of Steeple Morden. The captain and the only other occupant on board, a ground crew member, died, the former being buried in the Bassingbourn cum Kneesworth Cemetery.

Captain: NZ403425 Plt Off **John Edwin CASEY**, DFC, RNZAF - Age 23. 617hrs (547 solo - 450 on Wellington)

Sun 28/Mon 29 Jun 1942

MIDDLE EAST

Bombing attack over battle area

104 Squadron, RAF (LG 213 [Kabrit], Egypt - 236 Wing, 205 Group)

Wellington II Z8646 - shortly after take off at about 2350 suffered a starboard engine failure through a glycol leak. In preparation for an emergency landing the bomb load was jettisoned from 600-700 feet. They were fused for instantaneous detonation and the blast apparently caused Z8646 to dive straight into the Great Bitter Lake. There was only one survivor from the six crew, his comrades being buried at Fayid.

Captain: NZ402137 Flt Lt **Desmond John HAYDON**, RNZAF - Age 21. 456hrs. 23rd op.

2nd Pilot: NZ411416 Sgt **Ronald Frank LANDER**, RNZAF - Age 21.

BOMBER COMMAND

Circuits and landings

11 Operational Training Unit, RAF (Bassingbourn, Cambridgeshire, satellite at Steeple Morden - 91 Group)

Wellington IC X3173 - took off from Steeple Morden captained by Sgt J Elliott, RAF. The port engine reduction gear casing cut through, leading to loss of the propeller. While attempting to lower the undercarriage by emergency means X3173 orbited the airfield, but on the third circuit at 0410 stalled, crashed and caught fire near Ashwell, 2 miles to the SW. Two of the crew survived injured, including wireless operator-air gunner Plt Off **D P Wadey**, RNZAF. The three others died, two New Zealanders among them being buried in the Bassingbourn cum Kneesworth Cemetery.

Navigator(?): NZ405217 Sgt **Sydney ARMORY**, RNZAF - Age 26. 206hrs.
Navigator(?): NZ404361 Sgt **Keith Alexander HENDRY**, RNZAF - Age 25.
One of the two navigators was probably acting as air bomber for this flight.

Mon 29/Tue 30 Jun 1942

MIDDLE EAST

Bombing LG 05 [Sidi Barrani], Egypt

70 Squadron, RAF (Abu Sueir, Egypt - 231 Wing, 205 Group)

Wellington IC Z1170/O - captained by Flt Sgt J H Gunyon, RCAF, crashed and exploded on take off. Rear gunner Sgt **C N Harrop**, RNZAF, was the only one of the six crew to survive. His comrades are buried at Ismailia.

2nd Pilot: NZ41957 Flt Sgt **Herbert Bernard STEVENS**, RNZAF - Age 22.
Navigator: NZ405457 Flt Sgt **William Harry ROWLES**, RNZAF - Age 20. 275hrs. 5th op.
Front Gunner: NZ41587 Sgt **Selwyn Leonard HARRISON**, RNZAF - Age 21.

BOMBER COMMAND

Raid on Bremen, Germany (by 253 aircraft - 14 lost)

75 (NZ) Squadron, RAF (Feltwell, Norfolk - 3 Group)

Wellington III Z1616/D - took off at about 2320 captained by Plt Off R Bertram, RAF, but crashed and caught fire five minutes later near Red House Farm at Methwold, 3 miles NE of the base. The five crew died, all but the captain and navigator being buried at Feltwell.

Air Bomber: NZ404084 Sgt **Norman MITCHELL**, RNZAF - Age 25. 277hrs. 4th op.
Wireless Op: NZ404026 Sgt **Richard John GRENFELL**, RNZAF - Age 22. 192hrs. 14th op.

75 (NZ) Squadron, RAF (Feltwell, Norfolk - 3 Group)

Wellington III X3539 - took off at approximately 2325-2359 on same raid as the above, and shot down by a night-fighter over the Dutch coast, crashing into the Waddenzee between Ameland and the mainland. Four bodies were later recovered and buried at Holwerd, but that of the captain was lost without trace and he is commemorated on the Runnymede Memorial.

Captain: NZ411432 Plt Off **Walter Jack MONK**, RNZAF - Age 24. 326hrs. 11th op.
Navigator: NZ405454 Sgt **Douglas Haig RANDLE**, RNZAF - Age 24. 301hrs. 5th op.
Air Bomber: NZ411075 Sgt **John Gordon George HEGAN**, RNZAF - Age 23. 167hrs. 7th op.
Wireless Op: NZ411087 Sgt **Ernest Francis Sydney MONCRIEF**, RNZAF - Age 25. 152hrs. 10th op.
Rear gunner: NZ411079 Sgt **Murdoch Gordon McGREGOR**, RNZAF - Age 23. 178hrs. 6th op.

7 Squadron, RAF (Oakington, Cambridgeshire - 3 Group)

Stirling I N3706/S - took off at 2350 captained by Flt Sgt M G Bailey, RCAF, on same raid as the above, and brought down into the sea off the German island of Borkum at 0233, possibly after colliding with a night-fighter. Three of the crew, including the captain, survived to be taken prisoner. Three bodies were later recovered, two being buried in the Netherlands and the other in Germany. Of the navigator and eighth member no trace was found and they are commemorated on the Runnymede Memorial.

Navigator: NZ404543 Flt Sgt **Hugh Arthur Thomas WATSON**, RNZAF - Age 32. 322hrs. 6th op.
Watson's brother, Arthur Ernest Watson, died a day short of a year later while flying with 70 Sqn, RAF.

Thu 2 Jul 1942

FLYING TRAINING COMMAND

Training in wind speed and direction finding

14 (Pilots) Advanced Flying Unit, RAF (Ossington, Nottinghamshire - 21 Group)

Oxford II ED154 - spun into the ground from low altitude at about 1000 after structural failure and caught fire a $^1/_2$ mile south of Carlton on Trent, $2^1/_2$ miles east of the airfield. All three crew died, the pilot being buried at Orpington, Kent. Examination of the wreckage suggested that the gluing of the wooden structure by the manufacturer was sub-standard.

Pilot: NZ413489 Sgt **Frederick George SHEARS**, RNZAF - Age 21. 226hrs.

Thu 2/Fri 3 Jul 1942

MIDDLE EAST

Attack on enemy concentrations in the battle area around El Daba, Egypt

37 Squadron, RAF (Abu Sueir, Egypt - 231 Wing, 205 Group)

Wellington IC Z1179 - shot down from 2500 feet by a night-fighter and crashed approximately 10-15 miles west of the railhead at El Daba, 3-5 miles inland. The sole survivor was an RAF crew member who successfully baled out but was captured. His five comrades are commemorated on the Alamein Memorial.

2nd Pilot: NZ41886 Plt Off **Graham Lucas DONNE**, RNZAF - Age 26. 288hrs. 6th op.

BOMBER COMMAND

Raid on Bremen, Germany (by 325 aircraft - 16 lost)

12 Squadron, RAF (Binbrook, Lincolnshire - 1 Group)

Wellington II Z8579 - took off at 2335 captained by Sgt I C Dunn, RAAF and shot down over the Netherlands by a night-fighter, crashing at 0429 near Barchem, 17km east of Zutphen. Two RAF crew members survived to be taken prisoner, their three comrades being buried at Barchem.

Navigator: NZ402848 Sgt **Geoffrey Hugh BUTLER-WILLIAMS**, RNZAF - Age 24. 296hrs. 6th op.

Intruder attack on airfield at St Trond, Belgium

18 Squadron, RAF (Wattisham, Suffolk - 2 Group)

Blenheim IV Z7372 - took off at 0020 piloted by Sgt W E Furness, RAF, and lost without trace, its three crew being commemorated on the Runnymede Memorial.

Navigator: NZ411725 Plt Off **Robin William MORTON**, RNZAF - Age 27. 257hrs. 3rd op.

Fri 3 Jul 1942

MIDDLE EAST

Scramble over El Alamein area

213 Squadron, RAF (LG 154 [Amiriya area], Egypt - 243 Wing, 211 Group)

Hurricane IIB BN597 of 'A' Flight - took off at 1535 with four others, only one of which returned. Enemy fighters intercepted the Flight en route and shot down four of the Hurricanes, killing three pilots and wounding another. BN597 fell in the El Hamman area, where its pilot was buried, but later reinterred at El Alamein.

Pilot: NZ403502 Wt Off **Gordon Stuart BOLTON**, RNZAF - Age 25. 360hrs.

Attack on convoy off coast of Greece

217 Squadron, RAF (Luqa, Malta)

Beaufort I L9893 - took off at 1830 piloted by Sgt J Hutcheson, RAF, with two others of the Squadron, the three being led by a 39 Sqn Beaufort. During the attack south of Zante island two 217 Sqn aircraft were shot down into the sea by flak. Both crews were lost without trace, all eight airmen being commemorated on the Malta Memorial.

WOpAG: NZ404713 Sgt **Frank Stuart WEAVER**, RNZAF - Age 26.

Fri 3/Sat 4 Jul 1942

BOMBER COMMAND

'Gardening' - mining in the Great Belt area (by 6 aircraft - 2 lost)

61 Squadron, RAF (Syerston, Nottinghamshire - 5 Group)

Lancaster I R5488/F - took off at 2225 to lay mines in the Danish Sound off Verbena and presumed to have been brought down into the North Sea off the Danish coast. Two of the crew are commemorated on the Runnymede Memorial. The bodies of their five comrades later washed ashore along the Danish coast, including that of the captain, which was recovered near Rvidesande and is buried at Fourfelt near Esbjerg.

Captain: 40870 Sqn Ldr **Peter Wynn Mason WEST**, DFC, RAF - Age 25. 1140hrs. 52nd op.

West's brother, David George West, died on 23 February 1942 while flying with 41 OTU, RAF.

Sun 5 Jul 1942

FIGHTER COMMAND

Training

53 Operational Training Unit, RAF (Llandow, Glamorgan - 81 Group)

Spitfire IIB P8592 - collided with Spitfire IIA (P7504) of the same unit at 1045 and crashed at St Donat's, $5^1/_2$ miles SW of Cowbridge. Both pilots died, that of P8592 being buried at Llantwit Major.

Pilot: NZ413117 Sgt **Russell Stanley PEARCE**, RNZAF - Age 20. 241hrs.

Tue 7 Jul 1942

FIGHTER COMMAND

Dual check landings and aerobatics

55 Operational Training Unit, RAF (Annan, Dumfriesshire - 81 Group)

Master I T8411 - piloted by New Zealander, Act Flt Lt **D Rockel**, RAF, and at 1900 entered a dive from 6000 feet to demonstrate a loop. In the vertical position the port wing broke away, causing T8411 to carry out two upward flick rolls, stall and then enter a flat spin. Prior to baling out, the instructor saw the pupil pilot open his hood. Rockel, although injured, survived, but his pupil failed to bale out in time and died as the Master crashed at Kelsick, a mile west of Abbey Town, Cumberland. He is buried at Hazelrigg, 15 miles NW of Alnwick, Northumberland.

2nd Pilot: NZ412287 Sgt **Joel TREISTER**, RNZAF - Age 24. 222hrs (128 solo - 13 on Master)

Tue 7/Wed 8 Jul 1942

MIDDLE EAST

Attack on battle area, Egypt

70 Squadron, RAF (Abu Sueir, Egypt - 231 Wing, 205 Group)

Wellington IC Z8986/T - took off captained by Sgt J R Milligan, RCAF, but when at 75ft lurched downwards, striking the ground and burning out. Two crew survived, but the other four, including three New Zealanders, died and are buried at Ismailia.

Navigator: NZ41557 Plt Off **Reginald Francis WATSON**, RNZAF - Age 25. 291hrs.
Wireless Op: NZ404622 Sgt **Raymond Jeffery SMART**, RNZAF - Age 25.
Front Gunner: NZ41586 Sgt **Frank Owen Holder HARRISON**, RNZAF - Age 25.

Wed 8 Jul 1942

FIGHTER COMMAND

Formation flying practice

55 Operational Training Unit, RAF (Annan, Dumfriesshire - 81 Group)

Hurricane I Z7077 - took off at 0830 with two others for the exercise. About an hour later Z7077 collided with one of the other Hurricanes (AG125). They locked together and crashed near Drumburgh, Cumberland, 8 miles WNW of Carlisle. Both pilots are buried at Annan.

Pilot: 1375959 Sgt **Te Roki ROBIESON**, RAFVR - Age 19.

Wed 8/Thu 9 Jul 1942

BOMBER COMMAND

Cross country training to Port Douglas, Isle of Man, and return

19 Operational Training Unit, RAF (Kinloss, Morayshire - 91 Group)

Whitley V AD707 - took off at about 2315 and at 0340 was seen to crash in flames into the Sound of Jura, 7 miles south of Eilean a' Chuirn, Isle of Islay. All five crew died, the bodies of two later being recovered from the sea, while another washed ashore onto Rathlin Island, Northern Ireland. The captain and other member are commemorated on the Runnymede Memorial. A considerable amount of static activity was noted at the time of the accident and it was thought likely that the Whitley encountered lightning.

Captain: NZ404445 Wt Off **Edward Owen JOHNSON**, RNZAF - Age 27. 380hrs (125 solo - 36 on Whitley) 3 ops.

Raid on Wilhelmshaven, Germany (by 285 aircraft - 5 lost)

75 (NZ) Squadron, RAF (Feltwell, Norfolk - 3 Group)

Wellington III X3557 - took off at approximately 2340-0010 and shot down over the Dutch coast by a night fighter at 0130, crashing into the Waddenzee near the island of Rottumeroog. Four of the crew are commemorated on the Runnymede Memorial. The captain's body later washed ashore onto the island and is buried there. On 19 August, the air bomber's body washed ashore further to the SW, onto the island of Schiermonnikoog, and was buried there, but later reinterred on the mainland at Oldebroek, 15km SW of Zwolle.

Captain: NZ41953 Plt Off **Trevor Harry SMITH**, RNZAF - Age 24. 355hrs. 13th op.
2nd Pilot: NZ412267 Plt Off **Donald Norman POTTS**, RNZAF - Age 25. 286hrs. 1st op.
Navigator: NZ405367 Plt Off **Alan Stewart BROUN**, RNZAF - Age 32. 332hrs. 10th op.
Air Bomber: NZ405771 Sgt **George Anthony YOUNG**, RNZAF - Age 21. 138hrs. 15th op.
Wireless Op: NZ391085 Sgt **Frank Edwin McKENZIE**, RNZAF - Age 22. 225hrs. 15th op.
Rear Gunner: NZ402128 Plt Off **Jack Ralph GAVEGAN**, RNZAF - Age 30. 320hrs. 68th op.

Sixty of Gavegan's ops were undertaken in Defiant fighters prior to his joining Bomber Command.

Fri 10 Jul 1942

FERRY COMMAND

Delivery flight from Portreath, Cornwall, to the Far East

1444 Ferry Training Flight, RAF (Lyneham, Wiltshire - 44 Group)

Hudson IIIA FH319 - took off at 0711 from Gibraltar, piloted by Sgt W J Aitken, RAF, on the second leg of the journey, bound for Bathurst, Gambia. A few minutes later FH319 crashed into the sea a mile west of the airfield, killing all on board. The captain and one crew member are commemorated on the Malta Memorial. The body of the wireless operator is buried at Gibraltar, while that of the fourth member later washed ashore in Spain and is buried at Cadiz.

Wireless Op: NZ404604 Sgt **Richard John Lucas JONES**, RNZAF - Age 23.

Jones had been posted to the Far East on 2 July and was making his way there by means of the delivery flight.

MIDDLE EAST

Patrol

80 Squadron, RAF (LG 92 [7 miles SSE of Markaz Amriya] - 244 Wing, 211 Group, Western Desert)

Hurricane IIC BP503 - took off at 1330 and during an interception of enemy aircraft by the Squadron was shot down. Three days later the pilot's body washed ashore close to the Egyptian-Libyan border near Sollum, where it was buried by personnel of the 1st South African Division. A search for his grave after hostilities was unsuccessful, all trace having been obliterated, and he is now commemorated on the Alamein Memorial.

Pilot: NZ41373 Sgt **Ralph Stuart SYKES**, RNZAF - Age 23. 250hrs. 26th op.

BOMBER COMMAND

Cloud cover raids on Düsseldorf and Duisburg, Germany (by 8 aircraft - 1 lost)

75 (NZ) Squadron, RAF (Feltwell, Norfolk - 3 Group)

Wellington III X3720 - took off at approximately 1447-1530 to raid Düsseldorf. The aircraft on both raids were later recalled, all but X3720 safely returning. The lost bomber was brought down into the sea off the German-Netherlands coast. Three of the crew are commemorated on the Runnymede Memorial. The bodies of the wireless operator and air bomber washed ashore a few days later onto the German island of Borkum. They were buried there in the Lutheran Cemetery on the 15th, but later reinterred at Sage, 24km south of Oldenburg.

Captain: NZ411962 Sgt **John Charles WILMSHURST**, RNZAF - Age 25. 393hrs. 25th op.
Navigator: NZ411233 Sgt **James Douglas GOULD**, RNZAF - Age 21. 364hrs. 15th op.
Wireless Op: NZ405513 Sgt **Richard Edwin SHARP**, RNZAF - Age 23. 248hrs. 15th op.
Air Bomber: NZ403583 Flt Sgt **Peter Desmond LOWTHER**, RNZAF - Age 22. 229hrs. 15th op.
Rear Gunner: NZ404067 Sgt **Raymond John Finlay HIRST**, RNZAF - Age 22. 279hrs. 15th op.

Sat 11 Jul 1942

FLYING TRAINING COMMAND

Solo practice - precautionary landings and steep turns

2 (Pilots) Adavanced Flying Unit, RAF (Brize Norton, Oxfordshire - 23 Group)

Oxford II V3510 - took off at 2025 and collided at 75ft with another of the Unit's Oxfords (BF948) which had taken off on a converging course. The RCAF pilot of BF948 survived, injured, but the pilot of V3510 died and is buried at Black Bourton.

Pilot: NZ413496 Sgt **John Alexander SMITH**, RNZAF - Age 19. 169hrs (94 solo - 4 on Oxford)

Sat 11/Sun 12 Jul 1942

BOMBER COMMAND

Raid on U-boat yards at Danzig, Poland (by 44 aircraft - 2 lost)

97 Squadron, RAF (Woodhall Spa, Lincolnshire - 5 Group)

Lancaster I R5696/H - took off at 1642 and shot down by flak near the target area. The seven crew are buried in a collective grave at Malbork, 12km SE of Danzig. Danzig was the most distant target yet attacked by the command and involved a round trip of 1500 miles.

Captain: 36270 Flt Lt **Charles William MILLER**, DFC, RAF - Age 24. 1336hrs. 20th op.

Mon 13 Jul 1942

FIGHTER COMMAND

'Circus 199' - escort to 12 Bostons raiding Boulogne, France

602 Squadron, RAF (Redhill, Surrey - 11 Group)

Spitfire VB BM182 - took off at 1415 with 11 others and was last seen some 10 to 15 miles off Berck-sur-Mer, 37km NW of Abbeville, when about 25 enemy fighters were pursuing the Squadron. BM182 had fallen behind the others and was presumed to have been shot down into the English Channel. The pilot is commemorated on the Runnymede Memorial.

Pilot: NZ40769 Fg Off **Edgar Mostyn INNES-JONES**, RNZAF - Age 30. 971hrs.

Mon 13/Tue 14 Jul 1942

BOMBER COMMAND

Raid on Duisburg, Germany (by 194 aircraft - 10 lost)

97 Squadron, RAF (Woodhall Spa, Lincolnshire - 5 Group)

Lancaster I R5558 - after bombing, ditched in the sea at about 0430 four miles east of Wells-next-the-Sea, Norfolk. A lifeboat reached the floating wreckage several hours later, but an RAF airman was the only survivor of the seven crew. The others either died in the crash or shortly afterwards, two being lost without trace and commemorated on the Runnymede Memorial. The bodies of the four others were later recovered and three, including that of the captain, are buried at Great Bircham, 12 miles SW of Wells. The raiding force experienced electrical storms and cloud down to 300 feet, six bombers being lost over England or close to the coast.

Captain: NZ405221 Fg Off **John Thomas BLEASE**, RNZAF - Age 24. 398hrs. 9th+ op.

Tue 14/Wed 15 Jul 1942

MIDDLE EAST

Raid on Tobruk, Libya

104 Squadron, RAF (LG 213 [Kabrit], Egypt - 236 Wing[?], 205 Group)

Wellington II Z8649 - took off at about 2145 captained by Plt Off G Richards, RAF, being one of three which failed to return. At about 0200 was hit by flak and set on fire over the target area. It gradually became uncontrollable and at 8000 feet the six crew were forced to abandon the aircraft over the sea, about 20 miles east of Tobruk. Afterwards no trace could be found of the New Zealand wireless operator-air gunner and he is commemorated on the Alamein Memorial. His five comrades were captured, but the RAAF rear gunner died four days later and is buried at Tobruk.

WOpAG: NZ403573 Flt Sgt **Gordon McDonald FROSTICK**, RNZAF - Age 27. 193hrs.

104 Squadron, RAF (LG 213 [Kabrit], Egypt - 236 Wing[?], 205 Group)

Wellington II Z8658 - took off at 2146 captained by Plt Off R C Horton, RAF, on same raid as the above. While returning an engine began to falter and on nearing Sidi Barrani failed completely, resulting in a gradual loss of height. On attempting an emergency landing in the desert about 40 miles south of El Daba, Z8658 crashed and caught fire. The RAF rear gunner managed to struggle clear of the wreckage but was unable to assist any of his five comrades owing to the intense heat. He was picked up later in the day by friendly Arabs who escorted him to an advance unit of the 7th Armoured Division. Severe sandstorms later obliterated all trace of the wreckage and the remains of the dead, all of whom are commemorated on the Alamein Memorial.

Wireless Op: NZ40793 Sgt **William Cochrane HUTCHISON**, RNZAF - Age 37.

Wed 15 Jul 1942

MIDDLE EAST

Raid on Benghazi, Libya

159 Squadron, RAF (St Jean, Palestine - 242 Wing, 205 Group)

Liberator II AL566 - seen by another bomber crew to be attacked by enemy fighters in the target area and spiral into the sea. The body of an RAAF Sgt later washed ashore near Benghazi and is buried there. His six comrades are commemorated on the Alamein Memorial.

Air Gunner: NZ402118 Flt Sgt **Martin Costin FELL**, mid, RNZAF - Age 30. 440hrs. 56th op.

FIGHTER COMMAND

Army Co-operation exercise *LIMPET*

486 Squadron, RNZAF (Wittering, Northamptonshire - 12 Group)

Hurricane IIB BD725 - after completing a dummy attack on an Army column at 1200, struck trees when pulling up in a climbing turn and dived into the ground, catching fire 5 miles SW of Sudbury, Suffolk (at map reference M2955). The pilot is buried at Wittering.

Pilot: NZ402191 Plt Off **Ian Hamilton IRVINE**, DFM, RNZAF - Age 26. 572hrs (498 solo - 99 on Hurricane) 105+ ops.

COASTAL COMMAND

Anti-shipping patrol of Trost area, Norway

48 Squadron, RAF (Wick, Caithness-shire - 18 Group)

Hudson IIIA FH378/M - took off at 1150 and shot down at 1522 by enemy fighters off the Norwegian coast. On the 25th the captain's body was recovered near the Kya lighthouse, 120km NNW of Trondheim, and buried in the Saetervolden churchyard at Sör Froya, but later reinterred at Trondheim. His three comrades are commemorated on the Runnymede Memorial.

Captain: 41458 Sqn Ldr **Varley Allen PEDERSEN**, RAF - Age 28. 1602hrs. 126th op.

Thu 16 Jul 1942

BOMBER COMMAND

Training - beam approach and local flying

149 Squadron Conversion Flight, RAF (Lakenheath, Suffolk - 3 Group)

Stirling I R9299 - at 1635, while flying at 1500 feet SW of Newmarket, Suffolk, fire broke out in the starboard outer engine and spread to the inner. At 300 feet part of the starboard wing fell away, causing R9299 to spin into the ground and explode near Swaffham Bulbeck, Cambridgeshire. All six crew died, four New Zealanders being buried at Beck Row, 2 miles NW of Mildenhall. Investigators found that a front crank pin had fractured, resulting in oil being thrown out over the engine's exhaust ring. The Conversion Flight was affiliated to 149 Sqn, also based at Lakenheath.

Captain: NZ41481 Sgt **Henry Peter JOHNSON**, RNZAF - Age 24. 297hrs (160 solo - 23 on Stirling)
Front Gunner: NZ403625 Flt Sgt **Jack Napier NICOL**, RNZAF - Age 28. 200hrs.
Air Bomber: NZ39882 Sgt **Leslie Ernest POLE**, RNZAF - Age 26. 185hrs.
Rear Gunner: NZ392015 Sgt **Malcolm Vernon CATO**, RNZAF - Age 24. 132hrs.

Sun 19 Jul 1942

MIDDLE EAST

Raid on Tobruk, Libya

159 Squadron, RAF (St Jean, Palestine - 242 Wing, 205 Group)

Liberator II AL554 - took off from Fayid, Egypt, at 1340 captained by Flt Lt N E Sharp, DFC, RAF, and was attacked by a fighter over the target, although whether damaged during the engagement is unknown. On returning to Palestine, fuel began to run out and the number one engine cut. Unable to locate St Jean, a new course was set for Lydda, to the south near Tel Aviv, but while attempting an emergency landing en route at about 2255 AL554 crashed at Hadera, some 20 miles south of Haifa. Six of the crew were killed and another died of injuries the following day, leaving a seriously injured wireless operator as sole survivor. The dead are buried at Khayat Beach, immediately south of Haifa. Two other Liberators on the raid landed safely but with engines cutting as they did so.

Air Gunner: NZ401202 Plt Off **Henry Cumberland Stapleton COTTON**, DFM, RNZAF - Age 22. 512hrs. 42nd op.

Sun 19/Mon 20 Jul 1942

MIDDLE EAST

Raid on Tobruk, Libya

40 Squadron, RAF (Shallufa, Egypt - 205 Group)

Wellington IC HX399/S - took off at 2210, but on climbing from 3000 to 6000 feet about an hour later an explosion came from the starboard engine which then caught fire. Four of the crew baled out, while the captain and RAAF wireless operator carried out a belly landing at 2315 in an isolated area 10 miles west of Wadi Natrun, about 55 miles SSE of Alexandria. On striking the ground HX399 burst into flames, fatally injuring the two crew, who died shortly afterwards. They were buried close to the aircraft next day, a senior chaplain conducting the service. Both airmen are commemorated on the Alamein Memorial, presumably because the graves were subsequently obliterated by sand storms. Among the survivors was 2nd pilot Sgt **J A Tovey**, RNZAF.

Captain: NZ41482 Flt Sgt **Peter Reginald KINGSFORD**, RNZAF - Age 22.

Kingsford's father, Alfred Reginald Kingsford, was a First World War pilot who wrote two books based on his experiences in that conflict.

Tovey died on 10 December 1942 while travelling back to Britain in a 138 Sqn aircraft after completion of his tour with 40 Sqn.

BOMBER COMMAND

'Gardening' - mine laying off coast of France (by 19 aircraft - 1 lost)

12 Squadron, RAF (Binbrook, Lincolnshire - 1 Group)

Wellington II W5414 - took off at 2201 captained by Plt Off J P Muirhead, DFC, RAF, to lay mines in the 'Beeches' area (off St Nazaire) and brought down into the sea. The bodies of the five crew were later recovered along various parts of the coastline between St Nazaire and La Rochelle, the navigator and two of his comrades being buried near the latter, at Ars-en-Ré on Ile de Ré.

Navigator: NZ402000 Flt Sgt **Eric WAPLE**, RNZAF - Age 29. 377hrs. 19th op.

Raid on Vegesack, Germany (by 99 aircraft - 5 lost)

158 Squadron, RAF (East Kirby, Lincolnshire - 4 Group)

Halifax II W1162/D - took off at 2356 captained by Sgt P J Dillon, RAF, and brought down over the coast near the German-Dutch border. Three crew members are commemorated on the Runnymede Memorial. The bodies of their four comrades were recovered, that of the air bomber and at least two others washing up on the island of Borkum to be buried in the Lutheran Cemetery there. All four were later reinterred on the German mainland at Sage, 24km south of Oldenburg.

Air Bomber: NZ405482 Sgt **Herbert Edward GODFERY**, RNZAF - Age 23. 132hrs. 2nd op.

Mon 20 Jul 1942

COASTAL COMMAND

Low flying practice over the sea

7 (Coastal) Operational Training Unit, RAF (Limavady, Co. Londonderry - 17 Group)

Wellington IC DV772 - while attempting to establish its position in poor visibility and low cloud at 1610, flew into the ground and caught fire 200 yards NE of the hotel at Giant's Causeway, Co. Antrim, 40 miles ENE of Londonderry. The two-pilot crew is buried at Limavady.

Captain: NZ413913 Plt Off **Wilson TWENTYMAN**, RNZAF - Age 26. 294hrs (15 solo on Wellington)

Tue 21 Jul 1942

MIDDLE EAST

Patrol over El Alamein

80 Squadron, RAF (LG 92 [7 miles SSE of Markaz Amriya], Egypt - 244 Wing, 211 Group)

Hurricane IIB BN462 - took off at 1845 with others of the Squadron, 238 Sqn Hurricanes providing top cover. Forty-five minutes into the patrol both squadrons were attacked by enemy fighters and BN462 was shot down.

Believed to have crashed in the Matruh area, no trace of its pilot was found and he is commemorated on the Alamein Memorial.
Pilot: NZ403989 Flt Sgt **James Douglas ROBB**, RNZAF - Age 22. 313hrs. 53rd op.

Wed 22 Jul 1942

MIDDLE EAST

Formation practice
55 Squadron, RAF (LG Y [Qassassin, 6 miles NW Abu Sueir], Egypt)
Baltimore II AG824 - while flying low over the Landing Ground at 0730, struck the surface and burst into flames. All four crew are buried at Tel-el-Kebir.
Pilot: NZ41527 Plt Off **Cecil George RUSSELL**, RNZAF - Age 26. 199hrs.

Thu 23 Jul 1942

FIGHTER COMMAND

Sweep north of Brest, France
234 Squadron, RAF (Portreath, Cornwall - 10 Group)
Spitfire VB AR285 - took off at 1520 for a sweep between Plousecat and Guissény and one of four which failed to return. Flak and light machine gun fire were met near Landerneau and enemy fighters also encountered. AR285 was presumed to have been shot down into the English Channel, its pilot being commemorated on the Runnymede Memorial.
Pilot: NZ405351 Plt Off **John Dorset WEBSTER**, RNZAF - Age 24. 406hrs. 91st op.

Thu 23/Fri 24 Jul 1942

BOMBER COMMAND

Intruder operations over enemy airfields (by 8 aircraft - 1 lost)
18 Squadron, RAF (Wattisham, Suffolk - 2 Group)
Blenheim IV Z7428 - took off at 2245 piloted by Flt Lt F M Thorne, RAF, to bomb the German airfield of Vechta. Brought down over Germany and crashed at Goldenstedt, 12km NE of Vechta. All three crew were buried in the Municipal Cemetery at Vechta, but later reinterred at Sage, 24km south of Oldenburg.
Navigator: NZ40736 Plt Off **Bryan Manssen COOK**, RNZAF - Age 22. 600hrs. 25th op.

Raid on Duisburg, Germany (by 215 aircraft - 8 lost)
207 Squadron, RAF (Bottesford, Leicestershire - 5 Group)
Lancaster I R5632/N - took off at 0048 captained by Flt Sgt W G Hawes, RAAF, and brought down into the North Sea, probably off the Dutch coast, where the body of an RCAF crew member later washed ashore. His six comrades are commemorated on the Runnymede Memorial.
2nd Pilot(?): NZ404366 Sgt **Percival James HOOPER**, RNZAF - Age 21. 358hrs. 12th op.

207 Squadron, RAF (Bottesford, Leicestershire - 5 Group)
Lancaster I R5867/T - took off on same raid as the above, and brought down about 10km SE of the target at Krefeld. All seven crew were buried in the Principal Cemetery at Krefeld, but later reinterred at Reichswald Forest, 5km SW of Kleve.
Captain: NZ401470 Plt Off **John Justin Neville McCARTHY**, RNZAF - Age 27. 404hrs. 24th op.

Circuits and landing practice
25 Operational Training Unit, RAF (Finningley, Yorkshire - 92 Group)
Wellington IC T2701/L - while carrying out a circuit after an overshoot at 0355, collided at 800-1000 feet with another of the Unit's Wellington ICs (DV476). Both crashed and caught fire a mile NW of the airfield, killing the seven crew of the former and all six in the latter. The captain of T2701 is buried at Finningley along with three others. DV476 had earlier left for a two hour practice bombing sortie but reappeared at Finningley after only 30 minutes flying. As its bombs had not been released it was assumed that the Wellington had returned with a technical fault.
Captain: NZ402158 Plt Off **Richard Colvin Seddon BECK**, DFM, RNZAF - Age 30. 793hrs (660 solo - 121 on Wellington) 36 ops.

Fri 24 Jul/Sat 25 Jul 1942

MIDDLE EAST

Raid on Heraklion airfield, Crete
37 Squadron, RAF (Abu Sueir, Egypt - 231 Wing, 205 Group)
Wellington IC AD648/T - began take off at 1545 captained by Plt Off F J Pearce, but failed to lift off, struck a sandbank and caught fire, then blew up. All five crew were injured, including the wireless operator-air gunner who died later in the day and is buried at Tel-el-Kebir.
WOpAG: NZ412365 Sgt **William Alex Wylie SUTCLIFFE**, RNZAF - Age 22. 170hrs.

Raid on Tobruk, Libya
104 Squadron, RAF (Kabrit, Egypt - 205 Group)
Wellington II Z8568 - took off at 2308 and last heard from at 0300 when signalling its position as 3102N:2730E. Believed to have been brought down over the Burg El Arab area, one RAF crew member survived to be taken prisoner. His five comrades are commemorated on the Alamein Memorial.
Air Gunner(?): Aus403874 Sgt **Monte Ralph MOREY**, RAAF - Age 31.

Sat 25 Jul 1942

NEW ZEALAND

Solo flying practice
3 Elementary Flying Training School, RNZAF (Harewood)
Tiger Moth II NZ778 - while in the circuit over the south-western part of the airfield at 0845, collided at 800 feet with another of the School's Tiger Moths (NZ1411). The two camouflaged machines locked together and crashed to the ground, and caught fire. Both pilots died, that of NZ778 being buried at Westport. The accident card records that 'in an effort to avoid future accidents of a similar nature, all Tiger Moths at training schools are to be redoped yellow and a memo has been sent out to all units on the necessity for keeping an adequate look out.'
Pilot: NZ424449 LAC **Frank Hayman GOLDSMITH**, RNZAF - Age 20. 7hrs solo (all on Tiger Moth)

3 Elementary Flying Training School, RNZAF (Harewood)
Tiger Moth II NZ1411 - collided with Tiger Moth NZ778 as described above. The pilot is buried at Wellington.
Pilot: NZ42782 LAC **James Hylton MEAD**, RNZAF - Age 19. 16hrs solo (all on Tiger Moth)

SOUTH PACIFIC

Transit from Nausori, Fiji, to Tonga
4 Squadron, RNZAF (Nadi, Fiji)
Hudson III NZ2028 - took off from Nausori at 0919 to deliver two Army officers to Tonga and failed to arrive when scheduled 3hrs 25mins later. Late in the afternoon a message was received that NZ2028 had only an hour's fuel remaining and was lost owing to deteriorating weather. Contact was made with control at Tonga, who guided the Hudson to within a few miles of its destination. Unfortunately, the Hudson was confused by the American radar controller with US Navy aircraft from a carrier in the vicinity and failed to make a landfall. The last contact was timed at 1720 and the Hudson was believed to have ditched shortly afterwards south of Tonga, within 10 miles of the airfield. The Despite extensive searches by aircraft and naval vessels no sign of it was found. The four RNZAF and two army personnel on board are commemorated on the Bourail Memorial.
Captain: NZ404324 Fg Off **Donald Arthur ANDERSON**, RNZAF - Age 25. 423 solo (169 on Hudson)
Navigator: NZ411509 Fg Off **Clive Gaby IBBOTSON**, RNZAF - Age 19.
WOpAG: NZ40805 Sgt **Roland Harvey Winch WYBOURNE**, RNZAF - Age 21. 791hrs. 88 ops.
(Flt Mechanic): NZ402587 LAC **Arthur Norman CLAYTON**, RNZAF - Age 24.
Passenger: Maj Gen **Owen Herbert MEAD**, CBE, DSO, mid(3), NZ Infantry - Age 50.
Passenger: 8866 Lt **John Christie LESLIE**, NZ Infantry - Age 35.
Most unusually, CWGC register entries for Mead and Leslie simply record their deaths as occurring in 'July 1942' and '1942', respectively. Maj Gen Mead, the highest ranking New Zealander to lose his life during the war, was General Officer Commanding the Pacific Section of 2NZEF. By strange coincidence, J H Mead, who died in the flying accident at Harewood this day, was the only other New Zealand WWII air casualty of this surname. The two men are not thought to have been related.
The CWGC register incorrectly gives Wybourne's first name as Ronald, whereas he was named after his father, Roland. Clayton was on board to carry out the daily inspection on the Hudson prior to its return to Fiji.

FERRY COMMAND

Navigation and consumption test
Overseas Aircraft Preparation Unit, RAF (Filton, Gloucestershire - 44 Group)
Beaufort I DE121 of 4 Flight- took off at about 0900 from Lyneham, Wiltshire. Declared overdue by 1700. Lyneham control set about tracing the Beaufort and found it over the North Sea 300 or more miles off course. Bearings were given to bring it back on course and at 1830 it was seen approaching Mousehold, near Norwich, Norfolk, with undercarriage lowered. With engines cutting through lack of fuel DE121 spun into the ground at Mousehold and caught fire, killing the pilot, who is buried at Norwich.
Pilot: NZ412187 Plt Off **Colin William Bidwill BARTON**, RNZAF - Age 25. 226hrs.
Barton's last known posting, on 28 May, was to 5 (C) OTU. He may have been testing the Beaufort prior to its delivery overseas, but, if so, whether or not he was scheduled to also carry out this duty is unknown.

CANADA

Instrument training
12 Service Flying Training School, RCAF (Brandon, Manitoba - 2 Training Command)
Crane I 7776 - stalled at 1335 and spun vertically into the ground 6 miles north of Justice, about 18 miles NE of the airfield. A portion of the leading edge was seen to peel off from the wing during the Crane's fall. The two pilots and an RCAF air mechanic on board died, the New Zealander being buried at Brandon.
Pilot: NZ416538 LAC **Lionel Ralph PRIDAY**, RNZAF - Age 20. 182hrs (96 on Crane)
The other pilot was LAC R E Willis, RCAF, but it is unknown who had command at the time of the crash.

Sat 25/Sun 26 Jul 1942

MIDDLE EAST

Anti-submarine patrol between Port Said and Alexandria

230 Squadron, RAF (Fanara, Egypt - 201 Group)

Sunderland I L5806/Y - took off at 1600 captained by Flt Lt E H Brown, RAF, for a patrol 30 miles off the coast. Although a message was received seven minutes after take off, during a wireless test, nothing further was heard of L5806. All 12 crew are commemorated on the Alamein Memorial.

Navigator: NZ401450 Fg Off **Albert Gordon Boswell WILLIAMS**, RNZAF - Age 25. 913hrs. 32nd op.
Navigator: NZ402129 Fg Off **Jack Edward Sedgwick BENTLEY**, RNZAF - Age 27. 848hrs.

BOMBER COMMAND

Raid on Duisburg, Germany (by 313 aircraft - 13 lost)

150 Squadron, RAF (Snaith, Yorkshire - 1 Group)

Wellington III X3795 - took off at 0035 captained by Sgt D M Horsleman, RAAF. On the return journey it was damaged by flak and suffered an engine failure, resulting in a forced landing on the sea. Five of the crew survived to be taken prisoner, but a wireless operator-air gunner went down with the ditched Wellington and is commemorated on the Runnymede Memorial.

WOpAG: Aus403650 Flt Sgt **Richard Denison BEESTON**, RAAF - Age 33.

FIGHTER COMMAND

Intruder operation to Fécamp, France

43 Squadron, RAF (Tangmere, Sussex - 11 Group)

Hurricane IIC HL863/T - took off at 0130 and brought down over France, the pilot being buried at Le Havre, 33km SW of Fécamp.

Pilot: NZ41971 Sgt **Ronald Frederick WHITTEN**, RNZAF - Age 26. 332hrs. 2nd op(?)

Sun 26/Mon 27 Jul 1942

BOMBER COMMAND

Raid on Hamburg, Germany (by 403 aircraft - 32 lost)

75 (NZ) Squadron, RAF (Feltwell, Norfolk - 3 Group)

Wellington III X3714 - took off at approximately 2235-2330 and lost without trace, its six crew being commemorated on the Runnymede Memorial.

Captain: NZ404912 Flt Sgt **Colin Valentine McPHERSON**, RNZAF - Age 21. 17th op.
2nd Pilot: NZ41359 Flt Sgt **Stanley David ROSS**, RNZAF - Age 25. 291hrs. 3rd op.
Navigator: NZ404420 Plt Off **Philip Charles SPITTAL**, RNZAF - Age 26. 274hrs. 4th op.
Air Bomber: NZ402437 Flt Sgt **Louvain Trevor CAIRNS**, RNZAF - Age 25. 175hrs. 5th op.
Wireless Op: NZ391697 Sgt **Raymond Patrick COPPERSMITH**, RNZAF - Age 21. 149hrs. 10th op.
Rear Gunner: NZ40859 Flt Sgt **Norman Albert BRYSON**, RNZAF - Age 26. 465hrs. 48th op.
Cairns brother, Vickers Kitchener Cairns, died on 23 September 1943 while flying with 6 Sqn, RNZAF.

75 (NZ) Squadron, RAF (Feltwell, Norfolk - 3 Group)

Wellington III Z1596 - took off at approximately 2235-2330 on same raid as the above, and shot down by flak while flying away from the target at low level. Z1596 crashed near Dose, 10km west of Wilhelmshaven; the rear gunner was the sole survivor and taken prisoner. His four comrades were buried in the municipal cemetery at Wilhelmshaven, but later reinterred at Sage, 24km south of Oldenburg.

Captain: NZ404414 Plt Off **Ian James SHEPHERD**, RNZAF - Age 26. 380hrs. 24th op.
Navigator: NZ404907 Plt Off **Reginald Sidney LEES**, RNZAF - Age 26. 295hrs. 7th op.
Air Bomber: NZ412373 Sgt **James Francis WINSTANLEY**, RNZAF - Age 20. 127hrs. 7th op.
Wireless Op: NZ411101 Plt Off **Norman Bruce ROBERTSON**, RNZAF - Age 25. 216hrs. 10th op.

57 Squadron, RAF (Feltwell, Norfolk - 3 Group)

Wellington III Z1654 - took off at 2300 captained by Sgt K Hudson, RAF, on same raid as the above and brought down north of the target, crashing at Grossenaspe, 10km south of Neumünster. All five crew were buried in a collective grave at Neumünster, but later reinterred at Ohlsdorf, near Hamburg.

Navigator: NZ403985 Sgt **Arthur Walter PIERARD**, RNZAF - Age 22. 6th op.

142 Squadron, RAF (Waltham, Lincolnshire - 1 Group)

Wellington IV Z1289 - took off at 2305 captained by Flt Sgt N J Warfield, RAF, on same raid as the above. The navigator was killed in the first attack by a night-fighter on approach to the target. Though restricted to manipulating his damaged turret by hand, the rear gunner succeeded in shooting down the enemy in flames when it made a third firing pass. During these attacks the wireless operator had mistakenly baled out and the captain was wounded. The rear gunner assisted in piloting the damaged Wellington back to base. The navigator is buried at nearby Grimsby.

Navigator: NZ403787 Fg Off **Eric John Robert YATES**, RNZAF - Age 21. 467hrs.

142 Squadron, RAF (Waltham, Lincolnshire - 1 Group)
Wellington IV Z1376/M - took off at 2312 on same raid as the above and lost without trace, all five crew being commemorated on the Runnymede Memorial.
Captain: NZ403941 Flt Sgt **Noel Ernest Robertson CARREL**, RNZAF - Age 23. 433hrs. 14th op.

Mon 27 Jul 1942

NEW ZEALAND

Dual instruction
3 Elementary Flying Training School, RNZAF (Harewood)
Tiger Moth II NZ1429 - another of the School's Tiger Moths (NZ839) on a solo training flight overtook NZ1429 at 0945 from the port quarter as it made a left hand turn. They collided and crashed out of control a quarter of a mile east of Harris Field, an auxiliary landing ground at Eyreton, 5 miles north of Harewood. Both of NZ1429's pilots died, the instructor later being cremated at Wellington, while the pupil was buried at Hamilton. The pilot of NZ839, LAC **R B Knight**, survived his crash with serious injuries. Echoing a similar incident at the School only two days earlier, the accident report recorded; 'the fact that both aircraft were camouflaged is considered to have been a contributory factor' leading to the accident.
Pilot Instr: NZ40962 Fg Off **Kenneth Sinclair BLAIR**, RNZAF - Age 25. 1046hrs solo (990 on Tiger Moth)
Pilot: NZ421062 LAC **Leonard Matthew KEAT**, RNZAF - Age 29.
Knight died on 9 February 1945 while flying with 630 Sqn, RAF.

BOMBER COMMAND

Cloud cover raids over targets in northern Germany (by 8 aircraft - 2 lost)
57 Squadron, RAF (Feltwell, Norfolk - 3 Group)
Wellington III X3653 - took off at 1530 captained by Wg Cdr Peters-Smith, DFC, RAF, on a raid to Bremen and brought down near the target area. All five crew died, the navigator and two or more of the others being buried in the Walner Cemetery at Bremen, but later all five were reinterred at Becklingen, 17km south of Soltau.
Navigator: NZ403603 Fg Off **Nathaniel Edmund HODSON**, RNZAF - Age 25. 581hrs. 34th op.

Tue 28 Jul 1942

FIGHTER COMMAND

Air firing practice
32 Squadron, RAF (West Malling, Kent - 11 Group)
Hurricane IIC HL655 - took off at about 1545 and, on turning steeply in towards the target 15 minutes later, entered a spin. An overcorrection followed, but before it could be rectified HL655 crashed into the sea near Shoreham, Sussex. The pilot is buried at Brookwood in Woking, Surrey.
Pilot: NZ412664 Plt Off **Keith Henry Iveson EASTON** , RNZAF - Age 21. 275hrs (168 solo - 89 on Hurricane) 1 op.

Tue 28/Wed 29 Jul 1942

BOMBER COMMAND

Intruder operations in support of raid on Hamburg, Germany (by 43 aircraft - 3 lost)
107 Squadron, RAF (Great Massingham, Norfolk - 2 Group)
Boston III AL765 - took off at 2201 captained by Flt Sgt R A Russell, RCAF, for an attack on an airfield at Alkmaar, Netherlands. Shot down by flak at 2300, AL765 crashed at Egmond aan de Hoef, 6km WSW of Alkmaar, all four crew being buried 6km to the NE at Bergen.
Navigator: NZ404950 Sgt **Siddle Henry SCHOFIELD**, RNZAF - Age 22. 259hrs. 4th op.

Raid on Hamburg, Germany (by 256 aircraft - 33 lost)
16 Operational Training Unit, RAF (Upper Heyford, Oxfordshire - 92 Group)
Wellington IC R1450/Y1 - took off at 2214 captained by Plt Off F Lowe, RAF, and shot down at 11,500 feet over Germany by a night-fighter, crashing near Schleswig. Three of its crew survived to be taken prisoner, one of whom was murdered by the Gestapo in March 1944 after the 'Great Escape.' Their three comrades were buried in the Garrison Cemetery at Schleswig, but later reinterred at Kiel.
WOpAG: NZ411066 Sgt **Trevor Haughton CRAY**, RNZAF - Age 22. 79hrs. 1st op(?)
Cray's brother, Geoffrey Haughton Cray, died on 17 April 1944 while flying with 25 Sqn, RNZAF.

7 Squadron, RAF (Oakington, Cambridgeshire - 3 Group)
Stirling I W7533/G - took off at 2228 captained by Wt Off J M Black, RAF, on same raid as the above, and brought down over Germany, crashing at Klein Horsten, 16km SW of Wilhelmshaven, killing six of its crew. An RAF Sgt survived to be taken prisoner, as did the critically injured 2nd pilot, but he died on the 29th in the Marine Hospital at Varel, 14km SE of Klein Horsten. He was buried at Wittmund with some or all of the others, but all seven were later reinterred at Sage, 24km south of Oldenburg.
2nd Pilot: NZ411912 Sgt **Clarence Patrick LUND**, RNZAF - Age 26. 328hrs. 3rd op.
Lund's brother, Mervyn Sydney Lund, died just over a year earlier, on 25 July 1941 while flying with 103 Sqn, RAF.

75 (NZ) Squadron, RAF (Feltwell, Norfolk - 3 Group)
Wellington III BJ599 - took off at approximately 2235-2300 on same raid as the above, and brought down over Germany, crashing at Steinbeck, 4km south of Stade. All five crew were buried in the Municipal Cemetery at Stade, but later reinterred at Becklingen, 17km south of Soltau.
Captain: NZ41914 Flt Sgt **George Alister HUTT**, RNZAF - Age 25. 340hrs. 5th op.
Navigator: NZ411718 Plt Off **Murray Ellis CARNCROSS**, RNZAF - Age 19. 292hrs. 5th op.
Air Bomber: NZ411096 Sgt **William Clerken O'SHEA**, RNZAF - Age 28. 104hrs. 4th op(?)
Wireless Op: NZ405539 Sgt **James Gordon McMURCHY**, RNZAF - Age 31. 157hrs. 1st op.
Rear Gunner: NZ404620 Sgt **John Henry SAVAGE**, RNZAF - Age 33. 116hrs. 2nd op.

75 (NZ) Squadron, RAF (Feltwell, Norfolk - 3 Group)
Wellington III BJ661/X - took off at approximately 2235-2300 on same raid as the above, and shot down over the Netherlands by a night-fighter at 0305, crashing into the Ijsselmeer opposite Schellingwoude, Amsterdam. Three of the five crew died, the bodies of the captain and navigator being recovered from the wreckage and buried at Amsterdam. That of the rear gunner was recovered later some distance from the crash site and is buried at Harderwijk. The two survivors, RNZAF Sgts **R P Callahan** and **A W Rutherford**, were injured but floated free of the wrecked Wellington, to be rescued a few hours later and become prisoners of war. A very detailed account of this loss by Roy Nesbit may be found on pages 730-733 of the December 1988 issue of *Aeroplane Monthly* magazine.
Captain: NZ41894 Flt Sgt **John Edward GILBERTSON**, RNZAF - Age 22. 295hrs. 8th op.
Navigator: NZ404529 Flt Sgt **Martin John BYRNE**, RNZAF - Age 32. 351hrs. 10th op.

75 (NZ) Squadron, RAF (Feltwell, Norfolk - 3 Group)
Wellington III X3558 - took off at approximately 2235-2300 on same raid as the above, and brought down over Germany, crashing near Cuxhaven. All five crew were buried in the Central Cemetery at Brackeswalde, but later reinterred at Becklingen, 17km south of Soltau.
Captain: NZ405340 Flt Sgt **Alexander George SUTHERLAND**, RNZAF - Age 23. 369hrs. 19th op.
Wireless Op: NZ411104 Sgt **Adrian Oscar TABOR**, RNZAF - Age 25. 193hrs. 14th op.
Rear Gunner: NZ404623 Sgt **Ian Gordon STEWART**, RNZAF - Age 20. 230hrs. 19th op.
Tabor's brother, Lionel William Tabor, died on 15 February 1943 while flying with 90 Sqn, RAF.

75 (NZ) Squadron, RAF (Feltwell, Norfolk - 3 Group)
Wellington III X3664 - took off at approximately 2235-2300 on same raid as the above, and brought down over Germany at 0210, crashing at Fischbek, 28km ESE of Stade. All six crew were buried in a collective grave at Stade, but later reinterred at Becklingen, 17km south of Soltau.
Captain: NZ402475 Flt Lt **Peter John WILSON**, RNZAF - Age 22. 520hrs. 37th op.
2nd Pilot: NZ41970 Flt Sgt **Victor Kenneth WESTERMAN**, RNZAF - Age 24. 332hrs.
Navigator: NZ411204 Sgt **Robert Hylton BRISCO**, RNZAF - Age 26. 372hrs. 19th op.
Air bomber: NZ391857 Flt Sgt **Alan CAMPBELL**, RNZAF - Age 22. 410hrs. 52nd op.
Wireless Op: NZ403569 Flt Sgt **Ronald Fraser DAVIS**, RNZAF - Age 22.
Rear Gunner: NZ404016 Sgt **Gordon Edwin CAITCHEON**, RNZAF - Age 28. 234hrs. 32nd op.

75 (NZ) Squadron, RAF (Feltwell, Norfolk - 3 Group)
Wellington III Z1570/B - took off at approximately 2235-2300 on same raid as the above, and brought down over Germany near Lingen-on-Ems, 20km south of Meppen. All five crew were buried at Lingen-on-Ems, but later reinterred at Reichswald Forest, 5km SW of Kleve.
Captain: NZ41907 Flt Sgt **Arthur Grahame JOHNS**, RNZAF - Age 20. 321hrs. 17th op.
Navigator: NZ403458 Sgt **Mark KRALJEVICH**, RNZAF - Age 25. 321hrs. 27th op.
Air Bomber: NZ411753 Sgt **Laurie Albert FRAMPTON**, RNZAF - Age 20. 135hrs. 15th op.
Rear Gunner: NZ411785 Sgt **Leslie Ian STEWART**, RNZAF - Age 25. 184hrs. 20th op.

75 (NZ) Squadron, RAF (Feltwell, Norfolk - 3 Group)
Wellington III X3452 - took off at approximately 2235-2300 captained by Sgt **C Croall**, RNZAF, on same raid as the above. Over the target X3452 was caught in searchlights which were finally shaken off by diving to low level. While 'hedge-hopping' back towards the sea it was severely damaged by light flak and compelled to ditch. Croall and three other RNZAF crew members, Sgt **W H Bright**, Flt Sgts **J R Gratton** and **R S Harvey**, survived to be taken prisoner. The rear gunner, however, drowned, his body being recovered and buried in the Evangelical-Lutheran Cemetery at Tönning, 19km SSW of Husum, but later reinterred at Ohlsdorf, near Hamburg.
Rear Gunner: NZ405475 Sgt **Thomas Eric CRARER**, RNZAF - Age 21. 215hrs. 17th op.

115 Squadron, RAF (Marham, Norfolk - 3 Group)
Wellington III Z1605/R - took off at about 2300 on same raid as the above, and brought down over Germany, crashing at Hohner See, about a kilometre west of Hohn. All five crew were buried in the Hohn Cemetery, 10km west of Rendsburg, but later reinterred at Kiel.
Captain: NZ411487 Flt Sgt **Alan Campbell WILLIAMSON**, RNZAF - Age 21.
Air Gunner: NZ39097 Plt Off **Harold Douglas John FITZSIMONS**, RNZAF - Age 21.

115 Squadron, RAF (Marham, Norfolk - 3 Group)
Wellington III Z1624/D - took off at 2308 on same raid as the above and at 12,000 feet over the Netherlands underwent three night-fighter attacks. The first two caused fires which could not be extinguished, while the third hit the fuel tanks. Three of the crew baled out with burning parachutes but miraculously survived and were taken prisoner. The captain and the fifth crew member lost their lives as Z1624 fell into the Ijsselmeer, about 10km SW of Lemmer. They are commemorated on the Runnymede Memorial.
Captain: NZ411429 Plt Off **Leonard Martin MASON**, RNZAF - Age 25. 307hrs. 2nd op.

101 Squadron, RAF (Bourn, Cambridgeshire - 3 Group)
Wellington III X3668 - took off from Bourn at about 2315 captained by Sgt P G Teall, RAF, on same raid as the above and 15 minutes later collided with a Stirling I (N6121) of the Squadron's Conversion Flight which had taken off at 2311 from Oakington, $7^1/_2$ miles NE of Bourn. Circling overhead Oakington and climbing through 2300 feet to gain height, the Stirling pilot had sighted the approaching Wellington at the last moment and tried to turn sharply away to starboard. He was too late to prevent X3668 sheering off three-quarters of N6121's port tailplane and damaging the remainder, but all seven of its crew safely abandoned the now unstable aircraft. However, all five on board the Wellington died as both bombers crashed in the vicinity of Cottenham and Rampton, a few miles NE of Oakington. The wireless operator is buried at Cambridge.
Wireless Op: NZ412364 Plt Off **Robert James STUBBINGS**, RNZAF - Age 22. 168hrs. 4th op.

1651 Conversion Unit, RAF (Waterbeach, Cambridgeshire - 3 Group)
Stirling I N6069/G - took off at 2325 on same raid as the above and lost without trace, its eight crew being commemorated on the Runnymede Memorial.
Captain: NZ403465 Plt Off **John Dunlop McGREGOR**, RNZAF - Age 27. 543hrs. 35th op.
McGregor had completed an operational tour with 281 Sqn, RAF, and was instructing at the HCU when he undertook this additional op. His brother, Walter Young McGregor, died on 2 September 1943 while flying with 227 Sqn, RAF.

1651 Conversion Unit, RAF (Waterbeach, Cambridgeshire - 3 Group)
Stirling I N3655/T - took off at 2330 captained by Plt Off G C Bayley, DFC, RAF, on same raid as the above, and brought down over Germany, crashing on the east bank of the river Weser at Oslebshausen, in the NW district of Bremen. All seven crew were buried in Bremen's Walner Cemetery, but later reinterred at Becklingen, 17km south of Soltau.
Navigator: NZ41552 Sgt **Kenneth Stuart SPENCER**, RNZAF - Age 31. 263hrs. 2nd op.
Air Bomber: NZ411112 Sgt **Charles Easton WELLS**, RNZAF - Age 28. 124hrs. 2nd op.
Air Gunner: NZ405492 Sgt **Norman Robert KILPATRICK**, RNZAF - Age 25. 78hrs. 2nd op.
Wells brother, George Wells, died a year earlier, on 25 July 1941, while flying with 10 Sqn, RAF.

FIGHTER COMMAND

Intruder patrol to Noordwijk-Binnen, Netherlands
3 Squadron, RAF (Hunsdon, Hertfordshire - 11 Group)
Hurricane IIC HL858/Y - took off late evening and one of two which failed to return. It seems likely that both fighters were lost shortly after carrying out attacks on a night-fighter airfield at Gilze-Rijen, 55km SE of Rotterdam. Hurricane HL858 was presumed lost in the North Sea and its pilot is commemorated on the Runnymede Memorial.
Pilot: NZ402444 Flt Sgt **Peter Masters GAWITH**, DFM, RNZAF - Age 23. 360hrs. 37th op.

These 33 deaths, all but one within Bomber Command during a raid on Hamburg, represent the largest loss of New Zealand airmen for any single day or night, in peace or war, anywhere. No 75 (NZ) Sqn in particular was badly mauled, with 21 dead, having already lost 10 New Zealanders on another Hamburg raid only two nights earlier.

Wed 29/Thu 30 Jul 1942

BOMBER COMMAND

Raid on Saarbrücken, Germany (by 291 aircraft - 10 lost)
102 Squadron, RAF (Topcliffe - 4 Group)
Halifax II R9442/R - took off at 2306 and shot down over Belgium by a night-fighter at 0300, crashing near Corbais, 35km SE of Brussels. All six crew are buried in a collective grave at Corbais. The night-fighter, which had earlier destroyed the 149 Sqn Stirling detailed below, was shot down by return fire from the Halifax and one of its two crew also died.
Captain: NZ405338 Flt Sgt **Ronald Francis STONE**, RNZAF - Age 26. 398hrs. 14th op.
It will be recalled that Stone was a car passenger on 9 March 1941 when LAC Gerald Hewitt was accidentally killed in Christchurch, New Zealand. His brother, Robert James Stone, died a year and a day later, on 31 July 1943, while flying with 75 (NZ) Sqn, RAF.

61 Squadron, RAF (Syerston, Nottinghamshire - 5 Group)
Lancaster I R5737 - took off at 2359 on same raid as the above, and brought down over France, crashing at Vrizy, 4km NNW of Vouziers. An RAF Sgt was the sole survivor and taken prisoner. His six comrades are buried in a collective grave at Vrizy.
Captain: NZ405337 Plt Off **Jack Graham STEWART**, RNZAF - Age 28. 468hrs. 24th op.

149 Squadron, RAF (Lakenheath, Suffolk - 3 Group)
Stirling I R9161/T - took off at 0001 captained by Flt Lt F G Neate, RAF, on same raid as the above, and shot down over France by a night-fighter, crashing at Regniowez, 15km west of Revin. An RAF Sgt, the sole survivor, was taken prisoner. His six comrades were buried in a collective grave in the Communal Cemetery at Regniowez, but later reinterred at Choloy, 28km west of Nancy.
Rear Gunner: NZ412593 Sgt **Henry Woolls SAMPSON**, RNZAF - Age 34. 112hrs. 3rd op.
Sampson's brother, Richard Webb Sampson, died on 18 February 1944 while flying with 464 Sqn, RAAF.

Thu 30 Jul 1942

MIDDLE EAST

Scramble to meet incoming raid
1435 Flight, RAF (Luqa, Malta)
Spitfire VB EP344 - took off at 0755 with seven others and was attacked by enemy fighters while climbing, being shot down into the sea off St Paul's Bay. The pilot is commemorated on the Malta Memorial.
Pilot: NZ412780 Sgt **Colin Lethbridge WOOD**, RNZAF - Age 24.

FIGHTER COMMAND

Escorting bombers on raid to St Omer, France
122 Squadron, RAF (Hornchurch, Essex - 11 Group)
Spitfire VB BM242 - took off at 1820 with 11 others, as part of 'Blue' section, and one of four which failed to return, being shot down by flak at 1930 and crashed near Porsenle. The pilot is buried at Dunkirk.
Pilot: NZ411917 Sgt **Alfred Montgomery McPHERSON**, RNZAF - Age 25. 299hrs. 24th op.

Fri 31 Jul/Sat 1 Aug 1942

BOMBER COMMAND

Raid on Düsseldorf, Germany (by 630 aircraft - 31 lost)
16 Operational Training Unit, RAF (Upper Heyford, Oxfordshire - 92 Group)
Wellington IC DV736/F2 - took off at 2339 captained by Flt Sgt E H Vickers, RAAF, and brought down over Belgium at 0215, crashing at Dilsen, 44km NE of St Trond. All six crew were buried at St Trond, but later reinterred at Heverlee, 35km further west.
Wireless Op: NZ404085 Sgt **Francis Perrie George MOORE**, RNZAF - Age 24. 73hrs. 2nd op.

27 Operational Training Unit, RAF (Lichfield, Staffordshire - 93 Group)
Wellington IC DV552/N - took off at 2341 on same raid as the above, and brought down over Belgium at 0330, crashing at Huldenberg, 17km SSE of Brussels. All six crew are buried in the town's churchyard.
Captain: NZ40239 Fg Off **Mervyn George McNEIL**, RNZAF - Age 25. 748hrs. 19th op.
McNeil completed 16 of his ops with 218 Sqn before being posted to the OTU as an instructor in June 1941. This was his third op undertaken with the OTU.

101 Squadron, RAF (Bourn, Cambridgeshire - 3 Group)
Wellington III BJ841 - captained by Wt Off 1st Class A C Vautour, RCAF, on same raid as the above, and brought down about 20km WSW of the target, crashing at Mönchengladbach. All five crew were buried in the city's Municipal Cemetery, but later reinterred at Rheinberg, 12km south of Wesel.
Rear Gunner: NZ405746 Sgt **Aubrey Raymond DALLY**, RNZAF - Age 27. 87hrs. 2nd op.

Sat 1 Aug 1942

FLEET AIR ARM - FAR EAST

Reconnaissance
803 Squadron, FAA (HMS *Formidable* - Fleet aircraft carrier)
Fulmar II DR647 - presumed to have forced landed in the sea during a search over the Bay of Bengal near Ceylon. The pilot is commemorated on the RNZN Memorial at Devonport, Auckland, and his observer on the FAA Memorial at Lee-on-Solent.
Pilot: Sub-Lt (A) **Desmond Nelson ELWOOD**, RNZNVR - Age 24.

BOMBER COMMAND

Training - height test
103 Squadron Conversion Flight, RAF (Elsham Wolds, Lincolnshire - 1 Group)
Halifax II R9379 - captained by Sgt W Bagley, RAF, was turning on to final approach at 1108 when the port outer engine began to cut, apparently as a result of the loss of coolant. Stalling at 400 feet, R9379 spun into the ground

and caught fire, about 440 yards from the end of the runway. All twelve on board died, including the navigator, who is buried with four others at Brigg, 4 miles to the SW of the base. The Conversion Flight was affiliated to 103 Sqn, also based at Elsham Wolds.
Navigator: NZ405739 Sgt **James William KEANE**, RNZAF - Age 20. 139hrs. 1 op.

Sat 1/Sun 2 Aug 1942

FIGHTER COMMAND

Raid on Sotteville marshalling yards at Rouen, France
418 Squadron, RCAF (Bradwell Bay, Essex - 11 Group)
Boston III W8264 - took off at 2359 and brought down over France, crashing near Évreux, about 40km south of the target, killing all three crew. The New Zealand wireless operator-air gunner is buried at Overact, where it is likely the pilot and another RCAF airman were also at first laid to rest, prior to being reinterred in the Canadian War Cemetery near Bretteville-sur-Laize.
Pilot: J/6951 Fg Off **Ronald Garland BLAMIRES**, RCAF - Age 21.
WOpAG: NZ404602 Sgt **Harold Lewis GREEN**, RNZAF - Age 23. 218hrs. 1st op.

Sun 2 Aug 1942

MIDDLE EAST

Ferry flight to Khartoum
Communications Flight RAF Khartoum (Gordon's Tree, Khartoum, Sudan)
Blenheim IV Z7594. At about 0900 flew into the ground at a point '15 miles 230° West Station 2' (also placed 'near to' or 15 miles from Wadi Seidna and 19 miles west of Statur). The three crew were buried in the sand near the crash site. However, their graves were later effaced by the elements and they are now commemorated on the Alamein Memorial.
Pilot: NZ411856 Sgt **William Stead BURNS**, RNZAF - Age 29. 296hrs. 2 ops.
Burns is understood to have been serving at the time with 2 Middle East Ferry Control (headquartered at Wadi Seidna) but contemporary records also suggest 15 Sqn, SAAF, or the Communications Flight, although either of the latter may simply relate to ownership of the aircraft.

Scramble to intercept incoming raid
185 Squadron, RAF (Hal Far, Malta)
Spitfire VC BR321 - took off at 1400 with seven others and one of two shot down by enemy fighters during an engagement over the island of Gozo, BR321 going down in the sea. The pilot is commemorated on the Malta Memorial.
Pilot: NZ411916 Sgt **Matthew Leonard McLEOD**, RNZAF - Age 21. 289nrs. 24th op.

Returning from operational flight
24 Squadron, SAAF (LG 99 [15 miles SSE of Markaz Amriya], Egypt - 3 (SAAF) Wing, 211 Group)
Boston III AL696 - piloted by Lt W H Martin, SAAF. On approach at 2000 overshot and struck a stationary 37 Sqn Wellington IC (DV644) at the end of the runway. Both bombers burst into flames and DV644's bomb load exploded, setting off a chain reaction which caused 37 Squadron's nearby DV641, HF832 and HX480 to follow suit. The three Boston crew died, as did three members from 458 Sqn, RAAF, who were crewing DV644 while on attachment to 37 Sqn. All six are buried at Tel-el-Kebir.
WOpAG: NZ41622 Sgt **Albert Leslie HOPKINS**, RNZAF - Age 23. 89hrs. 35th op.

FIGHTER COMMAND

'Rhubarb' - sweep over Belgium to attack trains
485 Squadron, RNZAF (Kings Cliffe, Northamptonshire - 12 Group)
Spitfire VB BM247 (presentation machine named *Delhi IV - Central Public Works Department*) - took off at 1436 with others of the Squadron and after refuelling at Martlesham, crossed into Belgium slightly south of Ostend. BM247 was last seen on the return journey when still a few miles inland of the Belgian coast and was presumed to have gone down into the sea. The pilot is commemorated on the Runnymede Memorial.
Pilot: NZ412702 Sgt **Norman Mathew LANGLANDS**, RNZAF - Age 21. 290hrs. 15th op.

Mon 3 Aug 1942

COASTAL COMMAND

Squadron move to Wick, Caithness-shire (18 Group)
58 Squadron, RAF (St Eval, Cornwall - 19 Group)
Whitley VII Z9518 - while carrying out an overshoot at 1100 stalled and spun in from 300-400 feet at Wick. All six occupants on board died, the two pilots and an RCAF wireless operator being buried at Wick.
Captain: NZ41493 Flt Sgt **Malcolm William RANDALL**, RNZAF - Age 25. 468hrs (318 on Whitley) 23 ops.
2nd Pilot: NZ414605 Sgt **Kenneth Gordon DUNN**, RNZAF - Age 20.

FIGHTER COMMAND

'Rhubarb' - sweep over the Netherlands to attack trains

485 Squadron, RNZAF (Kings Cliffe, Northamptonshire - 12 Group)

Spitfire VB BM183/D - took off at 1325 with others of the Squadron as 'Red 2'. After refuelling at Martlesham the formation crossed the Dutch coast over Zeeland, two sections shooting up locomotives along the railway between Goes and Woensdrecht. After the attack Red 2 acknowledged to 'Red 1' that he was 'all right', but then disappeared. The pilot is buried at Vlissingen.

Pilot: NZ413156 Sgt **Ronald William VESSEY**, RNZAF - Age 19. 284hrs.

Mon 3/Tue 4 Aug 1942

FLYING TRAINING COMMAND

Circuit and landing training

1 Flying Instructors School, RAF (Church Lawford, Warwickshire - 23 Group)

Oxford I R6059 - while landing at 0030 on a relief landing ground at Warwick, 2 miles west of Leamington, bounced heavily and began an overshoot. At 150 feet the flaps were raised prematurely, causing R6059 to lose height and collide with a tree. In the crash which followed one crew member survived with injuries, but his companion died and is buried at Rugby.

Pilot: NZ413989 Sgt **Stanley Leonard PORTMAN**, RNZAF - Age 26. 282hrs.

The other pilot was Sgt E H Ridler, RAAF, but who was in command at the time is unknown.

Wed 5 Aug 1942

NEW ZEALAND

Interception practice

11 Operational Training Unit, RNZAF (Ohakea)

P-40E Kittyhawk NZ3001 - at 1430 began a tight left hand turn at 1200 feet which developed into a flat spin, failed to recover and crashed into the ground near Feilding. The pilot is buried at Petone.

Pilot: NZ39014 Sgt **Alfred Reginald Philip SAUL**, RNZAF - Age 22. 486hrs solo (6 on Kittyhawk)

Saul had served in the Far East during the Malayan campaign but it is not known if he flew on operations.

MIDDLE EAST

Anti-shipping strike off the Libyan-Egyptian coast

459 Squadron, RAAF (LG 'Z' [Mahsma], Egypt; detachment at LG 226 [Gianaclis] - 235 Wing, 201 Group)

Hudson IIIA FH302/C - took off at 0225 captained by Flt Sgt W L Staines, RAAF, and lost without trace, its four crew being commemorated on the Alamein Memorial.

Navigator: NZ405461 Sgt **Noel Crompton WOOLCOCK**, RNZAF - Age 27. 309hrs. 9th op.

Thu 6 Aug 1942

ARMY CO-OPERATION COMMAND

Army co-operation exercise

168 Squadron, RAF (Bottisham, Cambridgeshire - 34 Wing)

Tomahawk IIA AH891 - lost its port wing on striking trees while carrying out a low level dummy attack on gun emplacements at 1230 along the Thetford-Watton road near Thetford, Norfolk. AH891 flicked over once or twice before hitting the ground some 200 yards on and caught fire. The pilot is buried at Bottisham, Cambridgeshire.

Pilot: NZ412010 Plt Off **Robert Keith Anderson ANDREW**, RNZAF - Age 24. 270hrs.

CANADA

2 Instrument Link Training School, RCAF (Edmonton, Alberta)

Died after an unsuccessful brain tumour operation at Edmonton's University Hospital and is buried in that city.

(Link Trainer Instructor): NZ414339 Sgt **James Gough POLSON**, RNZAF - Age 29.

Thu 6/Fri 7 Aug 1942

BOMBER COMMAND

Raid on Duisburg, Germany (by 216 aircraft - 9 lost)

158 Squadron, RAF (East Kirby, Lincolnshire - 4 Group)

Halifax II W7750/M - took off at 0029 captained by Flt Lt J P M Haydon, RAAF, and brought down over Belgium, either by a night-fighter or through an accidental explosion in the aircraft, crashing at 0240 between Opglabbeek and Gruitrode, south of Bree. Of the five crew who successfully baled out, all but one managed to evade capture and return to Britain. The air bomber and rear gunner died, being buried at St Trond, about 40km to the SW, but later reinterred at Heverlee.

Rear Gunner: NZ405484 Sgt **Wilfred John GRAY**, RNZAF - Age 23. 157hrs. 6th op.

150 Squadron, RAF (Snaith, Yorkshire - 1 Group)
Wellington III X3698 - took off at 0040 captained by Flt Sgt R W Baxter, GM, RAAF, on same raid as the above, and seen to climb away, but at 200 feet veered to the right before crashing and catching fire about a mile west of the base. Four of the crew died in the crash or of injuries later in the day. The wireless operator suffered severe injuries and was admitted to the York Military Hospital, but died on the 15th and is buried at Selby with three of his comrades. Subsequent investigation revealed the propeller pitch on the starboard engine had suddenly coarsened as the aircraft climbed for height.
Wireless Op: NZ404068 Flt Sgt **John Herbert HODSELL**, RNZAF - Age 22. 289hrs. 21st op.

7 Squadron, RAF (Oakington, Cambridgeshire - 3 Group)
Stirling I R9154/F - took off at 0100 on same raid as the above, and brought down over Germany, crashing at Hüthum-Emmerich, about 55km NW of Duisburg. All seven crew were buried in the North Cemetery at Düsseldorf, but later reinterred at Reichswald Forest, 5km SW of Kleve.
Captain: NZ402541 Flt Sgt **Clifford George PULLEN**, RNZAF - Age 23. 405hrs. 11th op.

Fri 7 Aug 1942

NEW ZEALAND

RNZAF Station Wigram
Died of natural causes on his fifth day in the Christchurch Public Hospital and is buried in that city.
(Carpenter): NZ423527 AC1 **Frederick Charles BROWN**, RNZAF - Age 40.

Fri 7/Sat 8 Aug 1942

MIDDLE EAST

Raid on Tobruk, Libya
40 Squadron, RAF (Shallufa, Egypt - 205 Group)
Wellington IC DV663/U - took off at 1916 and one of two Squadron aircraft which failed to return. Believed to have collided with another 40 Sqn Wellington (HX431/O), the crew of which reported a collision after leaving the target and saw a Wellington fall in flames. Five of HX431's crew were taken prisoner, two after baling out and three on surviving a crash landing. The sixth died when baling out too low. No trace of DV663 was found and its six crew are commemorated on the Alamein Memorial.
Captain: NZ411412 Flt Sgt **Leonard Percival KERR**, RNZAF - Age 26. 420hrs.
Navigator: NZ404968 Flt Sgt **John Ivan THOMSON**, RNZAF - Age 27. 400hrs. 17th op.

Sat 8 Aug 1942

MIDDLE EAST

Interception of incoming raid
1435 Squadron, RAF (Luqa, Malta)
Spitfire VB EP336 - took off at 1020 with others of the Squadron as 'Blue 2' and shot down during an engagement with enemy fighters, crashing in the sea. An extensive search over a wide area failed to find any trace of the pilot and he is commemorated on the Malta Memorial.
Pilot: NZ39559 Sgt **Clarence Lloyd KELLY**, RNZAF - Age 21. 263hrs. 5th op.

FLYING TRAINING COMMAND

3 Personnel Reception Centre, RAF (Bournemouth, Hampshire - 54 Group)
Taken ill while on leave on 27 April and admitted to the Ballochmyle Hospital at Glasgow on 6 May with acute appendicitis, which proved gangrenous. His condition deteriorated and he died this day of Pylephlebitis, being buried at Auchinleck, Ayrshire.
(Pilot): NZ414696 Sgt **David Findlay STRACHAN**, RNZAF - Age 30. 147hrs.
Strachan had been posted to 15 (Pilots) Advanced Flying Unit on 16 May, but this was cancelled due to his illness.

Sun 9/Mon 10 Aug 1942

BOMBER COMMAND

Raid on Osnabrück, Germany (by 192 aircraft - 6 lost)
460 Squadron, RAAF (Breighton, Yorkshire - 1 Group)
Wellington IV Z1463/L - took off at 0010 captained by Flt Sgt J A Finlay, RAAF, and brought down over the Netherlands, crashing near Hengelo, 12km SE of Zutphen. All five crew are buried at Hengelo.
Wireless Op: NZ39099 Flt Sgt **Stephen Bryan GOORD**, RNZAF - Age 28. 332hrs. 26th op.

Tue 11 Aug 1942

COASTAL COMMAND

Torpedo strike on German pocket battleship *Lützow* off southern Norway
489 Squadron, RNZAF (Skitten, Caithness-shire - 18 Group)
Hampden TB I P1337/M - took off at 0037 captained by Sqn Ldr A C Triptree, RAF, with nine others, being the only one which failed to return. Lost without trace, its four crew are commemorated on the Runnymede Memorial.

After hostilities captured documentation from the *Lützow* suggested that P1337 may have located the ship on its radar at about 0500. No aircraft attacked or were seen by the ship, however.
WOpAG: NZ405490 Sgt **George Arroll JONES**, RNZAF - Age 30. 112hrs. 2nd op.

FIGHTER COMMAND

Shipping reconnaissance off Cherbourg, France
501 Squadron, RAF (Ibsley, Hampshire - 10 Group)
Spitfire VB EP128/J - took off at 0725 with others of the Squadron. When returning from France at 15,000 feet, the formation was attacked by four enemy fighters over the English Channel. The pilot of EP128 radioed that his aircraft had been hit and he was about to bale out. However, he was not seen to leave before the Spitfire spun into the sea. Aircraft searched during the day, but no trace of him could be found and is commemorated on the Runnymede Memorial.
Pilot: NZ404885 Plt Off **Stewart Gordon BRANNIGAN**, RNZAF - Age 23. 352hrs.

FLEET AIR ARM - USA

Training
VF-OTU5, US Naval Air Station (Miami, Florida - HMS *Saker II*)
Buffalo F2A-3 BuNo01590 - crashed 8 miles NE of the airfield at 1230 after engine failure. The pilot is buried at Miami.
Pilot: Sub-Lt (A) **Denis Beresford McLEAN**, RNZNVR - Age 22.

Tue 11/Wed 12 Aug 1942

BOMBER COMMAND

Raid on Mainz, Germany (by 154 aircraft - 18 lost)
75 (NZ) Squadron, RAF (Feltwell, Norfolk - 3 Group)
Wellington III X3646 - took off at approximately 2200-2255 and was badly damaged by flak over Cologne, the captain being severely wounded. When the Wellington had been forced down over the sea to 2000 feet, the crew were ordered to bale out but only an RAF Sgt survived and was taken prisoner. Two bodies were later recovered from the sea by the Royal Navy, that of an RAF Fg Off being brought ashore for burial, while the other was buried at sea. The latter, together with the three crew who were not found, is commemorated on the Runnymede Memorial.
Captain: NZ401954 Plt Off **George Edward Francis BRADEY**, RNZAF - Age 25. 435hrs.
Wireless Op: NZ402997 Flt Sgt **Cyril Vincent GREEN**, RNZAF - Age 21.

75 (NZ) Squadron, RAF (Feltwell, Norfolk - 3 Group)
Wellington III BJ625 - took off at approximately 2200-2255 on same raid as the above, and brought down over Germany, crashing near Düsseldorf. Rear gunner Sgt **J W F Atkin**, RNZAF, slightly wounded, was taken prisoner as the sole survivor. His four comrades were buried at Düsseldorf, but later reinterred at Reichswald Forest, 5km SW of Kleve.
Captain: NZ411358 Flt Sgt **Thomas Smith BARCLAY**, RNZAF - Age 22. 348hrs. At least 6th op.
Wireless Op: NZ411110 Sgt **Robert John WARRING**, RNZAF - Age 21. 164hrs. 5th op.
Air Bomber: NZ411758 Sgt **William Gordon Lloyd INGLIS**, RNZAF - Age 27. 150hrs. 7th op.

75 (NZ) Squadron, RAF (Feltwell, Norfolk - 3 Group)
Wellington III BJ767 - took off at approximately 2200-2255 on same raid as the above, and brought down over the Netherlands, crashing near Venlo, close to the German border. Two RAF Sgts and RNZAF wireless operator Sgt **J McQueen** survived to be taken prisoner. Their two comrades were buried at Venlo, but later reinterred in Jonkerbos War Cemetery, 4km SW of Nijmegen.
Captain: NZ401375 Flt Lt **Laurence St George DOBBIN**, RNZAF - Age 29. 728hrs. 3rd op.
Air Bomber: NZ411764 Sgt **Jack Leslie JURY**, RNZAF - Age 20. 130hrs. 4th op.

57 Squadron, RAF (Feltwell, Norfolk - 3 Group)
Wellington III Z1656 - took off at 2250 on same raid as the above, and brought down over Germany, crashing near Hagen, about 25km south of Dortmund. The five crew were buried at Hagen, but later reinterred at Reichswald Forest, 5km SW of Kleve.
Captain: NZ411449 Flt Sgt **Robert RIDDELL**, RNZAF - Age 24. 299hrs. 9th op(?)

106 Squadron, RAF (Coningsby, Lincolnshire - 5 Group)
Lancaster I W4109 - took off at 2300 captained by Plt Off B E Grein, RAF, on same raid as the above, and shot down over Germany, crashing at Waldesch, 10km SW of Koblenz. The seven crew were buried in the French Carthusian Cemetery at Koblenz, but later reinterred at Rheinberg, 12km south of Wesel.
WOpAG: NZ402981 Flt Sgt **Robin William CASKEY**, RNZAF - Age 27. 222hrs. 10th op.
WOpAG: NZ403565 Sgt **John Warren BADLEY**, RNZAF - Age 20. 232hrs. 12th op.

FIGHTER COMMAND

Night co-operation patrol with (Turbinlite?) Havoc
253 Squadron, RAF (Hibaldstow, Lincolnshire - 12 Group)
Hurricane IIC BN227 - while over the North Sea at 0205 was seen to lose height. The Havoc followed BN227 down until it was eventually lost to sight in the darkness off Spurn Head, Yorkshire. A search by sections of the

Squadron until late the following morning failed to find any trace of the aircraft or its pilot, who is commemorated on the Runnymede Memorial.
Pilot: NZ405318 Flt Sgt **Terence Barnes PEPPER**, RNZAF - Age 21. 366hrs.
Pepper's brother, Douglas Reid Pepper, died on 25 July 1944 while flying with 622 Sqn, RAF.

Wed 12/Thu 13 Aug 1942

BOMBER COMMAND

Raid on Mainz, Germany (by 138 aircraft - 6 lost)
150 Squadron, RAF (Snaith, Yorkshire - 1 Group)
Wellington III BJ588 - took off at 2205 captained by Sgt D J H Beaton, RAF, but returned early when it appeared that the starboard engine had caught fire. On attempting an emergency landing at 2245 about 17 miles SW of the base, 4 miles west of Scotter, Lincolnshire, it struck a small hillock. Three of the crew died in the crash or immediately afterwards, including the front gunner, who is buried at Selby. A fourth member died a week later, leaving an injured RAF Sgt as the sole survivor. Subsequent investigation, though revealing no engine fire, indicated that flames from a faulty exhaust stub may have conveyed an impression of one to the crew.
Air Bomber: NZ403768 Flt Sgt **Nelson Raymond ROWE**, RNZAF - Age 22. 220hrs. 11th op.

Thu 13 Aug 1942

MIDDLE EAST

Delivery of equipment to Teheran, Iran
162 Squadron, RAF (LG 208 [Bilbeis, Egypt]; detachment at Habbaniya, Iraq)
Bisley BA317 - took off at 0515 to deliver spares for Prime Minister Churchill's Liberator which had been delayed with engine trouble. The aircraft became airborne about a third of the way along the runway, but gradually swung away to starboard and struck the top of a bund, catching fire some 400 yards beyond the runway, just outside the airfield perimeter. All occupants were killed instantly and are buried at Habbaniya. Records disagree as to whether two or five died.
Pilot: NZ41891 Flt Sgt **William Arthur FERGIE**, RNZAF - Age 21.

Escort for *Pedestal* convoy en route to Malta
126 Squadron, RAF (Luqa, Malta)
Spitfire VB EP472 - took off at 0930 with two others, one of which experienced engine trouble soon after take off and returned to base. The remaining two carried on and engaged an enemy raid near the convoy, during which EP472 was shot down into the sea. The pilot is commemorated on the Malta Memorial.
Pilot: NZ41960 Flt Sgt **John Harold TANNER**, RNZAF - Age 22. 273hrs. 32nd op.

COASTAL COMMAND

Anti-submarine sweep and search for survivors from the *Letitia*
120 Squadron, RAF (Ballykelly, Co. Londonderry - 15 Group)
Liberator GR III LV341 - took off at 0533 and attacked a U-boat with six depth charges at 5320N:2242W between 1125 and 1138. At about 1515, while returning to base at 500 feet, a loud bang was heard and all four engines cut. With insufficient height to attempt a restart, LV341 ditched 37 miles WSW of Tory Island, off the coast of County Donegal, Eire. Two of the eight crew were lost as the Liberator sank and a further two, including the captain, drowned before a dinghy could be inflated and boarded. The other four airmen survived, their four comrades being commemorated on the Runnymede Memorial. Investigators concluded that the co-pilot may have inadvertently operated the inadequately protected master control switch, leading all four engines to cut. The *Letitia* has not been identified, but does not appear to have been a British vessel.
Captain: NZ40976 Plt Off **Victor Dennis JAMES**, RNZAF - Age 26. 575hrs solo (65 on Liberator)

Fri 14 Aug 1942

FIGHTER COMMAND

Interception practice
253 Squadron, RAF (Hibaldstow, Lincolnshire - 12 Group)
Hurricane IIC Z3261 - at 1530 made a steep climbing turn at 3000 feet and entered a spin. Recovery was made at 1000 feet, but a spin in the opposite direction immediately followed and Z3261 crashed into the ground 5 miles SE of Grimsby, Lincolnshire. The pilot is buried at Kirton-in-Lindsey.
Pilot: NZ404612 Plt Off **Lloyd Henry Roper McCARTHY**, RNZAF - Age 29. 384hrs (340 solo - 254 on Hurricane)

Aerobatics and spinning practice
55 Operational Training Unit, RAF (Annan, Dumfriesshire - 81 Group)
Hurricane I V6857 - made a downwind aileron turn at 1900, pulled out at about 4000 feet, rolled over and spun into the ground, catching fire near Longtown airfield, Cumberland, 11 miles east of Annan. The pilot is buried at Annan. It was thought that he may have blacked out after the aileron turn, when control appeared to be lost.
Pilot: NZ413822 Sgt **Derek Richard Sidney DIXON**, RNZAF - Age 20. 260hrs (166 solo - 66 on Hurricane)

CANADA

Air-to-splash-firing detail

31 Bombing & Gunnery School, RAF (Picton, Ontario - 1 Training Command)

Battle I L5427 - stalled off a steep left hand turn at 2000 feet while carrying out a dummy attack on another aircraft at 1145, flicked over and spun into Lake Ontario, near Soup Lake, about 10 miles south of the airfield. The staff pilot and two trainee air gunners on board died, the former being buried at Ottawa.

Pilot: NZ41385 Sgt **Blyth KEMPTON-WEROHIA**, RNZAF - Age 21. 311hrs (226 solo - 135 on Battle)

Sat 15 Aug 1942

FIGHTER COMMAND

Transit from Debden, Essex, to Llanbedr

41 Squadron, RAF (Llanbedr, Merionethshire - 9 Group)

Spitfire VB P8607/C (a presentation machine named *Palembang Oeloe II*) - at about 1745, dived out of low cloud and into the ground at or near Blounce Farm a few miles to the SE of the Weston Patrick, Hampshire. The pilot was buried in the village of Weston Patrick, but later reinterred at Brookwood in Woking, Surrey. A plaque commemorating his name was erected in the village cemetery in March 1994. Contemporary records state Blunt as the farm name but it is believed that Blounce was being referred to. The Squadron had been based at Debden until 8 July.

Pilot: NZ411981 Sgt **Bernard Bryn GOODALL**, RNZAF - Age 23. 312hrs. 41 ops.

BOMBER COMMAND

Flt Sgt **J H HODSELL** died of injuries received in a take off crash on the night of 6/7 August.

Sat 15/Sun 16 Aug 1942

BOMBER COMMAND

Raid on Düsseldorf, Germany (by 131 aircraft - 4 lost)

156 Squadron, RAF (Warboys, Huntingdonshire - [PFF] 3 Group)

Wellington III DF666 - took off at 0045 and shot down over the Netherlands by a night-fighter, crashing near Rozendaal, on the north-eastern outskirts of Arnhem. All five crew are buried in a private cemetery at Rozendaal.

Captain: NZ41994 Flt Sgt **Cyril Charles NEWLOVE**, RNZAF - Age 23. 316hrs. 6th op.

Newlove was the first New Zealand airman to lose his life with the recently formed Path Finder Force, although this particular raid was not in fact led by the Force.

Sun 16 Aug 1942

BOMBER COMMAND

Training

44 Squadron, RAF (Waddington, Lincolnshire - 5 Group)

Lancaster I R5489/X - captained by Sgt R Easom, RAF, and lost control while on approach, crashing at Cranston village, 3 miles NE of the base. The flight engineer and air bomber died but the other crew members were rescued from the wreckage by a local farmer and a butcher. The air bomber is buried at Lincoln.

Air Bomber: NZ405507 Sgt **David PULLINGER**, RNZAF - Age 30. 142hrs.

The CWGC register records Pullinger as belonging to 418 Sqn, RCAF, a unit he is not known to have served with.

Sun 16/Mon 17 Aug 1942

BOMBER COMMAND

'Gardening' - minelaying (by 56 aircraft - 2 lost)

207 Squadron, RAF (Bottesford, Leicestershire - 5 Group)

Lancaster I R5509/N - took off at 2131 to lay mines in the 'Willow' area (southern Baltic) and lost without trace, its seven crew being commemorated on the Runnymede Memorial.

Captain: NZ41631 Flt Sgt **Norman James SUTHERLAND**, RNZAF - Age 30. 390hrs. 11th op.

Mon 17 Aug 1942

FLYING TRAINING COMMAND

First solo on type

Central Gunnery School, RAF (Sutton Bridge, Lincolnshire - 25 Group)

Hampden I P1163 - took off at about 1630 but 10 minutes later collided with trees, crashed into a field and caught fire at Wingland, Terrington St Clement, about 4 miles east of the base. The pilot is buried at Sutton Bridge.

Pilot: NZ41381 Flt Sgt **Sidney Lionel VINTEN**, RNZAF - Age 26. 414hrs. 7 ops(?)

Mon 17/Tue 18 Aug 1942

BOMBER COMMAND

Raid on Osnabrück, Germany (by 139 aircraft - 7 lost)

214 Squadron, RAF (Stradishall, Suffolk - 3 Group)

Stirling I BF330/H - took off at 2133 captained by Sgt A Fleming, RCAF, and when returning at 0214, signalled that sufficient fuel remained for only ten minutes flying. A fix was obtained a minute later, placing BF330 about 70 miles east of the Wash, after which nothing further was heard. Five bodies later washed up onto the English coast, including that of the air bomber, who is buried at Manby, Lincolnshire, along with the captain and one of the others. The other two crew are commemorated on the Runnymede Memorial. In *Bomber Command Losses* BF330 is described as having been shot down by a night-fighter 30km north of Terschelling, but this seems unlikely in view of the above.

Air Bomber: NZ404024 Sgt **Adrian Harold Cooper GILL**, RNZAF - Age 24. 164hrs. 4th op.

FIGHTER COMMAND

Reconnaissance along part of French coast

3 Squadron, RAF (Hunsdon, Hertfordshire - 11 Group)

Hurricane IIB Z5257 - took off at 0050 and was later found to have come down into the sea off Hythe, Kent. The pilot is buried at Hawkinge, near Folkestone.

Pilot: NZ41511 Plt Off **Bruce Hamilton HAY**, DFC, RNZAF - Age 28. 373hrs. 50th op.

Tue 18 Aug 1942

NEW ZEALAND

Solo aerobatic practice

2 Service Flying Training School (ATS), RNZAF (Woodbourne)

Harvard II NZ964 - took off in fine weather at about 0800 for an hour-long practice flight and failed to return. Although an extensive search entailing 50 flights was conducted, no trace could be found. The pilot is commemorated on the Provincial Memorial at Dunedin.

Pilot: NZ417098 LAC **John Ronald Scott ORBELL**, RNZAF - Age 24. 97hrs solo (62 on Harvard)

Motor vehicle accident

RNZAF Station Levin

While walking towards the camp along the left hand side of Kimberly Road, was struck by an Austin 7 car heading in the same direction at 2330. Died instantly and is buried at Drury.

(Armourer): NZ422551 AC1 **Leonard Knox KEEN**, RNZAF - Age 21.

FIGHTER COMMAND

Sweep off Cherbourg, France

129 Squadron, RAF (Thorney Island, Sussex - 11 Group)

Spitfire VB BL934/J - took off at 0930 with 11 others as 'Red 3', being one of two which failed to return. Five miles NE of Cherbourg an enemy float plane flying just above the sea was intercepted by 'Red' Section. Immediately after the first attack some enemy fighters dived on to the Squadron and BL934 was seen to ditch in the sea and break into two or three pieces. After the engagement no sign of the pilot could be seen. On 27 August a destroyer recovered his body from the English Channel at position 5030N:0158W and brought it back to England for burial at Felixstowe, Suffolk on the 29th. The date he actually died is not known, but for official purposes is accepted as being on the 18th.

Pilot: NZ41642 Flt Sgt **Allan James EDWARDS**, RNZAF - Age 22. 432hrs. 105th op.

Formation practice

56 Squadron, RAF (Snailwell, Cambridgeshire - 12 Group)

Typhoon IA R7644/B - rear fuselage failed and broke away during a dive at 1245, the wreckage falling at Moulton, 4 miles east of Spalding, Lincolnshire. The pilot is buried at Sutton Bridge. Failure of the rear fuselage, of which this was the third incident in a relatively short time, was to plague the Typhoon throughout its service.

Pilot: NZ413424 Sgt **James Struthers JONES**, RNZAF - Age 23. 305hrs (215 solo - 43 on Typhoon)

CANADA

Gunnery exercises

1 Bombing & Gunnery School, RCAF (Jarvis, Ontario - 1 Training Command)

Battle I 1604 - at about 2500 feet in the vicinity of Evans Point, some 15 miles SE of the base, collided at 1700 with one of the School's Anson IIs (1604) which was engaged on low level bombing exercises. The Battle crashed into about 10 feet of water in Lake Erie and the Anson onto the shore. Both pilots died, as did their pairs of trainee air gunners and air bombers. The pilot of 1604 is buried at Jarvis.

Pilot: NZ411594 Sgt **John William WHITEHEAD**, RNZAF - Age 27. 251hrs (154 solo - 63 on Battle)

Tue 18/Wed 19 Aug 1942

BOMBER COMMAND

Raid on Flensburg, Germany (by 118 aircraft - 4 lost)

57 Squadron, RAF (Feltwell, Norfolk - 3 Group)

Wellington III X3371/G - took off at 2040 captained by Plt Off R P McLaren, RCAF, and lost without trace, the navigator and wireless operator being commemorated on the Runnymede Memorial. The body of the captain and rear gunner were later recovered and are buried on the Swedish island of Öckerö, about 18km to the west of Göthenburg. That of the air bomber washed ashore onto the German island of Sylt, at Kampen, and was buried further north at List, but later reinterred at Kiel.

Air Bomber: NZ411108 Sgt **Hector John WAKEMAN**, RNZAF - Age 23. 67hrs. 3rd op.
Wireless Op: NZ412309 Plt Off **Kivell Harold William ALDERTON**, RNZAF - Age 23. 177hrs. 3rd op.
Rear Gunner: NZ411230 Sgt **John Charles WILLIAMS**, RNZAF - Age 21. 147hrs. 3rd op.

218 Squadron, RAF (Downham Market, Norfolk - 3 Group)

Stirling I W7618/V - took off at 2107 captained by Plt Off I G McDaniel, RAF, on the same raid as above, and shot down by a night-fighter, crashing into the sea off the Danish west coast. Two of the crew are buried in Germany and two others, including a New Zealander, are commemorated on the Runnymede Memorial. The captain, wounded bomb aimer Sgt **C A Powell**, RNZAF, and one of the air gunners survived to be picked up by a Danish trawler after drifting in their dinghy for three days in rough weather. All three were taken prisoner when the boat was later stopped by a German patrol vessel.

WOpAG: NZ41597 Sgt **John Anderson MACPHERSON**, mid, RNZAF - Age 21. 150hrs. 2nd op.

Wed 19 Aug 1942

FIGHTER COMMAND

Operation JUBILEE - attack on gun defences and sea front buildings in support of combined operation against Dieppe, France

245 Squadron, RAF (Middle Wallop, Hampshire; detachment at Shoreham, Sussex - 10 Group)

Hurricane IIC BD766 - took off from Shoreham at 0445 with 10 others and one of three which failed to return. A successful attack was carried out, but smoke in the target area caused the Squadron to become broken up prior to making a second pass. Hurricane BD766 was lost without trace, its pilot being commemorated on the Runnymede Memorial. Testifying to the ferocity of the enemy defences, of the eight Hurricanes which did return to base only one remained serviceable.

Pilot: NZ402433 Plt Off **James Edgar BARTON**, RNZAF - Age 21.

ARMY CO-OPERATION COMMAND

Operation JUBILEE - tactical reconnaissance in support of Dieppe raid

26 Squadron, RAF (Gatwick, Surrey - 35 Wing)

Mustang I AL977 - took off at 0800 with one other, both failing to return. Brought down over enemy territory, AL977's pilot was taken prisoner. A participant in the 'Great Escape' from Stalag Luft III at Sagan, Poland, on the night of 24/25 March 1944, he was recaptured by the Germans at Flensburg and murdered, together with three of his comrades, by the Gestapo in a field near Kiel on the 29th. His body was later cremated at Kiel and the ashes returned to Sagan where, along with those of the other 49 'escape' victims, they were held in a communal vault in a cemetery about 500 yards from the camp. Later the remains were reinterred in a collective grave at Poznan, 40km NE of Sagan.

Pilot: NZ413380 Flt Lt **Arnold George CHRISTENSEN**, mid, RNZAF - Age 21. 300hrs. 1st op.

BOMBER COMMAND

Operation JUBILEE - laying smoke screen in support of Dieppe raid

226 Squadron, RAF (Swanton Morley, Norfolk; detachment at Thruxton, Hampshire - 2 Group)

Boston III AL710/Z - took off at 1015 from Thruxton piloted by Plt Off **W R Gellatly**, RNZAF, with three others and was hit by flak while over the target area, killing the wireless operator-air gunner and wounding the third crew member. A crash landing was carried out at Gatwick, Surrey, where the dead airman was removed for burial at Horley.

WOpAG: NZ411111 Plt Off **Laurence John Braithwaite WATERS**, RNZAF - Age 21. 127hrs. 1st op.

FIGHTER COMMAND

Operation JUBILEE - attack on the east headland at Dieppe

3 Squadron, RAF (Hunsdon, Hertfordshire; detachment at Shoreham, Sussex - 11 Group)

Hurricane IIB AM286 - took off at 1345 and led the Squadron on its third attack of the day. As the formation dived to 500 feet over Dieppe it was intercepted by about 15 enemy fighters. The leading Hurricane was seen to burst into flames and crash on a cliff top. The pilot's body later washed ashore at Boulogne and is buried at Dieppe.

Pilot: 43023 Sqn Ldr **Alec Edward BERRY**, DFC, RAF - Age 25.

Berry, who had enlisted as an AC1 in the RNZAF in 1937, gained a short service commission in the RAF in 1939. He is known to have flown 660hrs and completed 110 ops by the end of 1941.

FLYING TRAINING COMMAND

Tracer firing

1 (Observers) Advanced Flying Unit, RAF (Wigtown, Wigtownshire - 25 Group)

Botha I L6233 - while flying low over the sea at 1645, crashed into the water 3 miles SW of Abbey Head, five and a half miles SE of Kirkcudbright, killing all five on board. Only the body of an RAF airman was recovered, the others being commemorated on the Runnymede Memorial.

Pilot: NZ412523 Sgt **Brian Lundius GARLICK**, RNZAF - Age 18. 331hrs.

BOMBER COMMAND

Anti-submarine patrol off coast of Spain

61 Squadron, RAF (Syerston, Nottinghamshire; detachment at St Eval, Cornwall - 5 Group)

Lancaster I R5605 - took off from St Eval at 1055 captained by Sgt A Wotherspoon, RAAF, and failed to return from an attack on a [U-boat?] tanker off the north-western coast of Spain near Corunna. All seven crew are commemorated on the Runnymede Memorial. During July-August 1942 Squadron aircraft were detached at various times to St Eval for Coastal Command operations, including R5605 and its crew on this occasion.

Captain: NZ405742 Flt Sgt **Owen HAYNES**, RNZAF - Age 28. 516hrs. 28th op.
Air Gunner: 1382135 Sgt **Donald James KING**, RAFVR - Age 24. 205hrs. 15th op.

Thu 20 Aug 1942

AUSTRALIA

Air Force Headquarters, RAAF (Melbourne, Victoria)

Died at the Ascot Vale Hospital and cremated at Springvale, on the outskirts of Melbourne.

Administrative Officer: Aus5441 Plt Off **John Oliver EVA**, RAAF - Age 41.

BOMBER COMMAND

Anti-submarine patrol off coast of Spain

61 Squadron, RAF (Syerston, Nottinghamshire; detachment at St Eval, Cornwall - 5 Group)

Lancaster I R5543 - took off from St Eval at 1100 captained by Plt Off J H Harrad, RAF, and brought down over Spain, crashing near Cabanas, 43km WSW of Corunna. All seven crew were buried at Lornes, but later reinterred at Bilbao, about 480km to the east. During July-August 1942 some or all of the Squadron's aircraft were detached at various times to St Eval for Coastal Command operations, including R5543 and its crew on this occasion.

Air Gunner: NZ41566 Sgt **Thomas Lindsay BOLAND**, RNZAF - Age 26. 200hrs. 5th op.

Thu 20 Aug/Fri 21 Aug 1942

BOMBER COMMAND

'Gardening' - minelaying from Brest, France to Danzig, Poland (by 57 aircraft - 8 lost)

218 Squadron, RAF (Downham Market, Norfolk - 3 Group)

Stirling I BF319/C - took off at 2020 to lay mines off the entrance of Kiel Harbour and brought down 17km SW of Kiel, at Langwedel. All seven crew were buried 15km further south at Neumünster, but later reinterred at Ohlsdorf, near Hamburg.

Captain: NZ41868 Fg Off **Derek Vaughan BULLOCK**, RNZAF - Age 23. 451hrs. 29th op.

CANADA

Night navigation exercise

3 Air Observers School, RCAF (Regina, Saskatchewan - 4 Training Command)

Anson I 6877 - piloted by I Bowman and at 0150 crashed into Chapleau Lake, 5½ miles SW of Canadian Station and 60 miles east of Regina. The pilot, two RCAF trainee navigators and an RNZAF staff wireless operator all died. The latter is buried at Regina, along with one of the navigators. Bowman was a civilian contracted as a pilot under the Empire Air Training Scheme and is therefore not recorded in the CWGC register.

Wireless Op: NZ405766 Sgt **Leonard Allen INSALL**, RNZAF - Age 21.

Fri 21 Aug 1942

COASTAL COMMAND

A & E [Airframe & Engine?] test, transiting from Ballykelly to Nutts Corner, Co. Antrim

120 Squadron, RAF (Ballykelly, Co. Londonderry - 15 Group)

Liberator GR III LV340/X - captained by Plt Off M F Dear, RAF, while flying in cloudy weather about noon struck Limnalary Mountain, Gortnacory, Co. Antrim, killing all eight on board. A New Zealand wireless operator-air gunner is buried at Ballykelly, along with the 2nd pilot and passenger Gp Capt H K Goode, DSO DFC AFC. Goode, a First World War fighter ace, had retired from the RAF but, ironically, was at the time employed as a civilian in the Accident Branch of the service.

WOpAG: NZ405529 Sgt **Thomas William TAYLOR**, RNZAF - Age 20. 279hrs. 16 ops.

Sun 23 Aug 1942

NEW ZEALAND

Army Co-operation exercise with the Home Guard

2 Anti-Aircraft Co-operation Flight, RNZAF (Rongotai)

Tiger Moth II NZ1432 took off during the morning with a member of the Home Guard on board and failed to return. An extensive search by hundreds of personnel proved fruitless and three years passed before its fate was discovered. On the afternoon of 21 August 1945, a Te Marama Station farm hand looking for his dog in the Makara hills, a few miles NW of Karori, stumbled upon the wreckage and passenger remains at the edge of dense bush in a steep gully, about 2 miles off the South Makara Road. The pilot's remains were discovered three days later. He had been thrown out of the machine when it crashed. He is buried at Onehunga and the passenger at Karori. NZ1432 had apparently been caught by a down draught while flying up the gully in bad weather, crashed into trees beside a small stream and caught fire.

Pilot: NZ416102 Plt Off **Mason Irvine ELLIOTT**, RNZAF - Age 27. 229hrs.
Passenger: 5/2300 Lt **Harold Francis BOLLARD**, NZ Military Forces - Age 37.

COASTAL COMMAND

Navigation exercise

3 School of General Reconnaissance, RAF (Squires Gate, Lancashire - 17 Group)

Botha I L6318 - piloted by Sgt H L Pendal, RAF. In cloudy weather at 1000 flew into the side of 2000 foot-high Tal-y-fan Mountain, 4 miles SW of Conway, Caernarvonshire. The five crew died, the wireless operator and two New Zealand trainee navigators being buried at Caernarvon.

Navigator: NZ413468 Sgt **Robert Wilmore PATRICK**, RNZAF - Age 26. 134hrs.
Navigator: NZ412781 Sgt **John Bernard WOOD**, RNZAF - Age 26. 178hrs.

Mon 24/Tue 25 Aug 1942

BOMBER COMMAND

Raid on Frankfurt, Germany (by 226 aircraft - 18 lost)

150 Squadron, RAF (Snaith, Yorkshire - 1 Group)

Wellington III BJ831 - took off at 2101 and over the target was hit by flak, wounding the rear gunner, and on the return over Belgium shot down by a night-fighter, to crash near Tirlemont, about 40km west of Brussels. Two RAF Sgts survived to become prisoners, although one died in March 1945. The other three crew are buried in a collective grave at Kersbeek-Miskom, 10km NNW of Tirlemont.

Captain: NZ402912 Flt Sgt **Harry Malfroy THOMPSON**, RNZAF - Age 28. 543hrs. 12th op.

218 Squadron, RAF (Downham Market, Norfolk - 3 Group)

Stirling I W7562/R - took off at 2110 captained by Flt Sgt R Yates, RAF, on same raid as the above, and brought down over Belgium, crashing at 2350 between Thynes and Sovet, 6-10km NE of Dinant. All eight crew were buried in a collective grave at Thynes, but later reinterred at Gosselies, immediately north of Charleroi.

2nd Pilot: NZ41928 Flt Sgt **James Auld MORGAN**, RNZAF - Age 23. 300hrs. 3rd op.

61 Squadron, RAF (Syerston, Nottinghamshire - 5 Group)

Lancaster I R5662 - took off at 2120 on same raid as the above, and brought down over Germany, crashing near Cologne. All seven crew were buried in south Cologne, but later reinterred at Rheinberg, 12km south of Wesel.

Captain: NZ405303 Flt Sgt **Noel Roy MEYER**, RNZAF - Age 22. 500hrs. 28th op.

It will be recalled that Meyer was a car passenger on 9 March 1941 when LAC Gerald Hewitt was accidentally killed in Christchurch, New Zealand.

Tue 25 Aug 1942

MIDDLE EAST

Interception of incoming raid

249 Squadron, RAF (Ta Kali, Malta)

Spitfire VB EN976/C - took off at 1130 with nine others and during an engagement with enemy fighters north of Grand Harbour was shot down, crashing into the sea 500 yards off the coast near St Julian's Bay. The pilot is commemorated on the Malta Memorial.

Pilot: NZ41360 Fg Off **Reginald Pearson ROUND**, RNZAF - Age 28. 395hrs. 100th op.

COASTAL COMMAND

Conveying Air Commodore HRH The Duke of Kent to Iceland for a tour of RAF Stations

228 Squadron, RAF (Oban, Argyllshire - 15 Group)

Sunderland III W4026/M - took off at 1310 from Invergordon, Ross & Cromarty, captained by Flt Lt F M Goyen, RAF, and 32 minutes later, off track and in cloud at 700 feet, flew into a hillside at Eagle's Rock, 3 miles west of Dunbeath, Caithness-shire. Of the four passengers and 11 crew on board only the rear gunner survived, although badly injured. A New Zealand wireless operator-air gunner is buried at Pennyfuir, 2 miles north of Oban, Argyllshire, as are three others of the crew, including the Squadron's commanding officer, Australian Wg Cdr T L Moseley,

RAF. Informative analysis and comment on this controversial accident by Roy Nesbit appears in the January and February 1990, March 1992 and September 1996 issues of *Aeroplane Monthly* magazine.
WOpAG: NZ405467 Sgt **Edward Francis BLACKLOCK**, RNZAF - Age 30.

Thu 27/Fri 28 Aug 1942

BOMBER COMMAND

Raid on Kassel, Germany (by 306 aircraft - 33 lost)

156 Squadron, RAF (Warboys, Huntingdonshire - [PFF] 3 Group)
Wellington III X3367 - took off at 2026 captained by Flt Sgt M G Savage, RAF, and shot down near the German-Dutch border by a night-fighter at 2325, crashing onto the Laurenz Textile Factory at Epe, 4km south of Gronau. All six crew were buried in the Evangelical Cemetery at Rheine, 28km to the NNW, but later reinterred at Reichswald Forest, 5km SW of Kleve.
Navigator: NZ404484 Plt Off **Peter HAYES**, RNZAF - Age 20.

156 Squadron, RAF (Warboys, Huntingdonshire - [PFF] 3 Group)
Wellington III DF667 - took off at 2035 captained by Sgt R V James, RAF, on same raid as the above, and brought down over Germany, crashing near Krefeld, about 10km SW of Duisburg. All five crew were buried at Krefeld, but later reinterred at Reichswald Forest, 5km SW of Kleve.
Navigator: NZ403954 Flt Sgt **Richard Maxwell DARLING**, RNZAF - Age 28. 339hrs. 20th op.

156 Squadron, RAF (Warboys, Huntingdonshire - [PFF] 3 Group)
Wellington III Z1613 - took off at 2039 captained by Plt Off J W Longhurst, RAF, on same raid as the above, and shot down over the Netherlands by a night-fighter at 2323, crashing near Elsendorp, 25km NE of Eindhoven. All six crew are buried at Eindhoven.
2nd Pilot: NZ412201 Plt Off **Robert William BYERS**, RNZAF - Age 21. 255hrs. 2nd op.
Navigator: NZ412363 Sgt **Joseph William STUART**, RNZAF - Age 25. 253hrs. 2nd op.

101 Squadron, RAF (Stradishall, Suffolk - 3 Group)
Wellington III X3649 - took off at 2045 on same raid as the above, but an engine caught fire and then failed, forcing a return to base. Bombs and ammunition were jettisoned but X3649 continued to lose height until finally compelled to ditch at 2240 off the East Anglian coast at position 4222N:0206E. All five crew safely left the wrecked Wellington, but only three survived long enough to be rescued by a trawler an hour later. The two who died are commemorated on the Runnymede Memorial.
Captain: NZ411364 Flt Sgt **George Thirlwell BROWN**, RNZAF - Age 25. 386hrs. 9th op(?)

218 Squadron, RAF (Downham Market, Norfolk - 3 Group)
Stirling I R9160/G - took off at 2045 captained by Plt Off W R B McCarthy, DFC, RAF, on same raid as the above, and shot down over the North Sea at 0140 by a night-fighter. The bodies of two RAF crew members were recovered and buried in the Netherlands, while their six comrades are commemorated on the Runnymede Memorial.
WOpAG: NZ41591 Sgt **Louis Arthur JONES**, RNZAF - Age 30. 131hrs.
The 28th was Jones' 30th birthday.

75 (NZ) Squadron, RAF (Mildenhall, Suffolk - 3 Group)
Wellington III BJ708 - took off at approximately 2045-2110 captained by Flt Lt A F A Osborn, DFC, RAF, on same raid as the above, and brought down over Germany, crashing near Melsungen, 21km south of the target. All five crew were buried in the Korle Cemetery at Melsungen, but later reinterred at Hannover.
Navigator: NZ411378 Plt Off **Errol Thomas Paterson DALZELL**, RNZAF - Age 22. 279hrs. 1st op.
Wireless Op: NZ412331 Sgt **Denis Patrick HOGAN**, RNZAF - Age 23. 103hrs. 1st op.
Rear Gunner: NZ411788 Sgt **Victor Arthur TUNBRIDGE**, RNZAF - Age 28. 134hrs. 1st op.

214 Squadron, RAF (Stradishall, Suffolk - 3 Group)
Stirling I R9155/Q - took off at 2059 captained by Fg Off A R Skone, RAF, on same raid as the above, and brought down over Germany, crashing at Issum, about 25km WNW of Duisburg. All seven crew are buried at Reichswald Forest, 5km SW of Kleve.
Navigator: NZ404344 Plt Off **Raymond Harold DUNCKLEY**, RNZAF - Age 22. 404hrs. 9th op.

15 Squadron, RAF (Bourn, Cambridgeshire - 3 Group)
Stirling I W7624/E - took off at 2100 captained by Flt Sgt H Barton-Smith, RAF, on same raid as the above and believed to have been shot down by a night-fighter, being seen to burst into flames over the Netherlands at 0045 before crashing onto a farm at Bentelo, 5km SSW of Ambt-Delden. At least three of the seven crew were seen to bale out, but they were not heard of again. The rear gunner's parachute failed to deploy and he was killed as he struck the ground about 5km NE of the crash site. He is buried at Ambt-Delden, together with two of his unidentified comrades whose bodies were recovered from the wreckage. The four missing and two unidentified airmen are commemorated on the Runnymede Memorial.
WOpAG: NZ403556 Sgt **Leonard Edward MOSS**, RNZAF - Age 28.
Rear Gunner: NZ403032 Flt Sgt **Glen Allen SMITH**, RNZAF - Age 24. 187hrs. 6th op.
The CWGC register incorrectly gives Smith's age as 22.

44 Squadron, RAF (Waddington, Lincolnshire - 5 Group)
Lancaster I W4124/D - on same raid as the above, and shot down over Germany by a night-fighter at 0026, crashing at Haltern, 20km SW of Rheine. All seven crew were buried in the Evangelical Cemetery at Rheine, but later reinterred at Reichswald Forest, 5km SW of Kleve.
Captain: NZ411469 Fg Off **Russell Henry SUCKLING**, RNZAF - Age 26. 412hrs. 7th op.

FERRY COMMAND

Ferry flight from Canada to Britain
Ferry Command, RAF (Dorval, Quebec)
Ventura II AE917 - took off from Gander, Newfoundland, captained by Plt Off C de C Findlay, RAF, and last heard of when an SOS message was received at 0821 GMT, stating it was flying on only one engine. Presumed to have gone down into the sea, two of the Ventura's crew, including the captain, are commemorated on the Ottawa Memorial. The bodies of their two comrades were recovered, including that of the New Zealand wireless operator-air gunner, which washed ashore at Clooney, County Donegal, Eire. He is buried at Irvinestown, Co. Fermanagh, Northern Ireland.
WOpAG: NZ412292 Sgt **Francis Augustine WEAVER**, RNZAF - Age 27.
Weaver, previously with 31 OTU at Debert, Nova Scotia, is likely to have been attached to Ferry Command only for the purpose of the delivery flight, pending further posting on arrival in the UK.

Fri 28/Sat 29 Aug 1942

BOMBER COMMAND

Raid on Saarbrücken, Germany (by 113 aircraft - 12 lost)
408 Squadron, RCAF (Balderton, Nottinghamshire - 5 Group)
Hampden I AE227/D - took off at 1956 and brought down about 8km NNW of the target area, crashing near Dudweiler. All four crew were buried in the Dudweiler Cemetery overlooking the Sulz Valley, but later reinterred at Rheinberg, 12km south of Wesel.
Captain: NZ41897 Fg Off **William Herbert GOULD**, RNZAF - Age 20. 445hrs. 25th op.

Raid on Nuremberg, Germany (by 159 aircraft - 24 lost)
101 Squadron, RAF (Stradishall, Suffolk - 3 Group)
Wellington III X3754 - took off at approximately 2030-2115 and brought down over Belgium at 0030, crashing at Cheratte, about 7km NE of Liège. The navigator was the sole survivor and later captured. All four of his comrades were buried at St Trond, about 35km to the west, but later reinterred 33km further west at Heverlee.
Captain: NZ41319 Flt Sgt **James Douglas FOLLETT**, RNZAF - Age 21. 15th op.
Wireless Op: NZ411103 Sgt **Murray SWANN**, RNZAF - Age 27. 227hrs. 12th op.
Air Bomber: NZ411765 Sgt **Frank Ernest KEARNS** RNZAF - Age 20. 155hrs. 13th op.

75 (NZ) Squadron, RAF (Mildenhall, Suffolk - 3 Group)
Wellington III X3389 - took off at approximately 2040-2105 on same raid as the above, and shot down by flak over Germany, crashing in the northern outskirts of Frankfurt, a kilometre east of Niederhöchstadt. All five crew were buried in the Okerrad Cemetery at Frankfurt, but later reinterred at Durnbach, 16km east of Bad Tölz.
Captain: NZ411934 Flt Sgt **Eric PERKS**, RNZAF - Age 29. 299hrs. 2nd op.

149 Squadron, RAF (Lakenheath, Suffolk - 3 Group)
Stirling I N6081/H - took off at 2058 captained by Wg Cdr C Charlton-Jones, RAF, on same raid as the above, and shot down over Germany by a night-fighter, crashing at Airlenbach in den Odenwalke, 34km NNW of Mannheim. Two RAF Sgts taken prisoner were the sole survivors. Their six comrades are buried at Durnbach, 16km east of Bad Tölz.
2nd Pilot: NZ41980 Fg Off **Charles Dudley DOBSON**, RNZAF - Age 24. 350hrs. 1st op(?)

57 Squadron, RAF (Feltwell, Norfolk - 3 Group)
Wellington III BJ701 - took off at 2100 captained by Flt Sgt J H Rowney, RCAF, on same raid as the above, and brought down some 12km north of the target area, crashing near Kalchreuth. All five crew were buried in the New Cemetery at Kalchreuth in a collective grave, together with four crew members from a 49 Sqn Lancaster I (R5897) which also crashed during the raid in the same vicinity. All nine were later reinterred at Durnbach, 16km east of Bad Tölz.
Navigator: NZ404995 Flt Sgt **Kenneth Charles BROWN**, RNZAF - Age 22. 318hrs. 16th op.
Air Bomber: NZ402525 Sgt **Henry William PIKE**, RNZAF - Age 22. 193hrs. 13th op.
WOpAG: NZ411064 Sgt **David George CRAILL**, RNZAF - Age 24.

49 Squadron, RAF (Scampton, Lincolnshire - 5 Group)
Lancaster I L7567 - took off at 2111 on same raid as the above, and critically damaged by a night-fighter over the Netherlands near Venlo at about 2300, later crashing at Maasbracht, 42km SW of Venlo. The flight engineer and navigator survived and were taken prisoner. Their five comrades were buried in an unmarked grave near the crash site, but later reinterred in a collective grave at Mook, 11km SSE of Nijmegen.
Captain: NZ405290 Fg Off **John LOWRIE**, RNZAF - Age 21. 386hrs. 8th op.

Tue 1 Sep 1942

COASTAL COMMAND

Dual instruction

5 (Coastal) Operational Training Unit, RAF (Turnberry, Ayrshire - 17 Group)

Hampden I P1199 - while making a northerly approach at 1424, collided at 100 feet with one of the Unit's Beaufort Is (L9932), also on approach. Both crashed and burst into flames, killing the two Hampden crew, who are both buried at Dunure, about 7 miles NE of the airfield. Two of those in the Beaufort also died, the two others surviving injured.

Pilot: NZ401767 Wt Off **James Edward JONES**, RNZAF - Age 22. 364hrs. 3 ops.

Tue 1/Wed 2 Sep 1942

BOMBER COMMAND

Raid on Saarbrücken, Germany (by 231 aircraft - 8 lost)

218 Squadron, RAF (Downham Market, Norfolk - 3 Group)

Stirling I N3714/Q - took off at 2346 and on returning was shot down over Belgium at 5,000 feet by a night-fighter, crashing at 0315 at Sterrebeek, 11km east of Brussels. An RAF Sgt and navigator Plt Off **E G Ruegg**, RNZAF, were the only survivors, both being captured. Their five comrades are buried in a collective grave a few kilometres to the north, at Nossegem.

Captain: NZ41999 Flt Sgt **Keith Forbes RYAN**, RNZAF - Age 28. 373hrs. 9th op.
Air Bomber: NZ411778 Sgt **George Ernest REID**, RNZAF - Age 31. 170hrs. 3rd op.

Wed 2 Sep 1942

NEW ZEALAND

RNZAF Station Waipapakauri

Died of a self-inflicted gunshot wound to the head while on guard duty at the station and is buried at Hamilton.

(Aircrafthand): NZ425430 AC1 **Ronald Anthony MULCOCK**, RNZAF - Age 20.

Mulcock had enlisted for aircrew training and was serving on Aerodrome Defence Unit duty, pending entry to the Initial Training Wing at Rotorua.

COASTAL COMMAND

Night navigation training

Torpedo Training Unit, RAF (Abbotsinch, Renfrewshire - 17 Group)

Beaufort I N1180/S - veered slightly off track after wind strength increased during an unexpected weather change and at 2240 flew into a hill $1^1/_2$ miles east of Mull of Kintyre lighthouse, Argyllshire. All four crew died, a New Zealand air gunner and one other being buried at Kilkerran, near Campbeltown, while the pilot is buried at Brookwood in Woking, Surrey.

Pilot: NZ414287 Sgt **Albert Augustine HAYDON**, RNZAF - Age 26. 269hrs.
Air Gunner: NZ391207 Plt Off **Leonard Percy BOOKER**, RNZAF - Age 29. 90hrs.

Wed 2/Thu 3 Sep 1942

BOMBER COMMAND

Night training

11 Operational Training Unit, RAF (Bassingbourn, Cambridgeshire - 91 Group)

Wellington IC DV718 - flew into a hill and caught fire $2^1/_2$ miles SW of Angram Bridge, near Pateley Bridge, Yorkshire, after veering off track 19 miles to the west in rough and cloudy weather. Only one of the five crew survived, injured. The captain is buried 17 miles to the south, at Shipley. Although death is recorded as being on the 3rd, the actual time of this accident is unknown. If it occurred shortly after midnight, rather than before, then the event should appear under the night of 3/4 September. Another aircraft also crashed in the same area on the same night.

Captain: NZ414680 Sgt **George Francis RIDGWAY**, RNZAF - Age 20. 309hrs (185 solo - 53 on Wellington)

Thu 3/Fri 4 Sep 1942

BOMBER COMMAND

Raid on Emden, Germany (by 11 aircraft - 2 lost)

75 (NZ) Squadron, RAF (Mildenhall, Suffolk - 3 Group)

Wellington III X3396 - took off at approximately 2325-2330 captained by Sgt J Law, RAF, and brought down over Germany, crashing near Wittmund, 19km WNW of Wilhelmshaven. All five crew were buried at Wittmund, but later reinterred at Sage, 24km south of Oldenburg.

Navigator: NZ405254 Sgt **Horace Llewellyn GRANT**, RNZAF - Age 27. 250hrs. 2nd op.
Air Bomber: NZ405309 Sgt **Richard Alfred William NEWMAN**, RNZAF - Age 24. 248hrs. 2nd op.
Wireless Op: NZ412352 Sgt **Rupert Ernest RENTON**, RNZAF - Age 22. 183hrs. 2nd op.
Rear Gunner: NZ403362 Sgt **John Trevor Vivian GILL**, RNZAF - Age 27. 116hrs. 2nd op.

Fri 4 Sep 1942

BOMBER COMMAND

Accident to pedestrian

15 Operational Training Unit, RAF (Harwell, Berkshire - 91 Group)

Critically injured when knocked down by a bus at 1805 near the Great Western Railway Station at Reading, Berkshire. Admitted to hospital, but died the following day and buried at Harwell.

(Pilot): NZ414239 Sgt **Thomas Bernard BURROWES**, RNZAF - Age 19. 328hrs.

Fri 4/Sat 5 Sep 1942

BOMBER COMMAND

Raid on Bremen, Germany (by 251 aircraft - 14 lost)

214 Squadron, RAF (Stradishall, Suffolk - 3 Group)

Stirling I BF337/B - took off at 0004 captained by Flt Lt J G Smith, DFC, RCAF, and was badly damaged in a night fighter attack at 16,000 feet over the Netherlands near Groningen, later exploding in the air about 90km further south over Rouveen, 13km NNW of Zwolle. Four of the crew, including the captain, survived and were captured, while their three comrades are buried at Rouveen, 4km south of Staphorst.

Air Gunner: NZ411222 Plt Off **Arthur McCARTHY**, RNZAF - Age 23. 78hrs.

101 Squadron, RAF (Stradishall, Suffolk - 3 Group)

Wellington III BJ891 - took off at approximately 2330-0030 captained by Plt Off N I Gill, RAF, on same raid as the above and lost without trace, its five crew being commemorated on the Runnymede Memorial.

Air Bomber: NZ411763 Sgt **George Smith JENNINGS**, RNZAF - Age 24. 138hrs. 7th op.

Sat 5 Sep/Sun 6 Sep 1942

MIDDLE EAST

Raid on Tobruk, Libya

70 Squadron, RAF (Abu Sueir, Egypt - 231 Wing, 205 Group)

Wellington IC BB477/W - brought down near the target area. The 2nd pilot and front gunner are buried at Tobruk, while their four comrades are commemorated on the Alamein Memorial.

Captain: NZ402455 Flt Sgt **James Douglas GORDON**, RNZAF - Age 27. 485hrs. 31st op.
Front Gunner: NZ404059 Flt Sgt **Stanley Isaiah GALLAGHER**, RNZAF - Age 29. 358hrs. 30th op.

FLYING TRAINING COMMAND

Dual instruction in circuits and landings

5 (Pilots) Advanced Flying Unit, RAF (Tern Hill, Shropshire; satellite at Calveley, Cheshire - 21 Group)

Master III W8795 - took off from Calveley at about 0040 piloted by Sgt A Bradshaw, RAF, and believed to have suffered engine failure before crashing on striking a tree while attempting a forced landing at Brindley Common, about 3 miles south of the airfield. Both pilots are buried at Stoke-upon-Tern, Shropshire. There had already been two accidents in the area earlier in the night.

Pilot: NZ415346 Sgt **Euan William MACLAREN**, RNZAF - Age 20. 195hrs (97 solo - 7 on Master)

Sun 6 Sep 1942

BOMBER COMMAND

Night training

21 Operational Training Unit, RAF (Moreton-in-Marsh, Gloucestershire - 91 Group)

Wellington IC X9942 - took off at about 2115 and began a turn at 200 feet, but stalled, crashed and caught fire at Sezincote, 2 miles WSW of Moreton-in-Marsh. All five crew died, both pilots and two of their comrades being buried at Moreton-in-Marsh.

Captain: NZ414666 Sgt **Charles Ambrose O'HALLORAN**, RNZAF - Age 26. 334hrs.
2nd Pilot: NZ414682 Sgt **Edward Desmond SALE**, RNZAF - Age 20. 250hrs.

O'Halloran's brother, William Peter O'Halloran, died only eight days later, while flying with 1651 Conversion Unit, RAF.

The CWGC register records Sale's first name as Ernest but all official RNZAF documents, including his attestation form, quote Edward.

Sun 6/Mon 7 Sep 1942

BOMBER COMMAND

Raid on Duisburg, Germany (by 207 aircraft - 11 lost)

75 (NZ) Squadron, RAF (Mildenhall, Suffolk - 3 Group)

Wellington III X3867 - took off at approximately 0105-0135 and shot down over the Netherlands by a night-fighter, crashing at Riel, 5km SW of Tilburg. All five crew are buried in a collective grave at Tilburg.

Captain: NZ403822 Flt Sgt **William Ronald PARKES**, RNZAF - Age 31. 536hrs. 21st op.
Navigator: NZ404339 Sgt **Henry Varley Gibb CRAWFORD**, RNZAF - Age 28. 355hrs. 21st op.

Air Bomber: NZ403574 Flt Sgt **Alpheus Leslie GROVES**, RNZAF - Age 30. 277hrs. 19th op.
Wireless Op: NZ391713 Sgt **George Herbert ROSE**, RNZAF - Age 30. 268hrs. 17th op.
Rear Gunner: NZ403035 Flt Sgt **Eric Glover WILSON**, RNZAF - Age 27. 257hrs. 18th op.

75 (NZ) Squadron, RAF (Mildenhall, Suffolk - 3 Group)
Wellington III BJ765 - took off at approximately 0105-0135 on same raid as the above, and brought down over Germany, crashing at Hockshoefer. All five crew were buried in the Military Cemetery at Krefeld, but later reinterred at Reichswald Forest, 5km SW of Kleve.
Captain: NZ412746 Sgt **George William SHARMAN**, RNZAF - Age 27. 294hrs. 4th op.
Navigator: NZ411730 Sgt **Norman Priestley SAUL**, RNZAF - Age 30. 230hrs. 3rd op.
Air Bomber: NZ411769 Plt Off **George William Alfred MILLS**, RNZAF - Age 27. 106hrs. 4th op(?)
Wireless Op: NZ411722 Sgt **Claude Joseph Frederick JARVIS**, RNZAF - Age 22. 95hrs. 3rd op.
Rear Gunner: NZ40161 Sgt **Thomas Edward COLES**, RNZAF - Age 28. 66hrs. 2nd op.

7 Squadron, RAF (Oakington, Cambridgeshire - [PFF] 3 Group)
Stirling I W7629/Z - took off at 0111 captained by Flt Lt N A Bennitt, DFC, RAF, on same raid as the above, and brought down over Germany, crashing at the Baaler Bruch, 6km SSW of Goch. The seven crew were buried in a collective grave nearby, but later reinterred at Reichswald Forest, 5km SW of Kleve.
Rear Gunner: NZ403040 Flt Lt **Jack Menzies SMITH**, DFM, RNZAF - Age 25. 330hrs.
Jack Menzies Smith should not be confused with his similarly named brother who survived the war, Sqn Ldr James Menzies Smith, DFC.

Mon 7 Sep 1942

FLYING TRAINING COMMAND

Training
7 (Pilots) Advanced Flying Unit, RAF (Peterborough, Northamptonshire - 21 Group)
Master II DL838 - collided with another of the Unit's Master IIs (DL849) and crashed at Newton, Cambridgeshire, 5 miles SSW of Sutton Bridge, Lincolnshire. The two pilots in each Master died, those in DL838 being buried at Sutton Bridge.
Pilot: NZ413368 Plt Off **Noel Galbraith BAILEY**, RNZAF - Age 20. 232hrs.
The other pilot with Bailey was Sgt F N Andrew, RCAF, but it is not known who was in command at the time of the accident.

Tue 8 Sep 1942

MIDDLE EAST

Anti-submarine patrol
209 Squadron, RAF (Kipevu, Kenya; detachment at Pamanzi, Comoro Islands - 246 Wing[?], 207 Group, East Africa)
Catalina IIA VA713/L - crashed on take off at 0420, killing all ten crew. Five are buried at Diégo Suarez, Madagascar, while the others, including the captain, are commemorated on the Alamein Memorial.
Captain: 41093 Flt Lt **John Robertson WYLLIE**, RAF - Age 25. 599hrs. 106th op.

FLYING TRAINING COMMAND

Solo circuit and landing practice
7 (Pilot) Advanced Flying Unit, RAF (Peterborough, Northamptonshire; relief landing ground at Sibson - 21 Group)
Master II DL841 - after carrying out an overshoot at Sibson, 7 miles west of Peterborough, executed a series of turns and dives before diving into the ground at 0330. The pilot is buried at Peterborough.
Pilot: NZ415062 Sgt **Alan James CRAIG**, RNZAF - Age 20. 209hrs (116 solo - 2 on Master)

Tue 8/Wed 9 Sep 1942

FLYING TRAINING COMMAND

Dual check
15 (Pilots) Advanced Flying Unit, RAF (Leconfield, Yorkshire; satellite at Kirmington, Lincolnshire - 21 Group)
Oxford I BG270 - took off from Kirmington at about 0030, but failed to climb away and crashed a few minutes later into rising ground, catching fire SSE of the airfield. The two pilots are buried at Brigg, Lincolnshire. Subsequent investigation raised concern that the instructor had been overworked and that fatigue may have been a factor leading to the accident.
Pilot Instr: NZ413083 Sgt **Charles Glynn JONES**, RNZAF - Age 20. 450hrs (309 solo - 266 on Oxford)

Thu 10/Fri 11 Sep 1942

BOMBER COMMAND

Raid on Düsseldorf, Germany (by 429 aircraft - 38 lost)

78 Squadron, RAF (Middleton St George, Durham - 4 Group)

Halifax II DT491 - took off at 2014 captained by Plt Off C J Stevenson, RAF, and brought down near the target area. All seven crew were buried in the city's North Military Cemetery, but later reinterred at Reichswald Forest, 5km SW of Kleve.

Air Bomber: NZ404586 Sgt **Albert Edgar ARNOLD**, RNZAF - Age 23. 252hrs. 16th op.

158 Squadron, RAF (East Kirby, Lincolnshire - 4 Group)

Halifax II W7745 - took off at 2016 on same raid as the above, and brought down into the North Sea. Three of the crew are commemorated on the Runnymede Memorial. The bodies of their five comrades later washed up onto the Belgium and Dutch coasts, including that of the captain, who was buried at Ostend on 17 September.

Captain: 36274 Sqn Ldr **Peter Bettley ROBINSON**, DFC, RAF - Age 22. 1145hrs. 45 ops.

11 Operational Training Unit, RAF (Bassingbourn, Cambridgeshire - 91 Group)

Wellington IC DV890/G - took off at 2036 on same raid as the above, and brought down north of the target, crashing near Bismark, Gelsenkirchen. All five crew were buried in the East Cemetery at Bismark, but later reinterred at Reichswald Forest, 5km SW of Kleve.

Captain: NZ414344 Sgt **Norman John ROWE**, RNZAF - Age 29. 271hrs. 1st op.
Navigator: NZ41332 Sgt **Jack INSKIP**, RNZAF - Age 22. 198hrs. 1st op.
Air Bomber: NZ413308 Sgt **Walter Archibald O'MALLEY**, RNZAF - Age 21. 205hrs. 1st op.
Wireless Op: NZ411076 Sgt **Thomas Pearse HURLEY**, RNZAF - Age 26. 118hrs. 1st op.

11 Operational Training Unit, RAF (Bassingbourn, Cambridgeshire - 91 Group)

Wellington IC DV930/O(bar) - took off at 2105 on same raid as the above and on returning to England at 2350, crashed in the stack yard at Herne Hill Farm, Chediston, 2 miles west of Halesworth, Suffolk. Four of the crew were killed, a fifth, an RAF Sgt, dying a few hours later. The captain and navigator are buried at Ipswich.

Captain: NZ415005 Sgt **Thomas Fraser MUNRO**, RNZAF - Age 31. 308hrs. 1st op.
Navigator: NZ41553 Sgt **Jack Foden STANLEY**, RNZAF - Age 20. 228hrs. 1st op.

75 (NZ) Squadron, RAF (Mildenhall, Suffolk - 3 Group)

Wellington III BJ828 - took off at approximately 2109-2125 captained by Sgt E Lees, RAF, on same raid as the above and lost without trace, its five crew being commemorated on the Runnymede Memorial.

Navigator: NZ413336 Sgt **Douglas Howard JUDD**, RNZAF - Age 26. 160hrs. 1st op.

75 (NZ) Squadron, RAF (Mildenhall, Suffolk - 3 Group)

Wellington III BJ968/W - took off at approximately 2109-2125 captained by Flt Sgt F Burrill, RCAF, on same raid as the above, and lost without trace, its five crew being commemorated on the Runnymede Memorial.

Navigator: NZ411006 Plt Off **Randolph Cruickshank SMART**, RNZAF - Age 25. 234hrs. 1st op.

75 (NZ) Squadron, RAF (Mildenhall, Suffolk - 3 Group)

Wellington III BJ974 - took off at approximately 2109-2125 on same raid as the above, and brought down over Germany, crashing near Cologne. All five crew died, the New Zealanders being buried at South Cologne. Later the crew was reinterred at Rheinberg, 12km south of Wesel.

Captain: NZ414386 Sgt **Thomas Otto METCALFE**, RNZAF - Age 19. 336hrs. 4th op.
Navigator: NZ413334 Plt Off **Kevin John DEVLIN**, RNZAF - Age 26. 261hrs. 2nd op.
Air Bomber: NZ411908 Sgt **Walter Foch KELCHER**, RNZAF - Age 23. 176hrs. 2nd op.
Wireless Op: NZ412912 Sgt **Desmond James WALSHE**, RNZAF - Age 25. 177hrs. 2nd op.

The 11th was Devlin's 26th birthday.

218 Squadron, RAF (Downham Market, Norfolk - 3 Group)

Stirling I BF351/C - took off at 2125 captained by Flt Sgt G E Milligan, RCAF, on same raid as the above, and shot down over the Netherlands by a night-fighter at 2340, crashing at St Philipsland, about 17km NW of Bergen-op-Zoom. All seven crew are buried at Bergen-op-Zoom.

Air Gunner: NZ41611 Sgt **Maurice Varey TATE**, RNZAF - Age 28. 173hrs. 4th op.
Air Gunner: NZ404483 Sgt **Raymond Edgar GRAHAM**, RNZAF - Age 21. 167hrs. 1st op.

149 Squadron, RAF (Lakenheath, Suffolk - 3 Group)

Stirling I R9170/H - took off at 2141 captained by Flt Sgt A F Potts, RAF, on same raid as the above, and brought down over the Dutch island of Overflakkee, crashing at Ooltgensplaat, about 25km SSW of Rotterdam. An RAF Sgt baled out as the sole survivor and was taken prisoner. His six comrades are buried in the Netherlands, the New Zealanders at Rotterdam.

Air Bomber: NZ41610 Sgt **Reginald Frederick SWINDLEHURST**, RNZAF - Age 26. 194hrs. 12th op.
Wireless Op: NZ411236 Sgt **George Edward MARTIN**, RNZAF - Age 22. 219hrs. 11th op.
Rear Gunner: NZ405497 Sgt **James Flers LUMSDEN**, RNZAF - Age 25. 208hrs. 11th op.

Fri 11 Sep 1942

COASTAL COMMAND

Practice

5 (Coastal) Operational Training Unit, RAF (Turnberry, Ayrshire - 17 Group)

Beaufort II AW309 - port engine failed and its propeller flew off. On approaching the airfield for a forced landing the undercarriage would not lower, even using an emergency hand pump. When turning away to starboard at 0705, while attempting to fix the undercarriage, AW309 stalled and crashed into the sea off Maidens Harbour, about a mile to the north. Three crew members, two of them injured, escaped into a dinghy from the sinking Beaufort. The pilot could not, however, because prior to the accident his safety harness locking mechanism had been damaged and could not be fully released. He is commemorated on the Runnymede Memorial.

Pilot: NZ413013 Sgt **Norman James BARLOW**, RNZAF - Age 24. 381hrs (280 solo - 29 on Beaufort)

High level photographic exercise

8 (Coastal) Operational Training Unit, RAF (Fraserburgh, Aberdeenshire - 17 Group)

Spitfire IV BR655 - spun out of cloud over Montrose airfield at 1040 and crashed near the north end of the runway. The pilot is buried at Montrose.

Pilot: NZ413875 Sgt **Frederick Thomas MARTYN**, RNZAF - Age 21. 236hrs.

Sat 12 Sep 1942

COASTAL COMMAND

Anti-submarine patrol 'BP25'

502 Squadron, RAF (St Eval, Cornwall - 19 Group)

Whitley VII BD428/E - took off at 1255 captained by Plt Off J A Arbuckle, RCAF, and at 1715 signalled, 'am being attacked by enemy aircraft and am returning to base', after which nothing further was heard. All six crew died, only a dinghy and the body of a New Zealand wireless operator-air gunner being recovered by a lifeboat off the Scilly Isles. He is buried at St Mary's, while his comrades are commemorated on the Runnymede Memorial. It was later believed that the Whitley had been shot down over a convoy by 'friendly sea forces' in the vicinity of Lands End.

WOpAG: NZ404680 Flt Sgt **Leonard James Stafford BUSH**, RNZAF - Age 28. 322hrs.

ATLANTIC OCEAN

In transit from Middle East to Britain

HMT *Laconia*

The 19,695-ton armed liner *Laconia* was torpedoed at 2007 by U-156 about 500 miles south of Cape Palmas, Liberia. She was homeward bound from Suez via the Cape, carrying 2,725 crew, passengers and Italian prisoners of war. U-156 surfaced to take prisoners and, on hearing of the Italian prisoners, called in other U-boats and began a rescue. The operation was later halted by the arrival of an Ascension Island-based USAAF Mitchell bomber which proceeded to attack U-156. This action later led to German U-boat Commander-in-Chief, Grand Admiral Dönitz, issuing what became known as the *'Laconia Order'*, which decreed that the rescue of occupants of ships which had been sunk was to cease. Some 1000 survivors were later rescued by a Vichy French cruiser, while others drifted in lifeboats for many weeks before death or rescue. Among those lost at sea was a New Zealander in the Fleet Air Arm, who died on or about 1 October in a lifeboat which eventually landed in Liberia on the 10th. A New Zealander in the RAF died aboard another lifeboat, in which only four out of 51 occupants survived. His date of death is officially presumed to be 20 October, the day prior to the vessel being picked up. The airmen are commemorated on the Fleet Air Arm Memorial at Lee-on-the-Solent and the Runnymede Memorial, respectively.

(Pilot): Lt (A) **Peter Thurston MEDHURST**, RNVR - Age 26.

Medhurst had last served with 775 Sqn, FAA.

(Pilot): 42503 Flt Lt **Gilbert McLean HAYTON**, DFC, RAF - Age 25.

Hayton had last served with 89 Sqn, RAF.

Sun 13/Mon 14 Sep 1942

MIDDLE EAST

Raid on Tobruk, Libya, (by 101 aircraft) in support of a combined forces operation

70 Squadron, RAF (Abu Sueir, Egypt - 231 Wing, 205 Group)

Wellington IC Z8787/R - captained by Flt Lt E F W McCartan-Mooney, RAF, and shot down during the operation, all six crew surviving to be taken prisoner. Several later attempted an escape, but at 1930 on the 18th the captain, navigator and rear gunner lost their lives in a skirmish with an evening patrol. They were buried about three kilometres south of the 62km stone on the Egyptian road from Sidi Barrani to Sollum. Only the rear gunner's grave was located after hostilities and he is now buried at Halfaya Sollum, while his two comrades are commemorated on the Alamein Memorial.

Navigator: NZ403478 Fg Off **Nathan SPENCE**, RNZAF - Age 23. 450hrs. 32nd op.

Rear Gunner: NZ403588 Flt Sgt **Charles Cecil PIPER**, RNZAF - Age 24.

BOMBER COMMAND

Raid on Bremen, Germany (by 446 aircraft - 26 lost)

156 Squadron, RAF (Warboys, Huntingdonshire - [PFF] 3 Group)

Wellington III BJ600 - took off at 2245 and brought down over the Netherlands, close to the German border, crashing at 0313 near the village of Loozen, 3km south of Gramsbergen. All six crew are buried at Gramsbergen

Captain: NZ39082 Plt Off **Valentine George BROUGH**, RNZAF - Age 21. 365hrs. 17th op.

16 Operational Training Unit, RAF (Upper Heyford, Oxfordshire - 92 Group)

Wellington IC X9660 - took off on same raid as the above and last made contact at about 0500 when returning over the sea. Presumed to have ditched, its five crew are commemorated on the Runnymede Memorial.

Captain: NZ402859 Plt Off **Peter Stanley DUFF**, DFC, RNZAF - Age 21. 665hrs. 33rd op.

Duff had previously completed a tour of ops with 106 Sqn and was serving as an instructor with the OTU at the time of his death.

1651 Conversion Unit, RAF (Waterbeach, Cambridgeshire - 3 Group)

Stirling I BF332 - took off at about midnight captained by Wt Off G F H Sayers, RAF, on same raid as the above. On running in to the target its nose was hit by flak, a cannon shell killing the air bomber. BF332 and the rest of its crew, including air gunner Sgt **W J Fitzgerald**, RNZAF, safely returned to England, their dead comrade being buried at Cambridge on the 17th.

Air Bomber: NZ41352 Sgt **William Peter O'HALLORAN**, RNZAF - Age 23. 3rd op.

O'Halloran's brother, Charles Ambrose O'Halloran, died only eight days earlier while flying with 21 OTU, RAF.

1654 Conversion Unit, RAF (Wigsley, Nottinghamshire - 5 Group)

Lancaster I W4108 - took off at 2330 captained by Plt Off E F Dowdell, RAF, on same raid as the above was shot down over the Netherlands by a night-fighter, crashing at 0222 on the Kleistra brothers' farm at Buytenweg, Oudehorne, 10km east of Heerenveen. All seven crew are buried at Nieuwehorne, 9km east of Heerenveen.

Navigator: NZ391872 Plt Off **Patrick Aylmer VIVIAN**, RNZAF - Age 21. 300hrs. 32nd op.

Vivian had previously completed an operational tour with 83 Sqn, RAF.

Mon 14/Tue 15 Sep 1942

ATLANTIC OCEAN

In transit from Bathurst, Gambia, to Britain via Lisbon, Portugal

British Overseas Airways Corporation

S.30 Empire Flying-boat G-AFCZ (named *Clare*) - took off in the evening captained by Fg Off G B Musson, RAF, and little over an hour later signalled it was experiencing engine difficulties. About 20 minutes later there followed an emergency call of 'fire, fire, fire!' *Clare's* fate remained unknown until the 16th, when searching Catalina G-AGDA located wreckage and six bodies floating in the water off the Senegalese coast at position 1420N:1732W. *Clare's* six crew and 13 passengers all died; amongst the latter were at least eight tour-expired 37 Sqn personnel being repatriated to Britain, including a New Zealand navigator. The bodies of two RAF members of the Squadron were recovered and buried at sea, the remainder of their comrades and the *Clare's* crew being commemorated on the Alamein Memorial. The other five passengers were apparently Army, Navy or civilian personnel. Operating at the limit of their range, with minimum safety margins, the other S.30s were promptly withdrawn from the route following *Clare's* loss.

(Navigator): NZ404678 Plt Off **Terence Albert DIXON**, RNZAF - Age 20.

Dixon was returning to the UK, having completed an operational tour with 37 Sqn, RAF.

COASTAL COMMAND

U-boat search north of Shetland islands

489 Squadron, RNZAF (Skitten, Caithness-shire - 18 Group)

Hampden TB I AD795/K - took off at 1735 and signalled at 0217, 'short of petrol. May have to ditch', followed three minutes later by, 'will have to land in sea.' SOS signals continued to be sent between 0222 and 0227, when AD795 was plotted in a position some 60 miles east of Rattray Head, Aberdeenshire. Eight aircraft searched on the 15th but found nothing. Later the body of Sgt **Brown** was recovered on the German island of Sylt and buried at List, while that of Sgt **Newman** was found further south and buried at Husum. Both were later reinterred at Kiel. Their two comrades are commemorated on the Runnymede Memorial.

Pilot: NZ404925 Fg Off **Thomas Donald Gordon MURRAY**, RNZAF - Age 29. 336hrs. 3rd op.
Navigator: NZ404958 Sgt **Charles Henry BRANDON**, RNZAF - Age 24. 304hrs.
WOpAG: NZ41568 Sgt **Alexander George BROWN**, RNZAF - Age 22. 172hrs.
WOpAG: NZ41602 Sgt **Douglas NEWMAN**, RNZAF - Age 25. 173hrs. 4th op.

BOMBER COMMAND

Raid on Wilhelmshaven, Germany (by 202 aircraft - 3 lost)

218 Squadron, RAF (Downham Market, Norfolk - 3 Group)

Stirling I N3725/D - took off at 1955 captained by Fg Off J C Frankcombe, RAAF, but while over the target area suffered a starboard inner engine failure. On returning to the base at 0055, the starboard outer engine also cut, through lack of fuel, just as the undercarriage was being lowered while circling to land. N3725 then spun out of

control into the ground and caught fire a mile east of Stokeferry, 6 miles WNW of Downham Market. Six of the crew died, the captain, air bomber and rear gunner being buried at Marham. An RAF Sgt survived slightly injured, while Sgt **J J Marshall**, RNZAF, severely injured, lost both his legs. The cause of the inner engine failure was later established to be due to sleeve valve failure in the No 12 cylinder.
Air Bomber: NZ411776 Sgt **Ernest Trevor PELLOW**, RNZAF - Age 24. 215hrs. 11th op.
Rear Gunner: NZ412529 Sgt **Sidney Horace IVES**, RNZAF - Age 22. 219hrs. 10th op.

115 Squadron, RAF (Marham, Norfolk - 3 Group)
Wellington III BJ693/J - took off at 1958 on same raid as the above, but developed engine trouble on the return and was presumed to have ditched or crashed into the North Sea. Three bodies later washed ashore, two of them near the Danish town of Fourfelt, just north of Esbjerg, that of the 2nd Pilot being buried there on 12 October. The three others are commemorated on the Runnymede Memorial.
Captain: NZ411852 Plt Off **Frank Bruce BOADEN**, RNZAF - Age 24. 398hrs. 17th op.
2nd Pilot: NZ413162 Sgt **Kenneth John WARNER**, RNZAF - Age 20. 373hrs. 1st op(?)

Tue 15 Sep 1942

NEW ZEALAND

1 Personnel Dispatch Centre, Auckland City
Died of natural causes at the Alma Private Hospital at Auckland and was later cremated in that city.
(Clerk): NZ401548 Sgt **Albert John LESTER**, RNZAF - Age 32.
Prior to falling ill, Lester was serving with No 1 Stores Depot, Hobsonville.

FLYING TRAINING COMMAND

Glider towing training
4 Glider Training School, RAF (Kidlington, Oxfordshire - 23 Group)
Hector I K9706 - took off at about 0155 towing a General Aircraft Hotspur II glider (BT505) for a single circuit. About five minutes later in low cloud the combination flew into high ground near Wytham, 4 miles SSW of the airfield. The pilot of the Hector and both occupants of the glider died, the former being buried at Kidlington.
Pilot: NZ412180 Sgt **Alistair William ALLEN**, RNZAF - Age 20. 542hrs (485 solo - 362 on Hector)
Allen's brother, John Herbert Allen, died on 8 April 1941 while flying with 232 Sqn, RAF.

BOMBER COMMAND

Gunnery training
1484 Gunnery Flight, RAF (Driffield, Yorkshire - 4 Group)
Defiant I N3501 - took off at about 1000, but towards the end of the exercise the pilot unsuccessfully tried to inform control that the undercarriage would not lower. He then engaged in some aerobatic manoeuvres in an attempt to unlock it, but at 1137 control was lost and N3501 crashed at Hackness Park, about 5 miles west of Scarborough. The gunner safely baled out but the pilot died and is buried at Driffield.
Pilot: NZ401387 Wt Off **Derek Frederick Paterson JANNINGS**, RNZAF - Age 25. 539hrs. 14 ops.
Jannings 14 ops were undertaken with 40 Sqn, RAF, prior to his posting to the Flight as a staff pilot

COASTAL COMMAND

Anti-submarine patrol - 'BP13'
502 Squadron, RAF (St Eval, Cornwall - 19 Group)
Whitley VII Z9365/A - took off at 1305 and was forced to ditch in the sea 150 miles west of Brest, France, after an encounter with several enemy aircraft, during which the captain was wounded. Z9365 broke in two on impact and its large dinghy sank with the tail half. The 2nd pilot and an RCAF member of the crew drifted away from the wreckage and were not seen again. The other four clambered into a small dinghy, but the captain died on the 16th, another from injuries received in the crash, and a third from exposure. An RAF Sgt was the sole survivor, being picked up by a French fishing boat, only to be captured by the Germans on the 30th. His five comrades are commemorated on the Runnymede Memorial.
Captain: NZ40643 Plt Off **Alan Edward COATES**, mid, RNZAF - Age 30.
2nd Pilot: NZ414594 Sgt **Alexander Bridges COBURN**, RNZAF - Age 27. 382hrs.
Coburn's true family name was Cockburn, although he enlisted and served as Coburn.

CANADA

Precautionary landings and low flying instruction
12 Service Flying Training School, RCAF (Brandon, Manitoba - 2 Training Command)
Crane IA FJ261 - while low flying at 1600 in the vicinity of Souris, 23 miles SW of the airfield, struck a 20 foot-high straw stack and rolled over before striking the ground about 300 yards further south. The instructor and his RCAF pupil died, the former being buried at Brandon.
Pilot Instr: NZ415725 Plt Off **Clifford Harris THOMPSON**, RNZAF - Age 22. 384hrs.

Wed 16 Sep 1942

BOMBER COMMAND

Training

12 Operational Training Unit, RAF (Chipping Warden, Northamptonshire - 91 Group)

Wellington III BJ728 - starboard engine failed at 0610 while over Conway, Caernarvonshire. An hour later, while in the circuit for a precautionary landing at Little Rissington, Gloucestershire, the port engine was also thought to have failed, causing BJ728 to crash and burst into flames at Lower Farm, Upper Milton. The rear gunner was the sole survivor of the five crew, the captain and navigator being buried at Little Rissington.

Captain: NZ415072 Sgt **Robert George McCARTHY**, RNZAF - Age 27. 262hrs (166 solo - 47 on Wellington)

Navigator: 1387312 Sgt **William John FERGUSON**, RAF - Age 23. 150hrs.

Ferguson had enlisted in the FAA as a pilot but remustered six months later as a navigator on transfer to the RAF.

Wed 16/Thu 17 Sep 1942

BOMBER COMMAND

Raid on Essen, Germany (by 369 aircraft - 41 lost)

61 Squadron, RAF (Syerston, Nottinghamshire - 5 Group)

Lancaster I W4173 - took off at 2000 and brought down south of the target area, crashing near Heiligenhaus-Hösel, 14km NE of Düsseldorf. All seven crew were buried in the North Military Cemetery at Düsseldorf, but later reinterred at Reichswald Forest, 5km SW of Kleve. Nineteen of the 41 bombers lost were from training units, this being the last occasion on which these units were used to support Main Force raids in strength.

Captain: NZ404935 Wt Off **Aubrey William OSMAN**, RNZAF - Age 23. 423hrs. 18th op.

15 Operational Training Unit, RAF (Harwell, Berkshire - 91 Group)

Wellington IC X9920 - took off at 2002 captained by Sgt W J Nobbs, RAF, on same raid as the above, and shot down over the Netherlands by a night-fighter, crashing at 0012 on land owned by Mr Ligthart between Heiloo and Alkmaar. All six crew are buried at Bergen, a few kilometres to the north.

WOpAG: NZ404617 Flt Sgt **Donald MILLER**, RNZAF - Age 32. 239hrs. 1st op(?)

7 Squadron, RAF (Oakington, Cambridgeshire - [PFF] 3 Group)

Stirling I W7569/D - took off at 2005 on same raid as the above, and believed to have been damaged by flak before falling into the Ijsselmeer. All seven crew died, their bodies later washing ashore or being recovered from the water. The New Zealanders are buried at Lemmer and the others at Amsterdam.

Captain: NZ411377 Plt Off **Ben DALLENGER**, RNZAF - Age 23. 458hrs. 11th op.

Navigator: NZ41540 Flt Sgt **Donald Trevor LAMB**, RNZAF - Age 20. 294hrs. 5th op.

WOpAG: NZ411741 Sgt **Ronald Hugh CRABTREE**, RNZAF - Age 24. 215hrs. 9th op.

WOpAG: NZ41617 Sgt **William James ANDERSON**, RNZAF - Age 23. 160hrs.

Rear Gunner: NZ404547 Sgt **Harry Oldfield GODDARD**, RNZAF - Age 27. 187hrs. 5th op.

11 Operational Training Unit, RAF (Bassingbourn, Cambridgeshire - 91 Group)

Wellington IC DV612/L - took off at 2012 on same raid as the above, and coned by searchlights near the target, then shot down from 10,000 feet by flak over Duisburg. Three of the crew survived and were taken prisoner, while their two comrades are buried at Reichswald Forest, 5km SW of Kleve.

Captain: NZ41302 Fg Off **Victor Dominic BENEFIELD**, RNZAF - Age 28. 595hrs. 3rd op.

Air Gunner: NZ413277 Sgt **Ernest Joseph OUTEN**, RNZAF - Age 21. 16hrs. 1st op(?)

11 Operational Training Unit, RAF (Bassingbourn, Cambridgeshire; satellite at Steeple Morden, Hertfordshire - 91 Group)

Wellington IC Z1108/K - took off from Steeple Morden at 2029 on same raid as the above, and brought down at Dinslaken, about 15km north of Duisburg. All five crew were buried in the Military Cemetery at Düsseldorf, but later reinterred at Reichswald Forest, 5km SW of Kleve.

Captain: NZ414627 Plt Off **William Arthur HOSKINS**, RNZAF - Age 24. 253hrs. 1st op(?)

Navigator: NZ412225 Plt Off **Grahame Cyril Ralph GOODWIN**, RNZAF - Age 23. 194hrs. 1st op.

26 Operational Training Unit, RAF (Wing, Buckinghamshire - 92 Group)

Wellington IC DV941/G - took off at 2029 on same raid as the above, and brought down off the Dutch coast, killing the five crew. The bodies of an RCAF Sgt and the captain were recovered, the latter being buried at Noordwijk, 32km NNW of The Hague. Their three comrades are commemorated on the Runnymede Memorial.

Captain: NZ413505 Sgt **Lawrence William STREETER**, RNZAF - Age 21. 356hrs. 3rd op.

11 Operational Training Unit, RAF (Bassingbourn, Cambridgeshire - 91 Group)

Wellington IC DV480/N(bar) - took off at 2030 on same raid as the above, and brought down off the Dutch coast, killing the five crew. An RAF Sgt is buried in Belgium and another at Rotterdam. The bodies of the others washed ashore onto the island of Overflakkee and are buried at Ouddorp.

Captain: NZ414283 Plt Off **Ian Alexander GRANT**, RNZAF - Age 24.

Navigator: NZ411950 Plt Off **James Lyders SPEEDY**, RNZAF - Age 22. 226hrs. 3rd op.

106 Squadron, RAF (Coningsby, Lincolnshire - 5 Group)
Lancaster I W4179 - took off at 2030 captained by Sqn Ldr C M Howell, RAF, on same raid as the above, and brought down over Germany, crashing at Soers in the NE suburbs of Aachen. All seven crew were buried at Aachen on the 19th, but later reinterred at Rheinberg, 12km south of Wesel.
Air Gunner: NZ402983 Flt Sgt **Terence George CLARK**, RNZAF - Age 23. 252hrs. 12th op.

158 Squadron, RAF (East Kirby, Lincolnshire - 4 Group)
Halifax II W1222/S - took off at 2048 captained by Flt Sgt L J W Beckman, RAF, on same raid as the above, and brought down over Germany, crashing between Bösinghoven and Ossum, a few kilometres SE of Krefeld. At least five of the crew, including the wireless operator, were buried at Bockum, in the NW outskirts of Krefeld, but later all seven were reinterred at Reichswald Forest, 5km SW of Kleve.
Wireless Op: NZ403585 Flt Sgt **John William MURPHY**, RNZAF - Age 22. 262hrs. 7th op.

1654 Conversion Unit, RAF (Wigsley, Nottinghamshire - 5 Group)
Lancaster I W4138/U - captained by Sgt F H Huntley, RAF, on same raid as the above, and brought down over Germany, crashing at Willich. All seven crew were buried in the Krefeld-Bockum Cemetery, but later reinterred at Reichswald Forest, 5km SW of Kleve.
Navigator: NZ405364 Flt Sgt **Leslie Edwin BELL**, RNZAF - Age 23. 320hrs.

Fri 18/Sat 19 Sep 1942

MIDDLE EAST

Fg Off **N SPENCE** and Flt Sgt **C C PIPER** died while attempting to escape, after surviving the loss of their aircraft on the night of 13/14 September.

Sat 19 Sep 1942

FLEET AIR ARM - ENGLAND

Dummy deck landings
762 Squadron, FAA (Yeovilton, Somerset - HMS *Heron*)
Fulmar II N4126 - while turning at 400 feet on approach at 1458, collided with a 41 OTU Mustang I (AG438), both crashing near the north-eastern boundary of the airfield. The two pilots died, that of the Fulmar being buried at Yeovilton.
Pilot: Sub-Lt (A) **William HARRISON**, RNZNVR - Age 21.

Mon 21 Sep 1942

INDIA

Cross country from Asansol to Cuttack
353 Squadron, RAF (Cuttack, India - 221 Group)
Hudson III FH364 - on coming in to land at Cuttack at about 1000, collided with a 135 Sqn Hurricane IIB (BG973). The fighter pilot survived uninjured, while all four crew in the Hudson died and were buried at Cuttack, but later reinterred at Madras.
Pilot: NZ41901 Flt Sgt **Ronald HARGREAVES**, RNZAF - Age 21. 360hrs. 10 ops.

Tue 22 Sep 1942

BOMBER COMMAND

Conversion flying
460 Conversion Flight, RAAF (Holme-on-Spalding Moor, Yorkshire - 1 Group)
Halifax II W1272 - spun out of control from a considerable height at 1012 and crashed a quarter mile east of Middle Farm near Catterton, 19 miles WNW of the base. All four occupants died, the New Zealander being buried at Harrogate. It was thought that control might have been lost during a 'rudder stall' demonstration. The Conversion Flight was affiliated to 460 Sqn based nearby at Breighton .
Pilot Instr(?): NZ401031 Fg Off **John William PURCIVALL**, DFC, RNZAF - Age 30. 709hrs (482 solo - 67 on Halifax) 32 ops.

FLYING TRAINING COMMAND

Air gunnery training
7 Air Gunners School, RAF (Stormy Down, Glamorgan - 25 Group)
Defiant I T4075 - took off at about 1705 and 40 minutes later, after recovering from a spin off a steep left turn at 1000-1200 feet, dived into the sea a half mile SW of Port Talbot pier. The two crew are buried at Porthcawl.
Pilot: NZ41367 Flt Sgt **Thomas Wallace SHRIMPTON**, RNZAF - Age 25. 500hrs (416 solo - 190 on Defiant)

Wed 23 Sep 1942

FLEET AIR ARM - ENGLAND

Transit(?)

884 Squadron, FAA (HMS *Victorious* - Fleet Aircraft Carrier, Squadron disembarked/at Lee-on-Solent, Hampshire)

Spitfire VB AB940 - in poor visibility flew into Butser Hill, $2^1/_2$ miles SW of Petersfield, Hampshire, at 1350. The pilot is buried at Gosport.

Pilot: Sub-Lt (A) **Andrew Todd RAINES**, RNZNVR - Age 21

Wed 23/Thu 24 Sep 1942

BOMBER COMMAND

Raid on Wismar, Germany (by 83 aircraft - 4 lost)

9 Squadron, RAF (Waddington, Lincolnshire - 5 Group)

Lancaster I R5907/M - was shot down by flak near the target area, crashing at Wendorf, 3 miles NW of Wismar. All seven crew were buried at Wismar, but later reinterred at Berlin.

Captain: NZ41918 Plt Off **Cyril James McKEEN**, RNZAF - Age 21. 445hrs. 19th op.

9 Squadron, RAF (Waddington, Lincolnshire - 5 Group)

Lancaster I W4230/P - took off on the same raid as the above and brought down over Denmark, crashing into the sea north of Lolland Island, killing the seven crew. The bodies of two were later recovered, including that of Sgt **Henry**, which washed ashore on the island of Skalø, both being buried at Svinø, 29km NE of the island. Their five comrades are commemorated on the Runnymede Memorial.

Captain: NZ411003 Flt Sgt **Owen Beamish WHITE**, RNZAF - Age 21. 396hrs. 10th op.
Navigator: NZ41641 Fg Off **Harry Francis Burton SMITH**, RNZAF - Age 25.
WOpAG: NZ411742 Sgt **Noel Henry CROZIER**, RNZAF - Age 21. 136hrs. 3rd op(?)
Air Gunner: NZ405255 Sgt **James Cornwall GRAY**, RNZAF - Age 25.
Rear Gunner: NZ405266 Sgt **John Wallace HENRY**, RNZAF - Age 26. 124hrs. 3rd op.

The CWGC register records Smith as a member of 83 Sqn, a unit he appears never to have served with. His brother, Raymond Bernard Smith, died two years earlier, on 30 September 1940 while travelling as a passenger on an Oxford delivery flight in New Zealand.

Crozier's brother, Alexander Walter Crozier, died on 23 June 1943 while flying with 15 Sqn, RAF.

Raid on Vegesack, Germany (by 24 aircraft - 1 lost)

218 Squadron, RAF (Downham Market, Norfolk - 3 Group)

Stirling I R9187/A - took off at 2325 and shot down into the North Sea off the West Frisian islands by a night-fighter shortly after 0315. All seven crew are commemorated on the Runnymede Memorial.

Captain: 43710 Sqn Ldr **Cuthbert RAYMOND**, DFC, RAF - Age 25. 1097hrs. 51st op.

Raid on Flensburg, Germany (by 28 aircraft - 5 lost)

102 Squadron, RAF (Pocklington, Yorkshire - 4 Group)

Halifax II DT517/G - took off at 0015 and brought down near the target area. All eight crew were buried in the Garrison Cemetery, Friedenshugel, Flensburg, but later reinterred at Kiel.

Captain: NZ402559 Fg Off **Leo George MEE**, RNZAF - Age 26. 528hrs. 19th op.

102 Squadron, RAF (Pocklington, Yorkshire - 4 Group)

Halifax II W1055/F - took off at 0016 captained by Plt Off R C E Bassom, RAF, on same raid as the above, and brought down near the target area, crashing into the sea off Flensburg. Five bodies were recovered, including that of the air bomber, who was buried on the 28th in the Garrison Cemetery, Friedenshugel, Flensburg, but later reinterred at Kiel with the four others. Their two comrades are commemorated on the Runnymede Memorial.

Air Bomber: NZ405483 Sgt **Robert John GOULD**, RNZAF - Age 24. 209hrs. 7th op.

Thu 24 Sep 1942

FIGHTER COMMAND

Training(?)

53 Operational Training Unit, RAF (Llandow, Glamorgan - 81 Group)

Spitfire IIA P7319 - (a presentation machine named *Lahore City*) engine caught fire at 1500, but the pilot's parachute failed to deploy when he baled out near St Athan, about 5 miles SE of the base. He is buried at Llantwit Major.

Pilot: NZ39780 Sgt **Nelson Cremer TURNER**, RNZAF - Age 24. 238hrs.

Thu 24/Fri 25 Sep 1942

COASTAL COMMAND

Ferrying 119 Sqn aircraft from Pembroke Dock to Gibraltar for 202 Sqn, RAF

119 Squadron, RAF (Pembroke Dock, Pembrokeshire - 19 Group)

Catalina IB FP119 - captained by Plt Off T L Wright, RCAF. Encountered a severe electrical storm en route and at 0644 crashed into the sea off Cadiz, Spain, about 100km WNW of its destination. All seven airmen died, as did an

Army, Navy, and possibly civilian, passenger. Five crew bodies were later recovered, three being buried at sea and the others on Gibraltar. The navigator and his two comrades are commemorated on the Runnymede Memorial. It will be noted that by chance the Squadron and aircraft serial share the same number.
Navigator: NZ404013 Plt Off **Maurice Richard JEFFERY**, RNZAF - Age 28. 270hrs.

Fri 25/Sat 26 Sep 1942

MIDDLE EAST

Raid on Tobruk, Libya
40 Squadron, RAF (LG 213 [Kabrit], Egypt - 238 Wing, 205 Group)
Wellington IC HF846/P - took off at 2001 and brought down near the target area, being seen to crash in flames about a mile south of LG 146 (El Gubbi). All six crew are commemorated on the Alamein Memorial.
Captain: NZ41913 Flt Sgt **Gerald Fitzgerald LANGHAM**, RNZAF - Age 22. 278hrs. At least 20th op.
2nd Pilot: NZ41859 Fg Off **Donald Arthur ADAMS**, mid, RNZAF - Age 24. 372hrs. At least 10th op.
Navigator: NZ391432 Flt Sgt **John Miller WELSH**, RNZAF - Age 22. 499hrs. 32nd op.
WOpAG: NZ404593 Sgt **Richard Henry COOK**, RNZAF - Age 21. 341hrs. 32nd op.

BOMBER COMMAND

Cross country training over the Irish Sea
24 Operational Training Unit, RAF (Honeybourne, Gloucestershire - 92 Group)
Whitley V BD232 - took off at 1918, but veered off track 12 miles to the west and flew into the side of Foel Fras Mountain, 4 miles west of Dolgarrog, Caernarvonshire. The wreckage was spotted from the air two days later. All five crew died, the captain and an RAF Sgt being buried at Caernarvon.
Captain: NZ414356 Sgt **Charles Alexander STUART**, RNZAF - Age 22. 320hrs (159 solo - 38 on Whitley)
Stuart's brother, William Douglas Stuart, died on 24 Oct 1941 while flying with 40 Sqn, RAF.

Sun 27 Sep 1942

NEW ZEALAND

Plt Off **A C LONG** died of injuries received in an aircraft accident on 6 January 1941.

Sun 27/Mon 28 Sep 1942

MIDDLE EAST

Raid on Tobruk Harbour, Libya
148 Squadron, RAF (LG 237 [Jebel Hamzi], Egypt - 236 Wing, 205 Group)
Wellington IC DV520 - took off at 2010 captained by Sgt N G Beadle, RAF, and lost without trace, its six crew being commemorated on the Alamein Memorial.
Rear Gunner: 1343513 Sgt **John Manwaring Steward MOIR**, RAFVR - Age 19. 248hrs. 21st op.

Tue 29 Sep 1942

NEW ZEALAND

Air firing exercise
2 Squadron, RNZAF (Nelson)
Hudson III NZ2029 - while flying low over Tasman Bay at about 1555 some 5 miles off Motueka, the propellers struck the water, causing excessive vibration. NZ2029 climbed about 200 feet but lost height and crash landed on the water. All seven on board left the Hudson which sank in about 30 seconds. Excluding observer Wt Off **J A Colway**, most or all were wearing life jackets, but the dinghy, not being released, went down with the wreckage. The observer and two others struck out for the shore, an aircraft dropping a life jacket to Colway about 20 minutes later. He was picked up after about two hours in the water, but none of his comrades survived. The bodies of the Fg Off **Horrocks** and Sgt **Hill** were later recovered and are buried at Devonport and Christchurch, respectively. The others are commemorated on Provincial Memorials at Wellington (**Chapple, Stevens** and **Davis**) and Christchurch (**Bryson**).
Captain: NZ412234 Fg Off **Dean Rainsford HORROCKS**, RNZAF - Age 20. 409hrs (159 solo - 28 on Hudson)
WOpAG: NZ38248 Sgt **Vincent Charles HILL**, RNZAF - Age 23. 508hrs.
Air Gunner: NZ414856 Sgt **Jack Gilbert BRYSON**, RNZAF - Age 24.
Air Gunner: NZ414861 Sgt **Victor Newell CHAPPLE**, RNZAF - Age 21.
Air Gunner: NZ414894 Sgt **Gordon Francis STEVENS**, RNZAF - Age 27.
Armourer: NZ416787 AC1 **John Gordon DAVIS**, RNZAF - Age 32.
Horrocks served in the Far East during the Malayan campaign and is thought to have flown operationally there, although as navigator rather than pilot.
Davis's year of death is incorrectly given in the CWGC register as 1943.

Wed 30 Sep 1942

NEW ZEALAND

Solo training

1 Service Flying Training School (ITS), RNZAF (Wigram)

Oxford II NZ1280 - while carrying out steep turns at 400 feet, lost height, stalled and spun into the ground north of Oxford at 1440. The pilot is buried at Dunedin.

Pilot: NZ421749 LAC **Denis Charles Joseph MOONEY**, RNZAF - Age 24. 75hrs solo (34 on Oxford)

MIDDLE EAST

Search for enemy shipping

203 Squadron, RAF (LG X [Abu Sueir North], Egypt; detachment at LG226G [Gianaclis] - 247 Wing, 201 Group)

Baltimore I AG723 - took off in the afternoon and lost without trace west of Crete. Its four crew being commemorated on the Alamein Memorial.

Pilot: NZ411951 Fg Off **Kingi Te Aho Aho Gilling TAHIWI**, RNZAF - Age 27. 316hrs. 6th op.

Air Gunner: NZ403600 Sgt **Alan Dawson GLOVER**, RNZAF - Age 23. 261hrs.

The CWGC register gives Tahiwi's age as 33.

Wed 30 Sep/Thu 1 Oct 1942

BOMBER COMMAND

'Gardening' - minelaying off Texel and in the Frisian Islands (by 25 aircraft - 2 lost)

149 Squadron, RAF (Lakenheath, Suffolk - 3 Group)

Stirling I BF328 - took off at 1928 to lay mines in the 'Nectarines' area (Frisians) and brought down into the North Sea off the enemy coast. The captain's body washed ashore on 15 October and was buried in the Civil Cemetery on the German island of Heligoland. Post-war, however, his grave could not be located and he is now commemorated on a special memorial in the Hamburg Cemetery at Ohlsdorf. His six comrades are commemorated on the Runnymede Memorial.

Captain: NZ41969 Flt Sgt **Stuart Daniel WELLS**, RNZAF - Age 20. 345hrs.

Navigator: NZ411720 Sgt **Max Hilton ELLIOTT**, RNZAF - Age 21. 354hrs.

Thu 1 Oct 1942

ATLANTIC OCEAN

Lt (A) **P T MEDHURST** died in a lifeboat on or about this date after the sinking of the *Laconia* on 12 September.

Thu 1/Fri 2 Oct 1942

BOMBER COMMAND

Raid on Wismar, Germany (by 78 aircraft - 3 lost)

61 Squadron, RAF (Syerston, Nottinghamshire - 5 Group)

Lancaster I R5703 - took off at 1745, but on lifting off the dinghy inflated and released itself from the starboard wing storage compartment. Fouling the tail controls, it caused R5703 to enter a shallow dive, fly into the ground and explode a mile NE of Gunthorpe, about 3 miles SW of the base. All seven crew died, the captain being buried at Hemblington, Norfolk. Investigation revealed that the dinghy release had been caused by a short circuit in the immersion switch.

Captain: 41885 Sqn Ldr **George Ernest WESTON**, DFC, RAF - Age 23. 1230hrs. 44th op.

Weston was born in England but came to New Zealand at an early age. Selected for a short service commission in the RAF, he embarked for England in 1938. His New Zealand born brother, Godfrey Randal Weston, died on 20 April 1944 while flying with 1657 Heavy Conversion Unit, RAF. Both are buried in the Hemblington All Saints Churchyard.

61 Squadron, RAF (Syerston, Nottinghamshire - 5 Group)

Lancaster I R5759 - took off at 1805 on same raid as the above, and brought down over Germany, crashing at Sildemow, 4 kms south of Rostock. An RAF Sgt was the sole survivor and taken prisoner. His six comrades were buried in the New Cemetery at Rostock, but later reinterred at Berlin.

Captain: NZ41314 Flt Sgt **Geoffrey Ernest DALE**, RNZAF - Age 25. 492hrs. 31st op.

Raid on Flensburg, Germany (by 27 aircraft - 12 lost)

10 Squadron, RAF (Melbourne, Yorkshire - 4 Group)

Halifax II W7852/K - took off at 1804 and brought down south of the target area, crashing near Schleswig. An RAF Sgt and an RCAF Sgt survived to be taken prisoner, but their five comrades died. The captain and one or more of the others were buried in the Military Cemetery at Schleswig, but later all five were reinterred at Kiel.

Captain: NZ411924 Flt Sgt **John Hope MOLLER**, RNZAF - Age 22. 426hrs. 1st op.

76 Squadron RAF (Linton-on-Ouse, Yorkshire - 4 Group)
Halifax II W7812/B - took off at 1810 captained by Plt Off M W Atkinson, RCAF, on same raid as the above, and brought down over the target, crashing at Liehuus, 6km NNW of Flensburg. The air bomber and three or more of the others were buried in the Garrison Cemetery, Friedenshugel, Flensburg, but later all seven crew were reinterred at Kiel.
Air bomber: NZ403628 Sgt **Thomas William PEACOCK**, RNZAF - Age 21. 185hrs. 6th op.

Raid on Lübeck, Germany (by 25 aircraft - 3 lost)
15 Squadron, RAF (Bourn, Cambridgeshire - 3 Group)
Stirling I W7634/G - took off at 1915 and brought down off the coast of Denmark or Germany, killing the seven crew. The body of an RAF Sgt was recovered between the islands of Sylt and Rømø on the 2nd and buried on the former at List, but later reinterred at Kiel, while his six comrades are commemorated on the Runnymede Memorial. In *Bomber Command Losses*, W7634 is stated to have come down off Peenemünde, but this is some 300km east of the target and where the body was found the day following the raid.
Captain: NZ412252 Fg Off **James Lloyd MEREDITH**, RNZAF - Age 24. 478hrs. 20th op.

218 Squadron, RAF (Downham Market, Norfolk - 3 Group)
Stirling I W7613/N - took off at 1948 captained by Flt Lt J C F Du Toit, RAF, on same raid as the above, and brought down over Denmark, crashing on to the Skallingen, a low-lying tongue of land projecting towards the entrance of Esbjerg Harbour. All eight crew died, the bodies of four, including that of the wireless operator, being buried at Fourfelt, to the NW of Esbjerg. Of their four missing comrades, only three are commemorated on the Runnymede Memorial - the fourth, according to the CWGC registers, is commemorated on the Malta Memorial.
Wireless Op: NZ411774 Sgt **Alan Irwin PARKINSON**, RNZAF - Age 22. 226hrs. 18th op.

Fri 2 Oct 1942

FIGHTER COMMAND

'Circus 221' - escort to Boston bombers raiding Le Havre, France
486 Squadron, RNZAF (North Weald, Essex - 11 Group)
Typhoon IB R8683 - took off from Thorney Island at 1400 and was part of a formation attacked by enemy fighters soon after leaving the French coast. A Typhoon, possibly R8683, was last seen under attack while flying straight and level at 200 feet, about 50 miles south of Selsey Bill, Sussex. The pilot is commemorated on the Runnymede Memorial.
Pilot: NZ402893 Plt Off **Raymond Ignatius PHILLIPS**, RNZAF - Age 30. 465hrs. 19th op.

Fri 2/Sat 3 Oct 1942

BOMBER COMMAND

Raid on Krefeld, Germany (by 188 aircraft - 9 lost)
115 Squadron, RAF (Mildenhall, Suffolk - 3 Group)
Wellington III BK271/A - took off at 1845 captained by Sgt F C Adsett, RCAF, and brought down SW of the target area near Düsseldorf. All five crew were buried in the North Military Cemetery in Düsseldorf, but later reinterred at Reichswald Forest, 5km SW of Kleve.
Navigator(?): NZ412476 Sgt **George NELSON**, RNZAF - Age 21. 222hrs. 1st op.
Navigator(?): NZ411217 Sgt **Rex Nigel Hugh JONES**, RNZAF - Age 22.
WOpAG: NZ405302 Sgt **George Wilfred MARTIN**, RNZAF - Age 23. 180hrs.
Although both Nelson and Jones were navigators, one of them is likely to have acted as air bomber on this flight.

Sun 4 Oct 1942

FIGHTER COMMAND

Low flying and aerobatic practice
539 Squadron, RAF (Acklington, Northumberland - 13 Group)
Hurricane IIC BN382 - took off at 1525 and later dived out of low cloud into the ground 3 miles NE of Gilsland, Cumberland. The pilot is buried at Carlisle, 16 miles to the SW.
Pilot: NZ405355 Flt Sgt **Bernard Clive WILLIAMS**, RNZAF - Age 27. 400hrs. 8 ops.

Mon 5 Oct 1942

SOUTH PACIFIC

Motor vehicle accident
2 Aerodrome Construction Squadron, RNZAF (Fiji)
From a detached flight of the Squadron, as a passenger was seriously injured at 1815 when a Ford V-8 rolled over a bank at Viseisei, about 5 miles SW of Lautoka. Although admitted to the US 142nd General hospital, died at 1610 on the 9th. Buried at Lautoka, but later reinterred at Suva.
(Driver): NZ414164 LAC **Charles Frederick GREEN**, RNZAF - Age 30.

Mon 5/Tue 6 Oct 1942

MIDDLE EAST

Raid on Tobruk Harbour, Libya

148 Squadron, RAF (LG 237 [Jebel Hamzi], Egypt - 236 Wing, 205 Group)

Wellington IC DV649 - took off at 1919 and believed to have gone down in the target area, where three parachutes were seen descending. The remains of an RAF Flt Sgt were discovered but his grave was later effaced by the elements. No trace at all was found of his comrades and all six crew are commemorated on the Alamein Memorial.

Captain: NZ41998 Flt Sgt **Henry Sutherland RODGER**, RNZAF - Age 30. 499hrs. At least 23rd op.

WOpAG: NZ404635 Sgt **Royal Tuiri ERIA**, RNZAF - Age 24.

BOMBER COMMAND

Raid on Aachen, Germany (by 257 aircraft - 18 lost).

102 Squadron, RAF (Pocklington, Yorkshire - 4 Group)

Halifax II W7824 - took off at 1841 and was critically damaged when near the target area, crashing after seven of its eight crew had baled out. The captain is buried at Rheinberg, 12km south of Wesel. The raid was undertaken in poor weather, including thunderstorms and severe icing conditions - 10 of the 18 bombers lost crashed in England.

Captain: NZ404412 Wt Off **Frederick Arthur SCHAW**, RNZAF - Age 20. 247hrs. 13th op.

149 Squadron, RAF (Lakenheath, Suffolk - 3 Group)

Stirling I N3755/S - took off at 1908 on same raid as the above, but after running out of fuel on return to England at 0035, crashed and caught fire at Arnolds Oak Farm, Eastling, 13 miles west of Maidstone, Kent. Six of the crew baled out, but the captain was too low and died when his parachute failed to fully deploy in time. He is buried at Maidstone.

Captain: NZ412242 Fg Off **Ralph LONSDALE**, RNZAF - Age 21. At least 3rd op.

CANADA

Cross country

3 Service Flying Training School, RCAF (Calgary, Alberta - 4 Training Command)

Crane I 8133 - piloted by Plt Off G S McMonnies, RCAF, and on return at 0240 flew into a 400 foot-high hill 4 miles NW of the airfield. The pilot was severely injured, but the RAF pupil pilot and New Zealander on board as passenger both died, being buried at Calgary.

(Air Gunner): NZ416426 LAC **Edward Thomas HUGHES**, RNZAF - Age 19.

Tue 6 Oct 1942

NEW ZEALAND

Solo training

2 Elementary Flying Training School, RNZAF (Bell Block)

Tiger Moth II NZ881 - seen to circle the airfield at low altitude at 0810 before entering a flat diving turn, from which it failed to recover before striking the ground and bursting into flames. The pilot is buried at Timaru.

Pilot: NZ415268 LAC **Chester PROBERT**, RNZAF - Age 28. 9hrs solo (all on Tiger Moth)

Tue 6/Wed 7 Oct 1942

BOMBER COMMAND

Raid on Osnabrück, Germany (by 237 aircraft - 6 lost)

44 Squadron, RAF (Waddington, Lincolnshire - 5 Group)

Lancaster I W4188/G - captained by Sqn Ldr I A Stewart, DFC, RAF, and brought down over Germany, crashing at Osnabrück. All seven crew were buried in the Johannes Civil Cemetery, but later reinterred at Rheinberg, 12km south of Wesel.

Navigator: NZ401411 Fg Off **Patrick Hugh RYAN**, RNZAF - Age 24. 492hrs. 20th op.

Wed 7 Oct 1942

MIDDLE EAST

Circuits and landing practice

2 Middle East Training School, RAF (Aqir, Palestine - 205 Group)

Wellington II W5530/E - took off at about 2258 piloted by Flt Sgt **C A McDonald**, RNZAF, for practice at Yibna airfield 3 miles away. During a circuit at 600 feet five to ten minutes later the starboard engine failed. Height could not be maintained and with hilly terrain blocking a direct approach to Yibna W5530 headed back to Aqir, only to crash onto a hillside a half mile short of the base and catch fire. The two crew were both severely injured, McDonald survived but had to have his right foot amputated the following day. The 2nd pilot died almost immediately after the crash and was buried at Ramleh on the 9th. Investigation subsequently revealed the cause of the engine failure to be a strip of ferrous metal jamming the inlet valve, sabotage being suspected.

2nd Pilot: NZ41323 Flt Sgt **Bryan Joseph GROFSKI**, RNZAF - Age 29. 403hrs. 16 ops.

The 7th was Grofski's 29th birthday.

Thu 8 Oct 1942

ARMY CO-OPERATION COMMAND

Dual instruction

42 Operational Training Unit, RAF (Andover, Hampshire - 70 Group)

Blenheim I L1140 - piloted by Fg Off J E F Griffin, RAF, crashed at 1607 near Clatford, about 2 miles south of Andover, when part of the tailplane detached, possibly while carrying out a rolling dive. The two crew are buried at Andover.

2nd Pilot: NZ413095 Fg Off **Thomas James Albert LONG**, RNZAF - Age 23. 351hrs.

Sat 10/Sun 11 Oct 1942

FLEET AIR ARM - ENGLAND

'Gardening' - minelaying off Le Havre, France

816 Squadron, FAA (Thorney island, Sussex - HMS *Daedalus*)

Swordfish I W5968 - took off at about 2145 and lost without trace. The pilot is commemorated on the RNZN Memorial at Devonport, Auckland, and his two comrades on the FAA Memorial at Lee-on-Solent, Hampshire. The Squadron was operating under the direction of RAF Coastal Command's 16 Group at the time of the sortie.

Pilot: Sub-Lt (A) **Douglas Haig TAIT**, RNZNVR - Age 23.

BOMBER COMMAND

'Gardening' - minelaying off the Biscay ports and Frisian islands (by 47 aircraft - 1 lost)

149 Squadron, RAF (Lakenheath, Suffolk - 3 Group)

Stirling I BF348/P - took off at 1852 captained by Flt Sgt H J Hart, RAF, to lay mines off the mouth of the Gironde estuary, France. A decision to abort the sortie was made shortly afterwards when 'play' in the elevator and aileron controls arose. On approach to Watton for a precautionary landing at 2020, BF348 came in too low, struck trees, crashed and caught fire at Great Cressingham, about 4 miles WNW of the airfield. An injured RAF Sgt was the sole survivor of the seven crew, four New Zealanders of whom are buried at Watton. A note on the accident card states that the 'bad aerodynamic characteristics' of some Stirlings were under investigation.

Navigator: NZ41632 Sgt **Lewis John WARREN**, RNZAF - Age 24. 365hrs. 12th op.
Air Bomber: NZ404571 Sgt **Kenneth George McKENZIE**, RNZAF - Age 34. 346hrs. 13th op.
Wireless Op: NZ412372 Sgt **Hugh Clarence WHITWELL**, RNZAF - Age 26. 248hrs. 13th op.
Rear Gunner: NZ412887 Sgt **Harry Ian Forsyth CONNELLY**, RNZAF - Age 24. 196hrs. 14th op.

Sun 11 Oct 1942

MIDDLE EAST

Interception of incoming raid

126 Squadron, RAF (Luqa, Malta)

Spitfire VC BR183 - took off at 1630 with eight others to intercept a force of over 50 aircraft and shot down into the sea about 35 miles north of Comino Island. The pilot is commemorated on the Malta Memorial.

Pilot: NZ411493 Flt Sgt **David Derek MacLEAN**, RNZAF - Age 20.

CANADA

Aerobatic practice

31 Service Flying Training School, RAF (Kingston, Ontario - 1 Training Command)

Harvard II AJ590 - spun off the top of a loop at 1100 and dived into the ground 5 miles from Enterprise, about 25 miles NW of Kingston. The pilot is buried at Kingston.

Pilot: NZ3186 Leading Airman **Robert Ian BREINGAN**, RNZN, of HMS *Canada* (FAA) - Age 22.

Sun 11/Mon 12 Oct 1942

BOMBER COMMAND

'Gardening' - minelaying in many areas (by 80 aircraft - 3 lost)

75 (NZ) Squadron, RAF (Mildenhall, Suffolk - 3 Group)

Wellington III BK341 - took off at approximately 1800-1810 to lay mines in the Kattegat and brought down over the Danish coast, crashing at 2329 between the island of Langli and the Skallingen, north of the main entrance to Esbjerg Harbour. All five crew died, three being buried at Fourfelt, immediately NW of Esbjerg. The captain and fifth member are commemorated on the Runnymede Memorial.

Captain: NZ413897 Sgt **Charles John SHALFOON**, RNZAF - Age 22.

Mon 12/Tue 13 Oct 1942

COASTAL COMMAND

Night torpedo attack practice

Torpedo Training Unit, RAF (Abbotsinch, Renfrewshire - 17 Group)

Hampden TB I AN151 - took off in the late evening piloted by Sgt G Howell, RAF, to carry out dummy attacks in the Firth of Clyde off Troon. Dropped a flare towards 2340 preparatory to carrying out an attack on a trawler, but on turning for the run in struck the sea and crashed. Two bodies were recovered, while the pilot and the New

Zealand wireless operator-air gunner are commemorated on the Runnymede Memorial. AN151 was the third Hampden to be lost during practice attacks in the Firth of Clyde in less than a fortnight.
WOpAG: NZ411751 Sgt **Ernest William EVANS**, RNZAF - Age 26. 99hrs.

Tue 13/Wed 14 Oct 1942

BOMBER COMMAND

Raid on Kiel, Germany (by 288 aircraft - 15 lost)
75 (NZ) Squadron, RAF (Mildenhall, Suffolk - 3 Group)
Wellington III X3954 - took off at approximately 1820-1850 and shot down by flak over Germany, crashing in a field owned by Frau Nanny Meyer at Dahrenwurth, 7km NE of Tonning. All five crew were buried at Brunsbüttelkoog, but later reinterred at Kiel.
Captain: NZ413522 Sgt **Ventry WATTERS**, RNZAF - Age 22. 359hrs. 4th op.
Navigator: NZ412518 Sgt **Lewis Harry PARKINSON**, RNZAF - Age 20. 104hrs. 4th op.
WOpAG: NZ412218 Sgt **Ellison George FIRTH**, RNZAF - Age 19. 176hrs. 4th op.
Watters brother, Terrence Watters, died on 31 August 1943 while also flying with 75 (NZ) Sqn, RAF.

Wed 14 Oct 1942

MIDDLE EAST

Interception of incoming raid
1435 Squadron, RAF (Luqa, Malta)
Spitfire VB EP203/G - took off at 0940 with seven others and one of two which failed to return. An enemy force of about 28 aircraft was engaged some 30 miles north of Grand Harbour, EP203 and another Spitfire were shot down into the sea. Both pilots are commemorated on the Malta Memorial.
Pilot: NZ412270 Sgt **Ronald John ROE**, RNZAF - Age 22.

Wed 14/Thu 15 Oct 1942

MIDDLE EAST

Raid on Tobruk, Libya
104 Squadron, RAF (LG 213 [Kabrit], Egypt - 238 Wing, 205 Group)
Wellington II Z8594/T - captained by Plt Off E W Baldwin, RAF,- and may have been the aircraft seen by others to fall in flames from 4000-5000 feet over the target area. The six crew are commemorated on the Alamein Memorial.
2nd Pilot: NZ404441 Wt Off **David Mitchell KEANE**, RNZAF - Age 23.

Thu 15 Oct 1942

MIDDLE EAST

'Pipe' - anti-submarine patrol off the Egyptian coast
203 Squadron, RAF (LG X [Abu Sueir North], Egypt; detachment at LG 226 [Gianaclis] - 201 Group)
Baltimore II AG749 - took off from LG 226 at 1055 piloted by Flt Sgt A C M Crombie, RAF. During the patrol difficulty was experienced in switching fuel tanks to feed the port engine, which eventually cut. The starboard engine gradually overheated and also failed shortly before 1345, whereupon a belly landing was made on a beach 10 miles east of Rumana at position 3104N:3245E. Three of the four crew survived, injured, but the New Zealand wireless operator-air gunner died and is buried at Ismailia.
WOpAG: NZ405537 Sgt **Raymond Robert TODD**, RNZAF - Age 21. 145hrs. 4th op(?)

FIGHTER COMMAND

Height climb
52 Operational Training Unit, RAF (Aston Down, Gloucestershire - 81 Group)
Spitfire IIA P7966 (a presentation machine named *Manxman*) - dived into a shed of 7 Maintenance Unit, RAF, at Quedgeley, 15 miles NNW of Cirencester, Gloucestershire, setting it on fire. A female civilian worker died, while others were injured and considerable damage caused. The pilot also died and is buried at Cirencester. The diarist for 52 OTU despairingly recorded that this had been 'quite the worst day for fatal accidents in the history of the unit' to date - three other deaths also occurring in unrelated accidents.
Pilot: NZ416440 Plt Off **Keith Thomas BEANGE**, RNZAF - Age 21. 261hrs.

BOMBER COMMAND

Training
21 Squadron, RAF (Bodney, Norfolk - 2 Group)
Ventura I AE760 - piloted by Flt Sgt Williams, RAF, spun to port, crashed and caught fire about a mile west of the airfield at 1045 after an engine was feathered in order to carry out a single-engined approach. The three crew died, the navigator being buried at Watton. Records conflict as to whether the approach was being made to Bodney or nearby Watton.
Navigator: NZ411219 Flt Sgt **Douglas James KIRKLAND**, RNZAF - Age 22. 354hrs. 3 ops.

Cross country
11 Operational Training Unit, RAF (Westcott, Buckinghamshire - 92 Group)
Wellington IC DV717 - dived through cloud and into the ground at 2055, catching fire at Breck's Farm, Maplebeck, 5 miles SE of Ollerton, Nottinghamshire. The five crew died, four being buried at Ollerton. As a poignant footnote to the tragedy, the body of the wireless operator-air gunner **McCord** was repatriated to the land of his birth and buried in Ardagh Cemetery, County Longford, Eire - he had emigrated to New Zealand only four years earlier.
Captain: NZ414640 Sgt **Vincent Xavier KIRBY**, RNZAF - Age 23. 336hrs.
Navigator: NZ411969 Sgt **Richard John FULLER**, RNZAF - Age 21. 174hrs.
Air Bomber: NZ413438 Sgt **John Llewellyn LLOYD**, RNZAF - Age 22.
WOpAG: NZ413264 Sgt **Edward Alexander McCORD**, RNZAF - Age 26.
Air Gunner: NZ413282 Sgt **Ivan Raymond STYLES**, RNZAF - Age 27. 79hrs. 1 op.
In the preceding five weeks Styles had already experienced a crash landing and a take off crash.

Thu 15/Fri 16 Oct 1942

BOMBER COMMAND

Raid on Cologne, Germany (by 289 aircraft - 21 lost)
158 Squadron, RAF (East Kirby, Lincolnshire - 4 Group)
Halifax II W1108/E - took off at 1854 and shot down over the Netherlands by a night fighter at 2253, crashing at Rhaan, 5km north of Hellendoorn. The sole survivor was an injured RAF Sgt who was captured. His six comrades are buried at Hellendoorn.
Captain: NZ413831 Sgt **Weiland Fairfax ENSOR**, RNZAF - Age 29. 478hrs. 4th op(?)

103 Squadron, RAF (Elsham Wolds, Lincolnshire - 1 Group)
Halifax II W1213 - took off at 1856 on same raid as the above, and brought down near the target area. All seven crew were buried in the Military Cemetery at South Cologne, but later reinterred at Rheinberg, 12km south of Wesel.
Captain: 36247 Flt Lt **Graham Noel PARKER**, RAF - Age 23.
Parker was born in England, but emigrated with his parents to New Zealand in 1928. Selected for a short service commission in the RAF, he embarked for England in 1938 after preliminary training with the RNZAF.

106 Squadron, RAF (Syerston, Nottinghamshire - 5 Group)
Lancaster I W4195/W - took off at 1900 captained by Plt Off E F White, RCAF, on same raid as the above, and shot down over the Netherlands by a night fighter, crashing at 2245 near Valburg, 12km SW of Arnhem. Four of the seven crew survived and were captured, while their comrades, including the navigator, are buried at Uden, 30km to the SW.
Navigator: NZ404488 Plt Off **Raymond Keith MARSHALL**, RNZAF - Age 21. 391hrs. 31st op.

156 Squadron, RAF (Warboys, Huntingdonshire - [PFF] 3 Group)
Wellington III BK339 - captained by Sqn Ldr J Hobbs, RAF, on same raid as the above and lost without trace, its five crew being commemorated on the Runnymede Memorial.
Navigator: NZ404909 Fg Off **Russell Stronach LONGUET**, RNZAF - Age 26. 274hrs. 5th op.
Navigator(?): NZ412469 Plt Off **Alan Henry CUNNINGHAM**, RNZAF - Age 33.
Both Longuet and Cunningham were navigators but it seems likely that Cunningham was acting as air bomber on this flight.

149 Squadron, RAF (Lakenheath, Suffolk - 3 Group)
Stirling I W7526/V - took off at 1950 on same raid as the above, and shot down over the Netherlands by a night fighter, crashing near the villages of Ingen and Lienden, about 22km SSE of Amersfoort. All seven crew are buried at Oud Leusden on the southern outskirts of Amersfoort.
Captain: NZ405224 Flt Sgt **John Crawford BROCKET**, RNZAF - Age 21. 397hrs.
Wireless Op: NZ412351 Sgt **Trevor Charles Beckett PIKE**, RNZAF - Age 22. 227hrs. 4th op.
Air Gunner: NZ411773 Sgt **Francis William O'SULLIVAN**, RNZAF - Age 23. 185hrs. 6th op.

214 Squadron, RAF (Chedburgh, Suffolk - 3 Group)
Stirling I R9146/S - took off at 2000 on same raid as the above, and shot down over the Netherlands by a night-fighter at 2256, crashing about 1km south of the Oss railway station, 12km NW of Uden. The seven crew are buried at Uden.
Captain: NZ411932 Fg Off **John Mitchell PAAPE**, RNZAF - Age 28. 393hrs. 11th op.
Paape's brother, Arthur Mitchell Paape, died on 3 April 1943 while flying with 467 Sqn, RAF.

Fri 16 Oct 1942

SOUTH PACIFIC

Special emergency operational search patrol and then to land at *'Roses'* (Vila)
3 Squadron, RNZAF (Pallikulo [Bomber I], Espiritu Santo, New Hebrides)
Hudson IIIA NZ2055 - took off at 0325 as the second of 12 scheduled for the patrol and climbed away normally to about 300 feet. The climb became steeper, causing NZ2055 to stall and enter a spiral dive, before exploding and disintegrating on impact with the sea 125 yards out from a reef. All seven airmen are commemorated on the

Bourail Memorial. The emergency arose from a report that a 'large enemy naval force, including a carrier,' was within striking range, but this later proved to be a false alarm.

Captain: NZ391361 Fg Off **William James Little LAUCHLAN**, RNZAF - Age 31. 805hrs solo (7 solo at night)
2nd Pilot: NZ413015 Sgt **Stephen William George BISHOP**, RNZAF - Age 25.
Navigator: NZ411225 Fg Off **Lister Charles MULHOLLAND**, RNZAF - Age 28.
WOpAG: NZ411306 Sgt **Maxwell Herbert WALKER**, RNZAF - Age 23.
WOpAG: NZ412865 Sgt **William John CORNISH**, RNZAF - Age 23.
(Fitter IIE): NZ401483 LAC **James Bett FERGUSON**, RNZAF - Age 39.
(Fitter IIA): NZ411673 AC1 **Noel Louis FOWLER**, RNZAF - Age 21.

Ferguson and Fowler were travelling as passengers, being taken to 'Roses' *for ground staff duties.*
This was 3 Squadron's first day on operations in the South Pacific, NZ2055 becoming the RNZAF's first fatal operational loss outside New Zealand.

FIGHTER COMMAND

Standing patrol
486 Squadron, RNZAF (West Malling, Kent - 11 Group)
Typhoon IB R8698/Q - took off at about 1750 but about 35 minutes later the pilot reported to the section leader that his engine was on fire. Shortly afterwards R8698 fell in flames from 400 feet and crashed near Netherfield Road, immediately to the NNW of Battle, Sussex. The pilot is buried at Maidstone, Kent. The engine fire was apparently caused through excessive wear in a cylinder sleeve.
Pilot: NZ412787 Flt Sgt **David Buchanan CLARK**, RNZAF - Age 23. 338hrs (247 solo - 28 on Typhoon)

Fri 16/Sat 17 Oct 1942

BOMBER COMMAND

'Gardening' - minelaying of the Biscay ports (by 34 aircraft - 4 lost)
115 Squadron, RAF (Mildenhall, Suffolk - 3 Group)
Wellington III X3946/Q - took off at 1800 and brought down off the coast of France, killing the five crew. Two are commemorated on the Runnymede Memorial, while the bodies of three others washed ashore on to the Ile de Ré, off the port of La Rochelle, including that of the captain, who is buried at Le-Bois-Plage-en-Ré.
Captain: NZ413408 Sgt **Douglas Huntly GORDON**, RNZAF - Age 20. 420hrs. 7th op.

15 Squadron, RAF (Bourn, Cambridgeshire - 3 Group)
Stirling I R9312/C - took off at 1815 to lay mines off Bayonne and shot down over France by flak, crashing at Grandchamps-des-Fontaines, 17km from the centre of Nantes. Two RAF Sgts in the crew survived and were taken prisoner, while their five comrades are buried at Nantes.
Captain: NZ413148 Sgt **Arthur James TANNER**, RNZAF - Age 22. 329hrs. 4th op.
Air Bomber: NZ413309 Sgt **Noel Edward STONE**, RNZAF - Age 32. 271hrs. 5th op.

149 Squadron, RAF (Lakenheath, Suffolk - 3 Group)
Stirling I BF392/D - took off at 1830 captained by Flt Sgt J H Ekelund, RCAF, to lay mines in the Gironde estuary and brought down on the Ile d'Yeu. All seven crew are buried in a collective grave on the island at Port-Joinville.
Navigator: NZ404353 Fg Off **Morris Searle GILBERD**, RNZAF - Age 29.
Wireless Op: NZ405737 Sgt **Angus MARTIN**, RNZAF - Age 27. 317hrs. 23rd op.
Air Gunner: NZ402823 Flt Sgt **Ernest Hallright HOWELL** RNZAF - Age 21. 280hrs. 24th op.
Air Gunner: NZ404970 Sgt **Maurice Allan TORRANCE**, RNZAF - Age 30. 278hrs. 26th op.

Sat 17 Oct 1942

COASTAL COMMAND

'Flora 5' - anti-submarine patrol
48 Squadron, RAF (Sumburgh, Shetland Isles - 18 Group)
Hudson IIIA FH360/P - took off at 0808 piloted by Plt Off R W Hooper, RAF, and lost without trace, its four crew being commemorated on the Runnymede Memorial.
WOpAG: NZ411098 Sgt **Laurence James PEARCE**, RNZAF - Age 25. 135hrs. 7th op.

BOMBER COMMAND

Raid on Schneider factory at Le Creusot and transformer power-station at Montchanin, France (by 94 aircraft - 1 lost)
61 Squadron, RAF (Syerston, Nottinghamshire - 5 Group)
Lancaster I W4774 - took off at 1200 captained by Sqn Ldr W D Corr, DFC, RAF, as one of six Lancasters designated to bomb the transformer. At about 1700 W4774 was seen by ground witnesses to fly at very low altitude towards the Ecuisses railway station, 5km east of Montchanin, strike an electricity pylon and catch fire. It continued flying for several kilometres with one wing low, gradually losing height until striking the side of a slope and catching fire at the entrance to a wood. The sole survivor was an RAF Sgt who was badly injured on baling out from a little over 100 feet, and taken prisoner. His six comrades are buried at Ecuisses. Reports in *The Bomber Command War Diaries* and *Bomber Command Losses* that W4774 crashed into a building or buildings are not borne out by eyewitness statements from an aircrew member and the Mayor of Ecuisses.
Air Bomber: NZ403780 Plt Off **John Colin WEBSTER**, RNZAF - Age 21. 420hrs. 27th op.

Mon 19 Oct 1942

FLYING TRAINING COMMAND

Training

7 Flying Instructors School, RAF (Upavon, Wiltshire - 23 Group)

Magister I P6413 - took off at about 1600 and 30 minutes later, while engaged in low flying, stalled off a steep climbing turn to port before diving into the ground near Chirton, 4 miles WNW of Upavon. The two pilots are buried at Upavon.

Pilot: NZ415709 Plt Off **William Edward NILSSON**, RNZAF - Age 22. 342hrs.

The other pilot was Plt Off P R C Mars, RAF, but it is not known who was in command at the time of the accident.

Tue 20 Oct 1942

MIDDLE EAST

Low flying exercise

Air Delivery Unit, RAF (Cairo, Egypt - 206 Group)

Blenheim V BA162 - while low flying over northern Nigeria at 0630 struck a tree, crashed and caught fire on the Kano-Dawa (Dawaki?) road. The two crew are buried at Kano.

Pilot: NZ413112 Sgt **Edgar NEVE**, RNZAF - Age 29.

COASTAL COMMAND

Practice(?)

236 Squadron, RAF (North Coates, Lincolnshire - 16 Group)

Beaufighter VIC EL280 - broke up at about 1135 after recovering from a dive and crashed onto the beach on the airfield perimeter. Both the crew died, the pilot being buried at Belfast.

Pilot: 36191 Sqn Ldr **George Rodoway MacDONALD**, mid, RAF - Age 23. 1790hrs. Approx. 300 ops.

ATLANTIC OCEAN

Flt Lt **G M HAYTON** died in a lifeboat on or before this date after the sinking of HMT *Laconia* on 12 September.

Thu 22 Oct 1942

MIDDLE EAST

Passenger flight from Bilbeis East to Lydda, Palestine

267 Squadron, RAF (Bilbeis East, Egypt - 216 Group)

Lodestar II EW984 - while taking off at 1530 struck the top of a tent, turned over, crashed and burst into flames. All 13 occupants died and are buried at Heliopolis, including a New Zealander with the Squadron travelling as a passenger. Among those killed was the Squadron commanding officer, Act Wg Cdr J P S Smyth, DFC, RAF.

(Navigator): NZ404933 Flt Sgt **Brian Joseph O'CONNOR**, RNZAF - Age 26. 307hrs.

Fri 23 Oct 1942

NEW ZEALAND

Air firing and wireless telegraphy training

1 Operational Training Unit, RNZAF (Bell Block)

Oxford I NZ277 - took off at 0930 to undertake machine gun firing off the Taranaki coast, afterwards to proceed 50 miles north to Kawhia Harbour before returning to base. The last transmission from NZ277 was at 1041, placing it off the coast near New Plymouth. Extensive searches for some time afterwards failed to reveal any trace of the aircraft or its four crew, whose names were commemorated after the war on the Provincial Memorials at Auckland (Sgt **Martyn**), Christchurch, (Sgt **Martin**), and Wellington (Plt Off **Dandy** and Sgt **Dodson**). Their fate came to light 31 years later, when on 17 January 1974 Messrs **E Clince** and **D Rawson** stumbled upon the wrecked Oxford about 2000 feet up Mt Egmont, in a small valley in the Touakai Ranges on its northern side. The crew had been killed on impact and on the 18th their remains were recovered and taken to the airmen's home towns for burial. Investigation suggested that NZ277 had flown into the side of the valley at a low rate of descent with its trailing aerial still extended. A strong northerly wind was prevailing at the time, with cloud down to about 1000 feet. Immediately or shortly after the impact a six-foot diameter Rimu tree fell along the length of the machine, thus largely obscuring it from view.

Pilot: NZ417031 Plt Off **Rodney George DANDY**, RNZAF - Age 32. 530hrs solo (117 on Oxford)

WOp Instr: NZ439008 Sgt **Graham MacLeod MARTIN**, RNZAF - Age 26.

Air Gunner: NZ412319 Sgt **Edward Reid DODSON**, RNZAF - Age 26. 20hrs.

Air Gunner: NZ413271 Sgt **Douglas Gordon MARTYN**, RNZAF - Age 21.

BOMBER COMMAND

Raid on Hengelo, Netherlands (by 4 Mosquitoes - 1 lost)

105 Squadron, RAF (Marham, Norfolk - 2 Group)

Mosquito B IV DZ343 - took off at 1253 and was believed to have been shot down over the Netherlands by light flak, at Raard, 3km west of Dokkum. Dutch eyewitnesses said the Mosquito was on fire before it crashed and exploded. The two crew buried at Raard.

Pilot: 36147 Sqn Ldr **Joseph Cunningham SIMPSON**, mid, RAF - Age 25.

Fri 23/Sat 24 Oct 1942

BOMBER COMMAND

Raid on Genoa, Italy (by 122 aircraft - 9 lost)

158 Squadron, RAF (East Kirby, Lincolnshire - 4 Group)

Halifax II W7862/D - took off at 1742 captained by Sgt G S Hughes, RAF, and brought down into the English Channel, killing the eight crew. The bodies of an RCAF Plt Off and Sgt later washed ashore onto the French coast of the Pas de Calais, while their comrades are commemorated on the Runnymede Memorial. Six of the nine aircraft lost were written off in England.

Navigator: NZ404894 Flt Sgt **Lloyd Denzil CLARK**, RNZAF - Age 27. 325hrs. 9th op.

218 Squadron, RAF (Downham Market, Norfolk - 3 Group)

Stirling I R9184/U - took off at 1835 captained by Plt Off R A Studd, AFC, RAF, on same raid as the above, and brought down into the sea off the coast of France, killing the eight crew. The bodies of two RAF Sgts and the 2nd pilot later washed ashore near Dieppe and the mouth of the Somme, the latter being buried at Abbeville. The others are commemorated on the Runnymede Memorial.

2nd Pilot: NZ405366 Flt Sgt **Thomas Layton BOWKER**, RNZAF - Age 31. 520hrs. 1st op.

Sat 24 Oct 1942

NEW ZEALAND

Instrument flying practice

2 Service Flying Training School (ATS), RNZAF (Woodbourne)

Harvard II NZ962 - seen at 1100 to crash inverted into a hill near Ward, about 6 miles south of Lake Grassmere. The two crew are buried at Matamata and Waipukurau respectively.

Pilot Instr: NZ41956 Sgt **Graham Ballantine STEELE**, RNZAF - Age 21. 271hrs solo (40 on Harvard)

Pilot: NZ42458 LAC **Charles Arthur RICKEY**, RNZAF - Age 23. 172hrs.

Steele had served operationally in the Far East during the Malayan campaign.

BOMBER COMMAND

Raid on Milan, Italy (by 88 aircraft - 4 lost)

207 Squadron, RAF (Langar, Nottinghamshire - 5 Group)

Lancaster I W4121/B - took off at 1241 captained by Flt Sgt L R Wright, RAF, and shot down into the sea a few kilometres off the port of Cabourg, about 29km SW of Le Havre, France. The bodies of six crew were later recovered along the coast between Cherbourg and to the NE of Cabourg. The three others, including a New Zealand wireless operator-air gunner, are commemorated on the Runnymede Memorial. In *Bomber Command Losses* the crash site is given as in the sea off Blainville-sur-Mer, over 100km west of Cabourg. It seems likely, however, that this has been confused with the similarly named Blonville-sur-Mer, where a crew member is buried, which lies just a few kilometres NE of Cabourg.

WOpAG: NZ405489 Sgt **Albert John Victor HUNT**, RNZAF - Age 24. 222hrs. 15th op.

Sat 24/Sun 25 Oct 1942

BOMBER COMMAND

Raid on Milan, Italy (by 71 aircraft - 7 lost)

218 Squadron, RAF (Downham Market, Norfolk - 3 Group)

Stirling I R9241/L - took off at 1850 but about 35 minutes into the flight a fire broke out in the port engine and rapidly spread. Control was lost and R9241 broke up and crashed at Cattawade, 8 miles SSW of Ipswich, Suffolk. The sole survivor of the eight crew was an RAF Sgt who managed to bale out at the last moment. The captain and 2nd pilot are both buried at Ipswich.

Captain: NZ41329 Flt Sgt **Frederick Rex HIGGOTT**, RNZAF - Age 21. 473hrs. 11th op.

75 (NZ) Squadron, RAF (Mildenhall, Suffolk - 3 Group)

Wellington III Z1652 - took off at approximately 1855-1910 on same raid as the above, and brought down over France, crashing near the village of Ville-sur-Retourne, 13km SSE of Rethel. The captain and navigator died and share a joint grave in the village. RNZAF Sgts **J G Barnes** and **E Worsdale** survived, as did the rear gunner. Barnes was captured but the others evaded capture, Worsdale reaching Switzerland.

Captain: NZ414293 Sgt **Howard James HUGILL**, RNZAF - Age 21. 413hrs. 8th op.

75 (NZ) Squadron, RAF (Mildenhall, Suffolk - 3 Group)

Wellington III BJ725 - took off at approximately 1855-1910 on same raid as the above, and brought down over France, crashing near Valenciennes, 29km NE of Cambrai. All five crew buried in a pair of joint graves at Valenciennes.

Captain: NZ414646 Sgt **James Allison McCONNELL**, RNZAF - Age 21. 366hrs. At least 7th op.

Navigator: NZ41952 Sgt **Selwyn Clarence SMITH**, RNZAF - Age 29. 319hrs. 9th op(?)

Wireless Op: NZ413285 Sgt **Douglas Noel TONKIN**, RNZAF - Age 22. 176hrs. 9th op.

Rear Gunner: NZ412317 Sgt **Vallance Albert Oliver DIMOCK**, RNZAF - Age 22. 183hrs. 9th op.

115 Squadron, RAF (Mildenhall, Suffolk - 3 Group)
Wellington III BK306/K - took off on same raid as the above, and brought down into the sea off the southern Dutch coast, killing the five crew. An RAF Sgt is commemorated on the Runnymede Memorial, while another is buried with the captain at Rotterdam. The navigator was buried at Haamstede, on the island of Schouwen, but later reinterred with the air bomber at Bergen-op-Zoom.
Captain: NZ413988 Sgt **Donald Stuart OLDRIDGE**, RNZAF - Age 21. 335hrs. 4th op.
Navigator: NZ411422 Sgt **Robert Louis Andrew McFARLAND**, RNZAF - Age 20. 256hrs. 3rd op.
Air Bomber: NZ41389 Sgt **Kenneth Harry YEOMAN**, RNZAF - Age 27. 270hrs. 4th op.

Sun 25 Oct 1942

MIDDLE EAST

Interception of incoming raid
126 Squadron, RAF (Luqa, Malta)
Spitfire VC BR311/L - took off at 1517 with nine others and shot down into the sea, its pilot being commemorated on the Malta Memorial.
Pilot: NZ412727 Plt Off **Nigel Manfred PARK**, DFM, RNZAF - Age 21. 296hrs. 65th op.

FIGHTER COMMAND

Coastal patrol
486 Squadron, RNZAF (West Malling, Kent - 11 Group)
Typhoon IB R8814 - took off at about 0935 as leader of a section. About 30 minutes later, at 1000 feet, black smoke began to stream from the engine and the pilot signalled his intention to return to base. Complete engine failure followed shortly afterwards and the pilot advised he was baling out, but apparently the parachute caught on the cockpit structure as the fighter spun into the ground at Willesborough, immediately south of Ashford. He is buried at Maidstone.
Pilot: NZ40427 Plt Off **Jesse PEARSE**, RNZAF - Age 23. 391hrs (315 solo - 41 on Typhoon) 22nd op.

FLEET AIR ARM - SCOTLAND

Deck landing training
885 Squadron, FAA (HMS *Formidable* - Fleet Aircraft Carrier)
Spitfire VB AA872 - missed the arrester wires on approach at 1415, crashed over the bows and sank. The pilot is commemorated on the RNZN Memorial at Devonport, Auckland. At the time the Squadron was based at Machrihanish (HMS *Landrail*), Argyllshire, working up for *Operation TORCH*, the forthcoming invasion of NW Africa. They were temporarily using Spitfires, pending replacement by Seafires, and AA872 had been borrowed from 768 Sqn, FAA.
Pilot: Sub-Lt (A) **John David HOUSTON**, RNZNVR - Age 21.

Mon 26 Oct 1942

NEW ZEALAND

Army co-operation
Central Flying School, RNZAF (Tauranga)
Harvard II NZ901 - after a dummy attack on another aircraft at 1550, stalled on attempting to regain height too quickly while recovering from a steeply banked diving turn. NZ901 failed to recover and dived into the ground near the airfield, killing the pilot, who is buried at Levin.
Pilot: NZ39932 Flt Lt **Keith Stuart Strode PENNY**, RNZAF - Age 24. 1991hrs solo (1010 on Harvard)
Penny's brother, Russell William Strode Penny, died less than two months later, on 18 December, while flying with 486 Sqn, RNZAF. Although the family surname is usually shown as Penny in contemporary Air Force documents, the CWGC registers published in the 1950s record it as Strode-Penny.

MIDDLE EAST

Shipping strike against convoy 'TT' near Tobruk, Libya
47 Squadron, RAF (LG 214 [Shandur], Egypt; detachment at LG 226 [Gianaclis] - 201 Group)
Beaufort I DE115/G - took off at 1230 from Gianaclis piloted by Fg Off W Garriock, RAAF, with seven other torpedo bombers and one of two which failed to return. Accompanied by five Bisleys of 15 Sqn, SAAF, and nine RAF Beaufighters, the formation found and carried out several attacks on the ships at position 3220N:2342E shortly after 1530, but lost four of their number. Ten miles into the return journey, one of the Bisleys, evidently damaged or its pilot wounded, suddenly weaved erratically, climbed and collided with DE115. Locked together they plunged into the sea, killing the crews of both. The Beaufort's four are commemorated on the Alamein Memorial.
Navigator: NZ411228 Fg Off **Robert Wrey SIMPSON**, RNZAF - Age 30. 309hrs. 5th op.

Tue 27/Wed 28 Oct 1942

MIDDLE EAST

Raid on Máleme airfield, Crete

160 Squadron, RAF (Aqir, Palestine - 242 Wing, 205 Group)

Liberator II AL548/R - took off at 1918 with seven others and shot down over Crete in the early morning, being seen to fall south of the target. Seven of the crew were buried in a military cemetery at Galatas, about 10km east of Máleme, while an RAF Flt Sgt, was buried at Caneá. All eight were later reinterred at Suda Bay.

Captain: NZ401256 Flt Lt **Eric Clarence COX**, DFC, RNZAF - Age 28. 969hrs. At least 37th op.
Air Gunner: NZ40605 Plt Off **Carroll Ronald DURRANT**, RNZAF - Age 23. 307hrs.
Air Gunner: NZ40629 Plt Off **Charles Christopher REILLY**, RNZAF - Age 29. 365hrs.

Reilly and Durrant enlisted on the same day and became friends, sharing at least eight postings during their Air Force service, and were commissioned together.

BOMBER COMMAND

Night cross country

1651 Heavy Conversion Unit, RAF (Waterbeach, Cambridgeshire - 3 Group)

Stirling I BF314 - on returning at 0147 made a three-engined landing, bounced slightly and ran along the runway. Near the end an overshoot was begun and while climbing away steeply, BF314 stalled and spun into the ground, catching fire. Four died in the crash, while two others, including the captain, died later in the day at the RAF Hospital at Ely. The sole survivor was a seriously injured RAF Sgt. All the New Zealanders, along with an RCAF crew member, are buried at Cambridge.

Captain: NZ413808 Fg Off **John Hamilton BOMFORD**, RNZAF - Age 24. 354hrs (180 solo - 19 on Stirling)
Navigator: NZ411724 Fg Off **Albert William McLEAN**, RNZAF - Age 25. 260hrs. 2 ops.
WOpAG: NZ411475 Sgt **Ernest Henry TURNER**, RNZAF - Age 23.
Rear Gunner: NZ403760 Flt Sgt **Frederick Spendlove DEAN**, RNZAF - Age 23. 184hrs.

Wed 28 Oct 1942

NEW ZEALAND

Delivery flight from Rongotai to Harewood

3 Elementary Flying Training School (Harewood)

Tiger Moth II NZ1447 - a new aircraft destined for the School but piloted by Fg Off **J F Cruickshank**, a flying instructor with 1 Service Flying Training School - took off at 1015, but towards journey's end, at about 1230, struck exceptionally turbulent weather and flew towards the coast in an attempt to gain shelter. The throttle was opened fully to gain height, but NZ1447 finally became uncontrollable and crashed inverted into the sea 3 miles south of Motunau Island, some 35 miles NNE of Christchurch. Although suffering from shock and abrasions Cruickshank managed to swim to shore, climb cliffs and walk 3 miles to a house on Mt Vulcan Station to seek help. The passenger, however, died when his head struck the cockpit, his body being recovered the following afternoon and later cremated at Christchurch.

(Equipment Assistant): NZ39749 Sgt **Neill Ernest KING**, RNZAF - Age 27.

King was serving with the Transport Section at Air Headquarters, Wellington, and is believed to have been travelling home to Christchurch on leave.

Wed 28/Thu 29 Oct 1942

BOMBER COMMAND

Cross country

19 Operational Training Unit, RAF (Kinloss, Morayshire - 91 Group)

Whitley V AD710 - captained by Flt Sgt J E Williams, RAF. An unexpected deterioration in the weather took place after departure and while flying through an electrical storm AD710 was struck by lightning, setting the starboard engine on fire and causing it to fail. Two of the crew managed to bale out just before the Whitley crashed at 0055 near Dalcross, 7 miles NE of Inverness. Their five comrades died, however, the 2nd pilot being buried at Kinloss.

2nd Pilot: NZ414246 Fg Off **Walter Ian CHARTERIS**, RNZAF - Age 23. 231hrs.

Thu 29 Oct 1942

BOMBER COMMAND

Air test following rigging check

15 Squadron, RAF (Bourn, Cambridgeshire - 3 Group)

Stirling I BF386/Q - took off at 1108 captained by Sqn Ldr C G Fisher, RAF, and appeared to be flying normally when seen coming out of cloud about 40 minutes later. BF386 then appeared to stall off a steep turn between 500-1000 feet, before diving into the ground and catching fire in the vicinity of Nordelph and Salter's Lode, a few miles SW of Downham Market, Norfolk. All eleven occupants died, the 2nd pilot and five of the others being buried at Downham Market.

2nd Pilot: NZ413417 Sgt **Walter Huson HOOD**, RNZAF - Age 23. 383hrs (57 solo on Stirling) 1 op.

'Cloud cover' raid on Ruhr area of Germany (by 6 Wellingtons - 3 lost)
115 Squadron, RAF (Mildenhall, Suffolk - 3 Group)
Wellington III X3540/X - took off at 1400 to raid Essen and brought down into the North Sea, probably off the Dutch coast, where the captain's body was later recovered, to be buried at Vlissingen. His four comrades are commemorated on the Runnymede Memorial.
Captain: NZ405346 Flt Sgt **Reginald William URWIN**, RNZAF - Age 23. 460hrs. 14th op.
Wireless Op: NZ411768 Fg Off **Richmond Lovett MEREDITH**, RNZAF - Age 22. 204hrs. 15th op.
Rear Gunner: NZ412346 Sgt **Leslie Hunter MARTIN**, RNZAF - Age 33. 186hrs(?). 16th op.

Fri 30 Oct 1942

NEW ZEALAND

Anti-submarine patrol for USS *Crescent City* ('Force N') en route from Wellington to Guadalcanal
7 Squadron, RNZAF (Waipapakauri)
Vildebeest IV NZ122/B - took off at 1003 for outer patrol near the ship for three hours. The *Crescent City*, stated to be proceeding at 15 knots on course 330° true, then 341° true, could not be located by NZ122, which signalled so at 1430. Six minutes later it acknowledged an instruction to return to base, being expected back by 1530, but was not heard from or seen again. A search began soon after 1640 and that evening two unusual white lights were sighted about 15 miles off the coast and to the south of Cape Reinga. Shortly after midnight another white light was seen to rise from the sea's surface about 10 miles SW of Scotts Point, 10 miles SE of Cape Maria van Diemen. A search aircraft dropped a flare in the area soon afterwards but no further lights or activity were seen. On 5 December, a wheel from the Vildebeest washed ashore between the two capes. The three crew are commemorated on the Provincial Memorial at Wellington.
Pilot: NZ41193 Fg Off **Charles Henry GUNN**, RNZAF - Age 22. 444hrs. 20th op.
Navigator: NZ391325 Fg Off **Stanley Gordon BROMLEY**, RNZAF - Age 28.
WOpAG: NZ413741 Fg Off **Robert DENT**, RNZAF - Age 28. 137hrs.
Bromley had served on bombers in the Middle East and completed an operational tour prior to returning to New Zealand.

COASTAL COMMAND

Convoy escort
224 Squadron, RAF (Beaulieu, Hampshire - 19 Group)
Liberator III FK242 - captained by Fg Off G M Sellar, RAF, and on return at 2000 ran short of fuel. Amid some confusion as to its position in poor weather and a 400-500 feet cloud base, FK242's port wing tip struck tree tops and crashed at Fullamore Farm, near Horrabridge, about 8 miles north of Plymouth. One of the crew escaped with injuries but his six comrades died, the navigator being buried at nearby Buckland Monachorum.
Navigator: NZ403953 Fg Off **Victor Edward CROWTHER**, RNZAF - Age 24. 392hrs.

Sat 31 Oct 1942

MIDDLE EAST

Strafing along Sidi Barrani-Sollum road, Egypt
450 Squadron, RAAF (LG 175 [Amiriya area], Egypt - 211 Group)
Kittyhawk III FR270 - took off at 1340 with three others and while strafing a five-mile stretch of the road was forced to land 2 miles to the north (map ref 560364, Matruh Sheet). Seen to leave the cockpit, apparently uninjured, the pilot was taken prisoner. A participant in the 'Great Escape' from Stalag Luft III at Sagan, Poland, on the night of 24/25 March 1944, he was recaptured a few days later by the Germans near Liberec, Czechoslavakia. He and three of his fellow escapees were murdered by the Gestapo on the 29th somewhere between Liberec and Most, 105km to the WSW. His body was cremated at Most and the ashes returned to Sagan, where, together with those of the 49 other 'escape' victims, they were held in a communal vault in a cemetery about 500 yards from the camp. Later the remains were reinterred in a collective grave at Poznan, 40km NE of Sagan.
Pilot: 40652 Sqn Ldr **John Edwin Ashley WILLIAMS**, DFC mid, RAF - Age 23
Williams had been commanding the Squadron for just four days.

FIGHTER COMMAND

Standing patrol between St Catherine's Point, Isle of Wight, and Selsey Bill, Sussex
486 Squadron, RNZAF (Tangmere, Sussex - 11 Group)
Typhoon IB R8701 - took off at about 1615, but 55 minutes later the pilot reported that he was having to bale out. Before he could do so R8701 dived into the sea from a low altitude, about 6 miles SE of Selsey Bill. His 'No 2' circled the site for 15 minutes but all that could be seen was the pilot's dinghy, and he is commemorated on the Runnymede Memorial.
Pilot: NZ412294 Fg Off **Leslie Vincent WEIR**, RNZAF - Age 36. 427hrs. 21st op.

BOMBER COMMAND

'Cloud cover' raid on Emden, Germany (by 8 aircraft - 2 lost)
150 Squadron, RAF (Kirmington, Lincolnshire - 1 Group)
Wellington III BK360 - took off at 1040 captained by Flt Sgt A G Makay, RCAF, and believed to have been shot

down over the Netherlands by a fighter, crashing at De Polle, 6km NW of Steenwijk. All five crew are buried nearby at Willemsoord.
Wireless Op: NZ412322 Sgt **John Alexander FULLER,** RNZAF - Age 20. 229hrs. 16th op.

Sun 1 Nov 1942

FIGHTER COMMAND

Convoy patrol
616 Squadron, RAF (Westhampnett, Sussex - 11 Group)
Spitfire VI BR186 - took off at 1415 accompanied by another, but some 15 minutes later, just as they had just been recalled because of deteriorating weather, both crashed into a hillside at Nettlecombe, 3 miles west of Ventnor, on the Isle of Wight. The pilot of BR186 died and is buried at Tangmere, Sussex, but his companion survived with injuries to the hands and face.
Pilot: NZ413900 Sgt **Philip Shaw SMITH**, RNZAF - Age 22. 337hrs. 3rd op.

Sun 1/Mon 2 Nov 1942

NEW ZEALAND

Railway accident
RNZAF Station Milson
An airman's body was found in the Tawa Flat railway tunnel at about 0830. Due back on duty at Milson by 1500, he apparently fell from a train which had left Wellington for Palmerston North the evening before. He is buried at Taihape. Coincidentally, another airman, Sgt **F A Jennings**, lost his life in the same tunnel in similar circumstances only four months later, on 24 February 1943.
(Patrolman): NZ4212441 LAC **William Patrick CASEY**, RNZAF - Age 50.
Ironically, Casey, was employed by NZ Railways at the time of his enlistment in the RNZAF.

Tue 3 Nov 1942

MIDDLE EAST

Raid on battle area
162 Squadron, RAF (LG 208 [Bilbeis], Egypt - detachment at LG 86 [Amriya] - 205 Group)
Wellington IC DV944 - took off from LG 86 at 2305 captained by Plt Off J J Silcox, RCAF, but lost height as the undercarriage was retracted, causing the propellers to strike the ground. A further 100 yards on the port propeller flew off and DV944 crashed and burst into flames, three of its bombs exploding. The captain and 2nd pilot were killed and three others died over the following two days, all being buried at El Alamein. Only one crew member survived.
2nd Pilot: NZ411391 Plt Off **Patrick Joseph FARREN**, RNZAF - Age 23. 315hrs. 19th op.

Wed 4 Nov 1942

BOMBER COMMAND

Beam approach weather test
1518 Beam Approach Training Flight, RAF (Scampton, Lincolnshire - 5 Group)
Oxford I AT663 - with visibility reduced to 100 yards and cloud down to ground level at Scampton, the pilot reported trouble with his beam setting. Although advised to fly south to North Luffenham, Rutland, where the weather was clearer, the approach to Scampton was continued with, until at 1430 AT663 struck trees and crashed at Brattleby, about a mile north of the base. The two pilots are buried at Scampton.
Pilot Instr: NZ401751 Flt Lt **Charles Henry BUTT**, mid, RNZAF - Age 22. 903hrs solo (590 on Oxford)
Butt's number of operational flights is unknown, but he had completed an operational tour with Bomber Command prior to his appointment as an instructor with the BAT Flight.

Fri 6 Nov 1942

FLYING TRAINING COMMAND

Solo circuit and landing practice
6 (Pilots) Advanced Flying Unit, RAF (Little Rissington, Gloucestershire - 23 Group)
Oxford II AP473 - carried out an approach at 2150, but was instructed to overshoot and go round again when another of the Unit's Oxfords (BG551) attempted to land too closely behind. As they began to overshoot the Oxfords collided, crashed and caught fire. The pilot of each died, that of AP473 being buried at Little Rissington.
Pilot: NZ416472 Sgt **Hector William FARQUHARSON**, RNZAF - Age 19. 248hrs.

Fri 6/Sat 7 Nov 1942

BOMBER COMMAND

'Gardening' - minelaying between Lorient, France, and the Frisian Islands (by 65 aircraft - 4 lost)
7 Squadron, RAF (Oakington, Cambridgeshire - [PFF] 3 Group)
Stirling I W7620/L - took off at 1659 to lay mines in the 'Nectarines' area (Frisian Islands) and believed to have been shot down off the Dutch coast by flak, crashing at 1957 off the island of Vlieland. All seven crew are commemorated on the Runnymede Memorial.

Captain: NZ411473 Plt Off **David James TOTTMAN**, RNZAF - Age 24. 391hrs. 12th op.
Air Gunner: NZ40613 Fg Off **Peter Coulson WHITWELL**, DFM, RNZAF - Age 22.

15 Squadron, RAF (Bourn, Cambridgeshire - 3 Group)
Stirling I R9201/U - took off on similar raid to the above to lay mines in the 'Deodar' area (Gironde estuary, France) and brought down over France, crashing at St-André des Eaux, 9km WNW of St-Nazaire. All seven crew are buried at Escoublac-la-Baule, about 5km to the SW.
Captain: NZ412683 Flt Sgt **Wilfred James HANNAH**, RNZAF - Age 24. 353hrs. 6th op.
2nd Pilot: NZ411369 Flt Sgt **Kenneth Dudley CHAPMAN**, RNZAF - Age 24. 333hrs. 2nd op.
WOpAG: NZ411761 Sgt **Alfred William JARVIS**, RNZAF - Age 22. 189hrs. 2nd op.
Rear Gunner: NZ41140 Sgt **James Michael BURGESS**, RNZAF - Age 31. 209hrs. 2nd op.

Raid on Genoa, Italy (by 72 Lancasters - 3 lost)
83 Squadron, RAF (Wyton, Huntingdonshire - [PFF] 3 Group)
Lancaster I R5566/B - took off at 2210 captained by Plt Off T G Hackney, DFC, RAF, and brought down near the target area, crashing into the sea. The body of the New Zealand wireless operator-air gunner was recovered on the 8th and buried at Sori, a few kilometres SE of Genoa, but later reinterred at Genoa. His six comrades are commemorated on the Runnymede Memorial.
WOpAG: NZ403558 Sgt **Selwyn Charles ORME**, RNZAF - Age 24. 387hrs. 26th op.
Air Gunner: NZ404631 Flt Sgt **Claude Allan FOSTER**, RNZAF - Agc 28. 307hrs. 17th op.

Sat 7 Nov 1942

NEW ZEALAND

Solo practice
1 Elementary Flying Training School, RNZAF (Taieri)
Tiger Moth II NZ733 - port wing struck that of another of the School's Tiger Moth IIs (NZ788) which was climbing away from the Outram auxiliary landing ground at 1420 after a solo practice forced landing. NZ733 crashed a quarter of a mile away, killing the pilot, who is buried at Wyndham. The pilot of NZ788, LAC **R J Whisker**, baled out and survived with minor injuries.
Pilot: NZ421831 LAC **John Oughton MITCHELL**, RNZAF - Age 19. 63hrs (29 solo - 29 on Tiger Moth)

SOUTH PACIFIC

Formation practice
15 Squadron, RNZAF (Fua'amotu, Tongatapu, Islands of Tonga)
P-40E Kittyhawk 41-25103/8 - stalled off a steep turn at 100 feet at 1625 and struck the ground inverted, catching fire, after crashing through trees. The pilot was buried at Nuku'alofa, but later reinterred at Suva, Fiji.
Pilot: NZ415060 Sgt **Donald CHRYSTALL**, RNZAF - Age 29. 318hrs (252 solo - 48 on Kittyhawk)

Sat 7/Sun 8 Nov 1942

MIDDLE EAST

Raid on Decimomannu and Elmas airfields on the island of Sardinia
104 Squadron, RAF (LG 213 [Kabrit], Egypt - 238 Wing, 205 Group; detachment at Luqa, Malta)
Wellington II Z8590/D - took off at 2040, but a few minutes later crashed into high ground and caught fire near Jebel Chantar, 1½ miles WSW of Siggiewi. All six crew are buried in a collective grave on Malta.
Captain: NZ411376 Flt Sgt **Lincoln Leslie Arnold CRAIG**, RNZAF - Age 24. 478hrs. 28th op.
Navigator: NZ405371 Flt Sgt **Herbert Henry EARNEY**, RNZAF - Age 22. 413hrs. 19th op.
WOpAG: NZ411070 Sgt **Keith Langford, DONALD**, RNZAF - Age 23. 324hrs. 30th op.

Sun 8 Nov 1942

NEW ZEALAND

RNZAF Depot Hobsonville
Died of illness at the Auckland Public Hospital and is buried in that city.
(Medical Orderly): NZ423321 AC1 **John Henry MATHER**, RNZAF - Age 45

Mon 9 Nov 1942

COASTAL COMMAND

Search for missing 179 Squadron Wellington
612 Squadron, RAF (Wick, Caithness-shire - 18 Group)
Whitley VII Z9376 - took off at 0650 and while flying through a front at 0904, crashed into a hill at 1800 feet on the Faeroe Islands at Nakkur (position 6134N:0655W). Four of the crew were killed in the crash, while the captain died in hospital of his injuries the following day. All five are buried at Midvaag on the island of Vaagö. An RAF Sgt was the sole survivor of the accident, which was thought to have occurred when Z9376 veered off track

due to magnetic disturbance in the area, the restricted visibility preventing the crew from seeing the hill in time.
Captain: NZ401272 Fg Off **Harold John George HAEUSLER**, RNZAF - Age 24. 641hrs. 57th op.
The CWGC register gives the 9th as date of death for all those who died, although contemporary documents clearly record that Haeusler did not die until the 10th.

Mon 9/Tue 10 Nov 1942

BOMBER COMMAND

Raid on Hamburg, Germany (by 213 aircraft - 16 lost)
142 Squadron, RAF (Waltham, Lincolnshire - 1 Group)
Wellington III BJ711/Z - took off at 1740 captained by Plt Off W T Bent, DFM, RCAF, and brought down over Germany, most probably crashing near Lüneburg, about 45km SE of Hamburg. All six crew are buried at Ohlsdorf, near Hamburg. Three, including the New Zealand wireless operator-air gunner, share a collective grave with the seven crew of a 57 Sqn Lancaster (W4247) also lost this night, of which Flt Sgt **R O'Neil**, RNZAF, was a member (see below).
WOpAG: NZ403012 Flt Sgt **Isaac KENNEDY**, RNZAF - Age 23.

7 Squadron, RAF (Oakington, Cambridgeshire - [PFF] 3 Group)
Stirling I R9169/Y - took off at 1755 on same raid as the above and lost without trace, its seven crew being commemorated on the Runnymede Memorial.
Captain: NZ413065 Plt Off **Herbert Adams HARRIS**, RNZAF - Age 24. 479hrs. 8th op.

57 Squadron, RAF (Scampton, Lincolnshire - 5 Group)
Lancaster I W4165 - took off at 1758 captained by Sqn Ldr J N G Carter, RAF, on same raid as the above, and brought down over the German coast, most probably crashing into the sea close to shore near Cuxhaven. All seven crew were buried in the Brackeswalde district of the city, but later reinterred at Becklingen, 17km south of Soltau.
Navigator: NZ411870 Sgt **George Walter James EVERETT**, RNZAF - Age 26. 287hrs. 4th op.
Air Bomber: NZ412467 Sgt **Alan Frederick BADCOCK**, RNZAF - Age 23. 291hrs. 5th op.
Wireless Op: NZ412877 Sgt **Kenneth Frederick GARDINER**, RNZAF - Age 21. 188hrs.
Badcock's brother, David Herbert William Badcock, died on 6 September 1943 while flying with 149 Sqn, RAF.

218 Squadron, RAF (Downham Market, Norfolk - 3 Group)
Stirling I W7475/H - took off at 1755 captained by Sqn Ldr P F Hickling, DFM, RAF, on same raid as the above, and brought down over Germany, exploding on impact in the district of Lashorst-Huffe, about 30km ENE of Osnabrück. All seven crew were buried in the Protestant Cemetery at nearby Alswede, but later reinterred in a collective grave at Hannover.
Navigator: NZ404350 Sgt **Ian Ignatius FRIEND**, RNZAF - Age 27. 461hrs. 28th op.
Wireless Op: NZ411088 Sgt **William Wynne MORGAN**, RNZAF - Age 21. 350hrs. 28th op.
WOpAG: NZ41612 Sgt **Adam Haddon THOMSON**, DFM, RNZAF - Age 22. 324hrs. 28th op.

57 Squadron, RAF (Scampton, Lincolnshire - 5 Group)
Lancaster I W4247/X - took off at 1800 captained by Plt Off M E Walsh, DFC*, RAF, on same raid as the above, and brought down over Germany, crashing near Lüneburg, about 45km SE of Hamburg. All seven crew were buried at Lüneburg, but later reinterred at Ohlsdorf, near Hamburg. They share a collective grave with three crew members, including Flt Sgt **I Kennedy**, RNZAF, of a 142 Sqn Wellington (BJ711) also lost this night (see above).
Air Bomber: NZ411226 Flt Sgt **Ronald O'NEIL**, RNZAF - Age 20. 323hrs. 15th op.

COASTAL COMMAND

Navigation exercise
5 (Coastal) Operational Training Unit, RAF (Turnberry, Ayrshire - 17 Group)
Hampden I AE379 - took off at about 0029, but within a few minutes, for reasons unknown, lowered its undercarriage and began a left hand circuit with the intention of landing back at base. While turning AE379 dived into the ground and caught fire 500 yards from the Turnberry Hospital, near Kirkoswald Road, close to Milton Burn. All four crew are buried at Dunure, about 7 miles NE of the airfield.
Pilot: NZ412752 Sgt **Alan Charles SMART**, RNZAF - Age 22. 339hrs (181 solo - 41 on Hampden)

Tue 10 Nov 1942

BOMBER COMMAND

Attack against large merchant ship at Le Havre, France (18 Bostons - 2 lost)
107 Squadron, RAF (Annan, Dumfriesshire - 2 Group)
Boston III AL284 - took off at 1313 piloted by Plt Off W R N Davidson, RAF, and, on running in to the target from 4 miles out, collided with another of the Squadron's Bostons (Z2164) as one or both were taking evasive action to avoid enemy fire. Both plunged into the sea, their four-man crews being killed. The navigator of AL284 is buried at Le Havre, while the other seven airmen are commemorated on the Runnymede Memorial.
Navigator: NZ404908 Fg Off **James Allan LOGAN**, RNZAF - Age 29. 247hrs. 5th op.
WOpAG: NZ404023 Sgt **James Watt FENTON**, RNZAF - Age 25. 154hrs. 5th op.

COASTAL COMMAND(?)

'Bauxite' - anti-submarine patrol

53 Squadron, RAF (Edinburgh Field, Trinidad & Tobago, British West Indies)

Hudson III V9253/L - took off at 0945 captained by Flt Sgt R R Sillcock, RAAF, and failed to return. The fate of the aircraft was not established until after the war, when German naval records revealed that it had surprised the surfaced U-505 east of Trinidad and scored a direct hit with one of four depth charges dropped in an attack. Although seriously damaging the submarine and severely injuring two of its crew, the resulting explosion also caught V9253 and sent it crashing into the sea nearby, killing its four crew and a US Navy seaman 'passenger'. The crew are commemorated on the Ottawa Memorial. U-505 succeeded in limping back to Lorient, France, arriving on 12 December. However, in June 1944 it was captured by a US hunter-killer task group and can be viewed today as an exhibit at Chicago's Museum of Science and Industry.

Navigator: NZ411726 Sgt **Patrick George NELSON**, RNZAF - Age 21. 257hrs. 42nd op.

Tue 10/Wed 11 Nov 1942

BOMBER COMMAND

'Gardening' - minelaying off the southern Biscay coast and Frisian Islands (by 42 aircraft - 2 lost)

57 Squadron, RAF (Scampton, Lincolnshire - 5 Group)

Lancaster I W4262 - took off at 1721 captained by Flt Sgt R R Abercrombie, RAF, to lay mines in the 'Deodar' area (Gironde estuary, France). Because of poor weather at base on return a diversion was made to Tangmere, Sussex, but en route at 0030 W4262 collided with a hill at 400 feet at Burgh on Bain, about 20 miles ENE of Scampton. The seven crew died, the navigator and an RAAF member being buried at Scampton.

Navigator: NZ405130 Sgt **James Harold BARRY**, RNZAF - Age 22. 278hrs. 1st op.

Wed 11 Nov 1942

NEW ZEALAND

3 Elementary Flying Training School, RNZAF (Harewood)

Sixteen days after admission to Christchurch Public Hospital, died this day and is buried at Greymouth.

(Pilot): NZ416895 LAC **Jack MOSELEY**, RNZAF - Age 22.

COASTAL COMMAND

'East' - anti-submarine patrol

608 Squadron, RAF (Gibraltar)

Hudson V AM802/U - took off at 0816 piloted by Lt G W Brodziak, SAAF, on patrol over the Mediterranean and believed shot down at 1020 in error off Algiers at position 3800N:0230E, four Martlets of 893 Sqn from HMS *Formidable* mistaking AM802 for an Italian SM.84. The four crew are commemorated on the Malta Memorial.

Navigator: NZ411723 Sgt **William Herbert LIP-GUEY**, RNZAF - Age 20. 383hrs. 24th op.

Wed 11/Thu 12 Nov 1942

FERRY COMMAND

Delivery flight from Bermuda to Britain

Ferry Command, RAF (Dorval, Quebec)

Catalina IB FP209/V - took off at 1206 captained by Capt R H M Sandeman, and on arriving at 0530 in the Strait of Canso, Nova Scotia, crashed into the sea, catching fire near the CNR wharf at Pirate Harbour, Mulgrave. There were only two survivors from the six-man crew. Two civilians under contract to the Ferry Command, Capt Sandeman and a wireless operator, died in addition to an RCAF navigator and a New Zealand wireless operator. The latter two are commemorated on the Ottawa Memorial. An official accident summary states the above take off and crash times to be in GMT, resulting in a flight time of over 17 hours. The direct route is only 780 nautical miles, however, and this would normally be covered by a Catalina in about seven hours. Either one of the times is in error or the aircraft was diverted en route, possibly to carry out a search for a U-boat, or survivors from one of its victims.

Wireless Op: NZ414862 Sgt **Norman James CHENEY**, RNZAF - Age 33. 42hrs.

Cheney's last known posting was 31 OTU, RAF, at Debert, Nova Scotia. It seems likely that he had been posted to Britain and crewing the delivery flight was his means of travelling there.

Thu 12 Nov 1942

EASTERN AIR COMMAND

Interception

154 Squadron, RAF (Djedjelli, Algeria - 322 Wing, 242 Group)

Spitfire VB (serial?) - advance elements of the Squadron had only just arrived on the base when enemy aircraft appeared overhead near noon. The Spitfire made an emergency take off with only 25 gallons of fuel remaining in its tanks and was presumed to have been lost in one of the many combats that followed. The pilot is commemorated on the Malta Memorial.

Pilot: NZ411854 Sgt **Norman BRUCE**, RNZAF - Age 28. 360hrs. 24 ops(?)

FLEET AIR ARM - SCOTLAND

Solo Aerial Light Torpedo (ALT) simulated attack
785 Squadron, FAA (Crail, Fifeshire - HMS *Jackdaw*)
Albacore I N4170 - crashed into the Firth of Forth at 1115 while executing a simulated attack. The pilot is commemorated on the RNZN Memorial at Devonport, Auckland.
Pilot: Sub-Lt (A) **William Francis WARNER**, RNZNVR - Age 21.

Thu 12/Fri 13 Nov 1942

COASTAL COMMAND

Solo night landing practice
2 (Coastal) Operational Training Unit, RAF (Catfoss, Yorkshire - 17 Group)
Beaufighter VIC T5146 - shortly after take off, descended in a shallow dive and at 0149 flew into a cliff near Skipsea, 5 miles NNW of the base. The pilot is buried at Brandesburton.
Pilot: NZ414989 Sgt **Francis KINNON**, RNZAF - Age 21. 334hrs (197 solo - 31 on Beaufighter)

CANADA

Navigation exercise No 4 - second night cross country
31 Operational Training Unit, RAF (Debert, Nova Scotia - Eastern Air Command)
Hudson V AM892 - seen descending as it passed over the airfield on return at about 0500, and flew into high ground a few minutes later and caught fire 6 miles NE of the airfield. The four crew died, the pilot and RAF navigator being buried at Truro .
Pilot: NZ415809 Plt Off **Harold Frederick WILLIS**, RNZAF - Age 23.

Fri 13/Sat 14 Nov 1942

COASTAL COMMAND

Local night flying practice
5 (Coastal) Operational Training Unit, RAF (Turnberry, Ayrshire - 17 Group)
Beaufort II AW384 - took off at 2340 but crashed 19 minutes later in a meadow by Killochan Castle, 3 miles NE of Girvan. All four crew died, the pilot being buried at Dunure, about 7 miles NE of the airfield. Records appear equally divided as to whether this loss occurred on this night or the following, The CWGC register gives the pilot's date of death as the 13th.
Pilot: NZ413858 Fg Off **Maurice Carson JOLLY**, RNZAF - Age 25. 317hrs (181 solo 19 on Beaufort)

FLYING TRAINING COMMAND

Night navigation exercise: base-25 miles WSW of Bardsey Island-20 miles NE of Point Lynas-Anglesey-Thyl-Conway-base
9 (Observers) Advanced Flying Unit, RAF (Penrhos, Caernarvonshire; satellite at Llandwrog - 25 Group)
Anson I DJ628 - took off at 0124 from Llandwrog but did not respond to a general recall message sent at 0250 when weather deteriorated. Last heard from at 0132, DJ628 was later found to have crashed into the sea about 4 miles NW of Penrhos, killing the five crew. The bodies of three were recovered, while the pilot and another are commemorated on the Runnymede Memorial.
Pilot: NZ413016 Sgt **Trevor Morgan BLAKEMORE**, RNZAF - Age 21. 555hrs.

Sat 14 Nov 1942

FIGHTER COMMAND

Training
61 Operational Training Unit, RAF (Rednal, Shropshire - 81 Group)
Spitfire I X4621 (a presentation machine named *Martyn Evans Bevan*) - failed to recover from a dive and crashed near Ridley, Cheshire. The pilot is buried at Oswestry, Shropshire.
Pilot: NZ416123 Plt Off **Willie LEE**, RNZAF - Age 28. 282hrs.
Although born in New Zealand, Lee was educated in China and attended an Air Force flying school at Hangchow. He had completed about 100 flying hours before returning to New Zealand in 1938. It is not known if these hours are included in the above figure.

Sun 15 Nov 1942

NEW ZEALAND

Interception practice
14 Squadron, RNZAF (Masterton)
P-40E Kittyhawk NZ3033 - entered a dive at 7900 feet when breaking away from a dummy attack at 0925 and failed to recover, crashing 3 miles SW of Greytown. The pilot is buried at Picton.
Pilot: NZ42418 Sgt **Arthur Francis LAW**, RNZAF - Age 20. 137hrs solo (5 on Kittyhawk)

FLEET AIR ARM - ATLANTIC OCEAN

In transit from Gibraltar to Britain after *Operation TORCH* landings in North West Africa

HMS *Avenger*

At 0415 the 13,785-ton escort aircraft carrier was torpedoed by U-155 west of the Strait of Gibraltar at position 3615N:0745W. Hit in the bomb-room, *Avenger* blew up, 513 of her complement losing their lives and only 12 surviving. Three New Zealand naval pilots are commemorated on the RNZN Memorial at Devonport, Auckland. The sinking of *Avenger* represented the largest naval loss in connection with *TORCH*.

Of 802 Squadron, FAA (a Sea Hurricane unit embarked on board *Avenger*)

(Pilot): Sub-Lt (A) **Valentine Henry George COWPER**, RNZNVR - Age 25.

Of 883 Squadron, FAA (a Sea Hurricane unit embarked on board *Avenger*)

(Pilot): Lt (A) **Meliss Stuart LATTER**, RNZNVR - Age 24.

In transit

(Pilot): Sub-Lt (A) **Alexander David Richmond WEBBER**, RNZNVR - Age 20.

Webber was a Sea Hurricane pilot who had been embarked on the escort carrier HMS Dasher *for ferrying duties in connection with* TORCH. *He was returning on board* Avenger *as a passenger.*

Sun 15/Mon 16 Nov 1942

BOMBER COMMAND

Night cross country and photographic

101 Squadron, RAF (Holme-on-Spalding Moor, Yorkshire - 1 Group)

Lancaster I W4326/W - took off at 0100 captained by Sgt J W Spinney, RCAF. At 0400 observers on the ground heard an explosion and saw W4326 descend in flames from 12,000 feet to crash near Dolwen Hill, about 4 miles WSW of Llanerfyl, Montgomeryshire. The seven crew died, the navigator, captain and another RCAF airman being buried at Towyn, Merionethshire. Subsequent investigation considered that the explosion occurred through a photoflash igniting, which set the Lancaster on fire and destroyed the tail structure.

Navigator: NZ412468 Sgt **Henry William Amner COLLETT**, RNZAF - Age 20.

Mon 16/Tue 17 Nov 1942

NEW ZEALAND

Night flying practice

2 Operational Training Unit, RNZAF (Ohakea)

Harvard IIA NZ1011 - took off at about 2300 and 30 minutes later was seen carrying out aerobatics. After executing a half roll NZ1011 dived into the ground and caught fire near Mt Stewart, about 4 miles west of Feilding. The pilot is buried at Wellington.

Pilot: NZ417038 Sgt **John Edwin DORNBUSCH**, RNZAF - Age 23. 194 solo (152 on Harvard)

Tue 17 Nov 1942

EASTERN AIR COMMAND

Escorting a DC-3 to Oran

253 Squadron, RAF (Maison Blanche, Algeria - 323 Wing, 242 Group)

Hurricane IIC HV958 - disappeared on the return journey after safely escorting the transport aircraft to its destination. The pilot is commemorated on the Malta Memorial.

Pilot: NZ414279 Sgt **Harold Leonard Francis GOGGIN**, RNZAF - Age 22. 319hrs. 12 ops.

BOMBER COMMAND

Dual circuit and landing practice

11 Operational Training Unit, RAF (Westcott, Buckinghamshire - 92 Group)

Wellington IC X9614 - took off at 1130 but almost immediately an engine caught fire and failed. Attempting a forced landing directly ahead, X9614 stalled as it manœuvred to avoid trees and crashed at Woodside Farm, $2^1/_2$ miles NE of the airfield. A severely burnt rear gunner was the sole survivor. His four comrades died, the instructor and navigator being buried at Westcott. The engine failure was attributed to the fracture and detachment of a cylinder. Petrol vapour escaped, ignited and the fire rapidly spread from engine to airframe.

Pilot Instr: NZ404867 Fg Off **Francis Seymour HADEN**, RNZAF - Age 28. 602hrs (458 Solo - 218 on Wellington)

Tue 17/Wed 18 Nov 1942

BOMBER COMMAND

'Nickel' - leaflet dropping flights to France (by 14 aircraft - 1 lost)

158 Squadron, RAF (Rufforth, Yorkshire - 4 Group)

Halifax II W7863/V - took off at 1718 captained by Sqn Ldr P de G H Seymour, RAF, to drop leaflets on Nantes and brought down over France, crashing near Consigny, 22km ENE of Chaumont. An RAF Plt Off survived to be taken prisoner, but his seven comrades died and are buried at Consigny.

Navigator: NZ401492 Fg Off **Leonard John FAIRBAIRN**, RNZAF - Age 30. 444hrs. 27th op.

Wed 18 Nov 1942

NEW ZEALAND

Dual instruction

1 Service Flying Training School (ATS), RNZAF (Wigram)

Oxford II NZ1353 - crashed 2 miles south of Lincoln during the morning after a wing was seen to break up. The two crew are buried at Blenheim and Woodlands, respectively. Investigation indicated that gluing of plywood to the wing structure was faulty and had failed, apparently as an aerobatic manœuvre was being attempted.

Pilot Instr: NZ415393 Plt Off **Walter Keith WEBB**, RNZAF - Age 32. 378hrs solo (307 on Oxford)
Pupil Pilot: NZ424492 LAC **Ian Caldwell McINTYRE**, RNZAF - Age 23.

BOMBER COMMAND

'Nickel' - leaflet dropping flights to France (by 5 aircraft - 1 lost)

21 Operational Training Unit, RAF (Moreton-in-Marsh, Gloucestershire - 91 Group)

Wellington IC T2574 - took off at 1743 to drop leaflets over Nantes, but suffered an engine failure at 11,000 feet when outbound over the English Channel. Abandoning the sortie, T2574 returned to England, gradually losing height until an attempted forced landing at 1835 ended in a crash at Liss, 4 miles NNW of Petersfield, Hampshire. The rear gunner was the only one of the five crew to survive, although injured, while the captain is buried at Brookwood in Woking, Surrey.

Captain: NZ415700 Sgt **Rothesay Linton JONES**, RNZAF - Age 23. 328hrs (197 solo - 42 on Wellington) 1st op.

Thu 19 Nov 1942

MIDDLE EAST

Patrol over incoming convoy 100 miles SE of Malta

126 Squadron, RAF (Luqa, Malta)

Spitfire VB EP260 - took off at 0900 with five others and abandoned over the sea after engine failure. The pilot is commemorated on the Malta Memorial.

Pilot: NZ412269 Sgt **Henry ROBERTS**, RNZAF - Age 23. 294hrs. 52nd op.

FIGHTER COMMAND

'Rhubarb' - low level attack on canal barges near Furnes, Belgium

132 Squadron, RAF (Martlesham Heath, Suffolk - 11 Group)

Spitfire VB BL781 - took off at 1400 and during an attack 40 minutes later overshot the leader from underneath, possibly coming into its line of fire - either this or light flak causing a glycol leak. With engine failing BL781 disappeared into cloud as the pilot climbed for height to bale out. Two minutes later he was heard calling a 'Mayday,' but was not seen again and is commemorated on the Runnymede Memorial.

Pilot: NZ414321 Fg Off **Mervyn Jack MILLS**, RNZAF - Age 19. 273hrs. 16th op.

Thu 19/Fri 20 Nov 1942

MIDDLE EAST

Raid on Catania airfield, Sicily

104 Squadron, RAF (LG 213 [Kabrit], Egypt - 238 Wing, 205 Group; detachment at Luqa, Malta)

Wellington II W5583/E - took off from Luqa at about 1900 captained by Flt Lt J H D Mercer, RAF, and brought down over Sicily. All but one of the six crew are buried at Catania, the RCAF 2nd pilot being buried in the Canadian war cemetery at Agira.

WOpAG: NZ41585 Sgt **William Eric HAMMOND**, RNZAF - Age 23. 255hrs. 22 ops.

Fri 20 Nov 1942

MIDDLE EAST

Returning from El Adem to base

117 Squadron, RAF (LG 208 [Bilbeis], Egypt; operational element at El Adem, Libya - 216 Group)

Hudson VI EW945/M - failed to arrive at Bilbeis and later found burnt out near Qaret Somara, having crashed there at 0900. The pilot and sole occupant is buried at El Alamein. Although the first stage of the flight was made in good weather there was a severe dust storm approaching the destination at the time of the crash.

Pilot: NZ411384 Fg Off **Alastair Easson DINNIE**, RNZAF - Age 24.

BOMBER COMMAND

Cross country and exercises over the North Sea off the Lincolnshire coast

11 Operational Training Unit, RAF (Westcott, Buckinghamshire - 92 Group)

Wellington IC R1163 - took off at 1100 captained by Sgt R L D Phair, RAAF, and last heard of 15 minutes later. It was thought R1163's wireless failed but that in favourable weather the crew continued with the flight anyway. All five crew are commemorated on the Runnymede Memorial.

Navigator: NZ414249 Plt Off **Cedric William CLARE**, RNZAF - Age 22.

Fri 20/Sat 21 Nov 1942

COASTAL COMMAND

Torpedo anti-shipping strike off Ijmuiden, Netherlands

415 Squadron, RCAF (Thorney Island, Sussex - 16 Group)

Hampden TB I N9106/N - took off at 1709 piloted by Sgt A J Tippett, RAAF, and believed to have come down into the sea off the coast of NW France. Three bodies washed ashore, that of Sgt **Sleep** at Wissant, about 15km SW of Calais, and they are buried at Pihen-lès-Guines, 6km inland from Wissant. The pilot is commemorated on the Runnymede Memorial.

WOpAG: NZ411086 Sgt **Douglas Lloyd George MILLAR**, RNZAF - Age 23. 157hrs.
WOpAG: NZ411794 Sgt **Ronald Robert Theodore SLEEP**, RNZAF - Age 23.

BOMBER COMMAND

Raid on Turin, Italy (by 232 aircraft - 6 lost)

76 Squadron, RAF (Linton-on-Ouse, Yorkshire - 4 Group)

Halifax II DT571/M - took off at 1719 and brought down over Italy, crashing near Bardonécchia, about 75km west of the target city. All seven crew were buried at Bronchia, but later reinterred at Milan.

Captain: NZ412779 Flt Sgt **Bruce Alexander WISELY**, RNZAF - Age 28. 350hrs. 4th op.

214 Squadron, RAF (Chedburgh, Suffolk - 3 Group)

Stirling I W7584/D - took off at 1844 on same raid as the above, but experienced difficulties with the exactor controls which eventually led to failure of the port inner and starboard outer engines. Short of fuel on return to England, and with the base and nearby alternatives closed with low cloud, the captain instructed his crew to bale out. Navigator Sgt **A H Bradley**, RNZAF, and his five RAF comrades did so successfully, although one of them was injured. The captain, however, died when W7584 crashed at 0342 near Thurlow, 8 miles SW of Chedburgh, and is buried at Haverhill.

Captain: NZ412207 Flt Sgt **Edmund Murray CORLETT**, RNZAF - Age 33. 343hrs (193 solo - 66 on Stirling)

FERRY COMMAND

Returning from Gibraltar to Britain

511 Squadron, RAF (Lyneham, Wiltshire - 44 Group)

Albemarle V P1451 - took off at 0018 captained by Flt Lt G W Thorougood, RAF, but crashed in the sea off Gibraltar shortly afterwards. Only the captain's body was recovered. Buried at sea, he is commemorated on the Gibraltar Memorial, while his three comrades are commemorated on the Malta Memorial.

2nd Pilot: NZ405352 Flt Sgt **Leo Clifford WEBSTER**, RNZAF - Age 26. 274hrs.
Navigator: NZ405374 Flt Sgt **Denison Charles HENDERSON**, RNZAF - Age 25. 302hrs.

Sun 22 Nov 1942

MIDDLE EAST

Convoy escort 20 miles off Apollonia, Libya

450 Squadron, RAAF (Martuba No 4, Libya - 211 Group, 239 Wing, Western Desert)

Kittyhawk III FR303 - took off at 0858 with five others and not seen again after a severe rainstorm was encountered on the return. The pilot is commemorated on the Alamein Memorial.

Pilot: NZ411562 Flt Sgt **Ronald PAYNE**, RNZAF - Age 20. 234hrs. 5th op(?)

EASTERN AIR COMMAND

Enemy air raid

1 ALGS (Advanced Landing Ground Signals?), RAF (Souk el Arba, Tunisia - 333 Group)

Two RNZAF airmen of the unit were injured by shrapnel during an air raid attack on or near the airfield at Souk el Arba. Admitted to a French hospital at Souk el Arba, north of Le Kef, in north-western Tunisia, one died of his wounds on the 25th and was buried in Guelma Cemetery, about 60km south of Bone, where he was later reinterred. The other, LAC **A C Lobban**, recovered and was discharged from hospital six weeks later.

Wireless Operator/Mechanic: NZ405434 LAC **Hubert Edgar SUMNER**, RNZAF - Age 30.

CANADA

Local training

34 Operational Training Unit, RAF (Pennfield Ridge, New Brunswick - Eastern Air Command)

Ventura II AE868 - took off at 1655 piloted by Sgt D Smith, RAF, but the port engine began cutting intermittently. In trying to carry out an emergency crosswind landing, AE868 cartwheeled onto its back and burst into flames. The two crew are buried 10 miles west of the airfield at St George. Investigation revealed that depleted fuel tanks caused the engine to begin cutting immediately after take off. The pilot then inadvertently knocked the port engine switch into the 'off' position while attempting to operate the undercarriage control. In addition to ordering that a strict watch be kept on fuel consumption when short training flights were being undertaken in future, it was recommended that a guard be fitted to protect the port engine switch from accidental operation when a pilot was operating the undercarriage control by touch only.

WOpAG: NZ416039 Sgt **Gerald St George PUTT**, RNZAF - Age 23. 31hrs.

Sun 22/Mon 23 Nov 1942

BOMBER COMMAND

Raid on Stuttgart, Germany (by 222 aircraft - 10 lost)

44 Squadron, RAF (Waddington, Lincolnshire - 5 Group)

Lancaster I W4304/C - took off at approximately 1730-1840 captained by Plt Off S R Young, RAF, and lost without trace, its seven crew being commemorated on the Runnymede Memorial.

Wireless Op: NZ405531 Sgt **Edward Ashton WATERS**, RNZAF - Age 23. 371hrs. 31st op.

Tue 24 Nov 1942

INDIA

Dogfight practice

151 Operational Training Unit, RAF (Risalpur, India)

Hurricane I AG112 - when dogfighting at 0735 with another of the Unit's Hurricane Is (AG155), flown by Plt Off **S L E Weggery**, RNZAF, collided and crashed 5 miles west of Prang. The pilot was buried in Risalpur Cemetery, but later reinterred at Karachi. Weggery baled out without injury.

Pilot: NZ411905 Fg Off **Robert Morris JONES**, RNZAF - Age 23. 184hrs.

Wed 25 Nov 1942

EASTERN AIR COMMAND

LAC **H E SUMNER** died of injuries received during an air raid on the 22nd.

BOMBER COMMAND

'Cloud cover' raids to Germany (by 11 aircraft - 3 lost)

207 Squadron, RAF (Langar, Nottinghamshire - 5 Group)

Lancaster I R5694/F - took off at 1500 to raid Bad Zwischenahn, but was unable to complete the sortie and returned. Because of bad weather on arrival R5694 was instructed to circle at 2000 feet, pending diversion to another airfield, but for reasons unknown descended below this height and at 1820 flew into high ground 11 miles to the ESE near Eaton, Leicestershire. The eight crew died, two New Zealanders being buried along with three of their comrades at Bottesford, 6 miles north of Eaton. The crash site is given in *Bomber Command Losses* as near Easton, Lincolnshire, 8 miles east of Eaton, but this is contrary to contemporary records, which give the Leicestershire location.

Captain: 43035 Flt Lt **Raymund Joseph HANNAN**, DFC, RAF - Age 25. 924hrs solo (68 on Lancaster) At least 30th op.

WOpAG: NZ412885 Sgt **Bryant Leonard McKenzie JENKIN**, RNZAF - Age 24. 191hrs. 6th op.

Sat 28 Nov 1942

FIGHTER COMMAND

Attack on shipping in the Zuid Beveland Canal, Netherlands

485 Squadron, RNZAF (Kings Cliffe, Northamptonshire - 12 Group)

Spitfire VB BM232/Y - after refuelling at Martlesham, Suffolk, took off at 1120 with five others. An oil barge was attacked and left on fire in the Western Schelde, after which BM232 and another Spitfire climbed away into cloud. The former was not seen again and is believed to have been brought down by flak. The pilot is buried at Vlissingen, it not being clear if his body had been recovered from the sea or on land.

Pilot: NZ412724 Sgt **Francis William NORRIS**, RNZAF - Age 21. 304hrs. 7th op.

First solo dusk to dark training

51 Operational Training Unit, RAF (Cranfield, Bedfordshire; satellite at Twinwood Farm- 81 Group)

Havoc I BJ501 - while carrying out an overshoot at Twinwood Farm at 1830, climbed steeply away but stalled, crashed and caught fire at Rutter's Cottage Meadow near Thurleigh, a mile or two north of the airfield. The severely injured pilot was admitted to the Bedford County Hospital, where he died at 2350 without regaining consciousness, and is buried at Kempston, a few miles SW of Bedford.

Pilot: NZ414957 Sgt **Leonard James CASEY**, RNZAF - Age 21. 340hrs.

FLYING TRAINING COMMAND

Night navigation exercise

9 (Observers) Advanced Flying Unit, RAF (Penrhos, Caernarvonshire; satellite at Llandwrog - 25 Group)

Anson I DJ635 - piloted by Sgt D E Johnson, RCAF, was recalled due to an unexpected weather change. At 1929, on heading towards base - believed to be Llandwrog - DJ 635 strayed off track to starboard and flew into a mountain side near Bettws Garmon, to the NW of Lake Cwellyn. The five crew died, the captain and New Zealand pupil navigator both being buried at Caernarvon.

Navigator: NZ42303 Sgt **Clement Robert MORGAN**, RNZAF - Age 22. 146hrs.

BOMBER COMMAND

Circuit and landing practice

75 (NZ) Squadron, RAF (Newmarket, Suffolk; detachment at Oakington, Cambridgeshire - 3 Group)

Stirling I BF399/O - in poor visibility at 2229 was suddenly forced to execute a climbing turn, while with wheels and a third of flap down, to avoid a head on collision with another Stirling, but stalled and dived into the ground, catching fire at Trinity Farm, Milton, on Cambridge Road. The six crew died, five in the crash, and an RAF Sgt two days later. The captain and an RCAF Flt Sgt are buried at Cambridge, and the rear gunner at Chevington, Northumberland.

Captain: NZ39691 Sgt **Raymond Herbert John BROADY**, RNZAF - Age 28. 360hrs (solo 191 - 3 on Stirling) 6 ops.

Rear Gunner: NZ412891 Sgt **Alexander McISAAC**, RNZAF - Age 24. 151hrs. 4 ops.

Sun 29 Nov 1942

ARMY CO-OPERATION COMMAND

Low level photographic reconnaissance between The Hook and Westkapelle, Netherlands

2 Squadron, RAF (Sawbridgeworth, Hertfordshire - 34 Wing)

Mustang I AP164 - took off at 1409 and on completion of the reconnaissance was given a course to steer for home by the leader, but although seen following below five minutes earlier no reply was received. AP164 was not seen again and its pilot is commemorated on the Runnymede Memorial.

Pilot: NZ414260 Fg Off **Godfrey Melville CUNNINGHAM**, RNZAF - Age 25. 393hrs. 1st op(?)

Sun 29/Mon 30 Nov 1942

BOMBER COMMAND

Raid on Turin, Italy (by 36 aircraft - 3 lost)

149 Squadron, RAF (Lakenheath, Suffolk - 3 Group)

Stirling I R9202/K - took off at 0024 and brought down over Italy, crashing at Irasco-Finerola. The eight crew are buried at Milan.

Captain: NZ402450 Flt Sgt **Valance Trent BOWIE**, RNZAF - Age 25. 466hrs.

Mon 30 Nov 1942

EASTERN AIR COMMAND

Transfer from Maison Blanche to Bone

Aircrew Pool(?), RAF (Maison Blanche?, Algeria)

Spitfire (serial?) - collided with an aircraft on the ground and caught fire when landing in darkness at 1745 at Bone, where the pilot is buried. There is conflict in the records as to when the crash occurred. The CWGC register death date of 1 December indicates it may have been on the following day, assuming that the pilot died in the crash and not later from injuries.

Pilot: NZ414661 Sgt **Albert Lens MURRAY**, RNZAF - Age 24.

Wed 2 Dec 1942

FLYING TRAINING COMMAND

Solo night circuit and landing practice

18 (Pilots) Advanced Flying Unit, RAF (Church Lawford, Warwickshire - 23 Group)

Oxford II V3581 - took off at about 2110, but 20 minutes later crashed through a hedge 400 yards north of the airfield while carrying out an overshoot. The pilot was admitted to the Hospital of St Cross at nearby Rugby and three days later transferred to All Saints Emergency Hospital at Bromsgrove, Worcestershire, where he died on the 18th. He is buried at Wythall, 8 miles ENE of Bromsgrove. It was thought that premature raising of the flaps led to the crash.

Pilot: NZ416497 Sgt **William Francis IRWIN**, RNZAF - Age 24. 242hrs.

Wed 2/Thu 3 Dec 1942

BOMBER COMMAND

Raid on Frankfurt, Germany (by 112 aircraft - 7 lost)

115 Squadron, RAF (East Wretham, Norfolk - 3 Group)

Wellington III BK338/A - took off at 0141 captained by Wt Off II A M W McCrae, RCAF, and brought down over Germany, crashing near Monzernheim, 14km NW of Worms. All five crew are buried at Rheinberg, 12km south of Wesel, four of whom were members of the RCAF.

Wireless Op: NZ403440 Flt Sgt **John Francis GARLAND**, RNZAF - Age 21. 396hrs. 32nd op.

75 (NZ) Squadron, RAF (Newmarket, Suffolk - 3 Group)

Stirling I BK618/Q - took off at 0202 on same raid as the above, and set on fire after two fighter attacks ten minutes after leaving the target area, later to crash at Ida-Oberstein, just over 100km SW of Frankfurt. Six crew safely abandoned BK618 and were taken prisoner, including wireless operator Sgt **J R Smith**, RNZAF, and air

bomber Sgt **W K Knight**, RNZAF. One, an RAF Sgt, died in captivity on 10 April 1945 and is buried at Berlin. The captain and mid-upper gunner died in the crash and are buried at Rheinberg, 12km south of Wesel
Captain: NZ413484 Sgt **Alexander SCOTT**, RNZAF - Age 20. 447hrs. 15th op.

Thu 3/Fri 4 Dec 1942

MIDDLE EAST

Raid on Ragusa area, Sicily
40 Squadron, RAF (LG 237 [Jebel Hamzi], Egypt - 238 Wing, 205 Group; detachment at Luqa, Malta)
Wellington IC HX395 - took off at about 1800 captained by Plt Off V M Todd, RAF, but on touching down after the raid a hung-up bomb dropped, hit the runway, bounced back and exploded. Three of the crew survived, but the wireless operator and rear gunner died and are buried on Malta.
Wireless Op: 1064105 Sgt **Everard Aloysius ASPELL**, RAFVR - Age 25. 2nd op.

Raid on Marble Arch Landing Ground, Libya
70 Squadron, RAF (LG140 [Bir el Baheira No 1], Libya - 231 Wing, 205 Group)
Wellington IC BB464/Q - took off at 0100, but immediately suffered an engine failure and crashed, catching fire. Only one of the five crew survived, injured, his five comrades being buried at Tobruk. The engine problem was later traced to a reduction gear case failure.
Captain: NZ412179 Flt Sgt **John Edwin ALBERTSON**, RNZAF - Age 25. 28th op.

COASTAL COMMAND

Training
5 (Coastal) Operational Training Unit, RAF (Turnberry, Ayrshire - 17 Group)
Beaufort II AW286 - took off at 2251 but four minutes later was heard to crash, having been last seen over the sea, close inshore. The following morning a patch of oil was seen where AW286 was presumed to have gone down. All four crew died, the pilot being commemorated on the Runnymede Memorial.
Pilot: NZ414727 Sgt **Bernard MOORE**, RNZAF - Age 20. 292hrs.

Sat 5 Dec 1942

FLEET AIR ARM - SCOTLAND

Delivery flight(?)
Hatston Workshops, FAA (2 miles NW Kirkwall, Orkney Islands - HMS *Sparrowhawk*)
Sea Hurricane IB Z7089 - rolled over and dived into the sea close inshore at 1630 while executing a steep turn at 600 feet off Kirkwall Peerie. The pilot was buried at Kirkwall on the 7th.
Pilot: Lt (A) **Thomas Noel BUSH**, RNZNVR - Age 26.
Bush was serving with 771 Sqn based at nearby Twatt (HMS Tern*), 9 miles north of Stromness. His brother, Charles Roy Bush, died on 30 November 1948 while flying with the General Purpose Flight, RNZAF.*

FERRY COMMAND

Delivery flight from Dorval to Britain via Houlton, Maine, USA, and Gander, Newfoundland
Ferry Command, RAF (Dorval, Quebec, Canada)
Hudson VI FK687 - spun in off a stall at 1630 when attempting to go round again at Houlton. The four crew died, one of the three RCAF members being buried with the pilot at Houlton.
Pilot: NZ413842 Fg Off **George Newall HARRISON**, RNZAF - Age 23.

Sat 5/Sun 6 Dec 1942

FIGHTER COMMAND

Ground Controlled Interception (GCI) practice
488 Squadron, RNZAF (Ayr, Ayrshire - 13 Group)
Beaufighter IIF T3385 - on conclusion of the exercise at 0010, collided at 8000 feet with the other Beaufighter (V8162) engaged on it. Both caught fire before crashing 2 miles south of Heathfield, 6 miles west of Largs. The two crews died and all four airmen are buried at Ayr.
Pilot: NZ403049 Fg Off **Raleigh John Brandon PEACOCKE**, RNZAF - Age 26. 508hrs. 5 ops.

488 Squadron, RNZAF (Ayr, Ayrshire - 13 Group)
Beaufighter IIF V8162 - collided with Beaufighter T3385 as described above.
Pilot: NZ40772 Flt Lt **Athol Charles McKINNON**, RNZAF - Age 25. 671hrs (584 solo - 165 on Beaufighter) 15 ops(?)
Navigator: NZ40194 Flt Lt **Robert Ian McCHESNEY**, RNZAF - Age 29. 602hrs. 94 ops.
McChesney enlisted as an air gunner, 71 of his ops were undertaken in this role while flying in Blenheim and Defiant aircraft. He remustered in October 1942.

BOMBER COMMAND

Training
14 Operational Training Unit, RAF (Cottesmore, Rutland - 92 Group)
Wellington IC R1522/Q - took off at about 0008 but crashed immediately afterwards on the main Cottesmore-

Market Overton road near Barrow Village. The rear gunner was injured, but his four comrades died, the captain, RCAF navigator and an RAF wireless operator-air gunner being buried at Cottesmore. Investigators found the flaps had been raised instead of the undercarriage.
Captain: NZ416097 Sgt **Wilford John COLLINS**, RNZAF - Age 23. 231hrs (167 solo - 33 on Wellington)

Sun 6 Dec 1942

NEW ZEALAND

RNZAF Station Levin
Admitted a week earlier to the Palmerston North Public Hospital on account of illness, died this day and is buried at Waiuku.
(Armourer): NZ416678 AC1 **Everard Newlyn WRIGHT**, RNZAF - Age 21.

BOMBER COMMAND

Operation OYSTER - low level raid on Philips radio works at Eindhoven, Netherlands (by 93 aircraft - 16 lost)
487 Squadron, RNZAF (Feltwell, Norfolk - 2 Group)
Ventura II AJ196/C - took off at 1120 and shot down en route to the target by flak at 1225, crashing and exploding on Woensdrecht airfield. All four crew are buried a few kilometres to the north, at Bergen-op-Zoom.
Pilot: 29216 Wg Cdr **Francis Cecil SEAVILL**, RAF - Age 32.
Seavill was the Squadron's commanding officer.

Sun 6/Mon 7 Dec 1942

BOMBER COMMAND

Raid on Mannheim, Germany (by 272 aircraft - 15 lost)
115 Squadron, RAF (East Wretham, Norfolk - 3 Group)
Wellington III BK513/V - took off at 1732 and brought down over France, crashing near Montcornet, 48km north of Reims. All five crew are buried at Montcornet.
Captain: NZ414265 Fg Off **Maxwell Wintringham DIXON**, RNZAF - Age 22. 379hrs. 5th op.

Mon 7 Dec 1942

NEW ZEALAND

Drogue towing
1 Operational Training Unit, RNZAF (Bell Block)
Gordon NZ617 - took off at 1530 and failed to return. The pilot's body was recovered from the sea ten days later near Kawhia and is buried at Wellington. His comrade is commemorated on the Provincial Memorial at Wellington. On 12 March 1944 a wheel, thought to belong to the Gordon, was found washed up on Motunui Beach. It is believed that NZ617 crashed or ditched between Waitara, about 10 miles ENE of New Plymouth, and Taharoa, to the south of Kawhia Harbour.
Pilot: NZ412721 Sgt **Peter Albert MOREY**, RNZAF - Age 23. 236hrs solo (5 on Gordon)
Towed Target Operator: NZ39597 LAC **Thomas Patrick DAVIES**, RNZAF - Age 28. 132hrs.

COASTAL COMMAND

VH and DTC exercise
2 (Coastal) Operational Training Unit, RAF (Catfoss, Yorkshire - 17 Group)
Beaufighter VIC T5317 - took off at 0852 but eight minutes later spun out of cloud and crashed near Hunmanby, about 19 miles north of the airfield. Both crew died, the pilot being buried just to the SW of the base, at Brandesburton.
Pilot: NZ414306 Sgt **Ian John LAND**, RNZAF - Age 21. 287hrs (165 solo - 20 on Beaufighter)

Tue 8 Dec 1942

NEW ZEALAND

Crawl search for unidentified ship
8 Squadron, RNZAF (Gisborne)
Vincent NZ332/P - took off at 1422 as one of three to search for an unidentified merchant ship reported off the Wairarapa coast, the intercept point being estimated as 20 miles south of Cape Turnagain. The three split up to cover a greater search area on a parallel track search. Several wireless messages were received from NZ332, the last at 1644, all of which served only to indicate that its receiving set was out of order. When it was overdue at 1850 an air search was begun by another Vincent, with a full scale effort being undertaken over the next two days. Apart from sighting several inconclusive oil slicks no trace of the Vincent was found. The pilot is commemorated on the Provincial Memorial at Auckland and his two comrades on the Provincial Memorial at Wellington.
Pilot: NZ411413 Fg Off **Harry KINDER**, RNZAF - Age 27. 350hrs (251 solo - 140 on Vincent)
Navigator: NZ401448 Fg Off **Charles Neville TURNBULL**, RNZAF - Age 29.
WOpAG: NZ412909 Sgt **Nepia STEWART**, RNZAF - Age 31. 146hrs.
Kinder had previously served in Malaya but it is not known if he flew operationally there.
Turnbull's brother, George Watson Turnbull, died on 24 May 1943 while flying with 75 (NZ) Sqn, RAF.

ARMY CO-OPERATION COMMAND

Low level cross country exercise

296 Squadron, RAF (Andover, Hampshire - 38 Wing)

Whitley V BD541 - believed to have taken off at about 1420, but in low cloud crashed 70 minutes later into trees on the top of Sidown Hill, a mile west of Burghclere, and caught fire. The four crew died, the captain and an RAF wireless operator-air gunner both being buried at Andover.

Captain: NZ414383 Fg Off **Robert Leslie EWERT**, RNZAF - Age 26. 384hrs (232 solo - 73 on Whitley) 5 ops.

Wed 9 Dec 1942

FLEET AIR ARM - ENGLAND

Operational

816 Squadron, FAA (Thorney Island, Sussex - HMS *Daedalus*)

Swordfish II DK696 - piloted by Sub-Lt (A) **A W Robertson**, RNZVR, on return from the sortie flew into a hill on Harting Downs near Treyford and caught fire. All three crew were seriously hurt, the observer dying at 1100 two days later in hospital at Chichester. He is buried at West Thorney. At the time 816 Sqn was operating under the direction of RAF Coastal Command's 16 Group.

Observer: Sub-Lt (A) **John Mackenzie DOUGLAS**, RNZNVR - Age 23.

BOMBER COMMAND

Cross country exercise

11 Operational Training Unit, RAF (Westcott, Buckinghamshire - 92 Group)

Wellington IC AD592 - apparently went out of control at 1000 feet, rolled over and dived near vertically into the ground near Ot Moor, Oxfordshire, at 1945, killing the five crew. The captain is buried at Tankersley, Yorkshire, and the air bomber at Willesden, Middlesex. Investigation established that the captain was not strapped in at the time and speculated that he would almost certainly have lost his grip on the control column in the event of striking very heavy turbulence.

Captain: NZ413024 Flt Lt **Frank CHAMBERS**, RNZAF - Age 27. 313hrs (203 solo - 51 on Wellington)

Air Bomber: NZ413077 Fg Off **Charles Henry Ernest JAY**, RNZAF - Age 33.

Chambers and Jay were English-born, enlisting on the same day. Each is buried in the county of his birth.

Wed 9/Thu 10 Dec 1942

BOMBER COMMAND

In transit Cairo to Britain

138 Squadron, RAF (Tempsford, Bedfordshire - 3 Group)

Halifax II W1002 - took off from LG 224 (Cairo West) captained by Fg Off L Anderle, DFC, RAF, bound for Malta, and brought down over the Mediterranean, possibly shot down by a night-fighter. All 15 of its crew and passengers are commemorated on the Alamein Memorial. The 138 Sqn Halifax and its crew, a special duties unit normally based in England, had briefly been attached to 511 Sqn, RAF, in order to meet an urgent call for the transport of supplies to the Middle East. It was returning with five tour-expired bomber crew as passengers when lost.

(Pilot): NZ41524 Flt Sgt **George Geoffrey ORGAN**, RNZAF - Age 30.

(Pilot): NZ411474 Flt Sgt **John Alexander TOVEY**, RNZAF - Age 28. 35 ops.

Organ and Tovey had just completed operational tours with 108 and 40 Squadrons, RAF, respectively. Tovey, it will be recalled, was 2nd Pilot of the 40 Sqn Wellington in which Flt Sgt Peter Kingsford died on 19 July 1942.

Cross country exercise

11 Operational Training Unit, RAF (Westcott, Buckinghamshire - 92 Group)

Wellington IC DV777 - captained by Sgt F G James, RAF, and in drizzle, low cloud and gusty winds at 0212, flew into the ground, catching fire near Byfield, 9 miles NNW of Banbury, Oxfordshire. All five crew died, the air bomber being buried at Banbury. High winds on the approach of an unforecasted front had driven DV777 off track.

Air Bomber: NZ413091 Sgt **Ernest Samuel KIRKPATRICK**, RNZAF - Age 21. 175hrs.

Thu 10 Dec 1942

SOUTH PACIFIC

2 Aerodrome Construction Squadron, RNZAF (Nausori, Fiji)

Died of natural causes at Nausori and buried at Suva.

(Heavy Tractor Driver): NZ427149 LAC **George Wynn HUMPHREYS**, RNZAF - Age 38.

Fri 11 Dec 1942

FLEET AIR ARM - ENGLAND

Sub-Lt (A) **J M DOUGLAS** died of injuries received when his Swordfish crashed on the 9th.

BOMBER COMMAND

Training

12 Operational Training Unit, RAF (Chipping Warden, Northamptonshire; satellite at Turweston - 92 Group)

Wellington III BK243 - crashed during a low circuit at Turweston at 2303. Four of the crew survived, injured, while another died, either in the crash or shortly afterwards. The captain, also injured, died at the Northampton General Hospital next day and is buried at Banbury, 6 miles SSW of Chipping Warden.

Captain: NZ416122 Sgt **Robert Balfour LAPSLEY**, RNZAF - Age 20. 335hrs (179 solo - 27 on Wellington)

Sat 12 Dec 1942

NEW ZEALAND

Local flying practice

2 Operational Training Unit, RNZAF (Ohakea)

P-40E Kittyhawk NZ3004 - thought to have been lost in bad weather or run short on fuel while in a valley near Taihape, 42 miles NNW of Ohakea. NZ3004 was seen to circle a woolshed three times prior to lowering its undercarriage. On approaching to land at 1040 it undershot, striking a belt of trees before crashing inverted and catching fire. The pilot is buried at Hastings.

Pilot: NZ421103 Sgt **Blake Francis SCOTT**, RNZAF - Age 21. 162hrs solo (4 on Kittyhawk)

The CWGC register incorrectly gives Scott's service number as NZ421003.

Sat 12/Sun 13 Dec 1942

ARMY CO-OPERATION COMMAND

'Nickel' - leaflet raid on Roubaix, France

295 Squadron, RAF (Netheravon, Wiltshire - 38 Wing)

Whitley V BD437 - took off at 1653 captained by Plt Off C Wood, RAF, and brought down over France near Abbeville, about 110km SW of the target area. The five crew are buried at Abbeville.

Wireless Op: NZ411749 Fg Off **D'Arcy Clive Melville DUCKER**, RNZAF - Age 23. 160hrs.

Mon 14 Dec 1942

EASTERN AIR COMMAND

Raid on docks at Tunis, Tunisia

13 Squadron, RAF (Canrobert, Algeria - 326 Wing)

Blenheim V BA747 - took off at about 0500 piloted by Fg Off J Kruytbosch, RAF, but crashed on striking a hut some 400 yards from the airfield perimeter. All three crew are buried at Medjez-el-Bab, Tunisia.

WOpAG: NZ411743 Sgt **Charles Frederick DAY**, RNZAF - Age 20.

MIDDLE EAST

Anti-submarine patrol 120 miles NW of Alexandria

203 Squadron, RAF (LG 227 [Maryut], Egypt - 201 Group)

Baltimore I AG721 - took off at 1150 piloted by Fg Off M J Cowan, RAF, and lost without trace, its four crew being commemorated on the Alamein Memorial.

WOpAG: NZ404621 Sgt **Charles Edward Le Montais SMALL**, RNZAF - Age 28. 144hrs. 7th op.

COASTAL COMMAND

Anti-submarine sweep 'ONS 152 Convoy'

206 Squadron, RAF (Benbecula, Outer Hebrides - 15 Group)

Fortress IIA FL453/A - took off at 0937 captained by Fg Off J Owen, RAF, with two others and lost without trace. On failing to send a scheduled weather report on reaching the most distant point of its patrol, FL453 was assumed to have been lost some time prior to 1400. An extensive search next day was hindered by rough seas, turbulent cloud and lightning. All seven crew are commemorated on the Runnymede Memorial. 'ONS 152' stood for Outward-North Atlantic-Slow convoy No 152 - it lost one ship to U-boat attack during the crossing.

WOpAG: NZ411113 Sgt **Garfield Charles WILSON**, RNZAF - Age 21. 285hrs.

The CWGC register records Wilson as belonging to 98 Sqn, a unit he appears never to have served with.

Mon 14/Tue 15 Dec 1942

MIDDLE EAST

Torpedo strike on shipping west of Tripoli, Libya

38 Squadron, RAF (Shallufa, Egypt - 247 Wing, 201 Group)

Wellington VIII HF913/R - took off at 1915 captained by the commanding officer, Wg Cdr C V J Pratt, DSO, RAF, and sent a message at 0115, giving 0323 as estimated time of arrival back at base, but nothing further was heard. On 26 March 1943 the rear gunner's body was recovered washed up near El Jiya and buried at Gaza. His five comrades are commemorated on the Alamein Memorial.

2nd Pilot: NZ413121 Sgt **Dennis Sowden PROCTOR**, RNZAF - Age 21. 527hrs. 19th op.

Tue 15 Dec 1942

INDIA

Supply dropping over Sumprabum, Northern Burma

194 Squadron, RAF (Lahore, India; detachment at Tezpur)

Hudson IIIA FH431 - took off at 1000 from Tezpur captained by Plt Off W E Foot, RAF, and intercepted by three enemy fighters, being shot down within 15 miles of the target area to explode in jungle tree tops about 2 miles NW of Nhkang Ga (at position 2648N:9746E). No remains of the five crew could be found after hostilities and they are commemorated on the Singapore Memorial. Foot was attached from 353 Sqn to give instruction to 194 Sqn aircrew; an RCAF pilot from 353 Sqn was also on board at the time. It appears that the Hudson too had come from 353 Sqn, but whether it was on attachment or 'owned' by 194 Sqn is not clear.

2nd Pilot: NZ413910 Sgt **Leonard Keith TEMPEST**, RNZAF - Age 22.

COASTAL COMMAND

Anti-submarine patrol in Bay of Biscay

119 Squadron, RAF (Pembroke Dock, Pembrokeshire - 19 Group)

Sunderland III DV971/A - took off at 0859 captained by Fg Off A R McCallum, RCAF. Signalled an SOS at about 1215 from a position 4750N:0840W, WSW of the Scilly Isles, but nothing further was heard. All 11 crew are commemorated on the Runnymede Memorial.

Navigator: NZ402871 Fg Off **Edward Norman JACKSON**, RNZAF - Age 26. 494hrs. 7th op.

Wed 16 Dec 1942

CANADA

Navigation exercise and photographic reconnaissance over Georgia Strait

32 Operational Training Unit, RAF (Patricia Bay, Victoria Island, British Columbia - Western Air Command)

Anson I N9560 - took off at 0932 captained by Plt Off D J Robertson, RCAF, and did not return. An extensive search failed to reveal N9560's fate, but the wreckage was later chanced upon by a civilian in woods on Victoria Island, 7 miles inland from Nanaimo. Flying low in poor weather, the Anson had struck trees which severed the starboard tailplane before it crashed. All five crew members died, a New Zealander pupil navigator being buried at Victoria.

Navigator: NZ414549 Sgt **Andrew Joseph DOHERTY**, RNZAF - Age 20. 128hrs.

Wed 16/Thu 17 Dec 1942

BOMBER COMMAND

Raid on an aircraft depot at Diepholz, Germany (by 8 Wellingtons - 1 lost)

15 Squadron, RAF (Bourn, Cambridgeshire - 3 Group)

Stirling I R9168/T - captained by Plt Off F S Millen, RCAF, and, after being damaged by flak, was shot down over the Netherlands by a night-fighter at 2009, crashing at Gortel, 7km SW of Epe. An RAF Sgt was the sole survivor and taken prisoner. His six comrades are buried at Epe.

Air Bomber: NZ404389 Sgt **Robert Hugh McKILLOP**, RNZAF - Age 25. 296hrs. 6th op.

'Gardening' - mine laying in the Gironde and off Brest and Lorient, France (by 18 aircraft - 1 lost)

75 (NZ) Squadron, RAF (Newmarket, Suffolk - 3 Group)

Stirling I R9245/N - took off at 2211 to lay mines in the 'Deodar' area (Gironde estuary), but on lifting off in a cross wind swung, its starboard main wheel striking the 'Devil's Dyke' on the airfield perimeter moments later, causing the starboard inner oil tank to break away. Deprived of oil, the engine quickly seized and, unable to maintain height with a full load, R9245 crashed a mile from the airfield, where its mine load exploded. All seven crew died, the captain, navigator, wireless operator and flight engineer being buried at Newmarket. The English-born RNZAF air bomber is buried at Lakenham on the southern outskirts of Norwich, Norfolk. The preceding three aircraft had also swung to some degree on take off and after the crash further departures were cancelled.

Captain: NZ414277 Sgt **Benjamin Allan FRANKLIN**, RNZAF - Age 21. 420hrs (291 solo - 30 on Stirling) 8th op.

Navigator: NZ41561 Sgt **William Henry WHITCOMBE**, RNZAF - Age 32. 334hrs. 6th op.

Air Bomber: NZ41902 Sgt **Edgar William HARVEY**, RNZAF - Age 27. 286hrs. 7th op.

Wireless Op: NZ41709 Sgt **Harold Rangi WELCH**, RNZAF - Age 23. 239hrs. 8th op.

Cross country

57 Squadron, RAF (Scampton, Lincolnshire - 5 Group)

Lancaster I W4359 - crashed at 0030 and caught fire 5 miles NE of Predannack, Cornwall. All seven crew are buried at Helston, about 5 miles NW of the crash site. Investigation led to the belief that W4359 had caught fire in the air after a lightning strike exploded a photoflash.

Captain: NZ413102 Sgt **Thomas Irvine MANSON**, RNZAF - Age 23. 386hrs (251 solo - 17 on Lancaster) 1 op.

Thu 17 Dec 1942

NEW ZEALAND

Test flight

1 Squadron, RNZAF (Whenuapai)

Hudson IIIA NZ2077 - carried out four stall turns and gradually lost height before entering a shallow dive over the airfield from 500 feet, followed by a left hand turn. As its airspeed dropped NZ2077 entered a spin from which it failed to recover before crashing through a hangar and exploding. The captain and electrician are buried at Auckland, the fitter at Wellington, while the air gunner was cremated at Auckland. Also destroyed in the crash was Miles Magister NZ585 in the hangar.

Captain: NZ40227 Flt Lt **Graham Wellesley HAMLIN**, DFC, RNZAF - Age 24. 633hrs solo (96 on Hudson)
(Air Gunner): NZ401461 Plt Off **Ian McGREGOR**, RNZAF - Age 26. 321hrs.
(Fitter IIA): NZ40317 Cpl **John Vaughan BARRACLOUGH**, RNZAF - Age 29
(Electrician): NZ42787 AC1 **James MITCHELL**, RNZAF - Age 42

Hamlin had completed an operational tour with 75 (NZ) Sqn.
McGregor had served in bombers in the Middle East and with Bomber Command in Europe, completing an operational tour, possibly two.
Mitchell was born in Scotland and had served with the RFC and RAF during the 1914-18 war.

COASTAL COMMAND

Anti-submarine patrol over the Atlantic

10 Operational Training Unit, RAF (Abingdon, Berkshire; detachment at St Eval, Cornwall - 19 Group)

Whitley V EB363/Q - took off at 0725 from St Eval and shot down off the coast of north-western France by a fighter, crashing into the sea at 1240. The bodies of three of the six crew were recovered along the French coast, including that of the captain 40km NNW of Lorient, at Concarneau, where he is buried. Their comrades are commemorated on the Runnymede Memorial. The OTU was a Bomber Command unit based at Abingdon, Berkshire, but the detachment had been loaned to Coastal Command since August and came under its operational control.

Captain: NZ415064 Sgt **Maurice Thorpe DENHAM**, RNZAF - Age 31. 365hrs.

Thu 17/Fri 18 Dec 1942

BOMBER COMMAND

'Gardening' - mine laying from Denmark to Southern Biscay (by 50 aircraft - 2 lost)

101 Squadron, RAF (Holme-on-Spalding Moor, Yorkshire - 1 Group)

1 Group Lancaster I W4319/N - took off captained by Sgt M A Fussell, RAF, to lay mines in the Baltic and on return to Yorkshire at 2215 took three direct hits from 'friendly' fire from anti-aircraft batteries at Redcar, crashing about 4 miles to the SW near Grangetown. The seven crew died, all but the captain and flight engineer being buried at Thornaby-on-Tees. Investigators concluded that the Lancaster's IFF (Identification Friend or Foe) warning device was either not working or had not been switched on.

Navigator: NZ405298 Sgt **Monte Alan Gray McINTYRE**, RNZAF - Age 29. 291hrs. 9th op.
Air Bomber: NZ404392 Sgt **Sydney Stewart McLEAN**, RNZAF - Age 29. 218hrs. 7th op.
Wireless Op: NZ413278 Sgt **Michael James O'MALLEY**, RNZAF - Age 21. 220hrs. 11th op.

Raid on the Opel works at Fallersleben, Germany (by 22 aircraft - 8 lost)

115 Squadron, RAF (East Wretham, Norfolk - 3 Group)

Wellington III BK336/D - took off at 1733 captained by Sgt R P Crisp, RAF, and brought down over Germany, crashing at Grafeld, 19km WSW of Quakenbrück. All five crew were buried in the Evangelical Cemetery at Quakenbrück, but later reinterred at Rheinberg, 12km south of Wesel.

Navigator: NZ413710 Sgt **Charles Derrick LAMONT**, RNZAF - Age 34. 305hrs. 8th op.

75 (NZ) Squadron, RAF (Newmarket, Suffolk - 3 Group)

Stirling I BF396/X - took off at approximately 1745-1817 captained by Wg Cdr V Mitchell, DFC, RAF, on same raid as the above and lost without trace, its eight crew being commemorated on the Runnymede Memorial. Mitchell had been the Squadron's commanding officer since July.

2nd Pilot: NZ40640 Wt Off **Trevor Horace BAGNALL**, RNZAF - Age 26. 1015hrs. 30th op.

Bagnall had completed a tour with 40 Sqn, RAF, and was lost on his first op with the 'New Zealand' Squadron.

75 (NZ) Squadron, RAF (Newmarket, Suffolk - 3 Group)

Stirling I R9247/W - took off at approximately 1745-1817 captained by Flt Sgt H E Rousseau, RAF, on same raid as the above, and brought down over Germany, crashing near Vechta, killing the seven crew. With the possible exception of the captain, they were buried at Vechta, but all were later reinterred at Rheinberg, 12km south of Wesel.

Navigator: NZ404895 Plt Off **Mervyn Oliver CLARK**, RNZAF - Age 20.
Wireless Op: NZ412342 Sgt **Christopher James KENDAL**, RNZAF - Age 21. 293hrs. 16th op.
Rear Gunner: NZ411777 Sgt **Henry Welsh PULLAR**, RNZAF - Age 25. 213hrs. 3rd op.

218 Squadron, RAF (Downham Market, Norfolk - 3 Group)

Stirling I BF403/R - took off at 1802 on same raid as the above and en route to the target was shot down over the Dutch coast at 1915 by ship-borne flak, crashing into the sea in the vicinity of Den Helder and the island of Texel.

The captain's body was recovered and buried at Den Helder, but later reinterred at Bergen-op-Zoom. That of the navigator also washed ashore and is buried on Texel at Den Burg. Their five comrades are commemorated on the Runnymede Memorial.
Captain: NZ403994 Flt Sgt **Leonard Clarence SHEPHERD**, RNZAF - Age 29. 481hrs. 6th op.
Navigator: NZ405274 Sgt **Stanley HOLMES**, RNZAF - Age 29. 276hrs. 2nd op.

75 (NZ) Squadron, RAF (Newmarket, Suffolk - 3 Group)
Stirling I BF400/G - took off at 1817 on same raid as the above, and brought down over Germany, crashing near Ankum, 35km NNW of Osnabrück. The seven crew were buried in the Catholic Cemetery at Ankum, but later reinterred at Rheinberg, 12km south of Wesel.
Captain: NZ41333 Fg Off **Gerald Howard JACOBSON**, RNZAF - Age 27. 363hrs. 13th op.
Navigator: NZ403551 Plt Off **Walter Duncan McALPINE**, RNZAF - Age 30. 385hrs. 13thop(?)
Air Bomber: NZ41563 Sgt **Eric Lumley Durham ASHWIN**, RNZAF - Age 22. 193hrs. 9th op.
Wireless Op: NZ412362 Sgt **Wallace Frederick STOKES**, RNZAF - Age 27. 248hrs. 9th op.
Rear Gunner: NZ41717 Sgt **William George Henry WHITE**, RNZAF - Age 27.

Fri 18 Dec 1942

NEW ZEALAND

Dual flying training
1 Service Flying Training School (ITS), RNZAF (Wigram)
Oxford I NZ1324 - in the morning collided with another of the School's Oxford Is (NZ1286) at 100 feet above the airfield at Norwood, about 15 miles WSW of Wigram, crashed and caught fire. Both of NZ1324's crew died, the instructor being buried at Midhurst and the pupil at Kaikoura. Two anonymous members of the Fire Section were later commended for their efforts in extricating the pilots' bodies from the burning wreckage. LAC **K O Whitehouse**, flying solo in NZ1286, made a successful crash landing and escaped with slight injuries.
Pilot Instr: NZ40972 Fg Off **Cyril Griffith William HENDERSON**, RNZAF - Age 26. 1639hrs.
Pupil Pilot: NZ428813 LAC **Eric Mark BROADHURST**, RNZAF - Age 22.
Whitehouse died on 25 July 1944, while flying with 75 (NZ) Sqn, RAF.

Navigation exercise between Woodbourne and Ohakea
2 Service Flying Training School (ATS), RNZAF (Woodbourne)
Harvard II NZ949 - seen at 1335 descending through cloud in a spin, from which it failed to recover before crashing into the sea near Paraparaumu Beach. The two crew are commemorated on the Provincial Memorial at Auckland.
Staff Pilot: NZ421087 Plt Off **Gerald Temple ODLIN**, RNZAF - Age 26. 136hrs solo (101 on Harvard)
Pilot: NZ411034 LAC **William Walker JOHNSON**, RNZAF - Age 29.

SOUTH PACIFIC

Patrol - area 'A' search, 2V45
3 Squadron, RNZAF (Henderson Field [Bomber I] Guadalcanal, Solomon Islands)
Hudson IIIA NZ2054 - took off at 0550 and believed to have been brought down by flak at 0745 over the Japanese seaplane base at Reketa Bay, located at the northern end of the island of Santa Isobel. In August 1945 an RNZAF party visiting the area established that NZ2054 had crashed into a bay some 2 miles from Suavana Point, killing three of the crew. According to post-war accounts by two civilians previously held captive on Rabaul, the other two airmen, one severely injured, were taken prisoner by the Japanese. The wounded and unidentified airman died on the beach near the crash scene a few hours later, being buried there by the enemy. The surviving member, Flt Sgt **Newlands**, was later transported to New Britain via the Shortland Islands, reaching Rabaul in January. Later in 1943 he was taken from the camp by the Japanese and his fate is unknown. For official purposes his death is accepted as having occurred on 1 September of that year. After hostilities no trace of any of the crew could be found and all are commemorated on the Bourail Memorial.
Captain: NZ404401 Flt Sgt **Norman John NEWALL**, RNZAF - Age 21. 573hrs.
2nd Pilot: NZ413100 Sgt **Albert Patrick MAHONY**,RNZAF - Age 26. 382hrs.
Navigator: NZ41315 Sgt **Henry William Manfred DOWNARD**, RNZAF - Age 28. 247hrs.
WOpAG: NZ415910 Flt Sgt **David George NEWLANDS**, RNZAF - Age 20. 349hrs.
WOpAG: NZ411289 Sgt **Richard Talbot ANDREW**, RNZAF - Age 29. 490hrs.

MIDDLE EAST

Escort to convoy heading for Malta
227 Squadron, RAF (Ta Kali, Malta)
Beaufighter VIC X8133/M - took off at 0810 with three others and several hours into the flight was one of two vectored onto four Italian aircraft. During the encounter X8133 was seen to go down smoking and crash into the sea, about 150 miles SE of Malta. One of its two crew was seen to climb into a dinghy and paddle towards a group of Italian airmen grouped onto the wing of their machine, which had also been brought down. However, on reaching them he was seen from the other Beaufighter to be pushed away. The two crew are commemorated on the Malta Memorial.
Pilot: NZ412219 Sgt **Ford John FRANKLIN**, RNZAF - Age 20. 258hrs. 15th op.
Franklin and his navigator were 46 Sqn personnel on attachment to 227 Sqn, as their own unit was without aircraft on Malta at the time.

FLYING TRAINING COMMAND

Sgt **W F IRWIN** died of injuries received in a flying accident on 2 December.

FIGHTER COMMAND

Standing patrol between Selsey Bill and Shoreham

486 Squadron, RNZAF (Tangmere, Sussex - 11 Group)

Typhoon IB R8800 - took off at 1135 with one other and towards noon attacked an enemy bomber at about 500 feet, but either collided with the quarry or was shot down by return fire, crashing into the sea about 10 miles SE of Shoreham. The pilot is commemorated on the Runnymede Memorial.

Pilot: NZ404937 Flt Sgt **Russell William Strode PENNY**, RNZAF - Age 26.

Penny's brother, Keith Stuart Strode Penny, died less than two months earlier, on 26 October, while flying with the Central Flying School, RNZAF.

Although the family surname is usually shown as Penny in contemporary RNZAF and RAF documents, the CWGC registers published in the 1950s record it as Strode-Penny.

BOMBER COMMAND

Special wireless investigation flight in the Yarmouth area and up to 30 miles out to sea

1474 Special Duties Flight, RAF (Gransden Lodge, Cambridgeshire - 3 Group)

Wellington IC DV892/D - took off at 1421 captained by Plt Off F A Couper, RAAF, and prior to 1550 was seen circling Hopton, Suffolk, a few miles to the south of Great Yarmouth, Norfolk, being last seen at 1622 heading SE from Yarmouth. Nothing further was heard or seen of DV892, despite two days of searching. Later, the body of an RAF Plt Off was recovered from the sea 7 miles SE of Lowestoft, while that of the captain washed up on the Danish coast at Vejers Strand. Their five comrades are commemorated on the Runnymede Memorial. No 1474 Flight specialised in detecting the wavelengths of German radar. Cloppenburg, Germany, is given as 'target' on the Bomber Command Loss Card for DV892, and is repeated in *Bomber Command Losses*, but other records make no mention of intent to operate over enemy territory and simply indicate a 'local' flight as described above.

WOpAG: NZ404078 Flt Sgt **Edwin Francis McKENNA**, RNZAF - Age 23. 408hrs. 17th op.

Sat 18/Sun 19 Dec 1942

FERRY COMMAND

Ferry flight from Portreath, Cornwall, to Gibraltar

271 Squadron, RAF (Doncaster, Yorkshire - 44 Group)

Harrow II K7011 - took off at 0215 captained by Sqn Ldr G A Brown, RAF, and brought down over the Bay of Biscay, killing the four crew and six passengers (two of whom were non-Air Force). Four bodies later washed ashore along the French coast and were recovered, but of the others there was no trace, including a New Zealander, who is commemorated on the Runnymede Memorial.

(Function Controller(?): 118987 Fg Off **John Glen McCAULEY**, RAFVR - Age 30.

McCauley is believed to have been serving at RAF Station Digby, Lincolnshire, prior to the flight.

Sat 19 Dec 1942

BOMBER COMMAND

Night flying test with load

149 Squadron, RAF (Lakenheath, Suffolk - 3 Group)

Stirling I R9265/N - took off at about 1512 captained by Fg Off E A R Hunt, RAF, but at 1615 both wings detached following a structural failure, probably through pulling out of a dive after loss of control in the 10/10ths cloud prevailing down to 1000-2000 feet. R9265 crashed near Gransden, 6 miles WSW of St Neots, Huntingdonshire, killing the eight crew. Three New Zealanders and two others are buried at Brookwood in Woking.

Navigator: NZ411229 Flt Sgt **George Edward TAIT**, RNZAF - Age 29. 397hrs. 9 ops.

Air Bomber: NZ411083 Sgt **Irvine Norman John McNAUGHTAN**, RNZAF - Age 25. 309hrs. 12 ops.

WOpAG: NZ411734 Sgt **James Frederick ALPE**, RNZAF - Age 27. 316hrs. 9 ops(?)

COASTAL COMMAND

Navigation exercise

3 (Coastal) Operational Training Unit, RAF (Cranwell, Lincolnshire - 17 Group)

Wellington VIII Z8708/G - took off at 2200 and 90 minutes later was seen flying over HMS *Royal Arthur*, Skegness, at 60 feet, heading due east. There appeared to be a blue flame visible in the cockpit, and about 50 yards out Z8708 dived into the sea. A searchlight was immediately switched on and a lifeboat launched, but no trace of the six crew could be found at the time. At least two bodies later washed ashore, while the captain is commemorated on the Runnymede Memorial.

Captain: NZ412784 Fg Off **Frederick Harold ANGELL**, RNZAF - Age 24. 428hrs (282 solo - 35 on Wellington)

Sat 19/Sun 20 Dec 1942

MIDDLE EAST

'Gardening' - minelaying in Tunis Harbour, Tunisia

14 Squadron, RAF (LG 221 [Fayid], Egypt; detachment at Berka III [Benghazi], Libya - 201 Group)

Marauder I FK366 - took off from Berka III captained by Flt Lt J T Willis, RAF, with two others and one of two which failed to return, being brought down in the Gulf of Tunis. The captain survived to be taken prisoner, but his five comrades died. One, an RAF Fg Off, is commemorated on the Alamein Memorial, while the others are buried at Enfidaville (the 2nd pilot is incorrectly recorded in the CWGC register as having died on the 14th, because of an unintentional repetition of his squadron number for date).

Navigator: NZ402240 Fg Off **Eric William BARR**, mid, RNZAF - Age 25.451hrs. 37th op(?)

Sun 20 Dec 1942

NEW ZEALAND

RNZAF Station Onerahi

Had been swimming with friends in nearby Parua Bay shortly after lunch. Came out of the water feeling unwell and was taken to Onerahi for treatment, but died shortly afterwards and is buried at Dunedin.

(Flight Mech): NZ405085 AC1 **Kenneth Grigor Williamson AITKEN**, RNZAF - Age 23.

Sun 20/Mon 21 Dec 1942

BOMBER COMMAND

Raid on Duisburg, Germany (by 232 aircraft - 15 lost)

156 Squadron, RAF (Warboys, Huntingdonshire - [PFF] 3 Group)

Wellington III Z1660 - took off at 1802 and brought down over Germany, crashing near the target area. Three crew baled out and were captured, while the captain and bomb aimer were buried in the North Cemetery at Düsseldorf, but later reinterred at Reichswald Forest, 5km SW of Kleve.

Captain: NZ404847 Plt Off **Frank Vernon WATKINS**, mid, RNZAF - Age 20. 594hrs. 29th op.

A survivor's account of Watkin's refusal to abandon the Wellington and his badly wounded bomb aimer led to a recommendation for the award of the Victoria Cross. This was not awarded because of insufficient confirmatory evidence, but in recognition of his actions Watkins did receive a mention in despatches 'for gallant and distinguished service'. (The mid is the only alternative posthumous operational award to the VC.)

Mon 21/Tue 22 Dec 1942

BOMBER COMMAND

Raid on Munich, Germany (by 137 aircraft - 13 lost)

7 Squadron, RAF (Oakington, Cambridgeshire - [PFF] 3 Group)

Stirling I W7632/N - took off at 1706 captained by Fg Off J Davies, DFM, RAF, and brought down over France, crashing near Jussecourt-Minecourt, 16km NNW of Vitry-le-Francois. The seven crew are buried at Jussecourt-Minecourt.

Air Bomber: NZ405264 Sgt **Leo Daniel ENRIGHT**, RNZAF - Age 30. 358hrs. 13th op.

Wireless Op: NZ412350 Sgt **Bernard Leslie PETERSON**, RNZAF - Age 21. 305hrs. 12th op.

Mid-Upper Gunner: NZ403955 Sgt **John ELLIS**, RNZAF - Age 28. 234hrs. 12th op.

207 Squadron, RAF (Langar, Nottinghamshire - 5 Group)

Lancaster I W4191/Q - took off at 1715 captained by Sgt J R Walker, RAF, on same raid as the above, and shot down over the Netherlands by a night-fighter at 0002, crashing at Elst, 20km SSE of Amersfoort. The seven crew are buried at Oud Leusden on the southern outskirts of Amersfoort.

2nd Pilot: NZ414380 Fg Off **Douglas Robert BANNERMAN**, RNZAF - Age 30. 365hrs. 1st op(?)

Bannerman served for five years in the regular pre-war RNZAF as a flight rigger, his engagement concluding the day after war was declared. Two months later he applied for war service and was enlisted as a pilot on 27 July 1941. The CWGC register records his unit as 467 Sqn, RAAF, to which he was first posted six weeks earlier. Later he attended a Lancaster course with 1661 Heavy Conversion Unit, which concluded on 16 December. He was apparently on loan to 207 Sqn pending return to 467 Sqn.

Tue 22 Dec 1942

BOMBER COMMAND

Bombing training

26 Operational Training Unit, RAF (Wing, Buckinghamshire - 92 Group)

Wellington IC DV915 - took off at about 1924 and collided at 2030 with another of the Unit's Wellington ICs (Z8950) over the bombing range near Hogshaw, 8 miles west of Wing. Both crashed and caught fire, killing DV915's six crew and the four of Z8950, the pilot instructor of DV915 being buried at Aylesbury with two of the others.

Pilot Instr: NZ41382 Plt Off **Eric Gordon WAGSTAFF**, RNZAF - Age 29. 600hrs (323 solo - 159 on Wellington) 5 ops.

Wed 23 Dec/Christmas Eve 1942

FIGHTER COMMAND

Night flying practice

165 Squadron, RAF (Tangmere, Sussex - 11 Group)

Spitfire VB BM326 - took off at 2325, but 50 minutes later dived vertically into the ground and exploded 4 miles SE of Horsham. The pilot is buried at Tangmere. It was considered that he may have passed out through failure of the oxygen system.

Pilot: NZ415370 Sgt **Wilton McBride SAGE**, RNZAF - Age 25. 309hrs (196 solo - 89 on Spitfire)

Thu, Christmas Eve 1942

NEW ZEALAND

Motor vehicle accident

RNZAF Station Waipapakauri

Struck by a car while walking along the Kaitaia-Awanui road near Awanui during the evening. Rain was falling at the time and the driver, blinded by headlights of two oncoming vehicles, failed to see and avoid the airman. Admitted to the Kaitaia Hospital, he died on Christmas Day and is buried at Wellington.

(Wireless Op): NZ416873 AC1 **Keith Israel BLACKBURN**, RNZAF - Age 23.

Sat, Boxing Day 1942

EASTERN AIR COMMAND

(Duty?)

32 Squadron, RAF (Philippeville, Algeria; detachment at Maison Blanche - 323 or 325 Wing, 242 Group)

Hurricane IIC HW486 - crashed after a mid-air collision with a Dakota on coming in to land at Maison Blanche at 1145. The pilot was buried in the El Calia Cemetery at Maison Carrée, on the SE outskirts of Algiers, but later reinterred nearby at Dely Ibrahim. The fate of the Dakota is unknown but it appears that all of its occupants survived.

Pilot: NZ411972 Flt Sgt **Geoffrey Stanton LLOYD**, RNZAF - Age 23.

Sat, Boxing Day/Sun 27 Dec 1942

EASTERN AIR COMMAND

Bombing Mechanised Transport on the Tunis-Massicault road, Tunisia

114 Squadron, RAF (Canrobert, Algeria)

Blenheim V BA817 - took off at 0030 piloted Plt Off W R Berriman, RAAF, as one of four which took off at intervals to attack, being the only one to locate a target and bomb. On return Canrobert was covered with fog and after circling for some time BA817's port engine cut at 0510 through fuel starvation. The Blenheim flew into a hillock on attempting to locate the airfield for an emergency landing. Only two of the three crew survived, injured, the navigator being buried at Canrobert, but later reinterred at Medjez-el-Bab in Tunisia.

Navigator: NZ404417 Sgt **Kenneth Ewen SMITH**, RNZAF - Age 26.

Sun 27/Mon 28 Dec 1942

MIDDLE EAST

Raid on docks at Tunis, Tunisia

104 Squadron, RAF (LG 237 [Jebel Hamzi], Egypt; detachment at Malta)

Wellington II Z8908 - took off from Luqa at about 1815 with seven others and lost without trace, its six crew being commemorated on the Malta Memorial.

Captain: NZ39993 Sgt **Randall Tyrone COTTRELL**, RNZAF - Age 24. 489hrs.

Tue 29 Dec 1942

FIGHTER COMMAND

Standing patrol between Dungeness, Kent, and Beachy Head, Sussex

91 Squadron, RAF (Lympne, Kent - 11 Group)

Spitfire VB EN782 - took off at 1440 with another and shot down by an enemy fighter, crashing into the sea between Beachy Head and Hastings. The pilot's companion on the sortie, Plt Off **W G Mart**, RNZAF, and an air-sea rescue Lysander searched unsuccessfully for him. He is commemorated on the Runnymede Memorial.

Pilot: NZ404853 Plt Off **Irvin William DOWNER**, RNZAF - Age 22.

CANADA

1 (Y) Depot, RCAF (Halifax, Nova Scotia - 3 Training Command)

Assigned for embarkation to Britain, upon completion of training in Canada, died of a heart attack while marching from the Depot down to the docks and is buried at Halifax.

(Air Bomber): NZ416647 Sgt **Raymond Joseph SWINGLER**, RNZAF - Age 34. 160hrs.

Wed 30 Dec 1942

NEW ZEALAND

RNZAF Station Levin
Died of illness at the Palmerston North Hospital and is buried in that city.
(Aircrafthand): NZ4216471 AC1 **Daniel George Alexander RENNIE**, RNZAF - Age 33.
Rennie had enlisted as recently as 18 December and died on his 13th day of service.

MIDDLE EAST

Ferry flight from England to Karachi, India
159 Squadron, RAF (Salbani, India - 170 Wing, 221 Group)
Liberator II AL520 - took off from Palestine and in cloud over Transjordan flew into a hill near El Qutem (approximate position 3218N:3637E), near the Syrian border, ENE of Amman. All eight crew were buried at the crash site, but lack of access and inability to maintain the graves led to their names being commemorated on the Alamein Memorial. However, during the 1950s the crew was reinterred in a collective grave at Damascus, Syria.
Captain: NZ405244 Flt Sgt **Allan Carlyle FEISST**, RNZAF - Age 23.
WOpAG: NZ39876 Sgt **Gordon Vincent BROOKS**, RNZAF - Age 20(?)
WOpAG: NZ41620 Sgt **Ronald Herbert FITCHETT**, RNZAF - Age 21.
Air Gunner: NZ403449 Flt Sgt **Jack HENRY**, RNZAF - Age 20.

Thu, New Year's Eve 1942

NEW ZEALAND

Motor vehicle accident
1 Aerodrome Construction Squadron, RNZAF (Waipapakauri)
While directing trucks carrying shale to the airfield from a nearby quarry, was injured at 1430 on attempting to mount the running board of the last truck (MT1504) as it passed by, falling under its rear wheels on losing his grip. Died at 0100 on New Year's Day after admission to the Kaitaia Hospital, and is buried at Auckland.
(Grader Driver): NZW424797 LAC **Laurence Leo FORSMAN**, RNZAF - Age 34.

BOMBER COMMAND

Local bombing practice
16 Operational Training Unit, RAF (Upper Heyford, Oxfordshire - 92 Group)
Wellington III Z1665 - took off at 1815, but a few minutes later went out of control and crashed into a wood, catching fire at Barton Abbey, about 4 miles to the west of the airfield. The rear gunner survived with slight injuries but his six comrades died, the captain being buried at Middleton Stoney.
Captain: NZ402223 Flt Lt **Vernon Elton SUTHERLAND**, RNZAF - Age 30. 865hrs.
Sutherland's total number of ops is unknown but he had completed an operational tour with 405 Sqn, RCAF, prior to his appointment as a staff pilot at the OTU.

Afterword

'It is not the coat that counts'

Wed 15 Jul 1942

COASTAL COMMAND

Anti-shipping patrol of Trost area, Norway
48 Squadron, RAF (Wick, Caithness-shire - 18 Group)
Hudson IIIA FH378/M - took off at 1150 and shot down at 1522 by enemy fighters off the Norwegian coast. On the 25th the captain's body was recovered near the Kya lighthouse, 120km NNW of Trondheim, and buried in the Saetervolden churchyard at Sör Froya, but later reinterred at Trondheim. His three comrades are commemorated on the Runnymede Memorial.
Captain: 41458 Sqn Ldr **Varley Allen PEDERSEN**, RAF - Age 28. 1602hrs. 126th op.

The circumstances surrounding the loss of 'Jim' Pedersen and his crew had remained a mystery for over 30 years, when, after a long search, WWII aircrew researcher Arthur Arculus traced and made contact with Paul Schalk, the Luftwaffe pilot who shot down FH378 while serving in Norway as an Unteroffizier (Corporal) with Jagdgeschwader 5 (JG5). Writing from Frankfurt in March 1976, Schalk described the events of that now distant July afternoon:-

...We flew with two Mes and came across Pedersen's aircraft about 80-100km south of Trondheim, near the coast. First the Feldwebel [Sergeant] attacked and received some machine gun hits, probably in his radiator. He had to make an emergency landing on the sea and escaped from the aircraft. I observed him with one eye and Pedersen with the other. He could not escape from me. The Me 109F was too fast. The first burst of fire showed no results, after the second his right hand engine was burning. I had expected Pedersen would fly to the coast or make an emergency landing at sea to save the crew. He turned away in the direction of the sea. Suddenly his aircraft soared up and fell from about 30 to 50 metres down into the water. His engine was burning so I didn't fire any more at him but could observe the entire crash from close quarters.

I stayed a little while at the site, but only saw a rubber dinghy. After my landing at the base I only tanked quickly and set out again immediately to guide the air-sea rescue aircraft which had been alerted. We got the Feldwebel out (I cannot remember his name) and could only salvage the rubber dinghy...

It was later learnt that Schalk's leader and companion on the sortie was Feldwebel Ernst Koch. Six months after his ditching his luck too ran out.

On the afternoon of 21 January 1943, now flying a FW 190, 26-year old Koch took off on his last operational flight. During the sortie his engine suddenly failed at an altitude of about 1000 metres. Rather than bale out, he made for a small frozen alpine lake and attempted to carry out a forced landing. However, the lake proved too small and the fighter hit a steep cliff at the far end, killing Koch, who was buried in the Military Cemetery at Trondheim on the 29th.

During the war Ernst Koch's parents also lost their other two sons, Robert and Fritz. Their third child, daughter Anna, was the only one of four to survive the conflict. The Koch family would grieve for their loved ones no less than would those of Jim Pedersen living half a world away.

As Paul Schalk so fittingly said, in the closing paragraph of his letter to Arthur Arculus, 'it is not the coat that counts but the heart that beats within it.'

Appendix

Summary of Second World War deaths by memorial and country of burial

The following is a breakdown by CWGC memorial and by country of where the 4202 New Zealand airmen and women who died while serving during the Second World War are buried or commemorated.

Note: Ongoing research may result in minor adjustment to some figures.

Memorials by name

(No known grave - total of 1278)

The Runnymede Memorial, near London, England (660)

††
††
††
††
††
††
††

The Bourail Memorial, New Caledonia (232)

††
††
††††††††††††††††††††††††††††††††

The Alamein Memorial, Egypt (100)

††

The Malta Memorial (86)

††

The Royal New Zealand Naval Memorial, Devonport, Auckland, New Zealand (68)

††

The Singapore Memorial (63)

†††

The Ottawa Memorial, Canada (19)

†††††††††††††††††††

The Provincial Memorial, Wellington, New Zealand (18)

††††††††††††††††††

The Provincial Memorial, Auckland, New Zealand (15)

†††††††††††††††

Other memorials (less than 10 names on each - total 17)

†††††††††††††††††

Graves by country

(total of 2924)

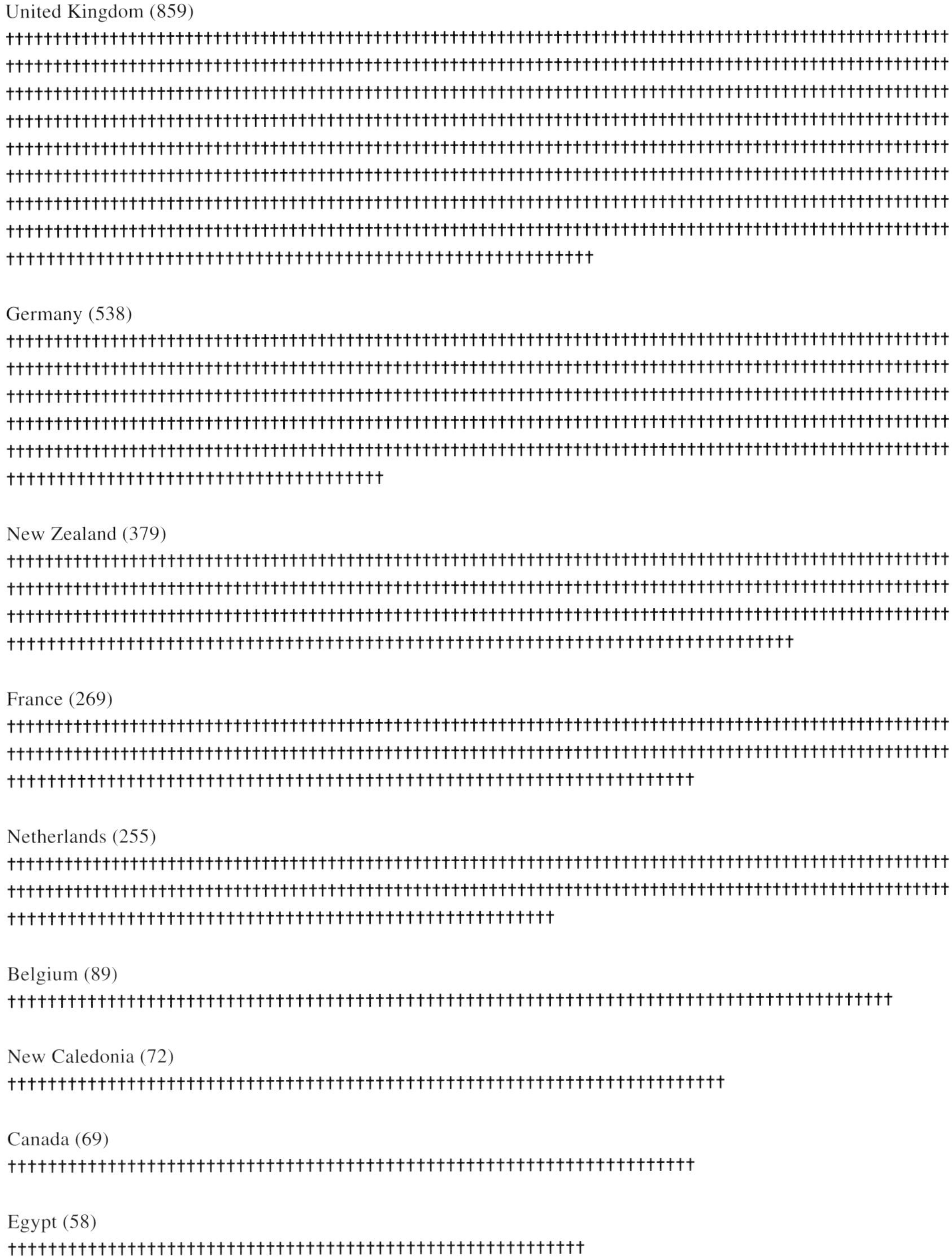

Italy (48)
††

Denmark (37)
†††††††††††††††††††††††††††††††††††††

India (including Pakistan) (33)
†††††††††††††††††††††††††††††††††

Libya (19)
†††††††††††††††††††

Norway (15)
†††††††††††††††

Singapore (14)
††††††††††††††

Tunisia (14)
††††††††††††††

Algeria (11)
†††††††††††

Burma (11)
†††††††††††

Greece (11)
†††††††††††

Fiji (10)
††††††††††

Malta (10)
††††††††††

Others countries (less than 10 graves each - total 103)
††
†††

Index to personnel

- Names of New Zealanders who died are shown in UPPER CASE (also includes other nationals who died while on attachment or secondment to any of the New Zealand air services).
- Names of New Zealanders other than deceased are shown in lower case (also includes other nationals on attachment or secondment to any of the New Zealand air services).
- Names of all others are shown in *lower case italic.*
- Surnames beginning with Mc or Mac are indexed in true alphabetical sequence.

Note: Volume Three, providing brief biographical details for each New Zealander who died, will also serve as a combined index to the two Fate volumes.

Remember